Ian Hamilton's March with Thomas Frankland's Account Annotated and Illustrated

Putting up a field telegraph.

Making a necessary join.

Volume 3 of the series, *The Last Horse War*

Series compiled, annotated, illustrated and expanded by Payton Fireman, JD.

ISBN: 978-1-7320124-5-5

Front Cover Photo Credits: With Hat,[10 11 12] Thomas Frankland.[13]
Back Cover Photo Credits: Arrival,[14] Rosebank.[15]
Title Page Photo Credits: Telegraph Line Photo Credits.[16 17]

[1] Please see the Bibliography and License Links sections regarding the use of material in this volume.

Original Cover.

Original Frontispiece, Ian Hamilton, by J.S. Sargent.[18]

Colonel Ian Hamilton.[19]

Another photograph taken at the same sitting.[20 2]

[2] Ian Hamilton (British Army officer) - Wikipedia

Contents

PF Preface

It has been my privilege to annotate and illustrate Winston's work. His writing is so dynamic and engaging. His insight is such that he knows where the good story is likely to occur and the intrepidness to go and see for himself on all occasions. He also has amazingly good fortune in his efforts.

It is also important to note that Winston was seriously engaged in supporting himself. When his father Randolph died, in 1895, his debts just about equaled his assets and his widow Jennie was left with just the 'entailed' property secured by her marriage settlement. These allowed her to live comfortably but Winston, as a young adult had to make his own way.[3] There was a purpose to his efforts as well as the abundantly evident enjoyment of his labours.[4] [5]

We should also remember that he wrote his despatches not at his desk but in a war zone where he was serving full time. In his words, "on the ground in a tent.....after a long days action or by the light which it was dangerous to use lest it drew fire—when I was tired and hustled amid other adverse circumstances..."[6]

The reason this work and the series is composed of a collection of first person accounts is that my background as an attorney encourages me to cite original sources and follow the best evidence rule.[7] These authors are the best witnesses for their experiences and I do not see any reason why they would need to have their tales retold by someone other than themselves.[8] I quote Churchill's other writings on the topic with his words since that is as close as we can come to his deposition too. My simply saying he, or they said something is merely hearsay.

These recollections are then augmented by the knowledge, facts, and circumstances of the other memorialists who participated, oftentimes in the same noteworthy events. To some degree they are synoptic, as each has their own perspective, and ancillary facts included in individual accounts. I have also done my best to fill in gaps in their knowledge that are now, in hindsight, available to be added.

As to annotation, the passage of time has obscured the social and political landscape of *who was who* and *what was what* that was as familiar to readers of his time as our own is to us. Many of the cultural and demographic references have fallen out of common use and knowledge, and many locations have changed their names as well. To get the most out his work, one really needs to be familiar with the personages and nomenclature then current. I hope the annotations and illustrations I have added will allow readers without a working knowledge of this time period to fully enjoy Winston's writing and observations. I also felt that illustrating the book would add context and atmosphere by putting faces to names, and images to places.

I've kept the text and spelling as intact as possible and sparingly inserted a few bracketed words for clarity. Oftentimes different writers will spell the same name or place differently. This is far too common to try and harmonize globally. Sometimes I will [bracket] the alternative spelling or footnote it. I have also mostly retained the rather idiomatic hyphenation of compound words that are now handled differently. Original footnotes are enclosed in brackets [ ]. I have taken the liberty of dividing up some of the longer paragraphs. Very rarely there are certain offensive words and phrases and I have declined to serve as a censor.

There are duplicates of pictures, maps and footnotes in the various sections and volumes based relevance to the topic at hand and space available. My desire was that each section should illustrate the topic under discussion and to some extent to be able to stand on its own as a separate work.

[3] See WSC My Early Life Ch. XVI 'I leave the Army' for a full description.

[4] Dr. Samuel Johnson famously stated that "No one but a blockhead ever wrote except for money." Churchill is quoted repeating this injunction in a letter to his mother Jennie. See Young Winston's Wars p. XV. Citing OBC p. 812.

[5] "The sales of the River War and of my two books of war correspondence from South Africa, together with the ten months' salary amounting to £2,500 from the Morning Post, had left me in possession of more than £4,000." Frederick Woods' Bibliography p. XXII citing My Early Life p. 374. He also wrote to his mother, accurately that: "...there is not one person in a million who at my age [25] could have earned £10,000 without any capital in less than two years." Letter to Jennie, January 1st, 1901 during his US speaking tour. Woods p. XXII and OBC p. 1225.

[6] Frederick Woods, Young Winston's Wars, p. XV Citing OBC p. 812.

[7] The best evidence rule applies first and foremost to documents, positing that an original is superior to a copy. My application of this concept conflates this rule with the rule regarding Hearsay that in broad brush terms states that first person accounts are more accurate than second hand one's.

[8] I don't think many will read what I think about this material in 50 years, but I'm sure that people will want to read the original texts for hundreds.

I realize that there are more footnotes, than truly necessary. Please feel free to read right past and through them. Many are destined for readers in decades to come when they may prove more useful than to us who are closer to the events described.[9] Some are solely for my own edification.

I reference Wikipedia[10] to identify a large number of topics. This is a deliberate choice since, they are already satisfactory for an introduction to the person, place or event, and once created are likely to be permanent and will continue to improve over time. Each volume contains an individual if somewhat abbreviated bibliography focusing on texts that are most relevant to that that volume. For the full bibliography please see Volume 5 containing primarily, A Century of Wrong.

I am particularly fortunate to have contacted the Churchill Archive Centre[11] and obtained the assistance of the archivist, Ms. Katharine Thomson, who has helped in many ways. I have also been fortunate to have made the acquaintance of the community that supports angloboerwar.com,[12] which is a comprehensive repository and forum of information on this time period.

Maps included in the original book are all reproduced herein.

Please see the South African Historical Background section of Volume 1; the Haldane, Frankland and Hofmeyr compilation, for an introduction to the time period and the events leading up to the conflict as well as for the series introduction and a discussion of the methodology for constructing the works in this series.

Space is not so tight in this book and more blank pages are included so that more chapters start on the right side.

Advertisement in New York Daily Tribune, Saturday Nov. 10th, 1900.[21]

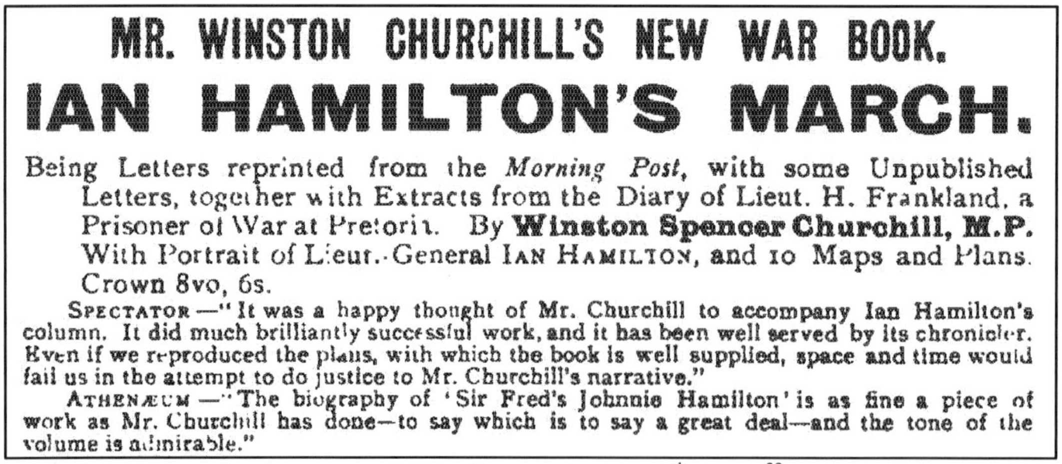

Advertisement carried in Navy & Army Illustrated Nov. 24th, 1900.[22]

[9] Footnotes and endnotes both have Arabic numerals. This was unavoidable since using Roman numerals for endnotes became unwieldy. In order to make some distinction I have used different fonts and font sizes to distinguish the notation type. It becomes easier to follow once the number sequencing diverges farther along into the book.

[10] Wikipedia, the free encyclopedia

[11] The Churchill Archives Centre – Churchill College (cam.ac.uk)

[12] Anglo Boer War - Home

Athenaeum Review of Ian Hamilton's March

This review from the Athenaeum covers Ian Hamilton's March, as well as Haldane's and Rosslyn's books.

Messrs. Longman publish Ian Hamilton's March, by Mr. Winston Churchill, M.P., a reprint of letters to the Morning Post, with alterations and additions.

The biography of "Sir Fred's Johnnie Hamilton" (pp. 108 to 136) is as fine a piece of work as Mr. Churchill has done—to say which is to say a great deal—and the tone of the volume is admirable, as witness the rebuke to "those tremendous patriots who demonstrate, but who do not fight"; but the style of many passages is the despair of admirers of the writer.

That the author of 'The Story of the 'Malakand Field Force' should be content to let his name go out on a book which contains "before we ever reached" for "before we reached," [&] "to chastise impalpable laagers," and continual repetition of the affected "stricken" for *hit*, is a sad testimony to the indifference begot of hasty production, as are slips like "Commander of the Bath" for Companion of the Bath.

We only scold because we know that the author can do work worthy of Napier when he tries.

Capt. Haldane's thrilling narrative of his escape from Pretoria, which has been appearing in Blackwood's, is now published by Messrs. Blackwood & Sons under the title How We Escaped from Pretoria. We can strongly recommend this little paper-covered book as the best work of adventure to which the war has as yet given rise.

A less admirable production, but one which Twice Captured, also published by Messrs. is also bright and readable, is Lord Rosslyn's Blackwood & Sons, a record of somewhat harum-scarum adventure, put together without much literary art.

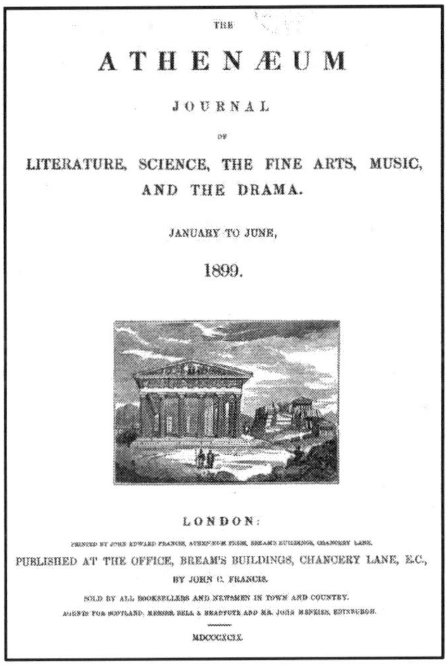
THE
ATHENÆUM
JOURNAL
OF
LITERATURE, SCIENCE, THE FINE ARTS, MUSIC, AND THE DRAMA.
JANUARY TO JUNE,
1899.
LONDON:
PUBLISHED AT THE OFFICE, BREAM'S BUILDINGS, CHANCERY LANE, E.C.,
BY JOHN C. FRANCIS.
SOLD BY ALL BOOKSELLERS AND NEWSMEN IN TOWN AND COUNTRY.
MDCCCXCIX.

Messrs. Longman publish *Ian Hamilton's March*, by Mr. Winston Churchill, M.P., a reprint of letters to the *Morning Post*, with alterations and additions. The biography of "Sir Fred's Johnnie Hamilton" (pp. 108 to 136) is as fine a piece of work as Mr. Churchill has done—to say which is to say a great deal—and the tone of the volume is admirable, as witness the rebuke to "those tremendous patriots who demonstrate, but who do not fight"; but the style of many passages is the despair of admirers of the writer. That the author of 'The Story of the Malakand Field Force' should be content to let his name go out on a book which contains "before we ever reached" for "before we reached," "to chastise impalpable laagers," and continual repetition of the affected "stricken" for *hit*, is a sad testimony to the indifference begot of hasty production, as are slips like "Commander of the Bath" for *Companion of the Bath*. We only scold because we know that the author can do work worthy of Napier when he tries.

Capt. Haldane's thrilling narrative of his escape from Pretoria, which has been appearing in *Blackwood's*, is now published by Messrs. Blackwood & Sons under the title *How We Escaped from Pretoria*. We can strongly recommend this little paper-covered book as the best work of adventure to which the war has as yet given rise.

A less admirable production, but one which is also bright and readable, is Lord Rosslyn's *Twice Captured*, also published by Messrs. Blackwood & Sons, a record of somewhat harum-scarum adventure, put together without much literary art.

See: Success in Journalism.[13] & The Scaffolding of Rhetoric.[14] Sample Cover of the Athenaeum.[15] Citation to the review.[16] William Francis Patrick Napier.[17] Pages 108-136 in the book are devoted to a biography of Ian Standish Monteith Hamilton.

[13] This grammatical of correction of, "before we ever reached", and others made by commentators, refined Churchill's writing. He eventually became a maven for split infinitives. See: Success in Journalism: The Inlander A Literary Magazine, Michigan University Vol. 11 No. 5 p. 168-170 (1900/1901) Catalog Record: The Inlander | HathiTrust Digital Library Location: https://babel.hathitrust.org/cgi/pt?id=mdp.39015071372380&seq=7 PF: Contra, see Oscar Wilde letter to The Times about pedant's.

[14] OBC Vol 1 Part II pages 816-821.

[15]The Athenæum was a British literary magazine published from 1828 to 1921. The Athenaeum (British magazine) - Wikipedia Internet Archive Catalog Record: https://archive.org/details/texts?tab=collection&query=%22The+Athen%C3%A6um%22&sort=date HathiTrust Catalog Record: Catalog Record: The Athenaeum | HathiTrust Digital Library

[16] Ian's March Review: The Athenaeum-No. 3808, Oct. 20th, 1900 p. 507. U. Illinois: #523 - The Athenaeum 1900:2. - Full View | HathiTrust Digital Library

[17] General Sir William Francis Patrick Napier (7 December 1785 – 12 February 1860) was a British soldier in the British Army and a military historian. See: William Francis Patrick Napier - Wikipedia

Original Preface

This book is a continuation of those letters to the *Morning Post*[18] newspaper on the South African war, which have been lately published under the title 'London to Ladysmith *via* Pretoria.' Although the letters had been read to some extent in their serial form, their reproduction in a book has been indulgently regarded by the public; and I am encouraged to repeat the experiment.

The principal event with which the second series deals is the march of Lieutenant-General Ian Hamilton's column on the flank of Lord Roberts's main army from Bloemfontein to Pretoria. This force, which encountered and overcame the brunt of the Boer resistance, which, far from the railway, marched more than 400 miles through the most fertile parts of the enemy's country, which fought ten general actions and fourteen smaller affairs, and captured five towns, was, owing to the difficulties of telegraphing, scarcely attended by a single newspaper correspondent, and accompanied continuously by none. Little has therefore been heard of its fortunes, nor do I know of anyone who is likely to write an account.

The letters now submitted to the public find in these facts their chief claim to be reprinted. While written in the style of personal narrative I have hitherto found it convenient to follow, they form a complete record of the operations of the flank column from the day when Ian Hamilton left Bloemfontein to attack the waterworks position, until he returned to Pretoria after the successful engagement of Diamond Hill.

Although in an account written mainly in the field, and immediately after the actual events, there must be mistakes, no care has been spared in the work. The whole book has been diligently revised. Four letters, which our long marches did not allow me to finish while with the troops, have been added and are now published for the first time. The rest have been lengthened or corrected by the light of after-knowledge and reflection, and although the epistolary form remains, I hope the narrative will be found to be fairly consecutive.

I do not want the reader to think that the personal incidents and adventures described in this book are extraordinary, and beyond the common lot of those who move unrestricted about the field of war. They are included in the narrative, not on account of any peculiar or historic interest, but because this method is the easiest, and, so far as my wit serves me, the best way of telling the story with due regard at once to detail and proportion.

In conclusion I must express my obligations to the proprietors of the Morning Post newspaper for the assistance they have given my publishers in allowing them to set up the copy as each letter arrived from the war; to the Duke of Marlborough,[19] to whom I am indebted for the details of the strength and composition of the force which will be found in the Appendix, and for much assistance in the attempt to attain accuracy; and thirdly, to Mr. [Thomas] Frankland, whose manly record of the heavy days he passed as a prisoner in Pretoria may help to make this book acceptable to the public.

Winston Spencer Churchill.

London: September 10th 1900.

[18] The Morning Post was a conservative daily newspaper published in London from 1772 to 1937, when it was acquired by The Daily Telegraph. The Morning Post - Wikipedia

[19] Charles Richard John Spencer-Churchill, 9th Duke of Marlborough, (13 November 1871 – 30 June 1934), was a British soldier and Conservative politician, and a close friend of his first cousin Winston Churchill. He was often known as "Sunny" Marlborough after his courtesy title of Earl of Sunderland. Charles Spencer-Churchill, 9th Duke of Marlborough - Wikipedia

Maps

Multiple maps are provided of various areas. Each has details or nomenclature that are not reproduced in others.

SA Political and Railway

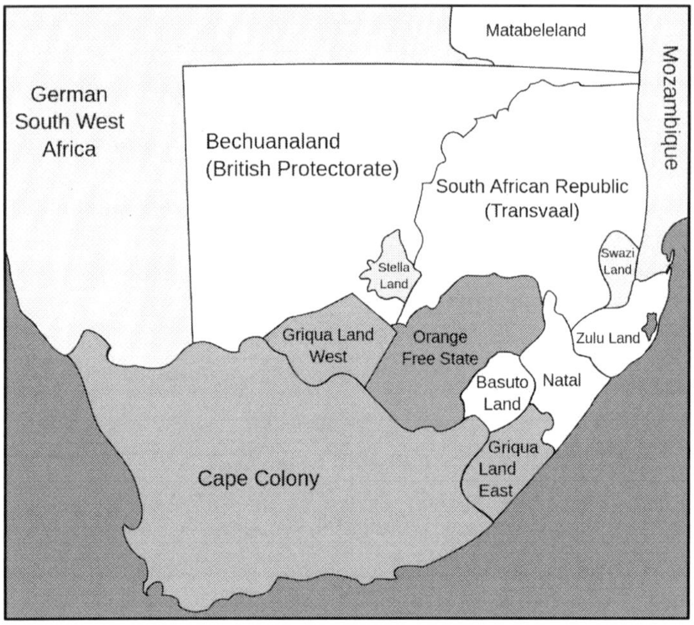

South Africa 1885.[23] [20]

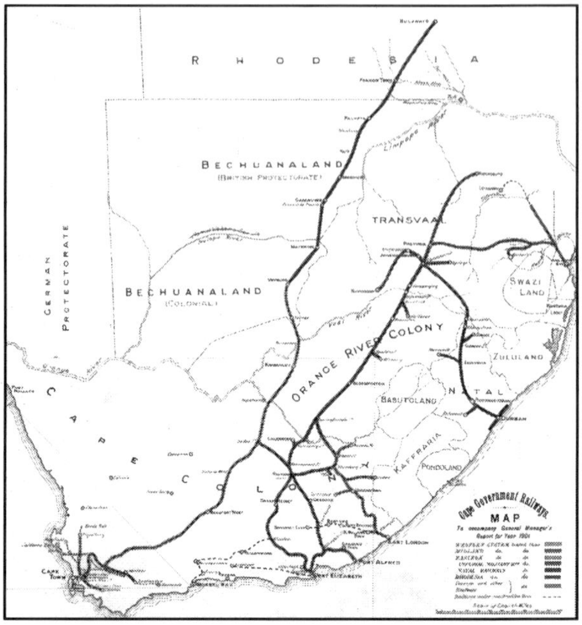

South African Railway System 1901.[24]

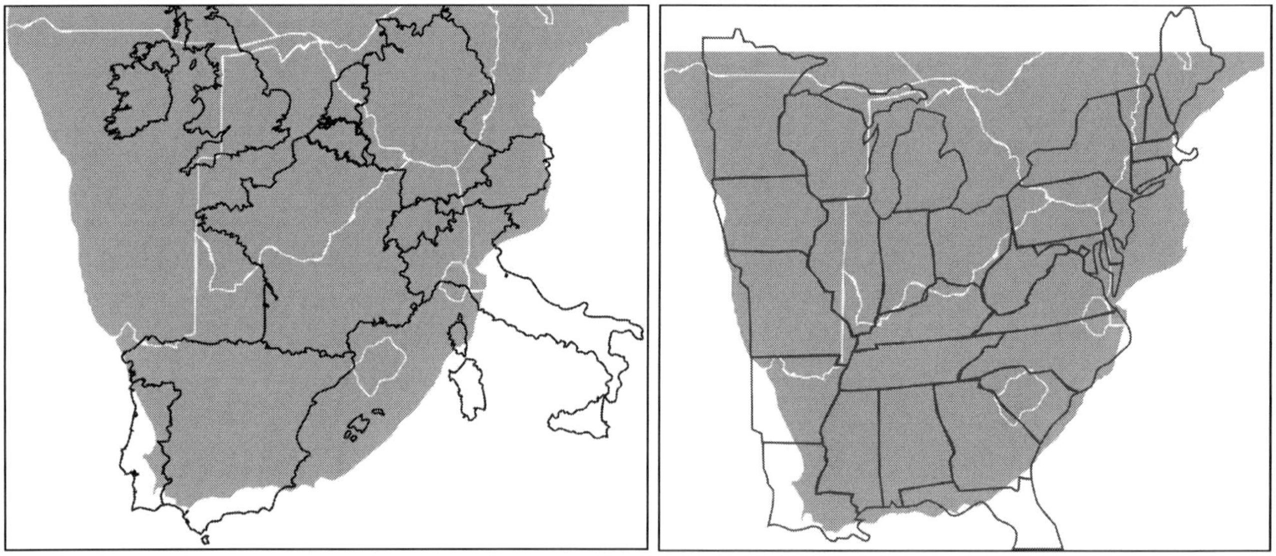

Relative size of Southern Africa compared to Europe and North America.[21]

[20] Basutoland - Wikipedia (Lesotho) Bechuanaland Protectorate - Wikipedia (Botswana) Cape Colony - Wikipedia German South West Africa - Wikipedia (Namibia) Griqualand East - Wikipedia Griqualand West - Wikipedia Matabeleland - Wikipedia Natal: Colony of Natal - Wikipedia Orange Free State - Wikipedia Portuguese Mozambique - Wikipedia Stellaland - Wikipedia Swaziland: Eswatini - Wikipedia South African Republic - Wikipedia Zulu Kingdom - Wikipedia File:SouthAfrica1885.svg - Wikimedia Commons

[21] Adapted from an original appearing in Hiley and Hassell's, The Mobile Boer, Grafton Press, New York, (c. 1902) "The above map, comparing South Africa with Europe was published by "C.I." as long ago as Jan 6th, 1900."

Theatre of Conflict

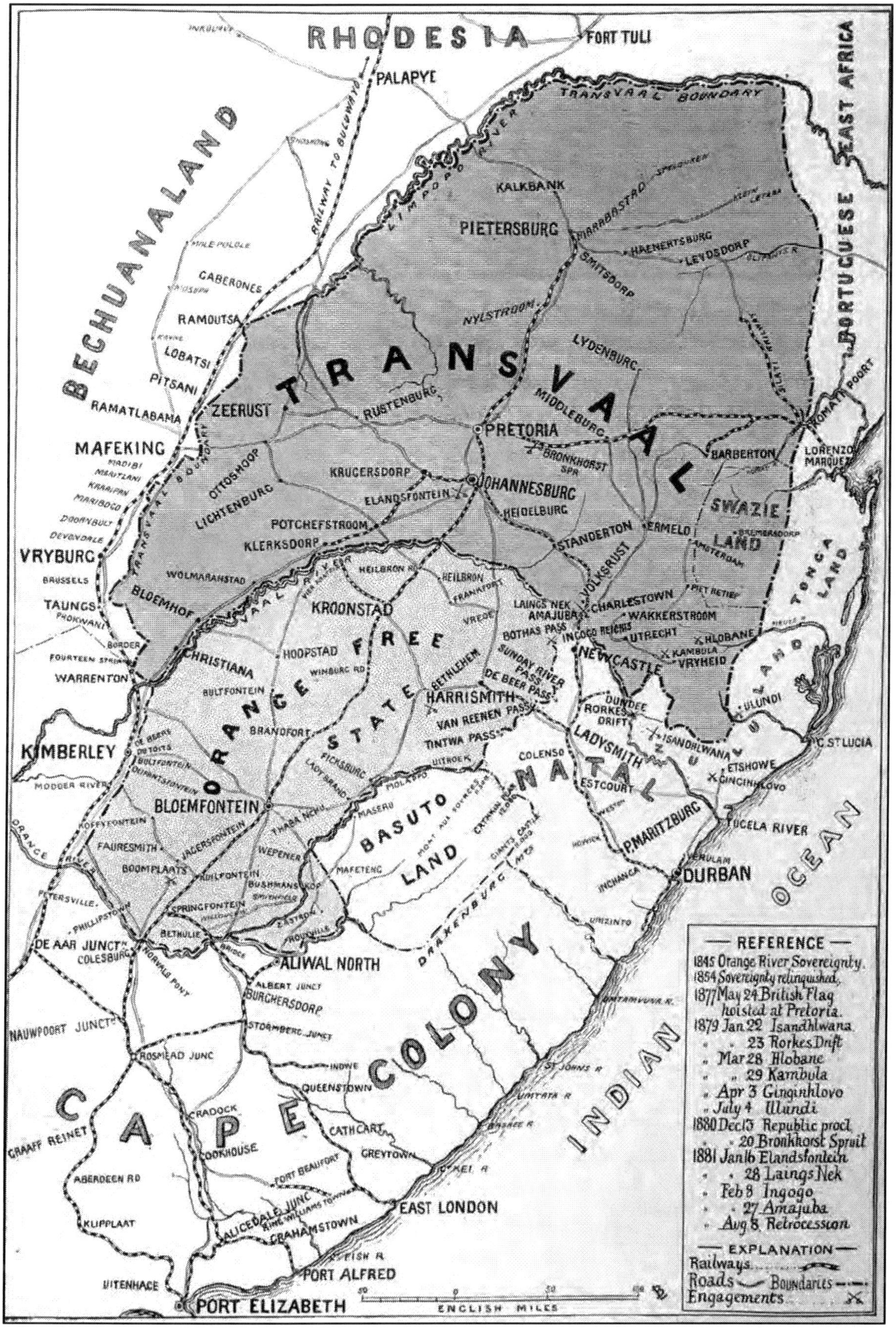

Theatre of Conflict.[25]

Maps 2

Hamilton's March Route.

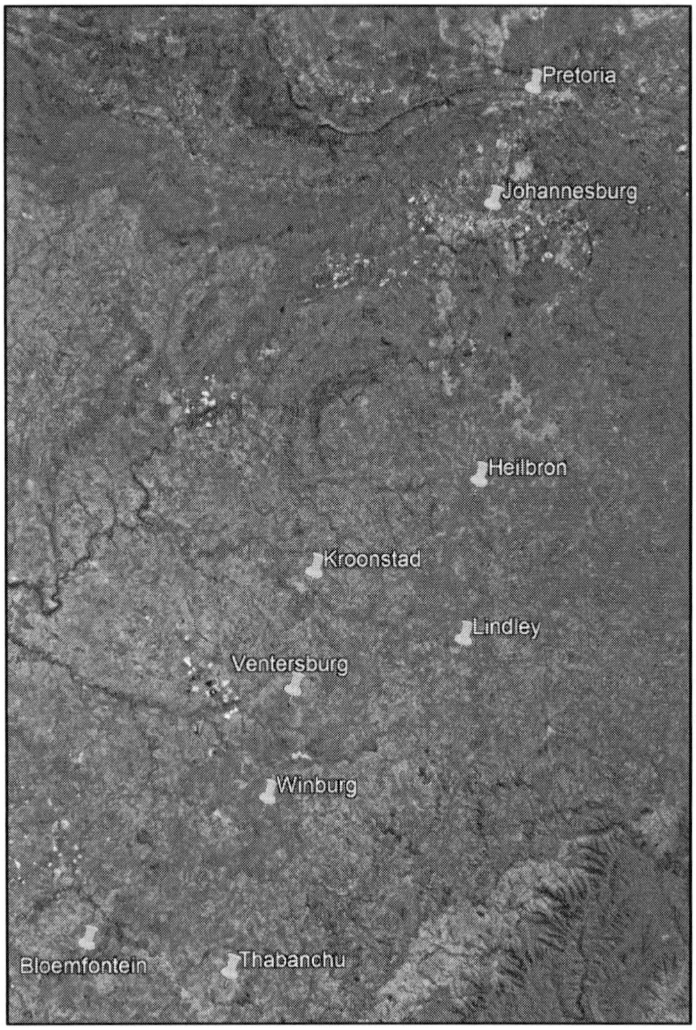

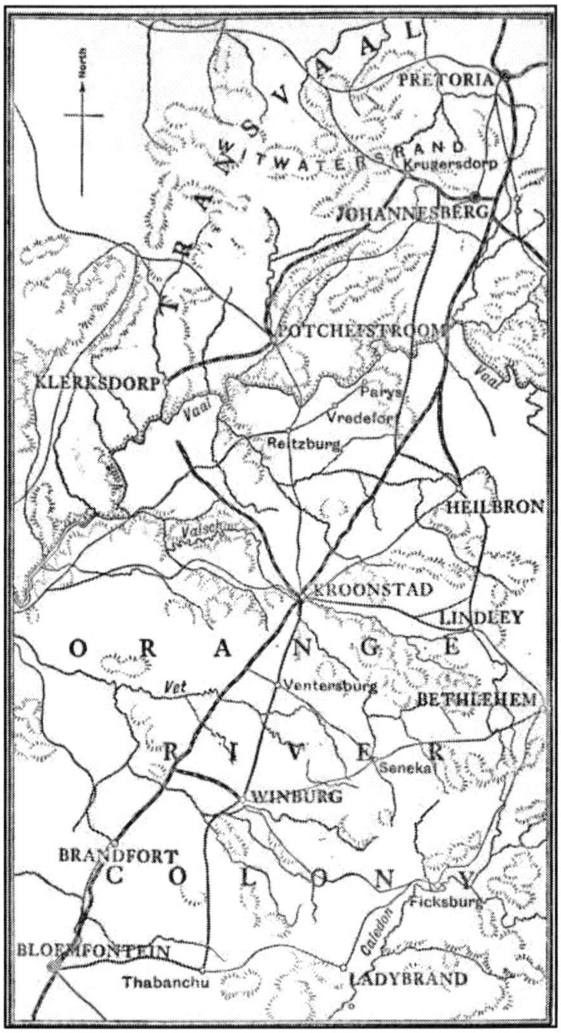

Map data: Google Image© 2023 Image Landsat/Copernicus. Bloemfontein to Pretoria.[26]

From Appendix: Distance traversed, 401 miles in a straight line. Time on the march, 45 days. Halts, 10 days.

Effective Fighting Strength
11,000 Men.
4,600 Horses.
8,000 Mules.
36 Field guns.
2 Five-inch guns
23 Machine guns.
6 Pom-poms.
[Departed] Bloemfontein, April 22.
Arrived at Pretoria on June 5.

General actions on nine days:
Israel's Poorte, April 25.
Houtnek, April 30 and May 1.
Welkom, May 4.
Sand River, May 10.
Affair of Lindley, May 20.
Doornkop (Florida), May 29.
Six Mile Spruit (Pretoria), June 4.
Diamond Hill, June 11 and 12.
Eighteen days' skirmishes.

Towns captured:
Thabanchu.
Winburg.
Ventersburg.
Kroonstadt.
Lindley.
Heilbron.
Johannesburg.
Pretoria.

Military Unit Sizes

Squad: 6 to 20 soldiers. Commanded by a non-commissioned officer. (NCO)

Platoon: 20-50 soldiers. Commanded by a Lieutenant.

Company: 100 to 250 soldiers. (3-7 Platoons) Commanded by a Major or Captain.

Battalion: 300 to 1,000 soldiers (2 or more Company's. Commanded by a Lieutenant Colonel.

Regiment: 800 or more soldiers (2 or more Battalions) Commanded by a Colonel.

Brigade: 2,000 to 4,000 soldiers (3-6 Battalions) Commanded by a Brigadier General.

Division: 6,000 to 25,000 soldiers (Regiments and Brigades) A combined arms unit capable of independent operations. Commanded by a Major General.

Corps: Two or more Divisions. (15,000 to 50,000 soldiers) Commanded by a Lieutenant General.

Army (Field Army): Two or more Corps. (80,000 to 200,000 soldiers) Commanded by a Lieutenant General or Army General.

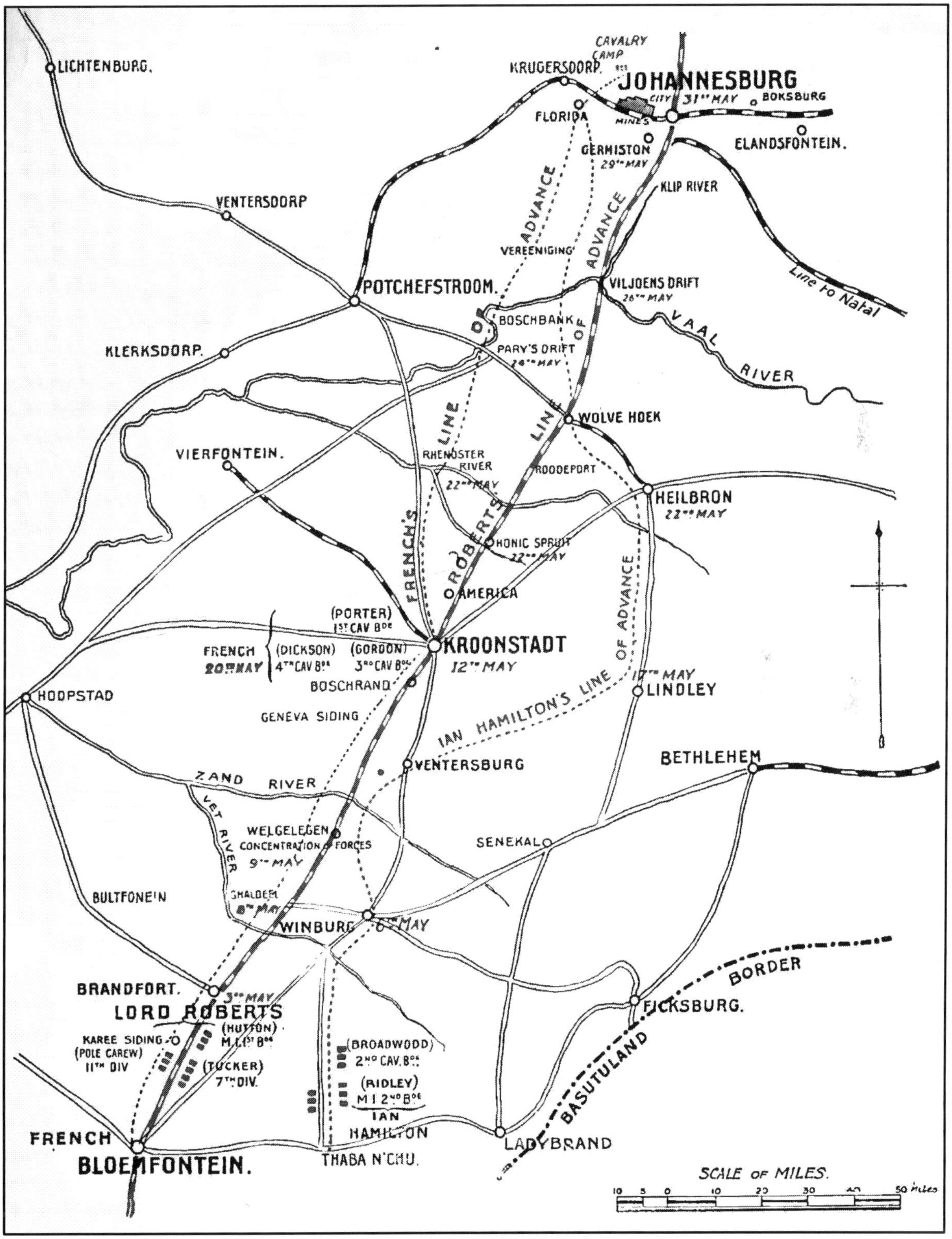

Hamilton's line of March with dates.[27]

Orange Free State Maps

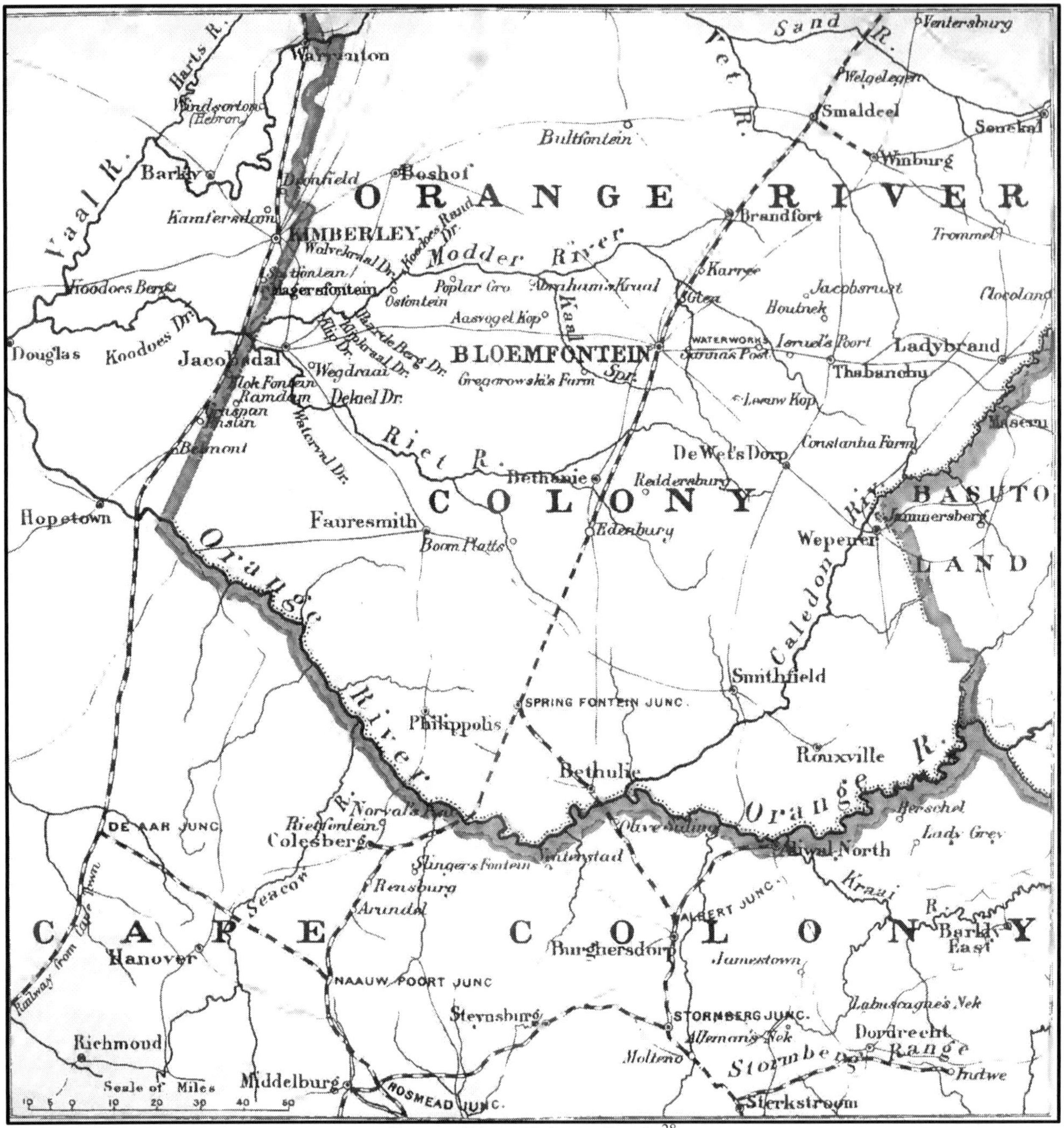

Orange River Colony South.[28]

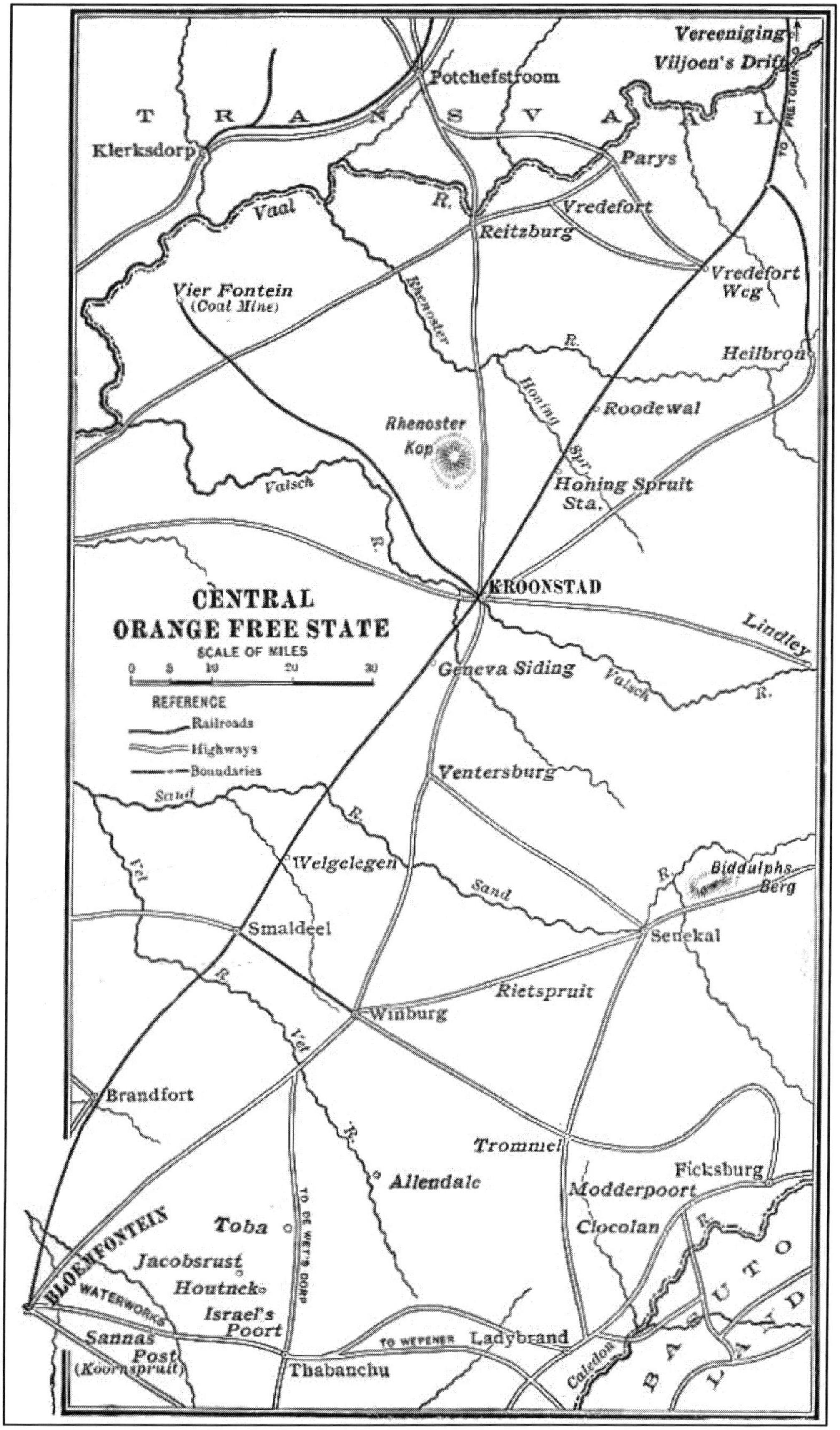

Central Orange Free State.[29]

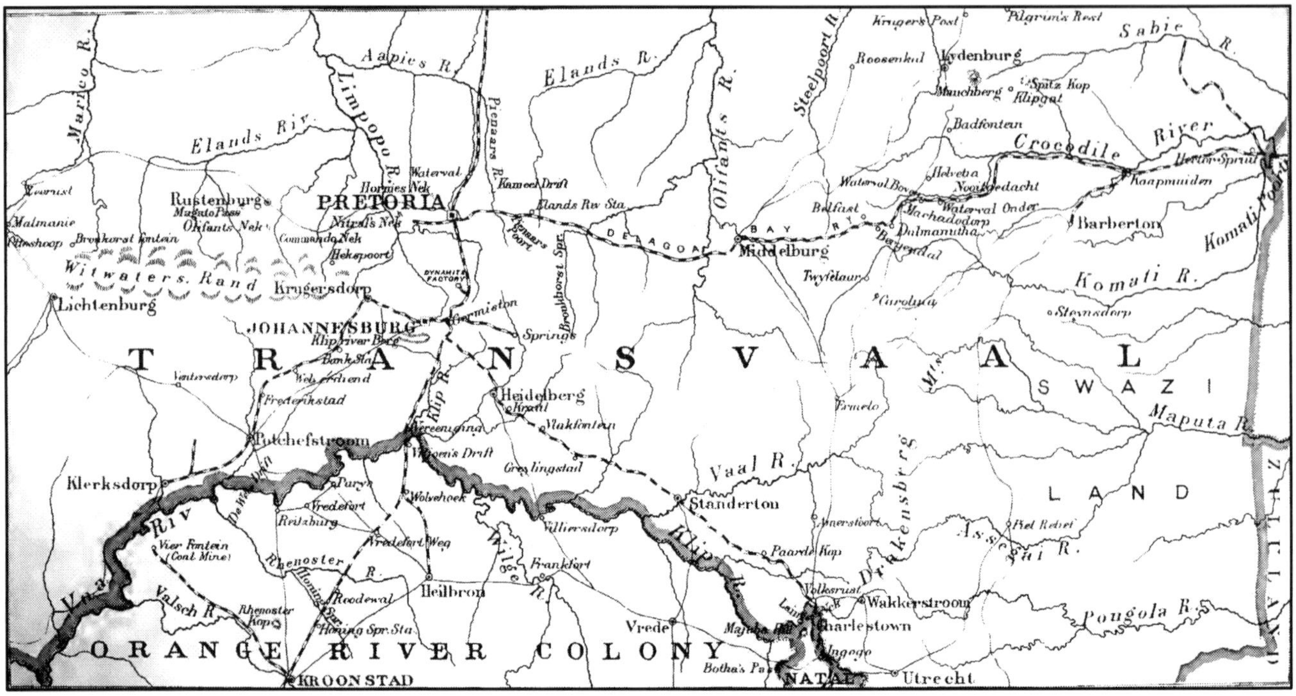

Southern Transvaal & Northern OFS.[30]

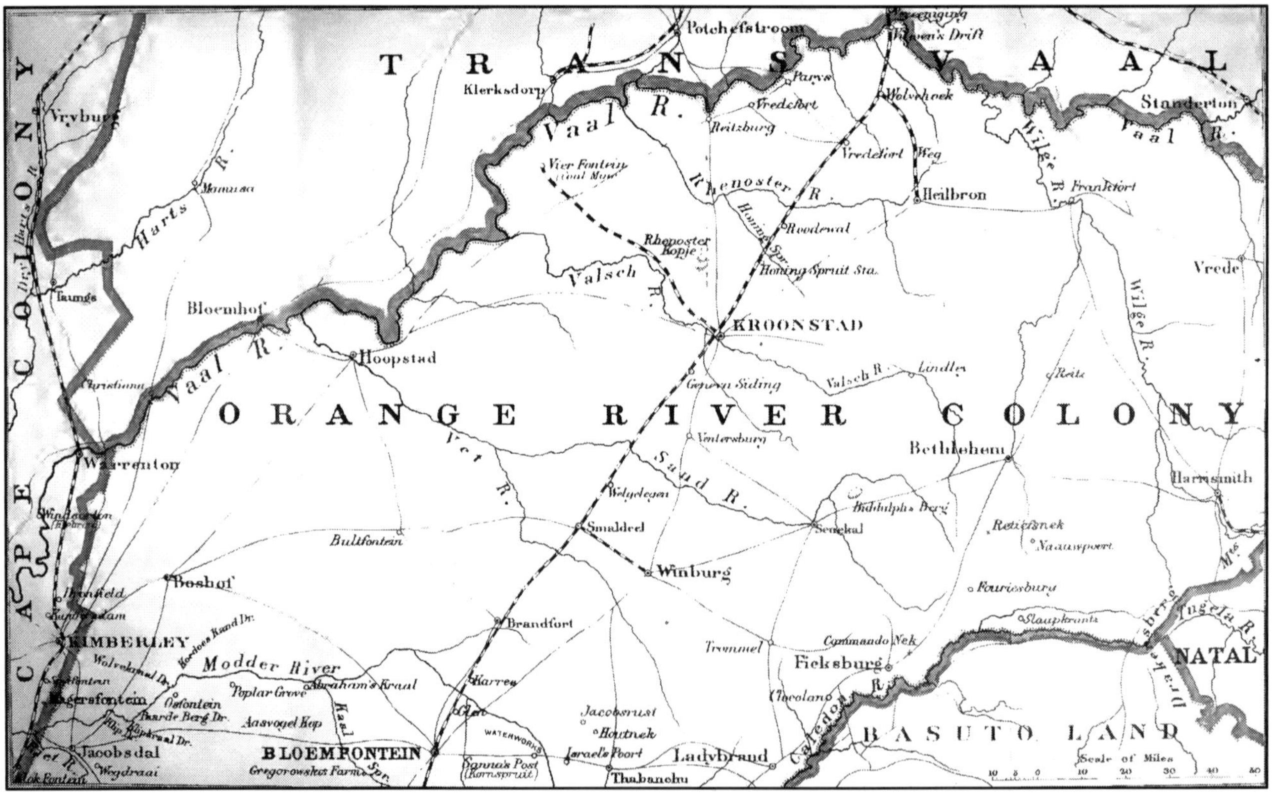

Orange River Colony North.[31]

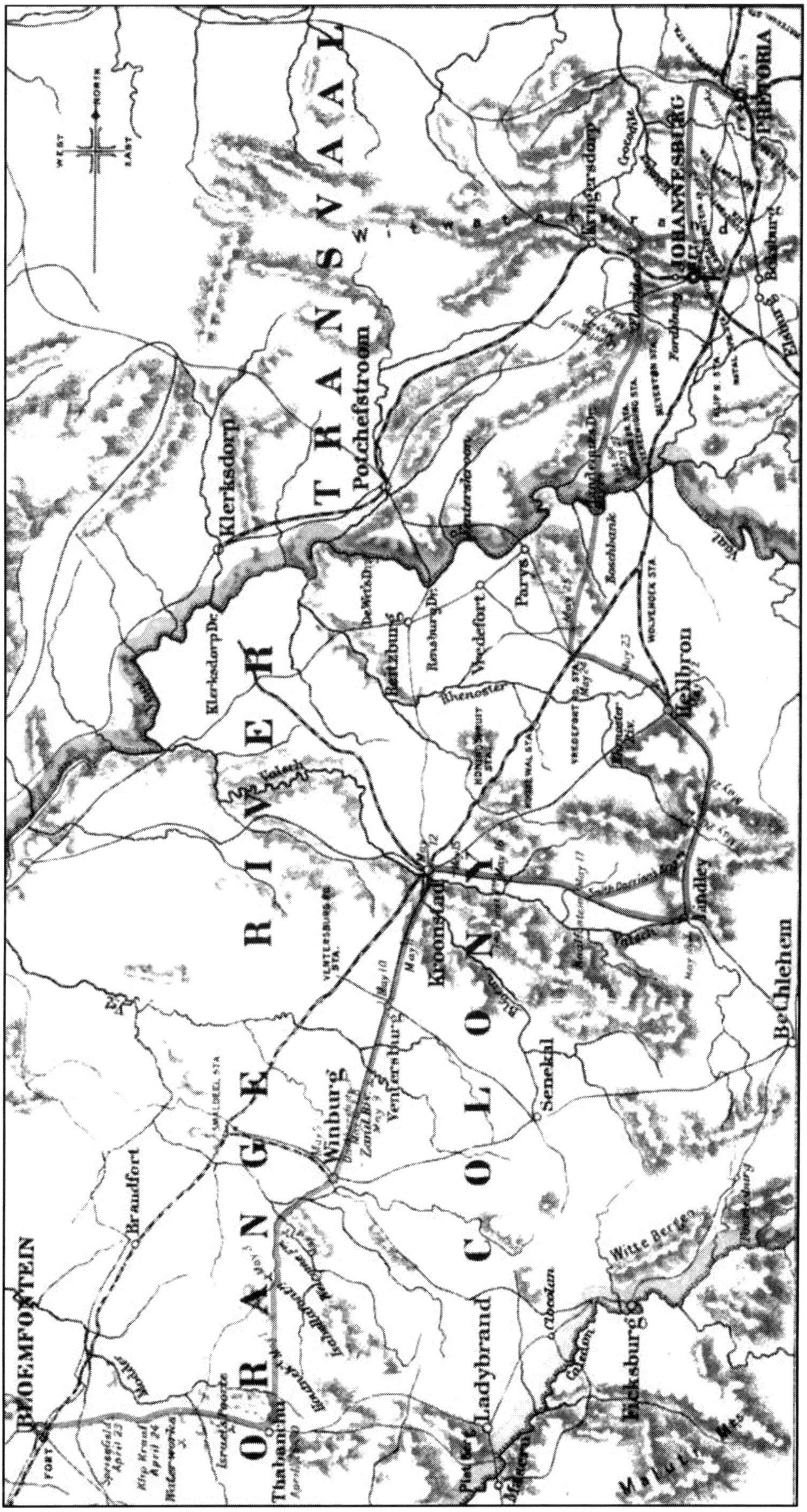

Map of March from Bloemfontein to Pretoria.[32] Original to book. North is top of page.

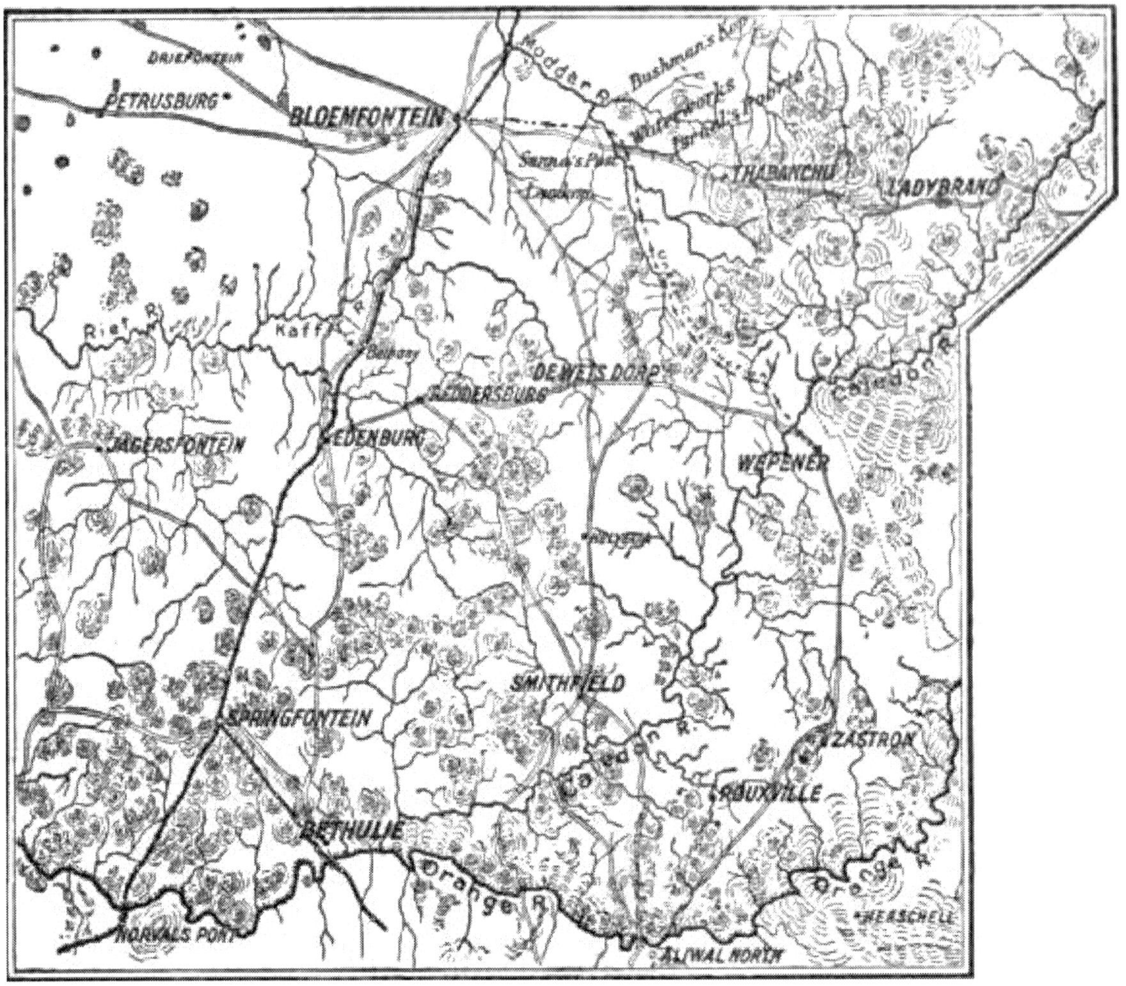

[Southern] Operations in the Orange Free State in April of 1900.[33]

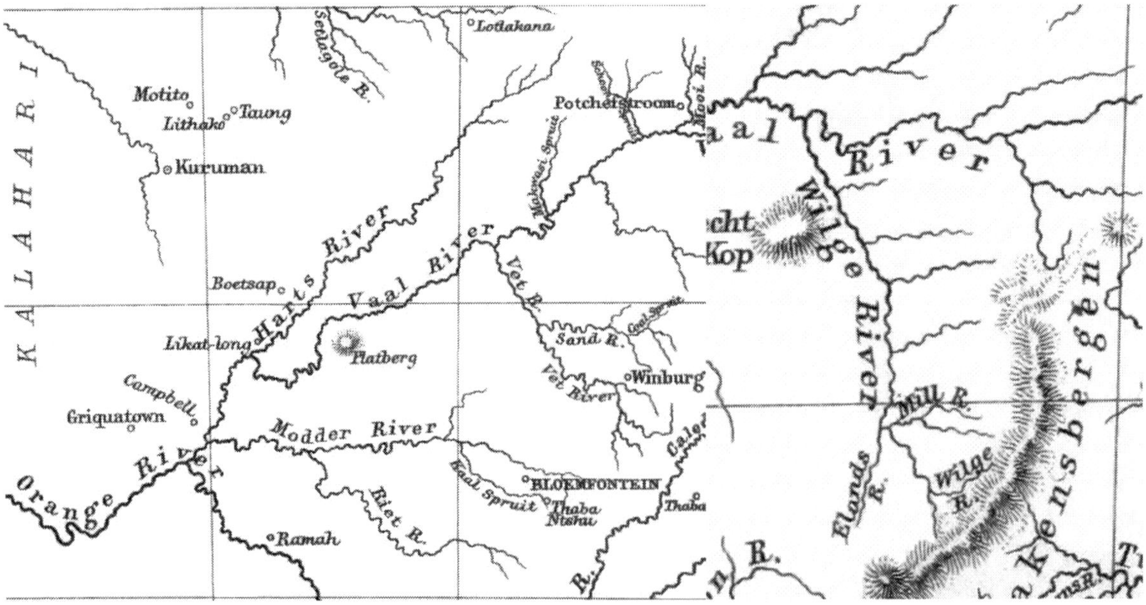

River system of the High Veld. A Composite of [34] and [35].

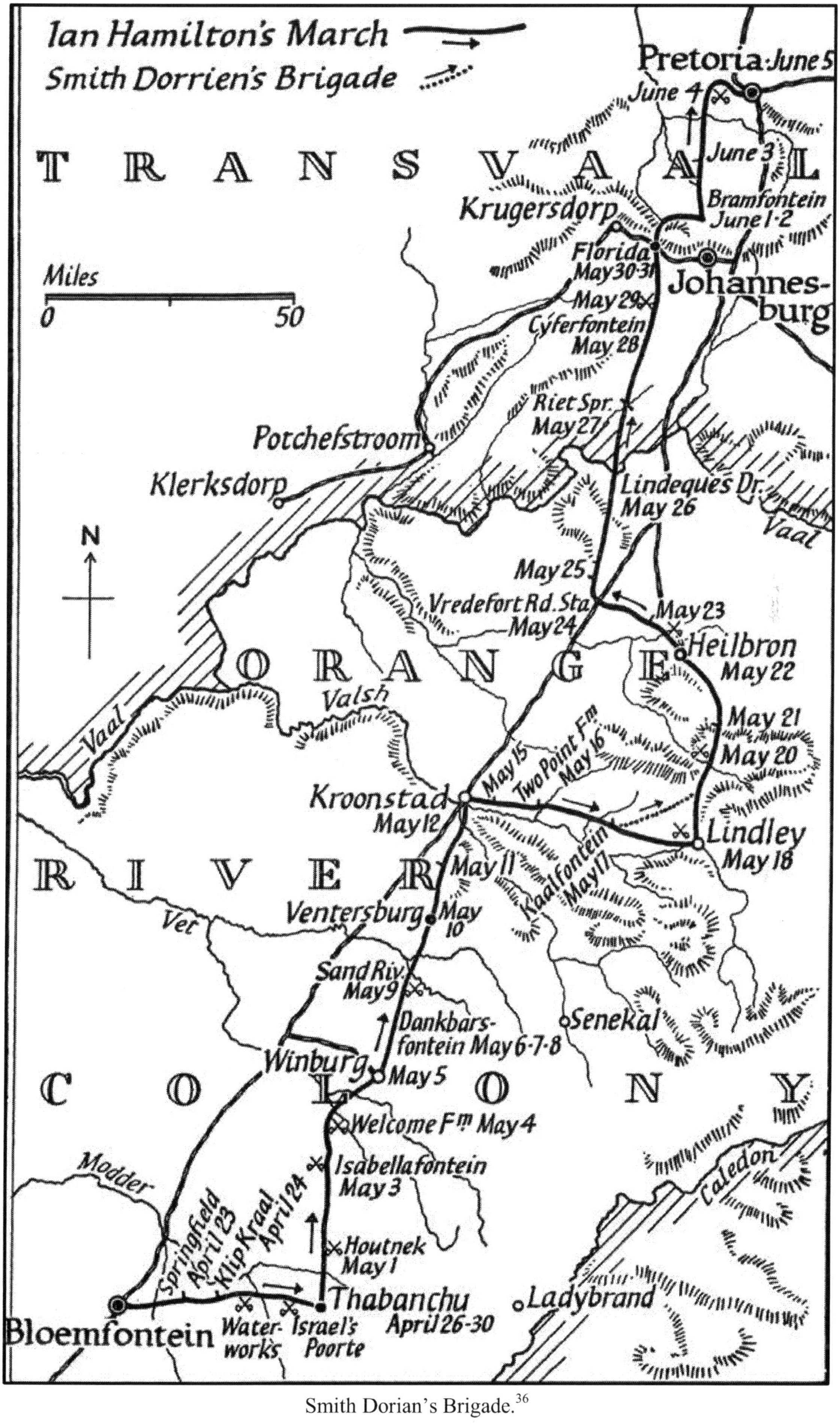

Smith Dorian's Brigade.[36]

Table of Distances

Table of Distances	Miles		Kilometers
By Sea			
London to Capetown	6,181		9,947
Capetown to Port Elizabeth	428		689
Port Elizabeth to East London	131		211
East London to Durban	253		407
Durban to Delagoa Bay	300		483
Capetown to Durban	812		1,307
By Rail	Miles		Kilometers
Capetown to De Aar Junction	501		806
Capetown to Kimberley	647		1,041
Capetown to Vryburg	774		1,246
Capetown to Mafeking	870		1,400
Capetown to Bulawayo	1,361		2,190
Capetown to Naauwpoort	570		917
Capetown to Norval's Pont	628		1,011
Capetown to Bloemfontein	750		1,207
Capetown to Johannesburg	1,014		1,632
Capetown to Pretoria	1,040		1,674
Port Elizabeth to Naauwpoort	270		435
Port Elizabeth to Bloemfontein	450		724
Port Elizabeth to Pretoria	740		1,191
Durban to P'Maritzburg	70		113
Durban to Ladysmith	189		304
Durban to Harrismith	249		401
Durban to Laing's Nek	301		484
Durban to Johannesburg	483		777
Durban to Pretoria	511		822
Lorenzo Marques to Komati Poort	58		93
Lorenzo Marques to Johannesburg	395		636
Larenzo Marques to Johannesburg	395		636

Table of Distances Source.[22]

[22] The Boer war: its causes and its interest to Canadians by Emerson Bristol Biggar. 3rd.Ed. 1899.

I. A Roving Commission

In the train near Pieters, Natal: March 31st 1900.[23]

Ladysmith, her garrison and her rescuers, were still recovering, the one from the effects of long confinement, the other from over-exertion. All was quiet along the Tugela except for the plashing of the waters, and from Hunger's Poorte[24] to Weenen[25] no sound of rifle or cannon shot disturbed the echoes.

Transport crossing the Tugela River. Note partially destroyed bridge in background.[37] Pontoon Bridge on Tugela River.[38] Pontoon Bridge on Tugela River. Note Falls in center distance and rapids in foreground.[39]

The war had rolled northward: the floods of invasion that had isolated—almost overwhelmed—Ladysmith and threatened to submerge the whole country had abated and receded, so that the Army of Natal might spread itself out to feed and strengthen at its leisure and convenience on the reconquered territory.

Knox's[26] (Ladysmith) Brigade went into camp at Arcadia, five miles west of the town. Howard's (Ladysmith) Brigade retired to the breezy plains south of Colenso. Clery's[27] Division—for the gallant Clery, recovered from his sickness, had displaced the gallant and successful Lyttelton[28]—moved north and encamped beyond Elandslaagte along the banks of Sunday's River.

[23] His last despatch included in London to Ladysmith was dated March 10th, 1900.

[24] Chuniespoort (recently Chuenespoort) is a mountain pass situated in the Limpopo Province, on the R37, the road between Polokwane and Lebowakgomo, South Africa. The river flowing through this gorge of the Strydpoort Mountains is known as the Chunies River, one of the tributaries of the Olifants River. Chuniespoort - Wikipedia

[25] Weenen (Dutch for "wept") [Near Durban] is the second oldest European settlement in Natal. It is situated on the banks of the Bushman River and was the site of Weenen Massacre.
The farms around the town grow vegetables, Lucerne, groundnuts, and citrus fruit. Weenen - Wikipedia

[26] Knox Eustace Chaloner Lieutenant Colonel He was the son of Lieutenant General Richard Knox, Honorary Colonel of the 18th Hussars, was born in March 1860, and served in the ranks for nearly three and a half years. He was promoted Lieutenant in the 18th Hussars, August 1882, Captain April 1886, Major May 1896, Lieutenant Colonel September 1900, and brevet Colonel the following November. Colonel Knox was adjutant of his regiment from 1886 to 1891. adjutant of the cavalry depot from 1894 to 1897. He served throughout the Nile Expedition of 1884-85 with the Light Camel Regiment, and took part in the operations of the Desert Column. In the South African War he took part in the action at Talana Hill, and the defence of Ladysmith, being mentioned in despatches by Lieutenant General Sir George White, March 23rd, 1900, and by General Sir Redvers Buller, November 9th 1900 (LG February 8th 1901). In December 1900, he was appointed to the command of the 2nd Cavalry Brigade in South Africa, which appointment he held till July 1901. He died in London of acute pneumonia February 18th 1902. He was buried at Brookwood. Source: Donner.

[27] Major-General Sir Cornelius Francis Clery (13 February 1838 – 25 June 1926) was a British Army officer who took part in the Anglo-Zulu War and later commanded the 2nd Division during the Second Boer War. Francis Clery - Wikipedia

[28] Lyttelton served in the Second Boer War as commander of the 4th Brigade in South Africa from 9 October 1899. He temporarily became general officer commanding the 2nd Division in February 1900, then commanded the 4th Division, and was involved in the Battle of Spion Kop in January 1900 and the Battle of Vaal Krantz in February 1900, leading to the Relief of Ladysmith later that month. Promoted to lieutenant general for distinguished service in the field on 22 March 1900. Lord Roberts in his despatch referred to Lyttelton as an officer "with great coolness under fire, and considerable tactical knowledge and resource...an excellent commander in the field." He was in command of the troops in Natal until June 1902, when he became Commander-in-Chief of the whole of South Africa following the end of the Second Boer War the previous month. In this role Lyttelton and his wife sought to repair relations with the Boer community. In the South Africa honours list published on 26 June 1902, Lyttelton was knighted as a Knight Commander of the Order of the Bath (KCB).Neville Lyttelton - Wikipedia

Hunter's[29] Division was disposed with one brigade at Elandslaagte and one at Tinta Inyoni. Warren,[30] whom it was no longer necessary to send to the Cape Colony, established himself and his two brigades north of Ladysmith, along the railway line to the Orange Free State. Brocklehurst,[31] with the remnants of what had once been almost a Cavalry Division, and now could scarcely mount three squadrons, occupied a neighbouring plain, sending his regiments one by one to Colenso, or even Mooi River, to be re-horsed; and around all this great Army, resting after its labours and preparing for fresh efforts.

Remount Camp at Stellenbosch. "The heavy mortality amongst the horses and mules during the campaign put the British forces to considerable inconvenience, and large numbers were purchased in North and South America, Spain and Italy."[40] Army Veterinary Department. Individual Inspection of doubtful horses brought for sale.[41]

A Group of Bengal Lancers at the remount depot at Kroonstad.[42] Landing mules for transport.[43]

The Cavalry brigades of Dundonald[32] and Burn-Murdoch[33] drew an immense curtain of pickets and patrols which extended from Acton Homes in the east, through Bester's Station right round to Wessels Nek and further still, and which enabled the protected soldiers within to close their eyes by night and stretch their legs by day.

[29] General Sir Archibald Hunter, (6 September 1856 – 28 June 1936) was a senior officer in the British Army. At the outbreak of the Second Boer War in October 1899, Major General Hunter (although actually Chief of Staff to General Sir Redvers Buller's 1st Army Corps) was on the staff of Sir George White's Natal Field Force during the Battle of Ladysmith in Natal and the subsequent 118-day siege On 8 December he successfully led a small raid against one of the Boers' Creusot "Long Tom" guns and a howitzer which they disabled with cotton charges who distinguished himself during the Boer War. He was Governor of Omdurman, in Sudan, and later of Gibraltar. Archibald Hunter - Wikipedia

[30] General Sir Charles Warren, GCMG, KCB, FRS (7 February 1840 – 21 January 1927) was an officer in the British Royal Engineers. He was one of the earliest European archaeologists of the Biblical Holy Land, and particularly of the Temple Mount. Much of his military service was spent in British South Africa. Previously he was police chief, the head of the London Metropolitan Police, from 1886 to 1888 during the Jack the Ripper murders. His command in combat during the Second Boer War was criticised, but he achieved considerable success during his long life in his military and civil posts. Charles Warren - Wikipedia

[31] Brocklehurst J F Colonel Entered 1874; Colonel, 1899. Staff service: DAA and QMG, Egypt, 1884-85; Equerry to the Queen, 1899; Major General, Cavalry Brigade, Natal, 1899. War Service: Egyptian Expedition, 1882 (medal with clasp; bronze star); Sudan Expedition, 1884-85 (Despatches; clasp; Brevet of Major); South African War, 1899-1900. Source: List of CB recipients.

[32] Lieutenant General Douglas Mackinnon Baillie Hamilton Cochrane, 12th Earl of Dundonald, 29 October 1852 – 12 April 1935), styled was a Scottish representative peer and a British Army general. He served in the Second Boer War and in November 1899 he was appointed Commander of the Mounted Brigade, part of the South Natal Field Force. He took part in the Relief of Ladysmith in February 1900, although his South African troops, unimpressed by his leadership, referred to him as "Dundoodle." Douglas Cochrane, 12th Earl of Dundonald - Wikipedia File:12th Earl of Dundonald.jpg - Wikimedia Commons

Sir Neville Gerald Lyttelton,[34] General Sir Charles. Warren[44] Douglas Cochrane Earl of Dundonald[45]

Meanwhile, the burghers had all retreated to the Drakensburg[35] and the Biggarsburg[36] and other refuges, from which elevated positions they defied intrusion or attack, and their scattered line stretched in a vast crescent even around our widely extended front from the Tintwa Pass,[37] through Waschbank[38] to Pomeroy.[39]

The Drakensberg.[46]

[33] J. F. Burn-Murdoch Lieutenant Colonel Born 1859, was the son of the Reverend Canon Burn-Murdoch. Entered 1878; Brevet-Colonel, 1st Dragoons, 1898. Staff service: Brigade-Major of Cavalry, 1890-91; Brigade-Major Cavalry Brigade, Aldershot, 1891-94; employed with Egyptian Army, 1894-95; Brigadier - General Cavalry Brigade, South Africa, February 1903. War service: Sudan Expedition, 1884-85 (medal with clasps; bronze star); Expedition to Dongola, 1896 (Despatches, Brevet of Lieutenant Colonel, Egyptian Medal with 2 clasps); South African War, 1899-1900; on Staff. Source: List of CB recipients.

[34] 1902 Times History War in S.A Vol 3 Photo by Gregory & Co.

[35] The Drakensberg is the eastern portion of the Great Escarpment, which encloses the central Southern African plateau. The Great Escarpment reaches its greatest elevation – 2,000 to 3,482 metres (6,562 to 11,424 feet) within the border region of South Africa and Lesotho. Drakensberg - Wikipedia

[36] Biggarsberg is a series of hills stretching south of Glencoe and Dundee in a north western/south eastern direction. It is named for Alexander Biggar. Biggarsberg - Wikipedia

[37] Endoumeni Pass, Is situated in the KwaZulu-Natal province of South Africa on the road between Dundee and Wasbank. Endoumeni Pass - Wikipedia

[38] Wasbank is a town on the Wasbank River, 25 km (16 mi) southwest of Dundee. Takes its name from the Wasbank River. Wasbank - Wikipedia

[39] Pomeroy is located 72 km north of Greytown and 56 km south-south-east of Dundee. The town was named Pomeroy after Sir George Pomeroy Colley who led the ill-fated British force during the Battle of Majuba Hill in 1881. Pomeroy, KwaZulu-Natal - Wikipedia

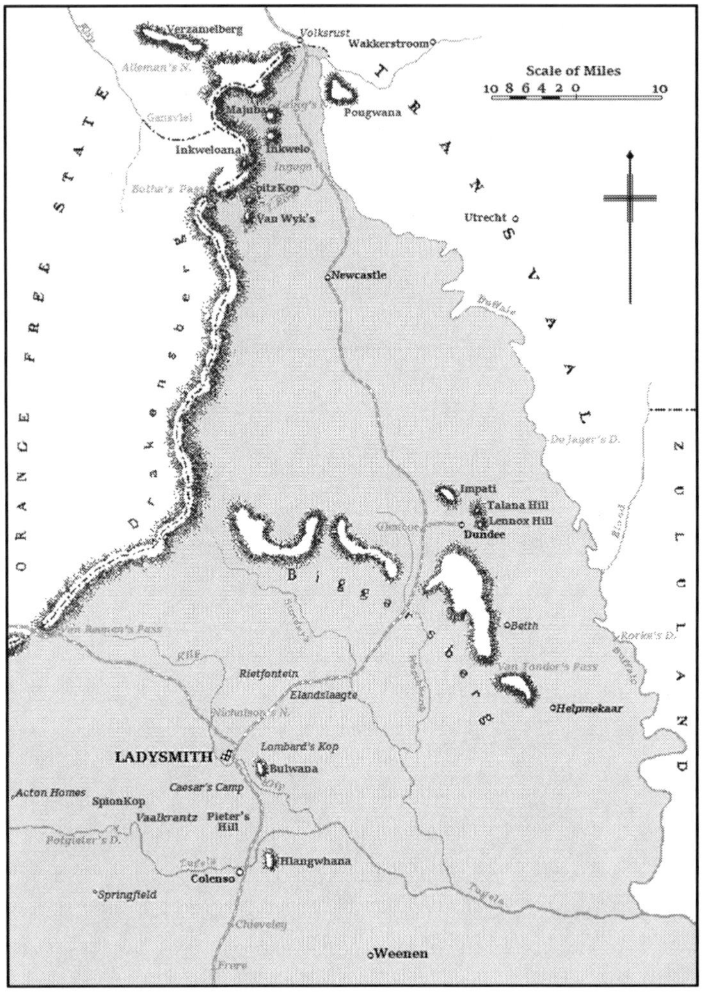

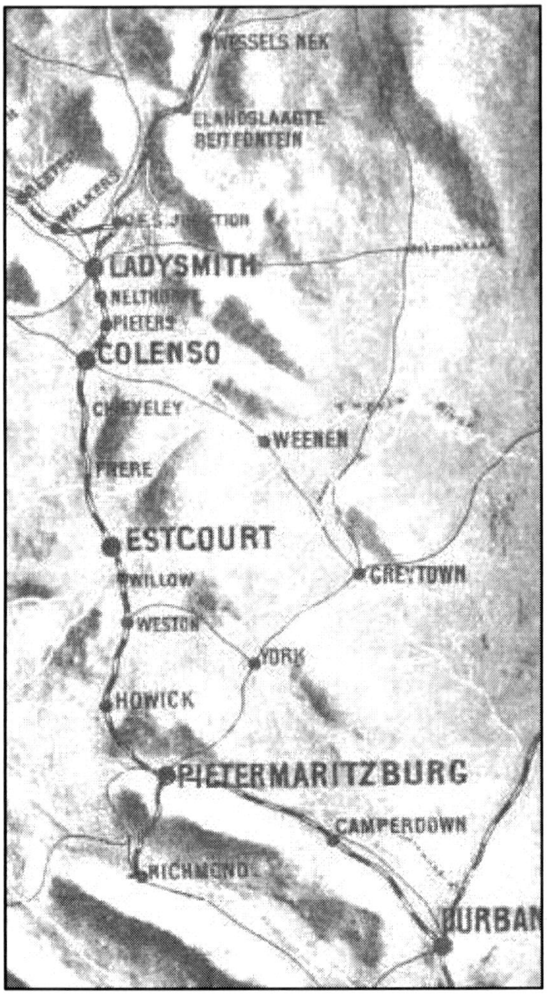

Northern Natal. Note Drakensberg all along the western frontier.[47] Country from Durban to Ladysmith.[48]

But with the exception of outpost skirmishes, wholly unimportant to those not engaged in them, a strange peace brooded over Natal, and tranquillity was intensified by the recollection of the struggle that was over and the anticipation of the struggle that impended. It was a lull in the storm.

All this might be war, but it was not journalism. The tempest for the moment had passed, and above the army in Natal the sky was monotonously blue. It was true that dark clouds hung near the northern horizon, but who should say when they would break? Not, at any rate, for three weeks, I thought, and so resolved to fill the interval by trying to catch a little of the tempest elsewhere.

After the relief of Ladysmith four courses offered themselves to Sir Redvers Buller. [1] To stand strictly on the defensive in Natal and to send Lord Roberts every gun and man who could be spared; [2] to break into the Free State by forcing Van Reenen's Pass[40] [3] or the Tintwa; [Pass] to attack the twelve thousand Boers in the Biggarsburg, clear Natal, and invade the Transvaal through the Vryheid district;[41] and, lastly, to unite and reorganise and co-operate with Lord Roberts's main advance either by striking west or north.

Which course would be adopted? I made inquiries. Staff officers, bland and inscrutable—it is wonderful how well men can keep secrets they have not been told—continued to smile and smile. Brigadiers frankly confessed their ignorance. The general-in-chief [Buller] observed pleasantly that he would 'go for' the enemy as soon as he was ready, but was scarcely precise about when and where.

[40] Van Reenen's Pass is a pass through the Drakensberg mountains in South Africa. The N3 freeway was constructed through the pass and is the main route from Durban to Johannesburg on the border between the provinces of KwaZulu-Natal and the Free State. The village of Van Reenen sits atop the pass and the two nearest town's are Ladysmith and Harrismith. Van Reenen's Pass - Wikipedia

[41] Vryheid is a coal mining and cattle ranching town in northern Natal. Vryheid is the Afrikaans word for "freedom." Vryheid - Wikipedia

It was necessary to go to more humble sources for truth, and after diligent search I learned from a railway porter, or somebody like that, that all attempts to repair the bridge across the Sunday's River[42] had been postponed indefinitely. This, on further inquiry, proved to be true.

An example of a bridge blown up by the Boers.[49] Temporary bridge over Sundays River.[50]

Now, what does this mean? It means, I take it, that no direct advance against the Biggarsburg is intended for some time; and as the idea of reducing the Natal Army to reinforce the Cape Colony forces has been definitely abandoned the western line of advance suggests itself.

It would be absurd to force Van Reenen's Pass[43] with heavy loss of life, when by waiting until the main Army has reached, let us say, Kroonstad,[44] we could walk through without opposition; so that it looks very likely that the Natal troops will do nothing until Lord Roberts's advance is more developed, and that then they will enter the Free State and operate in conjunction with him, all of which is strategy and common-sense besides. At any rate there will be a long delay.

Van Reenen's Pass.[45] Boer Soldiers holding Van Reenen's Pass.[51]

Therefore, I said to myself, I will go to Bloemfontein,[46] see all that may be seen there and on the way, and rejoin the Natal Army when it comes through the passes. Such was the plan, and the reader shall be a witness of its abandonment.

[42] Sundays river is a river in the Eastern Cape Province of South Africa. The source of the 250 kilometres (160 mi) long river is in the Sneeuberge (the highest mountain range in the former Cape Province) near Nieu-Bethesda. The river then flows in a general South/Southeasterly direction, before winding its way through the Zuurberg Mountains and then into the fertile Sundays River Valley. It empties into the Indian Ocean at Algoa Bay. Sundays River - Wikipedia

[43] Van Reenen's Pass is a pass through the Drakensberg mountains N3 freeway, the main road between Durban and Johannesburg, was constructed through the pass with the northern end of the pass lying at the border of the provinces of KwaZulu-Natal and the Free State. The village of Van Reenen sits atop the pass and the two nearest towns are Ladysmith and Harrismith, situated in the respective KwaZulu-Natal and Free State provinces. Van Reenen's Pass - Wikipedia

[44] Kroonstad (Afrikaans "Crown City") is the third largest city in the Free State and an important railway junction on the main line from Cape Town to Johannesburg. Kroonstad - Wikipedia

[45] File:Van Reenen Pass 1.jpg - Wikimedia Commons Author: Peter Black and Mario 1952.

[46] Bloemfontein, Afrikaans: lit. "fountain of flowers") is the present capital of the Orange Free State. Bloemfontein - Wikipedia

I left the camp of Dundonald's Brigade early in the morning of the 29th of March, and riding through Ladysmith, round the hill on which stands the battered convent, now serving as headquarters, and down the main street, along which the relieving Army had entered the city, reached the railway station and caught the 10 A.M. down train.

Ladysmith, 1900. Town Hall with damaged clock tower and cannon in foreground.[52]

We were delayed for a few minutes by the departure for Elandslaagte of a train load of Volunteers, the first to reach the Natal Army, and the officers hastened to look at these citizen soldiers. There were five companies in all, making nearly a thousand men, fine looking fellows, with bright intelligent eyes, which they turned inquiringly on every object in turn, pointing and laughing at the numerous shell holes in the corrugated iron engine sheds and other buildings of the station.

A few regulars—sunburnt men, who had fought their way in with Buller—sauntered up to the trucks, and began a conversation with the reinforcement. I caught a fragment: 'Cattle trucks, are they? Well, they didn't give us no blooming cattle trucks. No, no! We came into Ladysmith in a first-class doubly extry Pullman car. 'Oo sent 'em? Why, President ---- Kruger,[47] of course,' whereat there was much laughter.

I must explain that the epithet which the average soldier uses so often as to make it perfectly meaningless, and which we conveniently express by a ----, is always placed immediately before the noun it is intended to qualify. For instance, no soldier would under any circumstances say '---- Mr. Kruger has pursued a ---- reactionary policy,' but 'Mr. ---- Kruger has pursued a reactionary ---- policy.' Having once voyaged for five days down the Nile in a sailing boat with a company of Grenadiers, I have had the best opportunities for being acquainted with these idiomatic constructions, and I insert this little note in case it may be useful to some of our national poets and minstrels.

The train started across the well-known ground, and how fast and easily it ran. Already we were bounding through the scrub in which a month before Dundonald's leading squadrons, galloping in with beating hearts, had met the hungry picket line.

Intombi Spruit hospital camp was reached in a quarter of an hour. Hospital camp no longer, thank goodness! Since the bridge had been repaired the trains had been busy, and two days before I left the town the last of the 2,500 sick had been moved down to the great hospital and convalescent camps at Mooi River and Highlands, or on to the ships in the Durban Harbour.

[47] Paul Kruger is the current President of the Transvaal Republic. Paul Kruger - Wikipedia

British field hospital train collecting wounded soldiers.[53]

The Arrival of a Hospital Train.[54]

Getting on Board a Hospital Ship.[55]

Arrival at Cape Town of Wounded from Natal.[56]

Durban circa 1900. Note the size of ships and the small amount of docking space at the wharf.[57]

Nothing remained behind but 100 tents and marquees, a stack of iron cots, the cook houses, the drinking-water tanks, and 600 graves. Ghastly Intombi had faded into the past, as a nightmare flies at the dawn of day.

Estcourt camp in Natal before the Battle of Colenso.[58]

We sped swiftly across the plain of Pieters, and I remembered how I had toiled across it, some five months before, a miserable captive, casting longing eyes at the Ladysmith balloon, and vigilantly guarded by the Boer mounted escort.

The Balloon at Ladysmith.[59]

Observation Balloon at Ladysmith.[60]

Then the train ran into the deep ravine between Barton's Hill and Railway Hill, the ravine the Cavalry had 'fanned' on the day of the battle, and, increasing its pace as we descended towards the Tugela, carried us along the whole front of the Boer position. Signs of the fighting appeared on every side. Biscuit tins flashed brightly on the hillside like heliographs.

In places the slopes were honey-combed with little stone walls and traverses, masking the sheltering refuges of the Infantry battalions during the week they had lain in the sun blaze exposed to the cross fire of gun and rifle. White wooden crosses gleamed here and there among the thorn bushes. The dark lines of the Boer trenches crowned the hills. The train swept by—and that was all.

Heliograph in use.[61]

Boer Trenches at Colenso.[62]

Boer trench near the Bridle Drift at Colenso.[63]

Looking up the Trench. The Mauser Ammunition Was Left by the Boers on Monte Christo.[64]

I knew every slope, every hillock and accident of ground, as one knows men and women in the world. Here was good cover. There was a dangerous space. Here it was wise to stoop, and there to run. Behind that steep kopje a man might scorn the shrapnel. Those rocks gave sure protection from the flanking rifle fire. Only a month ago how much these things had meant. If we could carry that ridge it would command those trenches, and that might mean the hill itself, and perhaps the hill would lead to Ladysmith. Only a month ago these things meant honour or shame, victory or defeat, life or death. An anxious Empire and a waiting world wanted to know about every one of them—and now they were precisely what I have said, dark jumbled mounds of stone and scrub, with a few holes and crevices scratched in them, and a litter of tin-pots, paper, and cartridge cases strewn about.

The train steamed cautiously over the temporary wooden bridge at Colenso and ran into the open country beyond.

New Trestle Under Construction.[65] "Trestle Bridge Colenso with first train (Princess Christian[48] Ambulance Train) Passing Over it at 7 a.m. on March 19th, 1900."[66] Note fast flowing water downstream.

The British guns, captured at Colenso, arriving at Pretoria.[67]

Colenso Bridge demolished during the campaign.[68]

Colenso bridge.[69]

[48] Princess Christian was the third daughter and fifth child of Queen Victoria and Prince Albert. Princess Helena of the United Kingdom - Wikipedia

On we hurried past the green slope where poor Long's artillery had been shot to bits, past Gun Hill, whence the great naval guns had fired so often, through Chieveley Camp, or rather through the site of Chieveley Camp, past the wreck of the armoured train—still lying where we had dragged it with such labour and peril, just clear of the line—through Frere and Estcourt, and so, after seven hours' journey, we came to Pietermaritzburg.

Destroyed Bridge at Frere.[70]

New Bridge at Frere.[71]

The overturned ganger truck.[72]

Close-up of damage done to train.[73]

Battle of Colenso. X marks where the Long's guns were taken. Note low temporary bridge. Colonel Long's Artillery Horses, Battle Of Colenso.[74]

Pietermartizburg – Capital of Natal.[75]

Principal street of Pietermartizburg.[76]

An officer who was travelling down with me pointed out the trenches on the signal hill above the town. 'Seems queer,' he said, 'to think that the Boers might so easily have taken this town. When we dug those trenches they were expected every day, and the Governor, who refused to leave the capital and was going to stick it out with us, had his kit packed ready to come up into the entrenchments at an hour's notice.'

It was very pleasant to know that those dark and critical days were gone, and that the armies in the field were strong enough to maintain the Queen's dominions against any further invasion; yet one could not but recall with annoyance that the northern part of Natal was still in the hands of the enemy. Not for long, however, shall this endure.

After waiting in Pietermaritzburg long enough only to dine, I proceeded by the night train to Durban, and was here so fortunate as to find a Union boat, the *Guelph*,[49] leaving almost immediately for East London. The weather was fine, the sea comparatively smooth, and the passengers few and unobtrusive, so that the voyage, being short, might almost be considered pleasant.[50]

The captain took the greatest interest in the war, which he had followed with attention, and with the details and incidents of which he was extraordinarily familiar. He had brought out a ship full of Volunteers, new drafts, and had much to say concerning the British soldier and his comrades in arms.

The good news which had delighted and relieved everyone had reached him in the most dramatic and striking manner. When they left England Roberts had just begun his welcome advance, and the public anxiety was at its height. At Madeira there was an English cable to say that he was engaging Cronje, and that no news had arrived for three days. This was supplied, however, by the Spanish wire, which asserted with circumstantial details that the British had been heavily defeated and had fled south beyond the Orange River. With this to reflect on they had to sail. Imagine the doubts and fears that flourished in ten days of ignorance, idleness, and speculation. Imagine with what feelings they approached St. Helena. He told me that when the tug-boat came off no man dared hail them for news. Nor was it until the launch was alongside that a soldier cried out nervously, 'The war, the war: what's happened there!' and when they heard the answer, 'Cronje surrendered; Ladysmith relieved,' he said that such a shout went up as he had never heard before, and I believed him.

After twenty-four hours of breeze and tossing the good ship found herself in the roads at East London, and having by this time had quite enough of the sea I resolved to disembark forthwith.

South Africa Davitt Map, showing Winston's journey in detail.[77]

[49] SS Guelph, passenger ship for Union Steamship Co, launched 26 June 1894. List of ships built by Harland and Wolff (1859-1929) - Wikipedia

[50] Churchill always found sea voyages tedious. See the first chapter of London to Ladysmith.

Cape Town Background

In his 1930 autobiography, 'My Early Life,'[51] Winston provides very relevant background to his participation in the upcoming second phase of the War. What is notable is that publicly he reveals none of this in his reporting.

He pockets all of the slights, political difficulties, and outright opposition from the highest echelons of British society to his participation in and his reporting on the conflict. In his despatches he disregards, and almost universally ignores any hint of conflict with higher authority. He is unfailingly supportive of the command structure. Any criticism he makes is almost universally constructive. That's part of the problem too; he's so often correct in his assessments.

He is already an adept politician, as well as a reporter at twenty five years old. It is also of note that he does not trade on his name in print or even report on or about his contacts with the top tier of the Empire. No report whatsoever is made that while visiting Cape Town he went galloping over the veldt under Table Mountain with Milner, or that later he swung by Cecil Rhodes' house in July on the way out of town to talk to Cecil's brother Frank, who had edited the 'The River War.'

So, in 'Ian Hamilton's March', Winston does such a masterful job of reporting on his travel, the sights, sounds and intrigue of Cape Town, and his return to the front; that one does not notice he has been out of the action for over two weeks before his next despatch on April 13th. In particular he does not allude to the controversies he faced, two of which created the delay in his obtaining his press credentials. Thirty years later, he relates some of the circumstances.

First was the controversy he had engendered with an article regarding the treatment of the Boers subsequent to their military defeat. Winston was still in Natal on March 31st when he penned his Roving Commission despatch. Previously, on March 29th he published a letter in the Natal Witness.

Treatment of Rebels. Appeal for Leniency. From Winston Spencer Churchill, Ladysmith, March 24, 2.50 P.M. (Delayed by the Censor.) [52]

Bitter complaints are heard throughout the Colony at the lenient treatment apparently proposed for the rebels. The ridicule with which Lord Wolverton's speech has been received in London has given general satisfaction in Natal.

Whether, and, if so, how far, Lord Wolverton was authorised to make the statement attributed to him is not known here, but, in spite of the feeling of the loyal colonists who have fought so gallantly for the Empire, I earnestly hope and urge that a generous and forgiving policy be followed.

If the military operations are prosecuted furiously and tirelessly there will be neither necessity nor excuse for giving rebels who surrender a "lesson."

The wise and right course is to beat down all who resist, even to the last man, but not to withhold forgiveness and even friendship from any who wish to surrender.

The Dutch farmers who have joined the enemy are only traitors in the legal sense. That they obeyed the natural instinct of their blood to join the men of their own race, though no justification, is an excuse.

Certainly their conduct is less morally reprehensible than that of Englishmen who are regular Burghers of the Republics, and who are fighting as fiercely as the proper belligerents against their own countrymen. Yet even these Englishmen would deserve some tolerance were they not legally protected by their citizenship.

The Dutch traitor is less black than the renegade British-born Burgher, but both are the results of our own mistakes and crimes in Africa in former years.

On purely practical grounds it is most important to differentiate between rebels who want to surrender and rebels who are caught fighting.

Every influence should be brought to bear to weaken the enemy and make him submit.

On the one hand are mighty Armies advancing irresistibly, slaying and smiting with all the fearful engines of war; on the other, the quiet farm with wife and children safe under the protection of a Government as merciful as it is strong.

[51] Chapter XXVI.

[52] Treatment of Rebels 31 March: https://www.britishnewspaperarchive.co.uk/viewer/bl/0000174/19000331/028/000

The policy which will hold these two pictures even before the eyes of the Republican soldiers is truly "thorough," and therein lies the shortest road to "peace with honour."

Some Suggestions. Ladysmith, March 25, 5.30 p.m.

Further consideration tempts me to go more into detail on the question of the treatment of rebels.

Having affirmed the principle of generous treatment on the grounds alike of wisdom and mercy, and repudiated all desire of giving lessons or making examples, I consider it nevertheless necessary to discriminate between loyal and disloyal in an unmistakable manner.

This is demanded by public sentiment in the Colony and probably in England. A very convenient method suggests itself.

If a rebel continues to fight and is taken in arms he must stand his trial for high treason and suffer penal servitude, but if he makes surrender voluntarily he should be directed to return to his farm and remain there till the Queen's pleasure is further known. A Proclamation to this effect should be made.

This legally binds the Government to nothing, but morally amounts to an understanding that a repentant rebel will not suffer in life or limb. Where he may be fairly penalised is by being made to pay for damage done to the farms of loyalists. Here is a punishment which is disagreeable without being cruel, and which, moreover, to some extent involves reparation. It is not possible to imagine more perfect justice.

Take the example of a district invaded by an enemy. Half the farmers join the invaders, thus becoming rebels. The other half fly within the British lines. The rebels have their farms respected. The property of the loyalists is destroyed, the rebels very often taking the leading part in the destruction. After the war the rebel who is both wealthy and wicked has a farm flourishing, but must pay compensation to the poor but honest loyalist whose farm he has wrecked. The question of temporary suspension of civil rights might also be considered, and of course all who have committed dishonourable acts of war must be rigorously dealt with.

Letter to the Editor Natal Witness, 29 March 1900

Letter to the Natal Witness Published 29 March 1900.[53] [54] [Natal]

Sir,

I venture to make an appeal, through your columns, to the people of Natal, from whom I have received so many marks of kindness, and in the defence of whose beautiful country I have had the honour to bear a humble part.

The tide of war has turned at last in our favour. The invading armies of the Boers retreat, leaving the rebels who had counted on their success forlornly stranded. A cry, growing into a clamouring, which I can perfectly understand, has arisen that stem retribution should be meted out to these guilty and miserable people.

I read your newspapers and the reports of recent meetings industriously and all reveal the same spirit. 'Give them a lesson they will never forget.' 'Make an example.' 'Condign punishment.'[55] 'Our turn now.' These are the phrases or ideas which recur.

It is the spirit of revenge. It is wrong, first of all because it is morally wicked; and secondly because it is practically foolish. Revenge may be sweet, but it is also most expensive. Let me urge the material considerations first.

What is the first need and dearest desire of every man who has made South Africa his home? An honourable and lasting peace. There is only one means by which this happy issue can be assured. To make the Republican armies surrender and submit, and to annex both the Transvaal and the Orange Free State to the British Empire. That is the object for which we have been fighting, are fighting and are resolved to fight. Nothing less than this will satisfy us. It is the least we will take; but it must also be the most we demand.

We embarked on the stormy ocean of war to find true peace and the assurances of peace which we never had while the Boer menace clouded the sky. We did not seek vengeance. Perhaps it was because the idea of

[53] Natal Witness and OBC Vol 1 Part 2 P. 1162.

[54] Per Randolph Spencer Churchill: A few days later this letter was the subject of a lengthy and critical editorial in the Natal Witness.

[55] Fitting, appropriate, deserved, especially denoting punishment.

vengeance fell away from our minds that we have been allowed to clear the 27th of February from its load of shame.[56] But, at any rate, it was to win peace for ourselves, not to give others their deserts that we broke open the ammunition boxes. Let us do nothing that can delay the attainment of our object. Let us make the Republican armies submit as speedily as possible. To this end we must employ every influence we can command. While we continue to prosecute the war with tireless energy and remorselessly beat down all who resist -to the last man if necessary - we must also make it easy for the enemy to accept defeat. We must tempt as well as compel. On the one hand the Dutchman may see vast armies equipped with all the terrible machinery of war advancing irresistibly; on the other the quiet tin-roofed farm half buried in the trees far from the roar of conflict and the dread of death. That is the way to put an end to the Boer resistance quickly.

I take an example. There is a strong post of Thorneycroft's Mounted Infantry almost under the shadow of Spion Kop, and the officer in command has during the last fortnight been repeatedly approached by rebels who wish to desert the Boer forces in the Drakensberg and make their peace with the Imperial Government.

I readily admit they deserve to be shot. The question is not, however, what suits them but what suits us. If we say to these people, 'Although your lives and goods are justly forfeit, yet as we are strong so will we be merciful; return to your farms and live there upon your best behaviour until the Queen's pleasure concerning you be made known,' they will gladly embrace their only chance of safety and after the war is over, although their lives and liberties would naturally be respected, that being tacitly agreed, we may fairly lay upon them the burden of paying for a good proportion of the loss which their loyal neighbours have suffered and also suspend their civil rights so long as it may be convenient.

But if hardening our hearts, we show the handcuffs and say, 'Come and stand your trial for high treason and rebellion and may the Lord have mercy on your souls,' then these rebels will go back to the Boer lines and continue their practice of killing our friends and comrades with Mauser rifles.

Why would we be such fools as to catch [and imprison] General Joubert's deserters for him?

If I were President Kruger I should agree with every word of Mr. W. J. O'Brien's speech in Pietermaritzburg, and rejoice exceedingly whenever I heard that a surrendered rebel had been put in gaol. Indeed I would only ask that the Government should give no quarter to such treacherous creatures. Not being Mr. Kruger, I deprecate ferocity.

Beware of driving men to desperation. Even a cornered rat is dangerous. We desire a speedy peace and the last thing in the world we want is that this war should enter upon a guerilla phase. Those who demand 'an eye for an eye and a tooth for a tooth' should ask themselves whether such barren spoils are worth five years of bloody partisan warfare and the consequent impoverishment of South Africa.

But it is not only or even mainly upon these grounds that I urge generous counsels upon the people of Natal. See how great a reputation you have made for yourselves in the world by your courage and devotion during this war. The most distant countries have looked with interest to this little comer of South Africa and 'Natalian' has become almost a synonym for loyalty and pluck.

Fame has its responsibilities. Many eyes are upon you now, all admiring, some reluctantly admiring. Let not these watching multitudes be surprised and disappointed. Do not act or speak so that it may be said 'It is true, the Natal colonists have fought well; but they were drunk with racial animosity. They were brave in battle; but they are spiteful in victory.'

Peace and happiness can only come to South Africa through the fusion and concord of the Dutch and British races, who must forever live side by side under the supremacy of Britain. We have fought this war because it was certain that, so long as the Boer Republics existed, their malignant influence would prevent all union. The Boer Republics, and with them the dream of a Dutch Dominion will soon have vanished into the past. Already we may see a nobler and healthier South Africa, serene and prosperous in the sunlight of victorious peace.

Let no one try to make the golden days more distant.

Winston S. Churchill

[56] On the anniversary of the battle of Majuba Hill, 1881; General Cronje, with 44 commandants and other officers of all grades, and over 3500 men, surrendered unconditionally to Roberts. Also, Redvers Buller's force captured the Boer position at Pieters. This action opened the road to Ladysmith. Boers retired north to Ladysmith.

[Riposte] Treatment of Rebels. From Winston Spencer Churchill, Durban, March 30, 4.15 p.m.[57]

The Natal newspapers publish to-day inaccurate summaries of my telegrams advising clemency to the rebels.

They are much displeased with me, but rejoice that the *Morning Post* is "True to the Sons of the Empire."

I regret to learn that you cannot support my views in this matter. They are shared by many officers of distinction with special opportunities, through being constantly engaged in reconnaissance or intelligence work, of knowing the true situation in the disaffected districts.

Lord Dundonald and Colonel Thorneycroft, both of whom are closely acquainted with the local conditions, have pointed out to me the military advantages which would follow a Proclamation encouraging rebels and waverers to desert from the enemy's Army. Colonel Sandbach, the head of the Intelligence Department, favours the making of a definite pronouncement showing the rebels the advantages of voluntary surrender.

Moreover, I have reason to believe that General Buller does not desire to treat them severely, and I that he considers pecuniary punishment the most convenient form. It is strange that the soldiers in the field should hold more tolerant views than prevail at home. However, it is not, perhaps, the first time that victorious gladiators have been surprised to see the thumbs turned down in the Imperial box.

I am more than ever convinced of the importance of showing great generosity to submitting rebels, and surely the advantages of differentiating the treatment of the various classes must be admitted. I desire finally to put the situation before you as it strikes many Imperial officers here and as it impresses me.

Thereafter I will not allude to the subject in your columns.

Public Opinion in Natal.

The Natal Ministers are anxious that every returning rebel against whom there is a case shall be tried, and if convicted sentenced to some punishment, which they do not desire should be necessarily severe. They are not in a vindictive mood, nor do they want anyone hanged, but they do not consider that the rebellion can be allowed to pass *sub silentio*.

I believe I state their view fairly. In itself it is very reasonable. There is no cause to object to it in theory, but in practice it means that rebels who voluntarily surrender must first go to gaol and await trial for high treason.

This means that they will not surrender at all.

Secondly, I must earnestly protest against this spirit of revenge, which, though it may not animate the Ministers themselves, agitates the Colony, and may eventually, unless Imperial influence be exerted, carry all before it.

The First and Dearest Desire.

What is the first and dearest desire of every man who has made South Africa his home? An honourable and lasting peace. There is only one means by which this happy issue can be assured—to make the Republican Armies surrender and submit, and to annex both the Transvaal and the Orange Free State to the British Empire.

That is the object for which we have been fighting, are fighting, and are resolved to fight. Nothing less than this will satisfy us. It is the least we can take, but it must also be the most we demand.

We embarked on the stormy ocean of war to find true peace and assurances of peace which we never had while the Boer menace clouded the sky. We did not seek vengeance.

Perhaps it is because the idea of vengeance fell away from our minds that we have been allowed to clear the 27th of February from its load of shame.

But, at any rate, it was to win peace for ourselves, and not to give others their deserts, that we broke open the ammunition boxes.

"Tempt as well as Compel."

Do nothing that can delay the attainment of our object. Let us make the Republican Armies submit as speedily as possible. To this end we must employ every influence we can command—tempt as well as compel.

Let me take an example. There is a strong post of Thorneycroft's Mounted Infantry near Spion Kop, and the officer in command has during the last fortnight been repeatedly approached by rebels who wish to desert from the Boer forces in the Drakensberg and make their peace with the Imperial Government.

[57] Treatment of Rebels: Appeal for Leniency 27 March:
https://www.britishnewspaperarchive.co.uk/viewer/bl/0000174/19000327/063/0007

It is admitted that they deserve to be shot. The question is not, however, what suits them, but what suits us.

If we say: "Though your lives and goods are justly forfeit, yet, as we are strong, so we are merciful: return to your farms and live there on your best behaviour until the Queen's pleasure concerning you be made known," they will gladly embrace their only chance of safety, and, after the war is over, though their lives and liberties would naturally be respected—that being tacitly agreed—we may fairly lay on them the burden of paying for a good proportion of the loss which their loyal neighbours have suffered and also suspend their civil rights as long as may be convenient.

But if, hardening our hearts, we show them handcuffs, saying: "Come and stand your trials for high treason and rebellion, and may the Lord: have mercy on your souls," then these rebels will go back to the Boer lines and continue their practice of killing our friends and comrades with Mauser rifles.

Why be such fools as to catch the Boer general's deserters for him?

If I were President Kruger I should approve and rejoice exceedingly whenever I heard that surrendered rebels had been put into gaol. Indeed, I should only ask that the Government should give no quarter to such treacherous creatures.

By way of Warning.

Lastly, beware of driving men to desperation: even a cornered rat is dangerous. We desire a speedy peace, and the last thing we want is that this war should enter on a guerilla phase.

Those who demand "an eye for an eye and a tooth for a tooth" should ask themselves whether such barren spoils are worth five years' bloody partisan warfare. Peace and happiness can only come to South Africa through the union and concord of the Dutch and British races, who must for ever live side by side under the supremacy of Great Britain.

We have fought this war because it was certain that as long as the Boer Republics existed their malignant influence would prevent all fusion. The Boer Republics, and with them the dream of Dutch dominion, will soon have vanished into the shades of the past. Already we may see a nobler and wealthier South Africa, serene and prosperous in the sunlight of peace. Let no one try to make the golden days more distant.

PF: While the statesmen agreed with Churchill, the public did not. Even his brother Jack was concerned and said so in a letter to their mother, Jennie.[58] His first person contemporary account gives an indication of just how strong the current Winston is swimming against.

Jack to Lady Randolph

3 April 1900. Ladysmith.[59]

> . . . Winston is being severely criticised about his Peaceful telegrams—and everyone here in Natal is going against his views. They say that even if you are going to treat these Boers well after their surrender, this is not the time to say so. For knowing that they will lose nothing and may gain a lot, they may go on fighting. . . .[60]

This being less than a month after the relief of Ladysmith and while two Boer armies were still in the field, Winston's appeal was somewhat premature. Perspective like this reposed in statesmen like Joseph Chamberlain[61] and Alfred Milner,[62] both of whom Winston was in communication with. Joseph Chamberlain by letter and Milner in person when he arrives in Cape Town.

[58] OBC Vol 1 Part 2 P. 1165-66.

[59] Jack has recovered from his leg wound and is back with the SALH in Natal while his brother heads to the Orange Free State.

[60] This is in fact true in this conflict. Buller's Terms offered at Laings Nek in June, 1900. Kitchener's Terms offered at the Middleburg Conference in February of 1901 and eventually the Treaty of Vereeniging on May 31st, 1902 embodied honorable if unconditional surrender simply for the laying down of arms and the imposition of a new government. Knowing this the Boer forces, or rather their leadership and commanders in the field refused. Continuing to test the will of the British, for as long as possible. Eventually it was the deteriorating conditions in the concentration camps in India, Ceylon and Bermuda and particularly the women's camps with the associated high mortality rates, as much as the military situation, that compelled the Boer leadership to relent in their struggle.

[61] He was Secretary of State for the Colonies at the time. Joseph Chamberlain - Wikipedia

One might think that this was Winston being a controversialist and perhaps a little naive. It was not. These sentiments proved to be a lifelong policy.[63] [64] See Casablanca Conference section at the end of this work.

The Spearman's Camp Sermon

This leniency or terms for the cessation of hostilities issue was not the most serious issue he faced upon his arrival in Cape Town, his most serious issue was obtaining press credentials in order to enter the Orange Free State theatre of military operations. This was to prove a challenge. He had made a critique of one branch of the military and this had ruffled some very highly placed feathers.

> After several days had passed agreeably at Cape Town I began to wonder why no pass had reached me to proceed to Bloemfontein. When more than a week had elapsed without any response to the regular application which had been made, I realised that some obstacle had arisen.[65]

He had two good friends at Field Marshal, Frederick Roberts's Headquarters [In Bloemfontein]; Ian Hamilton, who had helped to defend Ladysmith during the siege, and General Nicholson,[66]—'Old Nick' of Lockhart's staff in Tirah.

Winston was informed by a telegram that his credentials were being held up by Roberts for two reasons.

First, Roberts' chief of staff was Herbert Kitchener; who had been offended by some of the passages that had appeared in 'The River War.' Kitchener had not wanted Winston to join that expedition. He even refused a request by the Prince of Wales[67] later Edward VII, to allow Winston to join the expedition. Winston eventually had managed to find a berth in the Sudan Expedition, and Kitchener's concerns were borne out in that he was not was he pleased with the result of his reporting. He wanted nothing more to do with him and there was no way that Winston was going to be allowed anywhere near where Kitchener was operating his wing of the Army.

But that was not the only issue; Roberts himself had taken umbrage to something Winston had written in his reporting.

> In a letter to the Morning Post written from Natal, I had criticised severely the inadequacy of a sermon preached to the troops on the eve of battle by a Church of England Army Chaplain. The Commander-in-Chief regarded this as a very unjust reflection on the spiritual ministrations of these devoted officials. He was, my friends said, 'extremely stiff.'[68]

The following despatch from Spearman's Hill that was published in the 'Morning Post' is what is causing the trouble.

[62] WSC. "Sir Alfred Milner was far more understanding, and spoke to me with kindliness and comprehension. His A.D.C. the Duke of Westminster, had organised a pack of hounds for his chief's diversion and exercise. We hunted jackal beneath Table Mountain, and lunched after a jolly gallop sitting among the scrub." My Early Life Ch. XXVI.

[63] [Many years later] Lord Birkenhead mentioned to me a Latin quotation which seems to embody this idea extremely well. 'Parcere subjectis et debellare superbos,' which he translated finely 'Spare the conquered and war down the proud.' I seem to have come very near achieving this thought by my own untutored reflections. The Romans have often forestalled many of my best ideas, and I must concede to them the patent rights in this maxim. Never indeed was it more apt than in South Africa. Wherever we departed from it, we suffered; wherever we followed it, we triumphed.

And not only in South Africa. I thought we ought to have conquered the Irish and then given them Home Rule: that we ought to have starved out the Germans, and then revictualled their country; and that after smashing the General Strike, we should have met the grievances of the miners. I always get into trouble because so few people take this line.

I was once asked to devise an inscription for a monument in France. I wrote: 'In war, Resolution. In defeat, Defiance. In Victory, Magnanimity. In peace, Goodwill.' The inscription was not accepted. ...... It is all the fault of the human brain being made in two lobes, only one of which does any thinking, so that we are all right-handed or left-handed; whereas if we were properly constructed we should use our right and left hands with equal force and skill according to circumstances. My Early Life Ch. XXVI.

As it is, those who can win a war well can rarely make a good peace, and those who could make a good peace would never have won the war. It would perhaps be pressing the argument too far to suggest that I could do both. My Early Life Ch. XXVI.

[64] "Parcere subiectis et debellare superbos" is a Latin phrase that means "spare the conquered and crush the proud." The phrase is from Virgil's Aeneid, VI, 8531.

[65] My Early Life Chapter XXVI.

[66] Probably: John Sanctuary Nicholson - Wikipedia

[67] Frederick Woods. Young Winston's Wars. P. XVII.

[68] My Early Life Chapter XXVI.

London to Ladysmith Quote:

A Fresh Effort and an Army Chaplain: Spearman's Hill: Saturday, February 4, 1900.[69]

The first gleams of daylight crept underneath the waggon, and the sleepers, closely packed for shelter from the rain showers, awoke. Those who live under the conditions of a civilised city, who lie abed till nine and ten of the clock in artificially darkened rooms, gain luxury at the expense of joy. But the soldier, who fares simply, sleeps soundly, and rises with the morning star, wakes in an elation of body and spirit without an effort and with scarcely a yawn.

There is no more delicious moment in the day than this, when we light the fire and, while the kettle boils, watch the dark shadows of the hills take form, perspective, and finally colour, knowing that there is another whole day begun, bright with chance and interest, and free from all cares.

Early morning on the veldt.[78]

Morning in camp, cooking a meal. .[79]

All cares—for who can be worried about the little matters of humdrum life when he may be dead before the night? Such a one was with us yesterday—see, there is a spare mug for coffee in the mess—but now gone for ever. And so it may be with us to-morrow. What does it matter that this or that is misunderstood or perverted; that So-and-so is envious and spiteful; that heavy difficulties obstruct the larger schemes of life, clogging nimble aspiration with the mud of matters of fact? Here life itself, life at its best and healthiest, awaits the caprice of the bullet. Let us see the development of the day. All else may stand over, perhaps for ever. Existence is never so sweet as when it is at hazard. The bright butterfly flutters in the sunshine, the expression of the philosophy of Omar Khayyám,[70] without the potations.[71] But we awoke on the morning of the 25th [January] in most gloomy spirits. ....

A Large British Camp.[80]

It is a solemn Sunday [January 29th], and the camp, with its white tents looking snug and peaceful in the sunlight, holds its breath that the beating of its heart may not be heard.

[69] London to Ladysmith Chapter XIX.

[70] Omar Khayyam (18 May 1048 – 4 December 1131) was a Persian polymath, mathematician, astronomer, historian, philosopher, and poet. As a mathematician, he is notable for his work on the classification and solution of cubic equations. There is a tradition of attributing poetry to him. This poetry became widely known to the English-reading world in a translation by Edward FitzGerald (Rubaiyat of Omar Khayyam, 1859), which enjoyed great success in the Orientalism of the fin de siècle. Omar Khayyam - Wikipedia

[71] The act of drinking.

On such a day as this the services of religion would appeal with passionate force to thousands. I attended a church parade this morning. What a chance this was for a man of great soul who feared God! On every side were drawn up deep masses of soldiery, rank behind rank—perhaps, in all, five thousand. In the hollow square stood the General, the man on whom everything depended. All around were men who within the week had been face to face with Death, and were going to face him again in a few hours. Life seemed very precarious, in spite of the sunlit landscape. What was it all for? What was the good of human effort? How should it befall a man who died in a quarrel he did not understand? All the anxious questionings of weak spirits. It was one of those occasions when a fine preacher might have given comfort and strength where both were sorely needed, and have printed on many minds a permanent impression. The bridegroom opportunity had come. But the Church had her lamp untrimmed.[72] A chaplain with a raucous voice discoursed on the details of 'The siege and surrender of Jericho.' The soldiers froze into apathy, and after a while the formal perfunctory service reached its welcome conclusion.

As I marched home an officer said to me:[73] 'Why is it, when the Church spends so much on missionary work among heathens, she does not take the trouble to send good men to preach in time of war? The medical profession is represented by some of its greatest exponents. Why are men's wounded souls left to the care of a village practitioner?' Nor could I answer; but I remembered the venerable figure and noble character of Father Brindle[74] in the River War, and wondered whether Rome was again seizing the opportunity which Canterbury disdained—the opportunity of telling the glad tidings to soldiers about to die.

In My Early Life he continues:

> Lord Roberts, a deeply religious man, ... felt that the Military Chaplains' Department had suffered unmerited aspersion, and the mere fact that outside assistance had now been proffered only seemed to aggravate the sting.[75] In these circumstances my prospects for several days seemed very gloomy, and I languished disconsolately amid the Capuan[76] delights of the Mount Nelson Hotel.[77]
>
> [Eventually] My pass was granted and I was free to proceed to Bloemfontein, with the proviso however that before taking up my duties as War Correspondent I should receive an admonition from the Military Secretary to the Commander-in-Chief against reckless and uncharitable criticism. This was good enough for me, and I started on the long railway journey that same night.
>
> But Lord Roberts maintained an air of inflexible aloofness. .... When one morning in the market place at Bloemfontein amid a crowd of officers I suddenly found myself quite unexpectedly within a few yards of him, he acknowledged my salute only as that of a stranger.[78 79]

This despite the fact Roberts made sure that Jennie's hospital ship Maine stayed in Cape Town harbour an extra day so that he could personally visit the vessel with her as his tour guide.

[72] PF: A used, carbonized, oil-lamp wick burns dimly and sooty.

[73] This may have been a subterfuge. It is not unheard of for Winston to relate a question or criticism originating from an anonymous source; but in fact was Winston's observation. His first letters from the Tirah expedition were all published unsigned. As if some friend of the anonymous officer had simply passed on a private letter to the newspaper. Everyone knew it was him and his mother, Jennie, made sure that important people knew. It was in Winston's words, a "pious fraud" as he allowed in a letter to Aylmer Haldane. Young Winston's Wars, p. XIX citing OBC p. 964 August 11th 1898 letter to Haldane.

[74] In My Early Life Winston gives more information that allows for the identification of this individual: Father Brindle: A well-known and honoured figure in the British Army in this period; and afterwards Bishop of Nottingham. Robert Brindle (4 November 1837 – 27 June 1916) was an English prelate of the Roman Catholic Church. He served as the Bishop of Nottingham from 1901 to 1915. See: Robert Brindle - Wikipedia

[75] "Several of the most eloquent divines, vacating their pulpits, had volunteered for the Front and were at this moment swiftly journeying to South Africa to bring a needed reinforcement to the well-meant exertions of the Army Chaplains Corps."

[76] A town in the province of Caserta, Campania, Italy.

[77] My Early Life Chapter XXVI.

[78] My Early Life Chapter XXVI.

[79] As we have seen, when Winston crosses through the lines to deliver Hamilton's despatches, there is a rapprochement in their relationship. Roberts was good friends with Winston's father and owed at least one promotion to Lord Randolph's efforts. See:
"Lord Roberts had been a great friend of my father's. Lord Randolph Churchill had insisted as Secretary of State for India in 1885 in placing him at the head of the Indian Army, thrusting on one side for this purpose the claims of Lord Wolseley himself. They had continued friends until my father's death ten years later; and I as a child had often met the General and could pride myself on several fascinating conversations with him." My Early Life, Ch, XXVI.

'The Maine' at Anchor.[81] General Roberts inspecting the Maine at Cape Town, January 26th, 1900.[82]

Also, at army headquarters in Bloemfontein Roberts' staff was coming under criticism from the press on the grounds that it was too top heavy with English nobility. At the time, serving in one capacity or another, were the Duke of Norfolk,[80] the Duke of Westminster,[81] and the Duke of Marlborough.[82]

Henry Howard.[83] Hugh Grosvenor, Duke of Westminster.[84] Charles Spencer-Churchill (Sunny), Duke of Marlborough.[85]

In order to appease the criticism Roberts selected the Duke of Marlborough, 'Sunny', for "retrenchment."

Sunny had been Winston's companion who had demanded the surrender of the garrison at the Officers Prison in Pretoria on June 5th. He was also Winston's cousin.[83]

Now deprived of his staff position Sunny was at loose ends and did not want to be left behind during the upcoming campaign.

[80] Henry Howard. At aged 53, he went in 1900 to South Africa for service in the Second Boer War as a lieutenant colonel in the Imperial Yeomanry, in the course of which he was wounded near Pretoria and invalided back to Britain. After the end of the war, he was promoted to Lieutenant-Colonel Commandant of his volunteer battalion. Henry Fitzalan-Howard, 15th Duke of Norfolk, (27 December 1847 – 11 February 1917) Henry Fitzalan-Howard, 15th Duke of Norfolk - Wikipedia

[81] Hugh Grosvenor. From February 1900 he served with the Imperial Yeomanry, as an ADC to Lord Roberts and Lord Milner. He served continuously and in December of 1901 was appointed captain of the Cheshire (Earl of Chester's) Imperial Yeomanry the following month. The war ended in May 1902. Hugh Richard Arthur Grosvenor, 2nd Duke of Westminster, (19 March 1879 – 19 July 1953) was a British landowner and one of the wealthiest men in the world. Hugh Grosvenor, 2nd Duke of Westminster - Wikipedia

[82] Charles Richard John Spencer-Churchill, 9th Duke of Marlborough, (13 November 1871 – 30 June 1934), was a British soldier and Conservative politician, and a close friend of his first cousin Winston Churchill. He was often known as "Sunny" Marlborough after his courtesy title of Earl of Sunderland. Charles Spencer-Churchill, 9th Duke of Marlborough - Wikipedia

[83] Winston's Grandfather John Spencer-Churchill, 7th Duke of Marlborough - Wikipedia had twelve children. Winston had a large extended family of cousins.

Eventually the ice thawed, but only that. On April 11th, 1900 Winston received his permit to accompany the Army.

From: Colonel Neville Chamberlain[84] to Winston S. Churchill.
April 11th, 1900,[85] Bloemfontein

My Dear Churchill,

Lord Roberts desires me to say that he is willing to permit you to accompany this force as a correspondent – for your father's sake.

Yours sincerely
Neville Chamberlain.

Privately Winston wrote acerbically to his mother regarding this permission slip: "You will form your own opinion, as to the justice of making me accept as a favour what was already mine as a right."[86]

Again his sound judgement prevails and he never reported any of this, nor cast a disparaging remark in either Roberts', Kitchener's, or French's direction during the campaign.[87] [88]

Frederick Roberts Writing Despatches.[89]

Herbert Kitchener of Khartoum.[86]

General Sir John French.[87]

[84] Colonel Neville Francis Fitzgerald of Oatlands, Castleknock, Dublin; and of the Kildare St, Dublin, and Naval and Military Clubs, was born in 1856, and is the only son of Lieutenant Colonel C F Falcon Chamberlain, CB. He joined the Army in 1873, being attached to the Devonshire Regiment until 1876, when he joined the Central India Horse; took part in the Afghan War in 1878-80, on the staff of Sir F Roberts, being present at the action at Kandahar (medal with four clasps and bronze star); and acted as ADC to Sir F Roberts from 1881-5, and from then until 1890 he acted as Persian interpreter; served in the Burmese Expedition in 1886-7 (medal and clasp), and acted as Private Secretary to Field Marshal Lord Roberts in Boer War. Sir Neville retired in 1900 from the Army, when he was appointed Insp. General of the Royal Irish Constabulary. He married, in 1886, Mary, daughter of Major General Hay.

[85] Telegram and quote, OB p. 507 and OBC Vol I Part 2 p. 1168.

[86] Telegram and quote, OB p. 507 and OBC Vol I Part 2 p. 1168.

[87] Roberts does reconcile with Winston after his daring to the point of foolhardy journey through Johannesburg to primarily reach the 'press censor' but also to carry Ian Hamilton's despatches to Army Headquarters. Winston will also write a review of Roberts' autobiography and publish it in The Worlds Work during his American Tour later in 1900.

[88] Winston also got the cold shoulder from General French. After April 21st when he had his adventure with McNeill. "I now attached myself to French's cavalry division and marched north with them. Here I found myself in a none too friendly atmosphere. It appeared that like a good many other Generals at this time, French disapproved of me. The hybrid combination of subaltern officer and widely-followed war correspondent was not unnaturally obnoxious to the military mind. But to these general prejudices was added a personal complication. I was known to be my old Colonel's partisan and close friend. [Babrazon] I was therefore involved in the zone of these larger hostilities. Even Jack Milbanke, Sir John Milbanke, 10th Baronet - Wikipedia French's aide-de-camp, now recovered from his wound and newly decorated with the V.C., was unable to mitigate the antagonism that prevailed. Although I was often with French's column in march and skirmish, the General completely ignored my existence, and showed me no sign of courtesy or goodwill. I was sorry for this, because I greatly admired all I had heard of his skilful defence of the Colesberg front, his dashing gallop through the Boer lines to the relief of Kimberley, and was naturally attracted by this gallant soldierly figure, upon whom fell at this moment the gleam of a growing fame. Thus during the South African war **I never exchanged a word with the General who was afterwards to be one of my greatest friends and with whom I was for many years to work at grave matters in peace and war.**" See My Early Life Ch. XXVI.

[89] Lord Roberts writing despatches in his wagon tent at Paardeberg. Cassell's History of the Boer War 1899-1902 No author listed. Originally published in the "Graphic"

Winston took his credentials, teamed up with his first cousin, 'Sunny', the Duke of Marlborough, and set out for Ian Hamilton's command where Sunny had secured a posting.[90]

Winston makes a point of mentioning Sunny's new appointment in a despatch to the 'Morning Post'.

The Duke of Marlborough. From Winston Spencer Churchill, Bloemfontein. April 13. 1.15 P.M.[91] [92]

Lord Roberts has appointed the Duke of Marlborough, who had been serving on General Brabazon's staff, assistant military secretary at headquarters.

Together they set out for their new posting.

> I inspanned my four-horsed wagon, and we [He and Sunny] started upon a forty-mile march to overtake the flanking column. We came through the Boer-infested countryside defenceless but safely, and caught up our friends on the outskirts of Winburg.[93]

Winston rarely interjects his personal affairs into his narratives and there is no mention by name mention of Sunny during his journey or in South Africa. The next mention of him is when they are departing in and Sunny is carrying despatches to Cape Town.[94]

> I travelled prosperously; came by rail to Edenburg, trekked from there in drenching rains, most unusual for this time of year, and greatly increasing the difficulties of supply; and, resting for the night at Reddersburg, caught up the marching column in its camp, about eleven miles from Dewetsdorp, on the night of the 19th. [April][95]

At that time: Wepener [Under Siege] making a brave defence; Brabant marching through Rouxville to bar their advance; Rundle, Chermside,[96] and Brabazon[97] striking east from Edenburg to shut the door behind them with two Infantry divisions, twenty-four guns, and 2,000 Yeomanry; and, further north, the great Bloemfontein Army—four Infantry divisions, [Ian] Hamilton's 10,000 mounted men, French's four Cavalry brigades, and many guns—is almost ready to move. ......'Qui vivra verra,'[98] and, if these letters continue, 'who runs may read,' for I purpose to journey *viâ* Edenburg[99] to Reddersburg to-morrow, and thence on to the point of collision, which must mark the climax of this extremely interesting event henceforward to be called 'The Operations in the Right-hand Bottom Corner of the Free State.'

[90] I proposed by telegram to Hamilton that he should take Marlborough upon his staff. The General agreed, and Lord Roberts, who never liked to treat anyone unfairly, gave a cordial approval. WSC My Early Life Ch. XXVII.

[91] Morning Post, The Duke of Marlborough 14 April https://www.britishnewspaperarchive.co.uk/viewer/bl/0000174/19000414/038/0005

[92] These despatches earned their author £250 a month, plus all expenses, four months guaranteed minimum of employment, and entire freedom of movement and opinion. 'Higher, I think, than any previously paid in British journalism to War Correspondents'. Frederick Woods' Quote from his Bibliography. Also WSC's observation. 'I received over £2000, apart from expenses, from the Morning Post'. (My Early Life, p. 244.)

[93] Winston rarely gets his facts wrong but in this case, Winburg is well north of where the campaign begins south of Thaba N'Chu. His first destination is Reddersburg and then DeWets Dorp; see map below.

[94] In a dramatic and a genuinely triumphant event, Sunny does compel the surrender of the British prisoner of war camp in Pretoria, but he is not mentioned by name.

[95] See also, My Early Life: It was the 8th Division, the last division of our regular army scraped together from our fortresses all over the Empire. It was commanded by Sir Leslie Rundle, later unkindly nicknamed 'Sir Leisurely Trundle,' whom I had known up the Nile. Brabazon's brigade was scouting on ahead. Rundle was affable and hospitable; and early the next morning I rode on to join Brabazon. He was delighted to see me, told me his grievances, and entertained me vastly with stories and criticisms of French, as well as of the war and the world in general. We abode together for some days. WSC My Early Life Ch, XXVI.

[96] General Sir Herbert Charles Chermside, (31 July 1850 – 24 September 1929) was a British Army officer who served as Governor of Queensland from 1902 to 1904. In 1899 Chermside was sent to South Africa to command the 14th Brigade and the 3rd Division during the Second Boer War.

[97] Brabazon John Palmer Major General Born 13 Feb 1843. He entered 1862; Colonel, January 1899. Staff service: Acting as Volunteer with rank of Captain, Ashanti Expedition, 1873-74; ADC (extra to Viceroy of India), 1877-79; Brigade-Major, Afghan Campaign, 1870-80; ADC to the Queen, 1889; Colonel on Staff; Commanding Cavalry Brigade, South-East District, 1899; Major General, Cavalry Brigade, South Africa, 1899-1900. War service: Ashanti, 1874 (medal with clasp); Afghan War 1878-80 (Despatches; March, November, 1879; January, May, December, 1880; Medal with 4 clasps; bronze star; Brevet of Major); Egyptian Expedition, 1884 (Despatches; medal with clasp; bronze star; Brevet of Lieutenant Colonel); Sudan, 1884-85 (clasp); South African War, 1899-1900 (Despatches). General Brabazon, born in 1843, was the son of the late Major Brabazon (late 15th Hussars), and the daughter of the late Sir W H Palmer, Baronet. Yeomanry. Imperial Yoemanry Article: https://www.angloboerwar.com/?option=com_content&view=article&id=1946

[98] Who(ever) lives will see.

[99] Edenburg is a town situated about 80 kilometres (50 mi) south of Bloemfontein. Edenburg - Wikipedia

Winston describes his camaraderie with his cousin and their work together:

> Every day we saw new country. Every evening we bivouacked—for there were no tents—by the side of some new stream. We lived on flocks of sheep which we drove with us, and chickens which we hunted round the walls of deserted farms. My wagon had a raised floor of deal boards[100] beneath which reposed two feet of the best tinned provisions and alcoholic stimulants which London could supply. We had every comfort, and all day long I scampered about the moving cavalry screens searching in the carelessness of youth for every scrap of adventure, experience or copy.[101] [102] [103]

Churchill Outside his tent.[88] [104] [105]

Unwelcome Visitors To [A] Tent.[89]

It is interesting to see that in My Early Life, Winston allows himself to write more like a youth. His prose resembles that of Deneys Reitz as we shall see in Commando. In the business of journalism, that is his brief, he keeps his reporter hat on while still providing vivid details, atmosphere, and drama. He does not over personalize nor invite his readership into many parts of his private life, either the travails or the personal details.

[100] Deal Board: A fir or pine board per Merriam-Webster Dictionary.

[101] As with other instances, if I paraphrase Winston's observations it will sound like I'm disparaging him. Let him recall these events and inform us about his prior narrative. There was also hardship and danger as we have and will see. WSC My Early Life Ch. XXVII.

[102] There was so much interest and excitement in everyday life, that there was little time to worry unduly about the displeasure even of so great a personage and so honoured a friend. [As General Roberts] Equipped by the Morning Post on a munificent scale with whatever good horses and transport were necessary, I moved rapidly this way and that from column to column, wherever there was a chance of fighting. Riding sometimes quite alone across wide stretches of doubtful country, I would arrive at the rear-guard of a British column, actually lapped about by the enemy in the enormous plains, stay with them for three or four days if the General was well disposed, and then dart back across a landscape charged with silent menace, to keep up a continuous stream of letters and telegrams to my newspaper. WSC My Early Life Ch XXVI.

[103] Recall that back in the UK, before his voyage; with his Morning Post contract in hand Winston set about equipping himself for his expedition with among other things laid in for the voyage and his reportage: 6 bottles of Vin d' Ay Sec (Champagne)
18 bottles of St. Emilion (Bordeaux wine)
6 bottles of Light Port
6 bottles of French vermouth
18 bottles of Scotch Whiskey (10 years old)
6 bottles of Very Old Eau de Vie landed 1886. (French clear colorless fruit brandy)
12 Rose's cordial lime juice. From Frederick Woods, Young Winston's Wars, p. XXII and other sources as well.

[104] This picture was taken at Bloemfontein. The arrangements of his quarters and person are made by Thomas Walden, Winston's Batman/Valet/Bodyman.

[105] "Batmen" A batman or an orderly is a soldier or airman assigned to a commissioned officer as a personal servant. Before the advent of motorized transport, an officer's batman was also in charge of the officer's "bat-horse" that carried the pack saddle with his officer's kit during a campaign. Batman (military) - Wikipedia

II. Exit General Gatacre

Bethany:[106] April 13th 1900.

If you go to sleep when the train leaves East London, you should wake, all being well, to find yourself at Queenstown.

Queenstown lies just beyond the high water-mark of war. The tide had flowed strong after [the battle of] Stormburg,[107] and it looked as if Queenstown would be engulfed, at any rate for a time. But Fortune and General Gatacre protected it. Sterkstroom entrenched itself, and prepared for daily attacks. Molteno was actually shelled. Queenstown suffered none of the horrors of war except martial law, which it bore patiently rather than cheerfully.

Martial law in Cape Colony: handing in bicycles to the military authorities At Graaff Reinet.[90] Five hundred bicycles handed in under military orders at Graaff Reinet.[91]

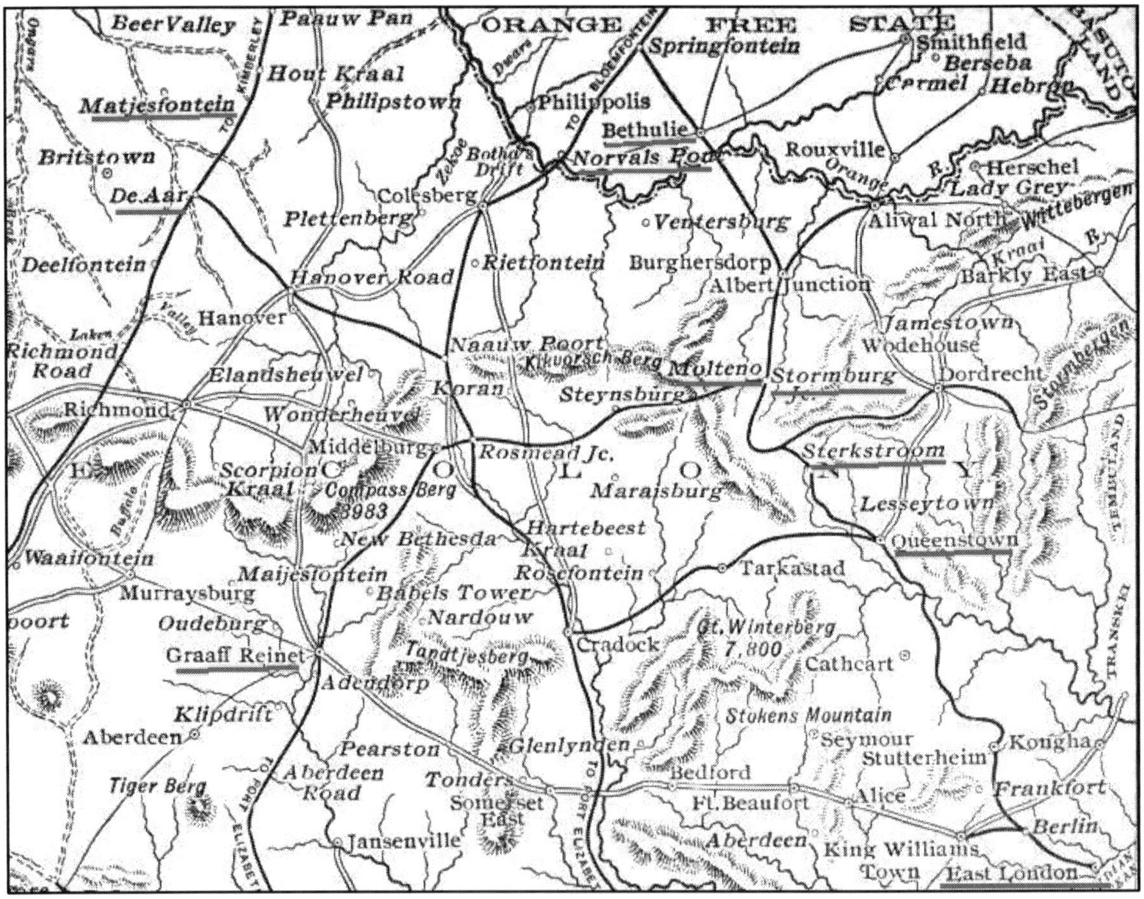

Northeastern Cape Colony.[92] Note locations he has mentioned.

[106] This town is just above Edenburg, and Norval's Pont along the railroad line. See Arthur Conan Doyle OFS map. Spelled: Bethulie and Bethanie on maps.

[107] The Battle of Stormberg (December 10th, 1899) (Gatacre Commanding) was the first British defeat of Black Week, in which three successive British forces were defeated by Boer forces in the Second Boer War. Battle of Stormberg - Wikipedia

Nothing in the town impresses the traveller, but at the dining-room of the railway station there is a very little boy, about twelve years old, who, unaided, manages to serve, with extraordinary despatch and a grand air, a whole score of passengers during the brief interval allowed for refreshments.

Five months earlier I had passed along this line, hoping to get into Ladysmith before the door was shut, and had been struck by this busy child, who seemed a product of America rather than of Africa. Much had happened in the meantime, not so far from where he lived. But here he was still—the war had not interfered with him, Queenstown was beyond the limit.

At Sterkstroom a line of empty trenches, the Red Cross flag over a hospital, and an extension to the cemetery enclosure filled with brown mounds which the grass had not yet had time to cover, showed that we had crossed the line between peace and war. Passing through Molteno, the last resting-place of the heroic de Montmorency,[108] the train reached Stormburg. Scarcely any traces of the Boer occupation were to be seen; the marks of their encampments behind the ridge where they had laagered—a litter of meat tins, straw, paper, and the like, the grave of Commandant Swanepoole and several nameless heaps, a large stone (in the station-master's possession) with the words engraved on it: 'In memory of the Transvaal commando, Stormburg, December 1899,' and that was all. The floods had abated and receded. This was the only jetsam that remained.

Trenches at Magersfontein.[93]

Magersfontien Battlefield.[109]

At Stormburg I changed my mind, or, rather—for it comes to the same thing and sounds better—I made it up.[110]

I heard that no immediate advance from Bloemfontein was likely or even possible for a fortnight. Therefore, I said, I will go to Cape town, and shelter for a week at 'The Helot's Rest.'[111] After all, what is the use of a roving commission if one cannot rove at random or caprice? So to Cape town I went accordingly—seven hundred miles in forty-eight hours of bad trains over sections of the line only newly reopened.

But to Cape town I will not take the reader. Indeed, I strongly recommend him to stick to the war and keep his attention at the front, for Cape town at this present time is not an edifying place. Yet, since he may be curious to know some reason for such advice, let me explain.

Cape town, which stands, as some writers have observed, beneath the shadow of Table Mountain, has been—and may be again in times of peace—a pleasant place in which to pursue business or health; but now it is simply a centre of intrigue, scandal, falsehood, and rumour.

[108] Captain the Honourable R De Montmorency, VC, 21st Lancers, in December 1899. Killed in action 23 Feb 1900 Place: Stormberg near Schoemans Kop. See: Raymond de Montmorency - Wikipedia

[109] Anglo-Boer War Album, Cape Town, D. Edwards & co. [1902?]

[110] PF: Ever politic and polite, Winston was probably informed that the local commanders, Roberts and Kitchener were not inclined to admit him to their theatre of military operations. He needed to look elsewhere, Cape town, for that permission.

[111] In Sparta, the helots were given a subordinate position in society more comparable to serfs in medieval Europe than the chattel slaves in the rest of Greece. Sparta - Wikipedia

Cape Town – Seat of Government, Cape Colony.[94]

The visitor stays at the Mount Nelson Hotel,[112] if he can be so fortunate as to secure a room. At this establishment he finds all the luxuries of a first-class European hotel without the resulting comfort. There is a good dinner, but it is cold before it reaches him; there is a spacious dining-room, but it is overcrowded; there are clean European waiters, but they are few and far between.

Mount Nelson Hotel, Cape Town.[95]

Mount Nelson Hotel - The Dining Room.[96 113]

Lawn and porch tables outside the Mount Nelson hotel.[97]

Cape town from the Mount Nelson Hotel, showing Signal Hill.[114]

[112] Belmond Mount Nelson Hotel - Wikipedia

[113] "Mount Nelson Hotel. –The Dining Room. – The dining-room measures some 80 feet in length, by 40 feet in width and has at the end a conservatory 20 feet in spread, which may be thrown into the saloon if need be. There are no pillars to impede the view, the roof being supported on massive girders from the main walls, which are paneled in oak to a height of 14 feet …. At the inner end is a gallery for musicians, flanked by two very fine plaster reliefs, representing the arrival of a Dutch fleet in Table Ban in the 17th century, while the arms of the United Kingdom, the British Colonies, and the South African States are displayed on shields in the ceiling."

[114] Rosslyn, Twice Captured, Blackwoods, 1900.

At the hotel, in its garden, or elsewhere in the town, all the world and his wife are residing—particularly the wife. We used to think, in the Army of Natal, that Lord Roberts's operations in the Free State had been a model of military skill and knowledge, and, in a simple way, we regarded [General] French as one of the first cavalry soldiers of the age.

All this was corrected at Cape town, and I learned with painful disenchantment that 'it' (the said operations) had all been a shameful muddle from beginning to end; that the field-marshal had done this and that and the other 'which no man in his senses,' &c., that French was utterly ... and as for Lord Kitchener, Cape town—let us be just, imported social Cape town—was particularly severe on Lord Kitchener.

It was very perplexing; and besides it seemed that these people ought to know, for they succeeded in making more news in the twenty-four hours than all the correspondents at the front put together. The whole town was overrun with amateur strategists and gossiping women. There were more colonels to the acre than in any place outside the United States, and if the social aspect was unattractive, the political was scarcely more pleasing.

Party feeling ran high. Some of the British section, those tremendous patriots who demonstrate but do not fight—not to be on any account compared with the noble fellows who fill the Volunteer corps—pot-house heroes, and others of that kidney, had just distinguished themselves by mobbing Mr. Schreiner[98] in the streets.

Hon. William Philip Schreiner.[99] "Crowded docks at Cape town." "The Majestic takes up a good deal of room."[100]

The Dutch section, some of them the men who, risking nothing themselves, had urged the Republics to their ruin, all of whom had smiled and rubbed their hands at the British reverses, sat silent in public, but kept a strict watch on incoming steamers for members of Parliament and others of more influence and guile, and whispered honeyed assurances of their devotion to the Empire, coupled with all sorts of suggestions about the settlement—on the broad general principle of 'Heads I win, tails you lose.'

British newspapers advocated short shrift to rebels— 'Hit 'em hard now they're down'; 'Give them a lesson this time, the dirty Dutchmen!' Dutch papers recorded the events of the war in the tone, 'At the end of the battle the British, as usual, fled precipitately, leaving 2,000 killed, *our* loss'—no, not quite that, but very nearly; everything, in fact, but the word 'our'—'one killed, two slightly wounded.'

Let no one stay long in Cape town now who would carry away a true impression of the South Africans. There is too much shoddy worn there at present.

Only at Government House did I find the Man of No Illusions, the anxious but unwearied Proconsul, [Milner] understanding the faults and the virtues of both sides, measuring the balance of rights and wrongs, and determined—more determined than ever; for is it not the only hope for the future of South Africa?—to use his knowledge and his power to strengthen the Imperial ties.

Government House in Cape Town.[115]

Sir Alfred Milner, High Commissioner for South Africa.[101]

All this time the reader has been left on a siding at Naauwpoort;[116] but does he complain of not being taken to Cape town? We will hasten back together to the healthier atmosphere of war.

Indeed, the spell of the great movements impending in the Free State began to catch hold of me before I had travelled far on the line towards Bloemfontein. Train loads of troops filled every station or siding.

Tommies on Train to Bloemfontein.[102]

A ceaseless stream of men, horses, and guns had been passing northwards for a fortnight, and on the very day that I made the journey Lord Kitchener had ordered that in future all troops must march beyond Springfontein,[117] because the line must be cleared for the passage of supplies, so that, besides the trains in the sidings, there were columns by the side of the railway steadily making their way to the front.

The one passenger train in the day stopped at Bethany[118] I got out. To go on was to reach Bloemfontein at midnight. Better, then, to sleep here and proceed at dawn.

'Are there many troops here?' I asked. They replied 'The whole of the Third Division.' 'Who commands?' 'Gatacre.' That decided me.

I knew the general slightly, having made his acquaintance up the Nile[119] in pleasant circumstances, for no one was allowed to pass his mess hungry or thirsty. I was very anxious to see him and hear all about Stormberg[120]

[115] Government House Cape Town. Seat of the Cape of Good Hope Government. A Winter Tour of South Africa, by F Young. Petherick & Co. London. 1890. File:Government House - Cape Town - Cape Colony.jpg - Wikimedia Commons

[116] Noupoort is a small town in the eastern Karoo region of South Africa. The town lies 54 km south of Colesberg Noupoort - Wikipedia

[117] Springfontein - Wikipedia

[118] This town is just above Edenburg, along the railroad line. See Arthur Conan Doyle OFS map. Spelled: Bethanie on map.

[119] During Winston's participation in the 1898 Sudan Campaign as chronicled by him in: *The River War*. Mahdist War - Wikipedia

and the rest of the heavy struggle along the eastern line of rail. I found him in a tin house close to the station. He received me kindly, and we had a long talk. The General explained to me many things which I had not understood before, and after we had done with past events he turned with a hopeful eye to the future. At last, and for the first time, he was going to have the division of which he had originally been given the command.

'You know I only had two and a half battalions at Sterkstroom[121] and a few colonial horse; but now I have got both my brigades complete.'

I thought him greatly altered from the dashing, energetic man I had known up the river, or had heard about on the frontier or in plague-stricken Bombay. Four months of anxiety and abuse had left their mark on him. The weary task of keeping things going with utterly insufficient resources, and in the face of an adroit and powerful enemy in a country of innumerable kopjes, where every advantage lay with the Boer, had bowed that iron frame and tired the strange energy which had made him so remarkable among soldiers. But when he thought of the future his face brightened.

Major General Sir William F. Gatacre.[103]

General Bindon Blood.[104 122]

The dark days were over. The broken rocky wilderness lay behind, and around rolled the grassy plains of the Free State. He had his whole division at last. Moreover, there was prospect of immediate action. So I left him, for it was growing late, and went my way. Early next morning he was dismissed from his command and ordered to England, broken, ruined, and disgraced.[123 124] I will not for one moment dispute the wisdom or the justice of his removal. In stormy weather one must trust to the man at the helm, and when he is such a man as Lord Roberts it is not a very hard thing to do. But because General Gatacre has been cruelly persecuted in England by people quite ignorant of the difficulties of war or of the conditions under which it is carried on in this country, it is perhaps not out of place to write a few words of different tenor.

Gatacre was a man who made his way in the army, not through any influence or favour which he enjoyed, but by sheer hard work and good service. Wherever he had served he had left a high record behind him.

[120] The first major defeat of British Forces on Dec. 10th in which 135 men were killed and 696 captured. Battle of Stormberg - Wikipedia

[121] The village is on the Hex River, at the southern foot of the Stormberg, 272 km north-west of East London. Sterkstroom - Wikipedia

[122] General Sir Bindon Blood GCB GCVO (7 November 1842 – 16 May 1940) was a British Army commander who served in Egypt, Afghanistan, India, and Southern Africa. Bindon Blood - Wikipedia Winston Churchill, who served under Blood on the North-West Frontier in 1897, dedicated his first non-fiction book, The Story of the Malakand Field Force (1898), to "Major-General Sir Bindon Blood, K.C.B., under whose command the operations therein recorded were carried out; by whose generalship they were brought to a successful conclusion; and to whose kindness the author is indebted for the most valuable and fascinating experience of his life."

[123] He returned to England and to his pre-war posting as General Officer Commanding Eastern District from June 1900 to December 1903. He retired in 1904.

[124] Lieutenant-General Sir William Forbes Gatacre KCB DSO (3 December 1843 – 18 January 1906) was a British soldier who served between 1862 and 1904 in India and Africa. He commanded the British Army Division at the Battle of Omdurman and the 3rd Division during the first months of the Second Boer War, during which time he suffered a humiliating defeat at the Battle of Stormberg. William Forbes Gatacre - Wikipedia

On the Indian frontier he gained the confidence of so fine a soldier as Sir Bindon Blood, and it was largely to his reputation won in the Chitral Expedition[125] that his subsequent advancement was due.

At Bombay in 1897 he was entrusted with the duty of fighting the plague, then first gripping its deadly fingers into the city. No one who is at all acquainted with the course of this pest will need to be told how excellent was his work. After the late Soudan campaign I travelled from Bombay to Poona with a Parsee gentleman, a wealthy merchant of the plague-stricken town, and I well remember how he dilated on the good which Gatacre had done.

'He was our only chance,' said the black man. 'Now he is gone, and the sickness will stay for ever.'

Gatacre's part in the Soudan campaign has been described at length elsewhere. His courage has never been questioned, because the savage critics did not wish to damage their cause by obvious absurdities. If I were to discuss his tactics in the Boer war here I should soon get on to ground which I have forbidden myself. It is sufficient to observe that Gatacre retained the confidence and affection of his soldiers in the most adverse circumstances.

When the weary privates struggled back to camp after the disastrous day at Stormburg they were quite clear on one point: 'No one could have got us out but him.' Two days before he was dismissed the Cameron Highlanders passed through Bethany, and the men recognised the impetuous leader of the Atbara charge;[126] and, knowing he had fallen among evil days, cheered him in the chivalry of the common man. The poor general was much moved at this spontaneous greeting, which is a very rare occurrence in our phlegmatic, well-ordered British Army. Let us hope the sound will long ring in his ears, and, as it were, light a bright lamp of memory in the chill and dreary evening of life.

Exit General Gatacre.[127] 'Now,' as my Parsee merchant remarked, 'he is gone'; and I suppose there are, here and there, notes of triumph. But among them I will strike a note of warning. If the War Office breaks generals not so much for incapacity as for want of success with any frequency, it will not find men to fight for it in brigade and divisional commands. Every man who knows the chances of war feels himself insecure. The initiative which an unsympathetic discipline has already killed, or nearly killed, in younger officers, will wither and die in their superiors. You will have generals as before, but they will not willingly risk the fruits of long years of service in damnable countries and of perils of all kinds. They will look at the enemy's position. They will endeavour to divide responsibility. They will ask for orders or instructions. But they will not fight—if they can possibly help it, and then only on the limited liability principle, which means the shedding of much blood without any result. Besides, as an irreverent subaltern remarked to me: 'If you begin with Gatacre, where are you going to end? What about poor old ----?'[128]

But I dare not pursue the subject further.

The Guards Brigade Entering Bloemfontein.[105]

[125] The Chitral Expedition was a military expedition in 1895 sent by the British authorities to relieve the fort at Chitral, which was under siege after a local coup following the death of the old ruler. An intervening British force of about 400 men was besieged in the fort until it was relieved by two expeditions, a small one from Gilgit and a larger one from Peshawar. Chitral Expedition - Wikipedia

[126] At the Battle of the Atbara River on 7 April 1898, British Army defeated Mahdist forces opening a line of march up the Nile.

[127] 4/11/1900 General Chermside promoted to command Third Division, vice General Gatacre, ordered home.

[128] I believe he is referring to General Buller.

III. At Half-Way House

Bloemfontein: April 16th 1900.

After a decent interval let the curtain rise on a new act. The scene and most of the characters are different, but it is the same play. The town—a town of brick and tin—stands at the apparent edge of a vast plain of withered grass, from whose inhospitable aspect it turns and nestles, as if for protection, round the scrub-covered hills to northward. From among the crowd of one-storied dwelling-houses, more imposing structures, the seats of Government and commerce, rise prominently to catch the eye and impress the mind with the pleasing prospect of wealthier civilisation. Here and there are towers and pinnacles, and, especially remarkable, a handsome building surrounded in the classic style by tall white pillars, and, surmounted by a lofty dome, looks like a Parliament House, but for the Red Cross flag which flies from the summit and proclaims that, whatever may have been its former purposes, the spacious hall within is at last devoted to the benefit of mankind. The dark hills—their uncertain outline marked at one point by the symmetrical silhouette of a fort—form the background of the picture: Bloemfontein, April, 1900.

General view of Bloemfontein.[106]

Changing of the Guard at the Presidency, Bloemfontein.[107 129]

Bloemfontein Post Office.[108]

It is five o'clock in the afternoon. The market square is crowded with officers and soldiers listening to the band of the Buffs. Every regiment in the service, every Colony in the Empire is represented; all clad in uniform khaki, but distinguished by an extraordinary variety of badges.

Each group is a miniature system of Imperial Federation. The City Volunteer talks to a Queensland Mounted Infantryman, who hands his matchbox to a private of the Line. A Bushman from New Zealand,[130] a Cambridge undergraduate, and a tea-planter from Ceylon stroll up and make the conversation general. On every side all

[129] Roberts occupied Bloemfontein on March 13th, 1900. His despatch ran:—"The British flag now flies over the Presidency vacated last evening by Mr. Steyn, late President of the Orange Free State. The inhabitants gave the troops a cordial welcome."

[130] New Zealand - 5th Contingent (Imperial Bushmen) https://www.angloboerwar.com/?option=com_content&view=article&id=418 Also: South Australian Citizen Bushmen - Wikipedia

kinds of men are intermingled, united by the sympathy of a common purpose and soldered together in the fire of war. And this will be of great consequence later on.

The inhabitants—bearded Burghers who have made their peace, townsfolk who never desired to make a quarrel—stand round and watch complacently. After all, there are worse things than to be defeated. Demand is keen, the army is wealthy, and prices are high. Trade has followed hard on the flag which waves from every building; and, whether it be for merchandise or farm produce, the market is buoyant.

The officers congregate about the pretentious building of the club, and here I find acquaintances gathered together from all the sentry beats of the Empire, for the regular army usually works like a kaleidoscope, and, new combinations continually forming, scatter old friends in every direction. But here all are collected once more, and the man we met on the frontier, the man we met 'up the river,' the man we met at manoeuvres with the comrade of Sandhurst, the friend or enemy of Harrow days, and the rival of a Meerut tournament,[131] stand in a row together. Merry military music, laughing faces, bright, dainty little caps, a moving throng, and the consciousness that this means a victorious British Army in the capital of the Free State, drive away all shadows from the mind.

One cannot see any gaps in the crowd; it is so full of animation that the spaces where Death has put his hand are not to be seen. The strong surges of life have swept across them as a sunny sea closes over the foundered ship. Yet they are not quite forgotten.

'Hullo, my dear old boy, I am glad to see you. When did you get up here? Have you brought ---- with you? Oh, I am sorry. It must have been a fever-stricken hole that Ladysmith. Poor chap! Do you remember how he .... Charlie has gone home. He can never play polo again—expanding bullet smashed his arm all to bits. Bad luck, wasn't it? Now we've got to find a new back .... and ---- was killed at Paardeberg .... spoiled the whole team.' The band struck into a lively tune. 'How long is it going to last?'

'With luck it ought to be over by October, just a year from start to finish.'

'I thought you said something about Pretoria the third week in March.'

'Ah, I must have meant May, or, perhaps, June.'

'Or August.'

'Who can tell? But I think this is the half-way house.'

The conversation stops abruptly. Everyone looks round. Strolling across the middle of the square, quite alone, was a very small grey-haired gentleman, with extremely broad shoulders and a most unbending back. He wore a staff cap with a broad red band and a heavy gold-laced peak, brown riding boots, a tightly-fastened belt, and no medals, orders, or insignia of any kind. But no one doubted his identity for an instant, and I knew that I was looking at the Queen's greatest subject, the commander who had in the brief space of a month revolutionised the fortunes of the war, had turned disaster into victory, and something like despair into almost inordinate triumph.

Other soldiers of career and quality mingle with the diversified throng. Macdonald sits on a bay pony near the club verandah talking to Martyr of the Mounted Infantry and of Central African repute. Pole-Carew,[132] who came to the Cape as Sir Redvers Buller's camp commandant, and passed at a bound to brigadier-general, and by another still greater leap to the command of the Eleventh Division, canters across the square. General French and his staff have just ridden up.[133] But the central figure holds all eyes, and everyone knows that it is on him, and him alone, that the public fortunes depend.[134]

[131] PF: I think he is referring to a polo tournament from his time in India. The city and district of Meerut lies 80 km (50 mi) northeast of the national capital New Delhi. Meerut - Wikipedia

[132] Reginald Pole-Carew. Reginald Pole-Carew (British Army officer) - Wikipedia

[133] Winston and General French are not on good terms at this time. "Although I was often with French's column ..... the General completely ignored my existence, and showed me no sign of courtesy or goodwill. I was sorry for this, because I greatly admired all I had heard of his skilful defence of the Colesberg front, his dashing gallop through the Boer lines to the relief of Kimberley, and was naturally attracted by this gallant soldierly figure, upon whom fell at this moment the gleam of a growing fame. Thus during the South African war I never exchanged a word with the General who was afterwards to be one of my greatest friends and with whom I was for many years to work at grave matters in peace and war." See My Early Life Ch. XXVI.

[134] Winston and General Roberts were not on good terms at this time. Roberts had only reluctantly granted Winston credentials because of Winston's disparagement of the minister who gave the lack luster sermon in Natal. So, here "When one morning in the market place at Bloemfontein amid a crowd of officers I suddenly found myself quite unexpectedly within a few yards of him, he acknowledged my salute only as that of a stranger." My Early Life Chapter XXVI.

Such was the scene on the afternoon of my arrival in Bloemfontein. What of the situation? The first thing to be done after the occupation of the town was to re-open the railway.

Norval's Pont Railway Bridge.[109]

The presence of a large army in their rear and the swift advance of Gatacre and Clements[135] compelled the invaders to withdraw from Cape Colony, so that Norval's Pont and Bethulie bridges were once more in British hands. Both were, however, destroyed or partially destroyed. Besides these, various other smaller bridges and culverts had been blown up. All these were forthwith repaired by the engineers, and through communication by rail was established between the advanced Field Army in the Free State and the sea bases at East London, Port Elizabeth, and Cape town.

In the meantime the Army at Bloemfontein lived on the reserve of rations it had carried from Modder River. When the railway was opened the line from Modder River was dropped.

Placing a water pipe across the river.[110] Troop train crossing a temporary bridge. This span is across the Modder River.[111]

A broad-gauge railway, even though it be only a single line, is usually capable of supplying an army of at least 50,000 men with considerable ease, and the reader may remember how the Natal Government Railway was able

[135] Clements took over command on the Colesberg front from Major General John French in late January and held it with lesser forces against the attacks of Boer commander Koos de la Rey. On 12 February, after the action at Pink Hill, Clements shortened his line in the face of the Boer advance by retreating to Rensburg siding. After the victories of Field Marshal Frederick Roberts on the western front, the Boers retreated and Clements took Colesberg on 28 February. Ralph Arthur Penrhyn Clements - Wikipedia

to support 30,000 men through January and February, to transport reinforcements and sick, and to run all its ordinary traffic in addition. But the repaired or provisional bridges on the Bloemfontein line caused so much delay that the carrying power of the railway was seriously diminished. When a permanent bridge has been blown up two alternatives present themselves to the engineers: a high level or a low level substitute. The high level bridge, such as was thrown across the Tugela after the relief of Ladysmith, takes much longer to build, but, when built, trains are run straight over it with very little diminution of speed. It is, moreover, secure against floods.

Modder River Bridge.[112]

Modder River Temporary Bridge.[113]

The low level bridge must be approached by zigzag ramps, which impose frequent shuntings, and cause great delay; and it is, of course, only to be trusted when there are no floods. But it has this inestimable advantage in military operations: speed in construction. The Army must be fed immediately. So the low level bridges were chosen; hence an early but reduced supply. When this was further minimised by the passage of reinforcements the commissariat depots could scarcely make headway, but must be content to feed the Army from day to day and accumulate at the rate, perhaps, of only one day in three, or even one in four. It was, therefore, evident that no offensive movement to the northward could be made for several weeks.

See how the stomach governs the world. By the rapid invasion of their territories, by the staggering blows which they had been dealt at Kimberley,[136] Paardeburg,[137] Poplar Grove,[138] and Dreifontein,[139] and by the bad news from Natal, the Boers in the Free State were demoralised.

Paardeberg Battlefield.[114]

Powder Barrels and Destroyed Wagons.[115]

Remains of burnt wagons.[116]

[136] The relief of Kimberly on February 15th, 1900. See: Siege of Kimberley - Wikipedia

[137] The Battle of Paardeberg or Perdeberg ("Horse Mountain", 18–27 February 1900) was a major battle during the Second Anglo-Boer War. It was fought near Paardeberg Drift on the banks of the Modder River in the Orange Free State near Kimberley. Battle of Paardeberg - Wikipedia

[138] The Battle of Poplar Grove took place on March 7th, 1900. General Sir John French's cavalry attacked the Boer force from the rear while mounted infantry and horse artillery attacked from the right flank. The Boers abandoned their positions in panic before the cavalry. The commander-in-chief of the Free State forces, Christiaan de Wet, in his book called the chapter on the subject "Wild Flight from Poplar Grove." Battle of Poplar Grove - Wikipedia

[139] The Battle of Driefontein on 10 March 1900 followed on the Battle of Poplar Grove. The Boer forces under the command of Christiaan de Wet were holding a 7-mile (11 km) line covering the approach to Bloemfontein. Lord Roberts subsequently ordered a division under Lieutenant General Thomas Kelly-Kenny to attack the position from the front, while Lieutenant General Charles Tucker's division moved against its left flank. The Boers were subsequently forced to withdraw losing 124 men killed and captured, while the British lost 82 killed and 342 wounded. Battle of Driefontein - Wikipedia

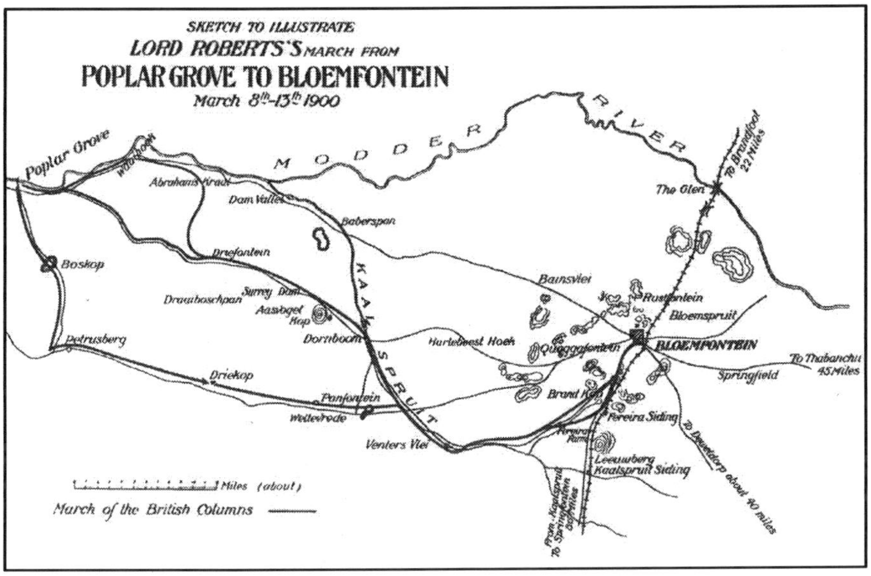

Poplar Grove to Bloemfontein.[117]

If we could have pressed them unceasingly the whole country would have been conquered to the Vaal River. Encouraged by Lord Roberts's Proclamation,[140] and believing that all resistance in the Southern Republic was at an end, great numbers of Free Staters returned to their homes, took the oath of neutrality, and prepared to accept the inevitable.

But while the Army waited, as it was absolutely forced to wait, to get supplies, to get horses—to get thousands of horses—to give the Infantry new boots, and all arms a little breathing space, the Boers recovered from their panic, pulled themselves together, and, for the moment, boldly seized the offensive.

Great, though perhaps temporary, were the advantages which they gained. The belief that the war in the Free State was at an end, which had led so many of the Burghers to return to their farms, was shared to some extent by the British commander, and loudly proclaimed by his colonial advisers. To protect the farmers who had made their peace the Imperial forces were widely extended. A line was drawn across the Free State from Fourteen Streams, through Boshof, Bloemfontein, and Thabanchu, south of which it was assumed that the country was pacified and conquered.

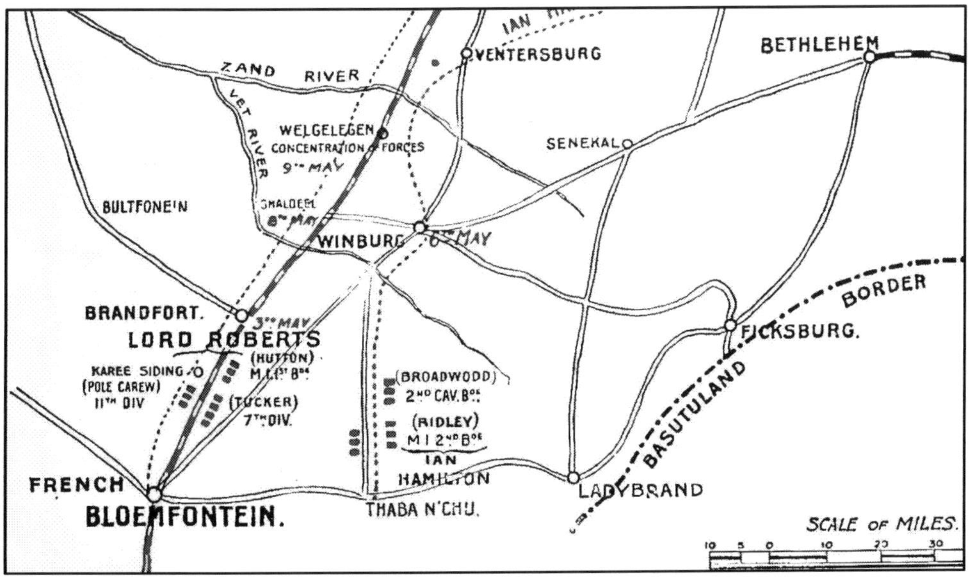

Note Thaba N'Chu East of Bloemfontein.

[140] Granting security of property and amnesty to those who would lay down their arms and take an oath of allegiance. See WSC infra.

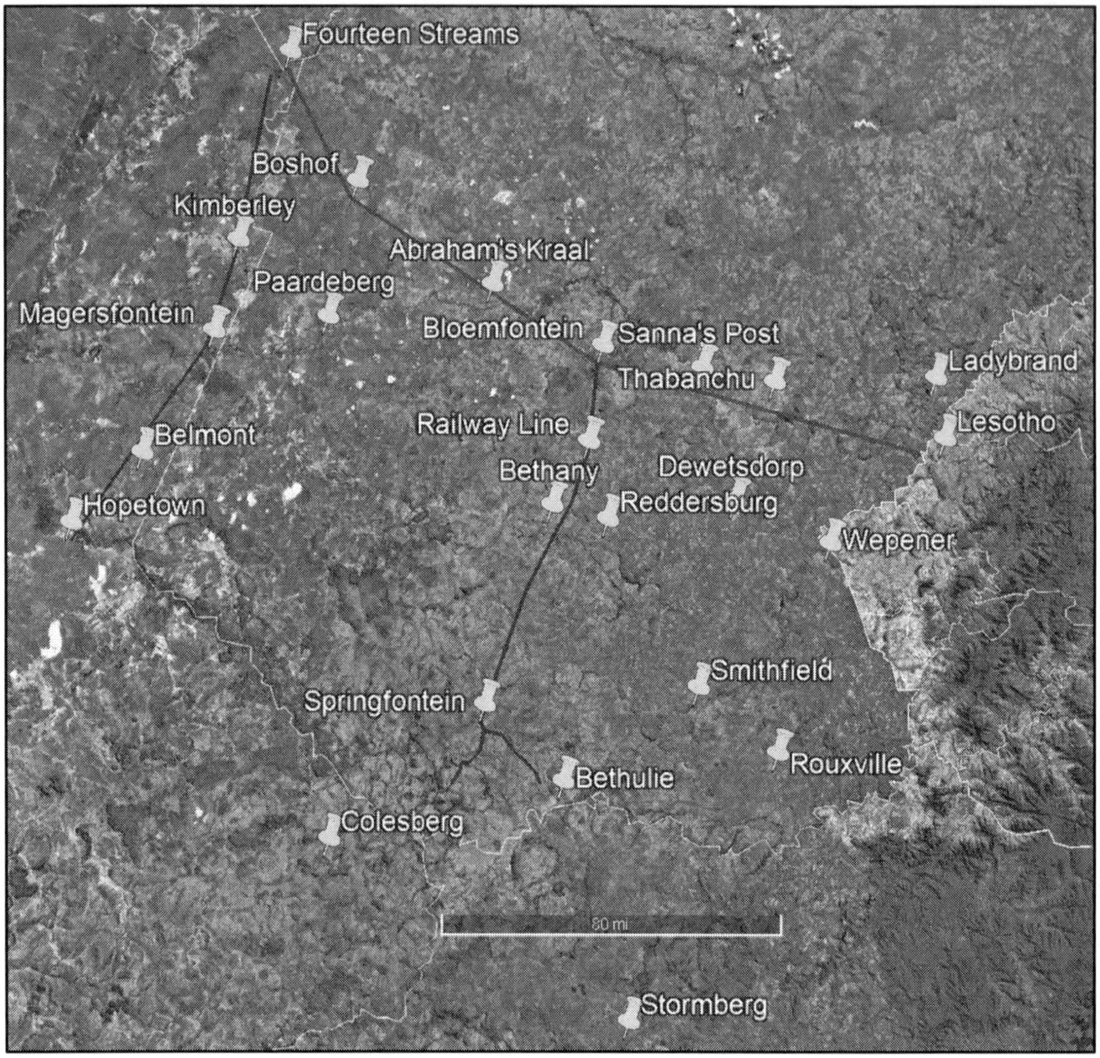

Area of Operations. Google Image© 2023 Image Landsat/Copernicus.

Meanwhile Olivier [141] and the southern commando, recalled from their operations in the Cape Colony, were making a hurried, and, as it seemed, a desperate march to rejoin the main Boer forces. They expected the attack of the same terrible Army which had already devoured Cronje; nor was it until they reached Ladybrand and found only Pilcher[142] with a few hundred men snapping at their heels that they realised that the bulk of the British troops were for the moment practically immobile at Bloemfontein. Then they turned.

Pilcher fled [westwards] warily before them, and fell back on Broadwood's Brigade, near Thabanchu. With renewed courage and strong reinforcements from their friends north of the line of occupation [De Wet] they pressed on. Broadwood[143] was compelled to fall back on the Ninth Division, which was camped west of the

[141] Jan Hendrik Olivier (1848 – 1930) His parents lived in the Burgersdorp area and later moved to Zastron. At the age of 15 he joined the Orange Free State border police. In 1865 he became a Field Cornet and fought in the last Basuto War. He is rewarded with the farm Olifantsbeen where he becomes prosperous. He became a member of the Volksraad in 1883 for the Caledon River ward.
He was in command of the Rouxville and Thaba Nchu commandos, and moved with his mother through the Colesberg, Barkly East and Dordrecht areas. He was in command of the Boer forces at the Battle of Stormberg in December 1899. Later he worked with General De Wet in the eastern Free State and Brandwater areas. After getting into an argument with De Wet, he decided to join the Transvaal forces. On the way there, he is captured. He was exiled to Ceylon as Boer prisoners of war along with three of his sons. J.H. Olivier - Wikipedia

[142] Thomas David Pilcher Lieutenant Colonel Entered 1879; Brevet Lieutenant Colonel, 1899. Staff service: DAAG, Dublin District 1895-97; Employed with W African Frontier Force, 1897-99; Special Service, South Africa, 1899. War service: W Africa, 1897-98 (Despatches; Brevet of Lieutenant Colonel); Boer War, 1899-1900. Commanding Corps Mounted Infantry. Bedfordshire Regiment. Thomas Pilcher - Wikipedia

[143] Major R. G. Broadwood Joined the 12th Lancers in 1881 and served in Egypt and the Sudan, before taking command of the 2nd Cavalry Brigade in the South African War. Fighting in the Orange Free State, he captured the town of Lindley and seized the Republican Government in Reitz, which was followed by operations in the Transvaal. From 1903 to 1904 he was in command in Natal. 12th Lancers. Robert George Broadwood - Wikipedia

waterworks. He made a twenty-mile march at night and laagered in the small hours of the morning, thinking, as most people would think, that pursuit was for the time being shaken off. Morning broke, and with it a Boer cannonade.[144]

Jan Hendrik Olivier.[118] Thomas Pilcher.[119] Robert G. Broadwood.[120]

I do not intend to be drawn into a detailed description of the action that followed.

For many reasons it deserves separate and detailed consideration, chiefly because it shows the Boer at his very best: crafty in war and, above all things, deadly cool. In a word, what happened was this: The shells crashed into the laager. Everyone said, 'Take the blasted waggons out of the shell fire. We will cover their retreat'; which they did most beautifully: Broadwood displaying all the skill which had enabled him to disentangle the reconnaissance of the 5th of April near the Atbara[145] from the clutches of the Dervishes.

The said waggons hurried out of the shell fire only to fall into the frying-pan of an ambuscade. Guns, prisoners, and much material fell into the hands of the Boers.[146 147]

The Ninth Division retreated suddenly—too suddenly, say the Army, with other remarks which it is not my business to transcribe—on Bloemfontein, and the force of the storm fell on Gatacre.[148]

Gatacre had a post at Dewetsdorp: three companies of the Royal Irish Rifles, two of Mounted Infantry. So soon as he heard of the retirement of the Ninth Division he sent orders by many routes for his post to fall back too. They fell back accordingly; but at Reddersburg[149] the net closed round them. Let us judge no man harshly or in ignorance. Fighting followed. With a loss of eight killed and thirty-one wounded, the retreating troops surrendered when relief was scarcely five miles away.[150]

Everything curled back on to Bloemfontein and the [North South] railway line, which it was *vital* to hold. Reinforcements were thrust to the front to meet the emergency: [Sir Leslie] Rundle,[151] with the Eighth Division, was diverted from Kimberley to Springfontein; Hunter, with the Tenth Division (our old friends the Irish and Fusilier Brigades), started from Natal, thus condemning Buller to the strict defensive, and the Boers swept southward.

144 3/30/1900 Colonel Broadwood with Cavalry Brigade and two batteries Royal Horse Artillery Retreat from Thabanchu to the waterworks pressed by Commandant Olivier.

145 In Sudan.

146 3/31/1900 Engagement at Sanna's Post (Koorn Spruit). Loss of convoy and six guns.

147 See: Sanna's Post - Wikipedia

148 [See Sanna's Post Engagement section of the Orange Free State Campaign Summary infra. Winston never names who was commanding the 'Ninth Division.']

149 Reddersburg is a small sheep and cattle farming 60 kilometres south of Bloemfontein. Reddersburg - Wikipedia

150 4/4/1900 Troops led by Christiaan de Wet attacked a convoy of 600 of the Royal Irish Rifles at Reddersburg on and obtained their surrender the next day.

151 General Sir Henry Macleod Leslie Rundle, (6 January 1856 – 19 November 1934) was a British Army general during the Second Boer War and the First World War. In January of 1900 he was appointed to the command of the 8th Division of the South African Field Force, with the temporary and local rank of lieutenant general. He left Southampton in March 1900 with the staff of the 8th division and 600 men of militia regiments. He served as commander until early March 1902, Leslie Rundle - Wikipedia

Now, in accordance with the terms of Lord Roberts's Proclamation, many farmers of the Free State, fighting men of the Boer Army—that is to say, who had thought that all was up: deserters, in other words—had come into the British posts, made their submission, taken the oath, and returned to their farms. The Boers were very angry with these people. What protection could we give them? Some, it is said—it may be a lie—were shot by the enemy. Most of them, from fear or inclination, rejoined their commandos.

The whole of the right-hand bottom corner of the Free State was overrun. Southward still hastened the Boer forces. Brabant was the next to feel the tempest. His garrison in Wepener was assailed, surrounded, fought well—perhaps is now fighting desperately. Other Boers approached the rebel districts of Cape Colony. The lately penitent rebels stirred, are stirring.

Sir Henry M. Leslie Rundle.[121] Edward Yewd Brabant.[122 152] John Brabazon.[123] Archibald Hunter.[124]

Mark, by the way, this sedition is not the result of misplaced generosity but of military misfortunes. No one expects beaten men to be grateful; but, under certain conditions, they will be loyal. An enemy at their throats is not one of those conditions. Southward still sweep the commandos *with empty carts*, for this is the most fertile of all the Republican territories; and, in the meanwhile, what are we doing? Divisions and brigades are being moved by a strong yet deliberate hand. The hope—general and special idea in one—is to catch these bold fellows who have thrust their heads thus far into the lion's mouth and enjoyed until now such immunity.

Wepener making a brave defence; Brabant marching through Rouxville to bar their advance; Rundle, Chermside, and Brabazon[153] striking east from Edenburg to shut the door behind them with two Infantry divisions, twenty-four guns, and 2,000 Yeomanry; and, further north, the great Bloemfontein Army—four Infantry divisions, [Ian] Hamilton's 10,000 mounted men, French's four Cavalry brigades, and many guns—is almost ready to move. Assuredly these Boers are in a dangerous place. Will they escape? Will they, perhaps, carry some part of the intercepting lines with them as a trophy of victory? 'Qui vivra verra,'[154] and, if these letters continue, 'who runs may read,' for I purpose to journey *viâ* Edenburg to Reddersburg to-morrow, and thence on to the point of collision, which must mark the climax of this extremely interesting event henceforward to be called 'The Operations in the Right-hand Bottom Corner of the Free State.'

[152] Major-General Sir Edward Yewd Brabant, (31 May 1839 – 13 December 1914) During the Second Boer War (1899–1902), he commanded the Colonial Division in 1900, and the Colonial Defence Force of Cape Colony in 1901. As a Brigadier General of the Eastern Cape troops, his command included: Cape Mounted Riflemen, the 79th Battery, RFA, the Kaffrarian Rifles, the Queenstown Volunteers, part of the 1st Battalion, Royal Scots, and Brabant's Horse. He occupied Jamestown and the Herschel district in February 1900. His units operated round the Queenstown/Dordrecht area and moved north to hold the Jammersburg Drift at Wepener, which they did under appalling rain and cold against a superior Boer force led by Christiaan De Wet. Edward Brabant - Wikipedia See Also: Anglo Boer War - Name search

[153] Brabazon John Palmer Major General Born 13 Feb 1843. He entered 1862; Colonel, January 1899. Staff service: Acting as Volunteer with rank of Captain, Ashanti Expedition, 1873-74; ADC (extra to Viceroy of India), 1877-79; Brigade-Major, Afghan Campaign, 1870-80; ADC to the Queen, 1889; Colonel on Staff; Commanding Cavalry Brigade, South-East District, 1899; Major General, Cavalry Brigade, South Africa, 1899-1900. War service: Ashanti, 1874); Afghan War 1878-80 Egyptian Expedition, 1884 Sudan, 1884-85 South African War, 1899-1900 General Brabazon, born in 1843, was the son of the late Major Brabazon (late 15th Hussars), and the daughter of the late Sir W H Palmer, Baronet. Imperial Yeomanry. _Imperial Yoemanry Article:
https://www.angloboerwar.com/?option=com_content&view=article&id=1946

[154] Who(ever) lives will see.

Close View of Area of Operations. Google Image© 2023 Image Landsat/Copernicus.

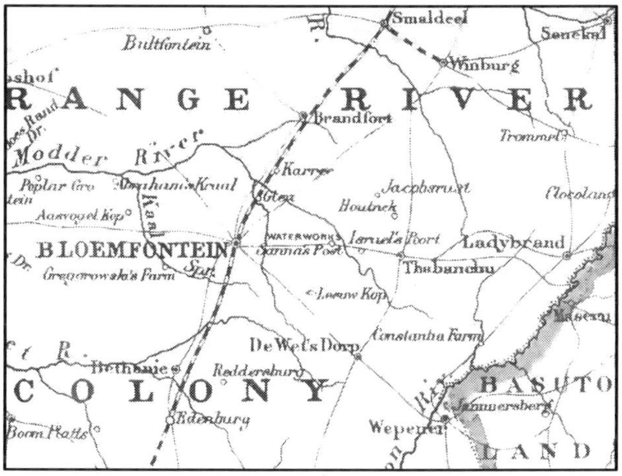

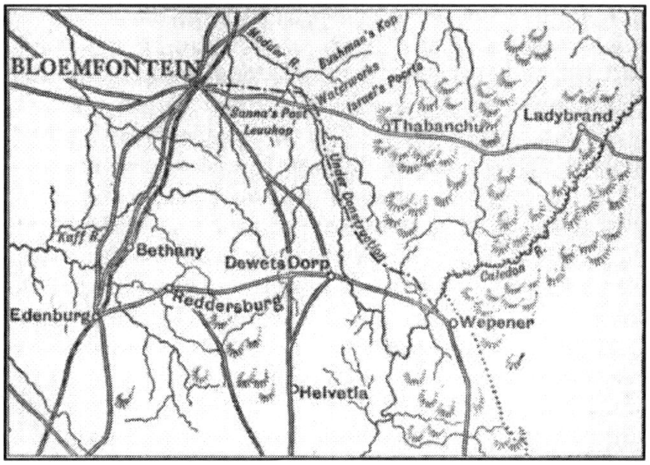

Orange Free State.[125] Note Poplar Grove, Abraham's Kraal, Waterworks and Sanna's Post and Israel's Poorte. Another View showing Sanna's Post and Israel's Poorte.[126]

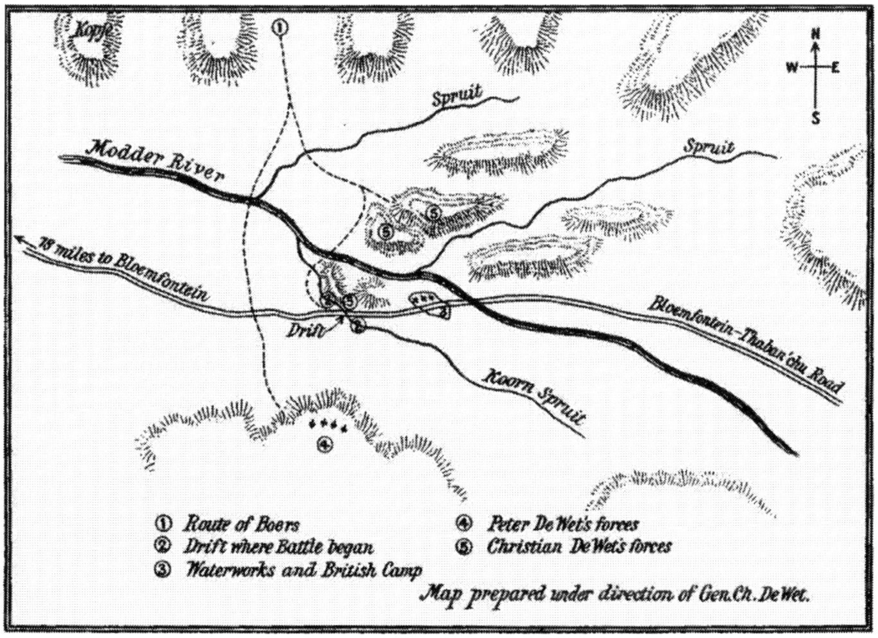

Map of Sanna's Post engagement.[127]

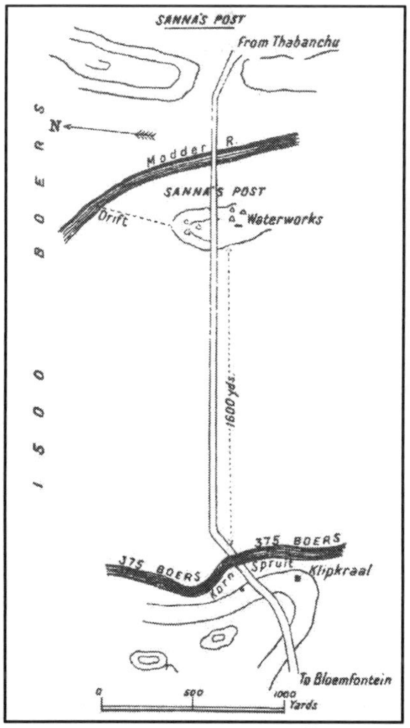

Sanna's Post Engagement.[128]

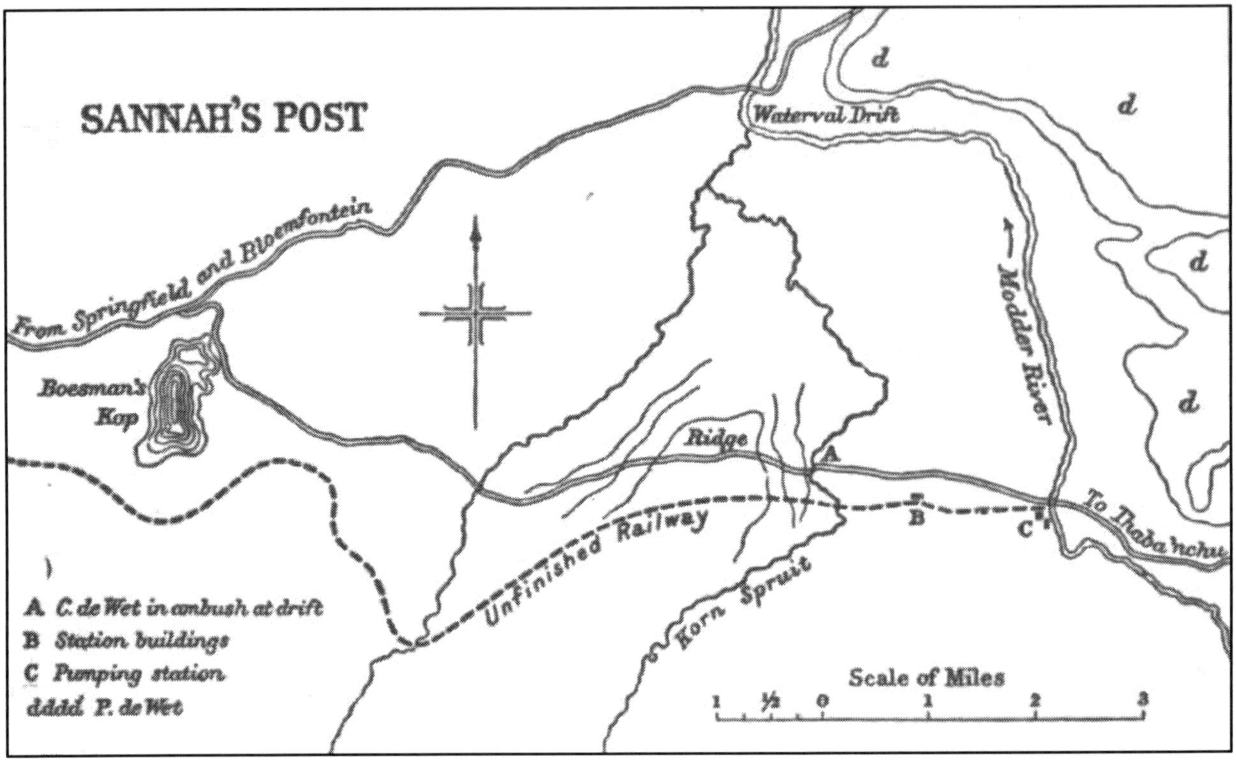

Sannah's Post.[155]

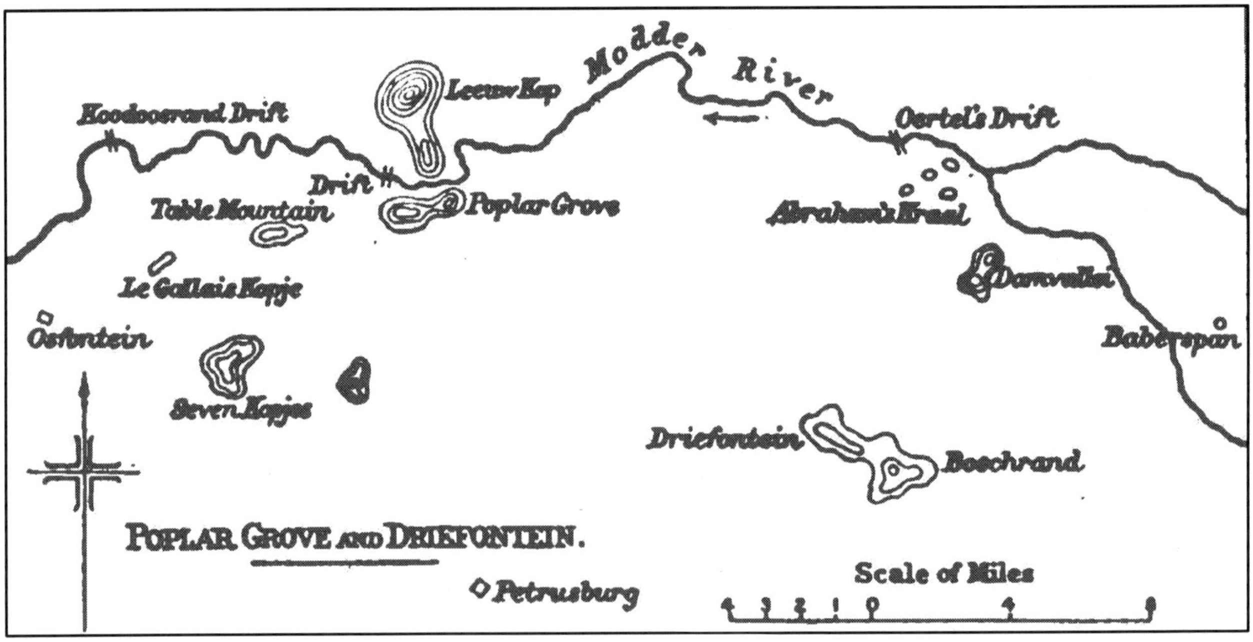

Poplar Grove.[156]

[155] A Handbook of the Boer War - Gale And Polden Limited, London 1910 Source: Anonymous ISBN 978-1-37-497455-5. https://babel.hathitrust.org/cgi/pt?id=uc1.l0061277489&seq=7

[156] A Handbook of the Boer War - Gale And Polden Limited, London 1910 Source: Anonymous ISBN 978-1-37-497455-5. https://babel.hathitrust.org/cgi/pt?id=uc1.l0061277489&seq=7

IV. Two Days with Brabazon

Before Dewetsdorp: April 21st, 1900.

When the incursion of the Boers into the recently pacified districts became known, the Eighth Division (Rundle) was diverted from Kimberley, whither it was proceeding, and concentrated at Springfontein. The Third Division (Chermside, in supersession of Gatacre) massed at Bethany. Still more troops were needed to guard the line and clear the country.

Lady and General Chermside in 1902 when he was Governor of Queensland.[157] [158]

Sir Redvers Buller was asked whether he could co-operate by forcing Van Reenen's Pass and bringing pressure on the enemy's line of retreat. His position in the centre of the triangle of Natal was, however, an inconvenient one. The strategic advantages possessed by the Boers in this scene of the war have before been noticed. But it may be worth while to explain them again.

The enemy possess the superiority of an enveloping frontier. If Sir Redvers Buller moves west through Van Reenen's Pass to make the diversion required in the Free State, down will come the Boers from the Biggarsburg on his communications and into South Natal. If he moves north to attack the Biggarsburg positions in order to clear Natal he will cut the Boers on his left flank and line.

According to the best information there are three thousand Boers on the Drakensburg Passes, and ten thousand on the Biggarsburg. Buller, therefore, would have preferred to mask Van Reenen's with the Ladysmith Division (Fourth, Lyttelton), which was getting well and strong again, and move northwards with the Second, Fifth, and Tenth Divisions. He did not consider until northern Natal should be cleared that he could safely move westward. On the other hand, the need in the Free State was urgent, and it was therefore arranged that the Tenth Division (Hunter)[159] should come by sea to East London—one brigade to replace the division diverted from Kimberley, one brigade to Bethulie, and that the rest of the Natal Field Army should remain strictly on the defensive until the situation was materially altered.

[157] Portraits of Sir Herbert Chermside (1850-1929), British Army officer and Governor of Queensland from 1902-1904., and his first wife, Lady Geraldine Katherine Webb. Illustrated page from the supplement to The Queenslander, March 15 1902, p. 609. File:Lady Chermside and Major-General Sir Herbert Chermside 1902.jpg - Wikimedia Commons

[158] Lieutenant-General Sir Herbert Charles Chermside, (31 July 1850 – 24 September 1929) was a British Army officer who served as Governor of Queensland from 1902 to 1904. In 1899 was sent to South Africa to command the 14th Brigade and the 3rd Division during the Second Boer War. Herbert Chermside - Wikipedia

[159] General Sir Archibald Hunter, (6 September 1856 – 28 June 1936) was a senior officer in the British Army who distinguished himself during the Boer War. He was Governor of Omdurman, in Sudan, and later of Gibraltar. Archibald Hunter - Wikipedia

East London, Cape colony. [160] The port of entry for the Orange Free State. Small steamers can cross the sand bar to enter the Buffalo river but the mail steamers must anchor outside. [129] Note lighter vessel bottom right. Northumberland Fusilliers Arriving in East London to Reinforce Gatacre. [130]

Practically, therefore, five brigades of troops were available for the operations in the right-hand bottom corner: Hart,[161] with a brigade of Hunter's Division at Bethulie, the Third and Eighth Divisions under Chermside and Rundle at Springfontein and Bethany. Besides these powerful bodies, which were quite independent of the communication troops or the Bloemfontein Army, there were fourteen hundred Yeomanry and Mounted Infantry under General Brabazon, and Brabant's Colonial Brigade, about two thousand five hundred strong.

It is scarcely necessary to follow all the movements in exact detail. Rundle formed a column at Edenburg, and, marching to Reddersburg, joined his force to part of Chermside's Division from Bethany, thus having under his immediate command eight battalions, four batteries, and Brabazon's Mounted Brigade. Another brigade was collecting at Edenburg under Campbell.[162] [163] Hart was moved north-east towards Rouxville, where was also Brabant with a thousand horse. The rest of Brabant's force, some fifteen hundred strong, were blockaded in Wepener by the enemy. Such was the situation when I left Bloemfontein on the morning of the 17th. [April]

I travelled prosperously; came by rail to Edenburg, trekked from there in drenching rains, most unusual for this time of year, and greatly increasing the difficulties of supply; and, resting for the night at Reddersburg, caught up the marching column in its camp, about eleven miles from Dewetsdorp, on the night of the 19th. [April][164]

Accompanied by his cousin Charles Spencer-Churchill (Sunny), Duke of Marlborough.[131] [165] [166] A General service wagon.[132] Note the brake shoes on the rear wheels and the screw handle (center) to apply them.

[160] East London, South Africa - Wikipedia

[161] Brigadier-General Arthur Henry Seton Hart-Synnot, (1870–1942) was a British Army general who saw service in the South Africa then in Japan, Hong Kong, Burma, and India; and finally in France in the First World War. Arthur Henry Seton Hart-Synnot - Wikipedia

[162] B. B. D. Campbell B B D Major General Commands & Staff. MID LG: 16 April 1901, page: 2604. Source: Field Marshal Roberts. 2 April 1901. Re: General mentions.

[163] David Campbell: David Campbell (British Army officer) - Wikipedia

[164] See also, My Early Life: It was the 8th Division, the last division of our regular army scraped together from our fortresses all over the Empire. It was commanded by Sir Leslie Rundle, later unkindly nicknamed 'Sir Leisurely Trundle,' whom I had known up the Nile. Brabazon's brigade was scouting on ahead. Rundle was affable and hospitable; and early the next morning I rode on to join Brabazon. He was delighted to see me, told me his grievances, and entertained me vastly with stories and criticisms of French, as well as of the war and the world in general. We abode together for some days. WSC My Early Life Ch, XXVI.

The position of the various troops was then as follows: Rundle, with eight battalions, four batteries, and fifteen hundred horse at Oorlogs Poorte, about twelve miles from Dewetsdorp; Campbell, with two battalions and a battery near Rosendal,[167] marching to join him; the Grenadier Guards double marching through Reddersburg to catch up the main force; Hart, with four battalions in Rouxville; Brabant, with one thousand horsemen eight miles north of Rouxville; Dalgety, with a garrison of fifteen hundred men, holding Wepener. [Under siege]

So far as could be learned the enemy had about seven thousand men with twelve guns south of the Bloemfontein-Thabanchu line under Commandants Olivier and De Wet, and with this force, which made up in enterprise and activity what it lacked in numbers or material, they were attempting to blockade and attack Wepener, to bar the road of Rundle's column to Dewetsdorp, and to check Brabant and Hart at Smithfield.[168] Besides proposing this ambitious programme, the Boers sent their patrols riding about the country commandeering all pacified farmers under threats of death.

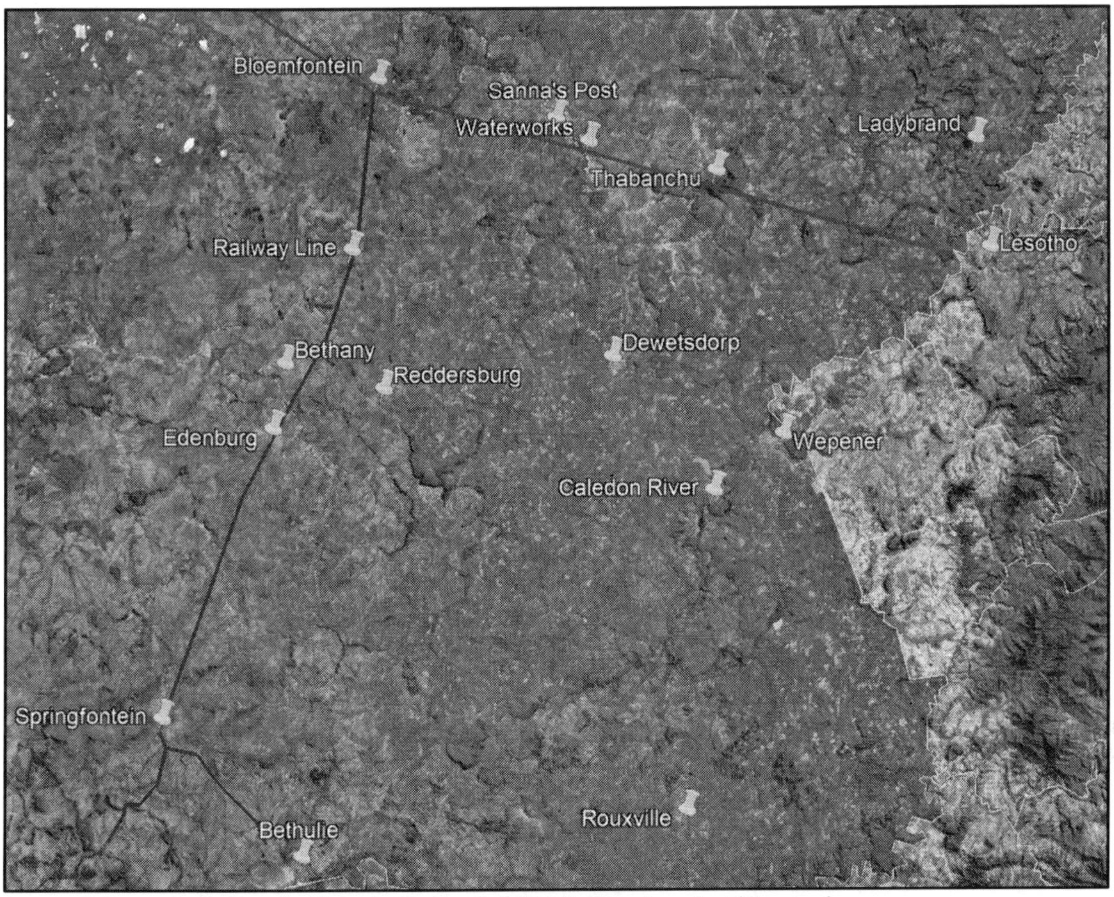

Area of Operations South. Google Image© 2023 Image Landsat/Copernicus.

We had a very pleasant ride from Reddersburg, and it was evening when we rounded the shoulder of a grassy hill and saw the camp of the main British column before us. It lay about the foot of a prominent knoll rising from a broad plain, which was in striking contrast to the mountains of Natal, and seemed to promise ample opportunity to the regular soldier. 'Camp' is, perhaps, an inaccurate description, for there were scarcely any tents to be seen, and the rolling ground was littered with swarms of grazing horses and oxen, and overspread with an immense canopy of white smoke from the hundreds of gleaming grass fires lighted to cook the soldiers' suppers.

[165] I proposed by telegram to Hamilton that he should take Marlborough upon his staff. The General agreed, and Lord Roberts, who never liked to treat anyone unfairly, gave a cordial approval. WSC My Early Life Ch. XXVII.

[166] I inspanned my four-horsed wagon, and we [He and Sunny] started upon a forty-mile march to overtake the flanking column. We came through the Boer-infested countryside defenceless but safely, and caught up our friends….. WSC My Early Life Ch. XXVII.

[167] This town is well north of the line of control. Rosendal, South Africa - Wikipedia

[168] Smithfield, South Africa - Wikipedia

A British Camp.[133]

General View of Mounted Infantry Camp.[134]

Officers Having Breakfast.[135]

I presented myself to Sir Leslie Rundle, who received me courteously, and briefly explained the outlines of the situation. We had arrived in the nick of time. The whole force would march at dawn. The scouts had exchanged shots during the day. The Kaffir[169] [170] spies reported that the enemy would fight on the morrow. What could be better? So with much satisfaction we went to bed.

There was a biting chill in the air when the first light of dawn began to grow in the sky, nor was I the only one who searched a modest kit for some of those warm clothes which our friends at home have thoughtfully been sending out. The South African winter was drawing near. But the sun soon rose, and we shivered no longer. The Cavalry were early astir. Indeed their mounted squadrons in silhouette against the morning sky was my first waking impression, and by half-past five all were in motion. I started a little later, but it was not long before I overtook them. Though the command was not a large one it presented several interesting features.

For the first time I saw the Imperial Yeomanry[171] in the field. Trotting across the beautiful green pasture land in a most extended formation, to which they seemed readily to adapt themselves, were seven hundred Yeomen, all good men and true, who had volunteered to fight because they understood the main causes of the quarrel, and from personal conviction earnestly desired to be of some assistance to the State, and who were, moreover, excellently mounted on smart, short-docked cobs,[172] which they sat and rode like the sportsmen they mostly were.

A large group of horsemen of the Imperial Yeomanry galloping over a plain.[136] A typical Imperial Yeoman on campaign.[137]

[169] During the Dutch and English Colonial Period the word Kaffir was used to describe monotheistic peoples (Nguni ethnic groups in particular) of South Africa, who were not of a Christian or of an Islamic religious background, without derogatory connotations, until the early twentieth century. It appears in many historical accounts by anthropologists, missionaries and other observers, as well as in academic writings. See : Kaffraria and British Kaffraria (Queen Adelaide's Province) were historical regions along the coast north or Port Elizabeth. See: File:Eastern Frontier, Cape of Good Hope, ca 1835.png - Wikimedia Commons Subsequently it became a racial slur. Kaffir (racial term) - Wikipedia See also the British military unit Kaffrarian rifles which were white and black units which were still organized as late as 1947. East London, 1 March 1947. King George VI inspecting 'Kaffrarian Rifles' guard of honour. - Atom site for DRISA And: Kaffrarian Rifles AKA: Buffalo Volunteer Rifles Buffalo Volunteer Rifles - Wikipedia Also: a historical region in Afghanistan and the setting of most of Rudyard Kipling's famous 1888 novella "The Man Who Would Be King." Kafiristan - Wikipedia And: an Arabic and Islamic term which, in the Islamic tradition, refers to a person who disbelieves in God as per Islam. Kafir - Wikipedia PF Note: I have no authority to change the wording used by original authors. There are instances where a pejorative or admiring modifier is used. While this is racist in and of itself, to identify someone by their complexion, in this narrative, identifiers such as Dutch, Boer, English, Uitlander, Indian, and many others are used as means of identification sometimes with pejorative or admiring modifiers.

[170] See also: Racial views of Winston Churchill - Wikipedia

[171] The original proposal to organise regiments of Yeomanry for service in the Boer War was made by Lord Chesham and other Yeomanry officers in Oct 1899. On 13 Dec 1899, the War Office decided it would allow a contingent of volunteer forces based on the standing yeomanry regiments to serve overseas. Under a Special Army Order of 4 Jan 1900, a committee of Yeomanry officers was constituted to administer the force. The standing Yeomanry regiments were asked to provide service companies of around 121 men each. The new Imperial Yeomanry (IY) were to be raised on a county basis with the core being the men of the existing volunteer units. The remaining numbers were to be recruited from individuals that met the necessary service criteria. 66 Volunteer Service Companies (7,427 men) served during the Boer War. https://www.angloboerwar.com/?option=com_content&view=article&id=1946

[172] A cob is traditionally a draft type pony. Typically of a stout build, with strong bones, large joints, and steady disposition, it is a body type of horse rather than a specific breed. Historically, in the United Kingdom and, to a lesser extent, the eastern United States, a 'cob' may be a common horse used for everyday riding but in the past was used for driving carts. Cob (horse) - Wikipedia

We were moving along in a wide formation, which secured us against all possibilities of surprise, when suddenly I noticed that the scouts far in front were halted.

'Tit-tat, tit-tat': two shots from a high plateau to the right. Shots fired towards you, I must explain, make a double, and those fired away from you a single, report.

We had flushed one of the enemy's outposts. Riding nearer, I could see their figures—seven in all—exposed on the skyline. This showed they were only an outpost, and wished to make us believe they were more. When the Boer is in force he is usually invisible. Still, the position was a strong one, and it is always a possibility worth considering with the Boer that he may foresee your line of thought, and just go one step further, out of contrariness. General Brabazon therefore halted his centre squadrons and detached a turning force of three companies of Yeomanry to the right.

We waited, watching the scouts exchange shots with the Boer picket, and watching—for it was a very pretty sight—the Yeomanry spread out and gallop away to the flank like a pack of hounds in full cry, each independent, yet the whole simultaneous. In a quarter of an hour they were scrambling up the steep sides of the plateau almost in rear of the obstructive picket, which hurriedly departed while time remained. Then the centre swung forward, and the whole Cavalry force advanced again, the greater part of it moving on to the plateau, where a running fight with the Dutch outposts now commenced at long range.

Several times we thought that we had unmasked their main position, and that the Cavalry work for the day was over; but each time Brabazon's turning movement on the right, the execution of which was entrusted to Colonel Sitwell,[173] a very dashing officer of Egyptian note, compelled them to fall back. After an hour of this sort of thing we were in possession of practically the whole of the plateau, which turned out to be of large extent.

Beyond it, commanding it, essential to it, yet not of it, was a steep rocky kopje. The swift advance and the necessity of pressing the enemy had left the Infantry a long way behind. The General felt, however, that this point must be secured. [Angus] McNeill[174] [175] made a dash for it with the scouts. The Yeomanry galloped off to the right again, as if about to surround it, and the Boers allowed themselves to be bounced out of this strong and important position, and scampered away to a smooth green hill a mile in rear. Brabazon made haste to occupy the captured kopje in force, and did so just in time, for as soon as the turning force—two companies (I am going to call them squadrons in future) of Yeomanry and a company of Mounted Infantry—approached the green hill, the musketry suddenly grew from an occasional drip into a regular patter, and there was the loud boom of a field gun. We had found the main Boer position, and the Cavalry came to a standstill. The enemy now directed a very sharp fire on the captured kopje, which, it seems, they originally intended to hold had they not been hustled out of it as has been described. They also shelled the Yeomanry—who were continuing the flank movement—rather heavily as they retired, inflicting some loss.

We had now to wait for the Infantry, and they lagged on the road. The Boer fire began to take effect. Several soldiers were carried wounded off the top of the hill—one poor fellow shot through both cheekbones. Others had to lie where they were struck because it was not possible to move them while the fire was so accurate.

On the reverse slope, however, there was good cover for man and horse. Some of the men were engaged for the first time, and though their behaviour was excellent, the General [Brabazon] thought it necessary to walk along the firing line and speak a few words here and there.

The Infantry still lagged on the road, but at about two o'clock Sir Leslie Rundle himself arrived. The firing about the kopje had been loud, and a rumour—who starts these tales?—ran back along the marching columns that the Cavalry were hard pressed, were running short of ammunition, and that the Boers were turning both flanks. At any rate, I found anxious faces in the divisional staff.

Rundle considered that the retention of the kopje was of first importance, and Sir Herbert Chermside, his second in command, fully agreed with him. But the Infantry of the advanced guard were alone near enough. It was

[173] W. H. Sitwell Colonel Entered 1880; Brevet Lieutenant Colonel, 1898. Staff service: Employed with Bechuanaland Border Police, 1891-93; DAAG, Guernsey, 1895-97; Special Service, Ashanti, 1895-96; Employed with Egyptian Army, 1897-99. War Service: Afghan War, 1880 (medal); Ashanti Expedition 1895-96 (star); Nile Expedition 1897; Nile Expedition 1898, wounded (Despatches; Brevet of Lieutenant Colonel; Egyptian medal with 2 clasps; medal); Boer War, 1899-1900.

[174] A. J. McNeill Captain MID LG: 16 April 1901, page: 2610. Source: Field Marshal Roberts. 2 April 1901. See: Montmorency's Scouts at: https://www.angloboerwar.com/?option=com_content&view=article&id=431 189 Records.

[175] In The River War Angus McNeill (Seaforth Highlanders) had contributed a dozen illustrations to both volume I and II.

decided to push them on. At this moment a reassuring message arrived from Brabazon engaging that he could hold his own, and hoping the Infantry would not be hurried so as to lose their breath.

Everyone was very cheerful after this, and when at last the leading battalion the Worcester Regiment[176] marched to the kopje all were able to admire the fine cool way in which they crossed the dangerous ground behind it; and I myself saw three pom-pom shells strike all around a young officer, who waved his rifle thereat in high delight, and shouted out loudly, 'By the left!' an order the purport of which I am as uncertain as the reader, but which, doubtless, was encouraging in spirit. When the Infantry had relieved the mounted men the latter withdrew to safer positions, and as the evening was drawing on the action came to an end—by mutual consent and by the effective intervention of the British Artillery.

Field Artillery Deployed.[138]

Maxim Automatic Machine-Gun (the "Pom Pom")[139] [177]

Pom Pom Maxim[140]

The events of the next day, though according to the scale of the war unimportant, were nevertheless instructive from the military point of view, and, so far as they concerned me, sufficiently exciting to require, if not to deserve, a letter to themselves.

[176] Worcester Regiment: https://www.angloboerwar.com/?option=com_content&view=article&id=672

[177] During the Second Boer War, the British used Maxim machine guns, and the Boers used a modified Maxim, a belt-fed, water-cooled machine gun that fired explosive 1-pound (0.45 kg) rounds (smokeless ammunition) at 450 rounds per minute, which became known as the "pom-pom." Maxim Nordenfelt Guns and Ammunition Company - Wikipedia And: A converted Maxim machine gun, firing a one-pound percussion-fused shell, was adopted for service by the Boers. Known as the "Pom Pom" (from the sound the gun made when firing) it was the first use of automatic light artillery in land warfare. The Pom-Pom was simply a large caliber, belt-fed machine gun firing explosive rounds. It was hard to locate on the battlefield as the smokeless ammunition (some of the first of its kind) produced little or no firing signature. Canvas belts held 25 brass-cased rounds carrying explosive charges with their percussion fuses. First-hand accounts describe the Pom-Pom as being very effective; standard artillery, mostly, could still be avoided by quickly taking cover in the interval between the flash indicating the firing of the shell and its arrival. The Pom-Pom, on the other hand, could keep up a continuous stream of fire, to devastating effect. South Africa's National Museum of Military History (smallarmsreview.com) See also this 15-minute video on the Forgotten Weapons channel describing the weapon. Maxim Pom-Pom 37mm Machine Gun - YouTube See also: QF 1-pounder pom-pom - Wikipedia

V. Two Days with Brabazon (Continued)

Camp before Dewetsdorp: April 22nd 1900.

Whether I am to see the white cliffs of Dover again I know not, nor will I attempt to predict. But it seems that my fortunes in this land are to be a succession of adventures and escapes, any one of which would suffice for a personal experience of the campaign. I acquit myself of all desire to seek for these. Indeed, I have zealously tried to avoid all danger except what must attend a War Correspondent's precarious existence.[178][179] This I recognise as a necessary evil, for the lot of the writer in the field is a hard and heavy one. 'All the danger of war and one-half per cent, the glory': such is our motto, and that is the reason why we expect large salaries.[180] But these hazards swoop on me out of a cloudless sky, and that I have hitherto come unscathed through them, while it fills my heart with thankfulness to God for His mercies, makes me wonder why I must be so often thrust to the brink and then withdrawn.

However, I will tell the tale of the doings of the Army, and what happened to me shall fill its proper place, so that the reader may himself be the judge of the matter.

The night of the 20th [April] passed quietly, but the Boers were awake with the sunrise and saluted us with discharges of the 'pom-pom,' which, as far as I could see, did no harm to anyone. We could not press the attack on the previous day because the Infantry were tired out and the enemy's position of sufficient natural strength to make an assault a serious business. In the night the Dutchmen had been busy, and the black lines of entrenchments marked the hillsides. When I inquired whether there would be a battle or not that day, staff officers pointed over the veldt to a column of dust which was coming slowly nearer.

General Campbell, with three battalions (including two of her Majesty's Guards) and a battery, was marching to join the main column. It was necessary, in view of the entrenchments and the approaching reinforcements, to wait until the force was complete. The event would be decided on the morrow, [22nd] and meanwhile Brabazon and the mounted troops—Cavalry, I shall call them—were to make a reconnaissance of the Boer left.

Troops of one of the British mounted regiments on the move.[141] Column of troops of one of the British mounted regiments on the move. They are probably servicemen of the Lothians and Berwickshire Yeomanry.[142]

The brigade, which included the Mounted Infantry, and was about a thousand strong, moved southward behind the outpost line and, making a rapid and wide circuit, soon came on the enemy's left flank. Here we waited while patrols were pushed out and while Brabazon was clearing his own right by a still wider turning movement. The patrols soon drew the fire of the Boer pickets, and the rifle shots began to ring out in the clear cool air of the morning. Presently a party of a dozen Boers appeared in the distance, galloping down towards a farm whence they might fire on the gradually advancing Cavalry. The General asked the subaltern in charge of

[178] Whether to award QSAs to correspondents was discussed in 1901 with the decision was made to issue medals under certain circumstances. There are 194 records in this article, from both sides. Many with extensive references. https://www.angloboerwar.com/?option=com_content&view=article&id=1861

[179] See: List of journalists killed during the Mahdist War - Wikipedia Seven individuals in this conflict by enemy action and friendly fire. Also: Mr. George W. Steevens, the most famous war correspondent at the time, died of disease in Ladysmith during the siege. G. W. Steevens - Wikipedia

[180] Winston's contract with the *Morning Post* was for £1,000 pounds for the first four months work, with £200 per month thereafter. He was also reimbursed for expenses and retained copyright on his work. These were very generous terms but he was an experienced, energetic and astute reporter as we see. Considering the courage, intrepidness and tenacity he has demonstrates in London to Ladysmith and this work I think he earned every shilling. For more detail see: Sandys, p. 13 and WSC My Early Life.

our two guns whether they were within range. The young officer was anxious to try. We watched the experiment with attention.

The practice was extremely good. The first shell burst in the middle of the Boer horsemen, who at once spread into a looser formation. The next exploded in front of them, and all the seven shells that were fired fell within measurable distance of someone.

For the first time in this war I saw the Boers show what I consider cowardice; for without anyone being killed or wounded the whole party turned back and, abandoning their intention or duty, scurried away to cover behind the long swell of ground over which they had come. The Boer Army in Natal was not thus easily dissuaded from its objects.

Meanwhile the flanking movement was in progress, and as the ground to our right was gradually made good and secured by Colonel Sitwell,[181] Brabazon pushed his centre forward until McNeill's scouts were cantering all over the slopes where the Boers had just been shelled, and hunting such of the enemy as tarried to safer and more remote positions. At last we arrived at the edge of the swell of ground. It fell steeply towards a flat basin, from the middle of which rose a most prominent and peculiar kopje. Invisible behind this was Dewetsdorp. Round it stood Boers, some mounted, some on foot, to the number of about two hundred.

Our rapid advance, almost into the heart of their position, had disturbed and alarmed them. They were doubtful whether this was reconnaissance or actual attack. They determined to make certain by making an attempt to outflank the outflanking cavalry; and no sooner had our long-range rifle fire compelled them to take cover behind the hill than a new force, as it seemed, of two hundred rode into the open and passing across our front at a distance of, perhaps, 2,000 yards, made for a white stone kopje on our right.

Angus McNeill ran up to the General. [Brabazon] 'Sir, may we cut them off? I think we can just do it.' The scouts pricked up their ears. The General reflected. 'All right,' he said, 'you may try.'

'Mount, mount, mount, the scouts!' cried their impetuous officer, scrambling into his saddle. Then, to me, 'Come with us, we'll give you a show now—first-class.'

A few days before, in an unguarded moment, I had promised to follow the fortunes of the scouts for a day. I looked at the Boers, they were nearer to the white stone kopje than we, but, on the other hand, they had the hill to climb, and were probably worse mounted. It might be done, and if it were done—I thought of the affair of Acton Homes[182]—how dearly they would have to pay in that open plain. So, in the interests of the '*Morning*

[181] William Henry Sitwell, Major and Brevet Lieutenant Colonel, was born at Benares 20 November 1860, eldest son of Major Francis Henry Massey Sitwell, of Barmoor Castle. He was educated at Harrow, and the Royal Military College, Sandhurst, and gazetted to the 16th Foot as Second Lieutenant, 14 January 1880; became Second Lieutenant, 5th Foot, 31 January 1880. He served in the Afghan Campaign of 1880 (Medal); was promoted to Captain 10 April, 1889; was employed as Adjutant to the Bechuanaland Border Police 20 March 1891 to 9 May 1893.

Served in the Nile Expedition of 1898; was severely wounded at Shebalia; took part in the capture of Shendy, and in the Batties of the Atbara arid Khartoum. was given the Brevet of Lieutenant Colonel 16 November 1898; received the Medal, and three clasps to the Egyptian Medal. He served also in the Nile Expedition of 1899, commanded the 14th Sudanese in the first pursuit of the Khalifa, and received a clasp to the Egyptian Medal. In South Africa, January to 31 August 1900, he commanded the 9th and 10th Mounted Infantry, 3rd Division, and afterwards raised the 22nd Mounted Infantry and commanded a Column of All Arms to the end of the war (March, 1901 to 31 May 1902).

He was present in the operations in the Orange Free State, February to May 1900; operations in Orange River Colony, May to 29 November 1900; operations in Cape Colony, south of Orange River, 1899—1900; operations in the Transvaal, Orange River Colony and Cape Colony, 30 November 1900 to 31 May 1902.

Major and Brevet Lieutenant Colonel, Northumberland Fusiliers. In recognition of services during the recent operations in South Africa." There is no official account of his DSO, but it is understood to have been awarded for capturing a hill in the attack on Dewetsdorp in April, 1900, two Victoria Crosses being awarded to Colonel Sitwell's command on the same occasion.

Lieutenant Colonel Sitwell was given the Brevet of Colonel 2 July 1904. He commanded the 4th Northumberland Fusiliers, and on its disbandment commanded the 2nd East Lancashire Regiment from 1906 to 1908. He was promoted to Colonel 17 February 1908, and was created a CB on King Edward VII's birthday, 1908. Colonel Sitwell was AAG and GSO1, 3rd Lahore Division, 14 March, 1909 to 14 March, 1913, and received the Delhi Durbar Medal, 1910. He commanded the Quetta 1st Infantry Brigade, 1913-14, and became Temporary Brigadier General, November 1913. He served in the European War, and commanded the 34th Infantry Brigade at the landing in Suvla Bay, 6 August 1915.

[182] See Chapter XVI of London to Ladysmith. [Reggie] "Barnes, adjutant of the Imperial Light Horse, joyful, with a breathless horse; he explained how they had seen two hundred Boers moving towards distant hills, to make sure of their line of retreat by the Acton Homes road into the Free State; galloped to cut them off; reached the hills first, with just five minutes to spare; dismounted, commanding the road, and waited.

Post,' I got on my horse and we all started—forty or fifty scouts, McNeill and I, as fast as we could, by hard spurring, make the horses go. It was from the very beginning a race, and recognised as such by both sides. As we converged I saw the five leading Boers, better mounted than their comrades, outpacing the others in a desperate resolve to secure the coign[183] of vantage. I said, 'We cannot do it'; but no one would admit defeat or leave the matter undecided. The rest is exceedingly simple.

A Reconnoitre in Force.[143]

Reconnoitring in Force.[144]

We arrived at a wire fence 100 yards—to be accurate 120 yards—from the crest of the kopje, dismounted, and, cutting the wire, were about to seize the precious rocks when—as I had seen them in the railway cutting at Frere, grim, hairy and terrible—the heads and shoulders of a dozen Boers appeared; and how many more must be close behind them?

There was a queer, almost inexplicable, pause, or perhaps there was no pause at all; but I seem to remember much happening. First the Boers—one fellow with a long, drooping, black beard, and a chocolate-coloured coat, another with a red scarf round his neck. Two scouts cutting the wire fence stupidly. One man taking aim across his horse, and McNeill's voice, quite steady: 'Too late; back to the other kopje. Gallop!'

Then the musketry crashed out, and the 'swish' and 'whirr' of the bullets filled the air. I put my foot in the stirrup. The horse, terrified at the firing, plunged wildly. I tried to spring into the saddle; it turned under the animal's belly. He broke away, and galloped madly off. Most of the scouts were already 200 yards off. I was alone, dismounted, within the closest range, and a mile at least from cover of any kind.

One consolation I had—my pistol.[184] I could not be hunted down unarmed in the open as I had been before.[185] But a disabling wound was the brightest prospect. I turned, and, for the second time in this war, ran for my life on foot from the Boer marksmen, and I thought to myself, 'Here at last I take it.' Suddenly, as I ran, I saw a scout. He came from the left, across my front; a tall man, with skull and crossbones badge, and on a pale horse. Death in Revelation, but life to me.

The Boers admitted afterwards that they thought that the squadrons visible on the other hills two miles back were the head of our column, and they also blamed their scouts, particularly one, an Austrian. 'It all comes of trusting these cursed foreigners! If we had only had a veldt Boer out we should never have been caught.' Caught, however, they undoubtedly were. The Carabineers and the Imperial Light Horse held their fire until the scouts walked into their midst, and then let drive at the main body, 300 yards range, mounted men, smooth open grass plain. There was a sudden furious, snapping fusillade the Boer column stopped paralysed; then they broke and rushed for cover. The greater number galloped fast from the field; some remained on the ground dead or wounded. Others took refuge among the rocks of the kopjes and apparently proposed to hold out until dark, and hence the arrival of Lieutenant Barnes demanding reinforcements, 60th Rifles, Mounted Infantry, and anything else, so as to attack these fellows in flank and 'bag the lot.'

[183] A projecting corner or angle; a cornerstone.

[184] His Mauser C-96. The one that caused so much anxiety and later acrimony after he was captured during the armoured train ambush.

[185] He had left the pistol in the train engine while he endeavored to organize the clearing of the tracks.

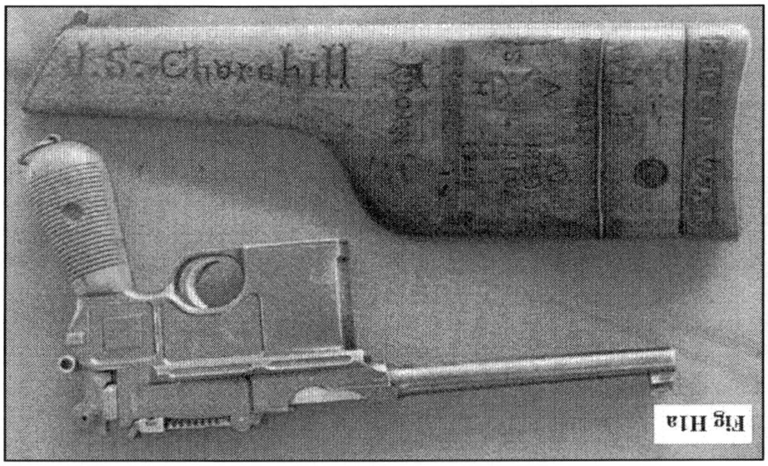

Winston's brother, John 'Jack' S. Churchill carried the same model of Mauser C96.[145] When Jack left the conflict, after being wounded, he presented the weapon to a brother officer, W. L. Edwards. This would make sense since it would be way more useful in the field than in the U.K. Cap badge of the 17th Lancers.[146]

I shouted to him as he passed: 'Give me a stirrup.' To my surprise he stopped at once. 'Yes,' he said, shortly. I ran up to him, did not bungle in the business of mounting, and in a moment found myself behind him on the saddle.

Then we rode. I put my arms around him to catch a grip of the mane. My hand became soaked with blood. The horse was hard hit; but, gallant beast, he extended himself nobly. The pursuing bullets piped and whistled—for the range was growing longer—overhead.

'Don't be frightened,' said my rescuer; 'they won't hit you.' Then, as I did not reply, 'My poor horse, oh, my poor ---- horse; shot with an explosive bullet. The devils! But their hour will come. Oh, my poor horse!'

I said, 'Never mind, you've saved my life.' 'Ah,' he rejoined, 'but it's the horse I'm thinking about.' That was the whole of our conversation.

Judging from the number of bullets I heard I did not expect to be hit after the first 500 yards were covered, for a galloping horse is a difficult target, and the Boers were breathless and excited. But it was with a feeling of relief that I turned the corner of the further kopje and found I had thrown double sixes again.[186]

The result of the race had been watched with strained attention by the rest of the troops, and from their position they knew that we were beaten before we ever reached the wire fence. They had heard the sudden fierce crackle of musketry and had seen what had passed. All the officers were agreed that the man who pulled up in such a situation to help another was worthy of some honourable distinction. Indeed, I have heard that Trooper Roberts[187] [188]—note the name, which seems familiar in this connection—is to have his claims considered for the Victoria Cross. As to this I will not pronounce, for I feel some diffidence in writing impartially of a man who certainly saved me from a great danger.

Well satisfied with my brief experience with the scouts, I returned to General Brabazon. While we had been advancing deeply into the Boer flank, they had not been idle, and now suddenly, from the side of the solitary kopje behind which they had collected, three guns came into action against us. For ten minutes the shell fire was really hot. As these guns were firing with black powder, the smoke springing out in a thick white cloud from the muzzle warned us whenever a projectile was on its way, and, I think, added to the strain on the nerves. You could watch the distant artillery. There was the gun again; four or five seconds to wonder whether the shell would hit you in the face; the approaching hiss rushing into a rending shriek; safe over; bang! right among the horses a hundred yards behind. Here comes the next—two guns fired together this time. Altogether, the Boers fired nearly thirty shells—several of which were shrapnel—on this small space of ground. But fate was in a

[186] The probability of throwing double sixes is 1 in 36.

[187] Clement (Clem) Roberts Corporal Source: Nominal roll in WO127 Montmorency's Scouts Article: https://www.angloboerwar.com/?option=com_content&view=article&id=431

[188] In 1907 Churchill had the case reopened and "Trooper Roberts received for his conduct on this occasion the Distinguished Conduct Medal." Fn. My Early Life Ch. XXVI. See also OB p. 509-510.

merciful mood that day, for we had but one man killed and five or six—including the General's orderly—wounded by them.

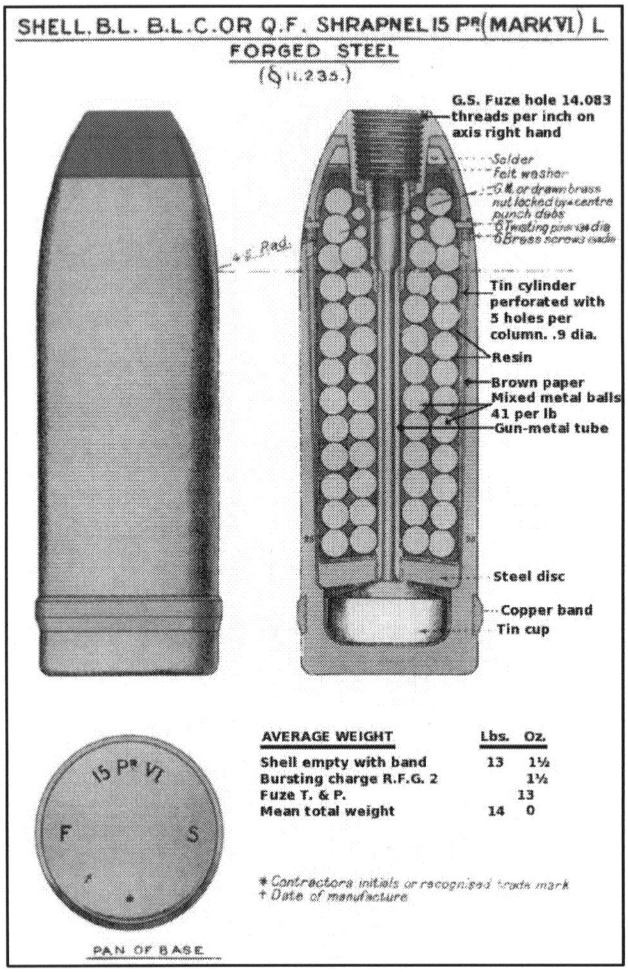

Diagram of a shrapnel shell.[147] Shell exterior with driving band (Bottom) and Timed Fuse (Top).

It was, however, evident that this could not endure. Brabazon had not cared to bring his own two guns into such an advanced position, because they were not horse guns, and might not be able to get away safely if the Boers should make a strong counter attack. Indeed, so long as the loss of guns is considered a national disaster instead of only an ordinary incident of war, Cavalry officers will regard them rather as sources of anxiety than as powerful weapons.[189]

Horse guns crossing the Klip River.[148] Not a Horse Gun. One of the 4.7's in action.[149]

Without guns it was useless to stay, and as, moreover, Sir Leslie Rundle's orders were that the Cavalry were not to be severely engaged, Brabazon decided to withdraw the reconnaissance, and did so most successfully, after

[189] He is referring to Colonel Long and the loss of 10 of the 12 forward guns lost at Colenso. Lord Roberts' son Frederick died in the attempt to retrieve these pieces and was awarded a posthumous Victoria Cross.

an instructive little rear-guard action. He had penetrated far into the enemy's position; had compelled him to move his guns and disturb his frontal dispositions; had reconnoitred the ground, located the laagers, and come safely away with the loss of little more than a dozen men. Had there been on this day an Infantry support behind the Cavalry we should have hustled the enemy out of his whole position and slept triumphantly in Dewetsdorp.

Sir Leslie Rundle was much impressed by the vigour and success of the Cavalry, whose fortunes were watched from the plateau, and as evening came the report spread through the camp that a general engagement would be fought on the next day. [22nd] He also decided to entrust the direction of the actual turning attack to General Brabazon, who, besides his Cavalry force, was to have twelve guns and an Infantry brigade under his command.

With every feeling of confidence in the issue the Army went to bed, impatient for the dawn. But in the dead of night a telegram arrived from Lord Roberts, instructing Rundle not to press his attack until he was in touch with Pole-Carew and other reinforcements; and it thus became evident that the operations had grown to an altogether larger scale.

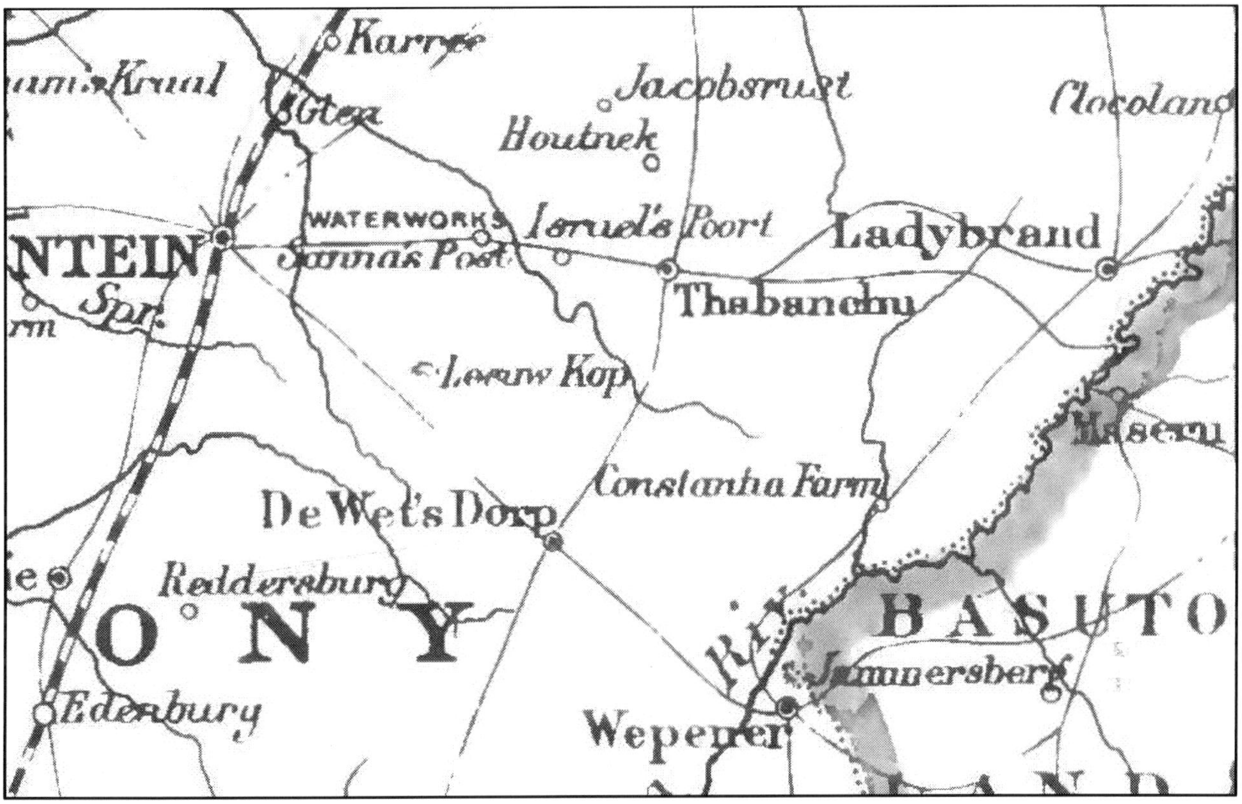

Orange Free State South.[150]

VI. The Dewetsdorp Episode

Bloemfontein: May 1st, 1900.

Sometimes it happens that these letters are devoted to describing small incidents, and often personal experiences in a degree of detail which, if the rest of the campaign were equally narrated, would expand the account to limits far beyond the industry of the writer or the patience of the reader. At others many important events must be crowded into a few pages. But though the proportions of the tale may vary, I shall not deserve criticism so long as the original object of conveying a lively impression of the war is strictly pursued; nor should the reader complain if, for his instruction or amusement, he is made one day to sit with the map of the Orange Free State spread before him, and move little flags to show the course of the operations, and on another day is invited to share the perils of a scout's patrol or try the chances of a cavalry skirmish. To-day there is much to tell, and we must remain almost beyond the sound of the cannon watching a distant panorama.

The object of the operations was in any case to relieve Wepener, and to clear the right-hand bottom corner of the Orange Free State of the Boers, and, if the enterprise prospered and the fates were kind, to cut off and capture some part of their forces. In all five columns were in motion.

There were to be demonstrations along the east of the railway line, increasing in earnestness according as they were nearer the south, and the lowest columns were to actually push the matter through.

Ian Hamilton, with 2,000 Mounted Infantry, was ordered to demonstrate against the waterworks position. French, supported by Pole-Carew, was instructed to move on Leeukop.

Rundle, in conjunction with Hart and Brabant from the southward, was to force his way to Dewetsdorp and to relieve Wepener.

Southern Area. Map data: Google Image© 2023 Image Landsat/Copernicus.

The attack on the Boer position in front of Dewetsdorp[190] had not been made on the 20th [afternoon] because Sir Herbert Chermside pointed out that the Infantry were fatigued with marching.[191]

190 A Dorp is Dutch for A village or small town; a town considered provincial.

191 From My Early Life. [That evening] I was admitted to the council. Brabazon was all for battle. All preparations were made for a regular attack next day. However, very early in the morning the leading Brigadier, Sir Herbert Chermside, made representations to our chief commander [Rundle] upon the gravity of the enterprise. In 1878—twenty-two years before—Chermside had been in the Russo-Turkish war. He therefore spoke with high authority. He declared that the Boers now held positions as formidable as those of, Plevna and that it would be imprudent without gathering every man and gun to launch an assault which might cost thousands of lives.
It was therefore resolved to await the arrival of a third brigade under General Barr-Campbell, containing two battalions of Guards, who were already on the march from the railway and should arrive by night.

Weary Troops Asleep under Arms.[151]

The next morning [21st] the smooth hills were crowned with entrenchments, and it was thought better to wait for Campbell's Brigade, which would arrive at sundown.[192] [193]

Meanwhile Sir Leslie Rundle had telegraphed to Lord Roberts describing the horseshoe position of the enemy, and its strength, explaining that with the small mounted force at his disposal any attack which he might make would develop into something very like a frontal attack, and would be costly.

A strong memorandum had previously been circulated among divisional and brigade commanders condemning, almost prohibiting, frontal attacks,[194] and the General, not unnaturally, wished to assure himself that the price of victory would not be grudged.

When this telegram reached Bloemfontein it was apparently misunderstood. 'Rundle is hung up,' they said. 'He can't get on'; and hence the reply which arrived in the dead of night, and prevented the attack of the 22nd. 'Wait till you get into touch with Pole-Carew,' or words to that effect. [195]

So the powerful force—almost equal in strength to that with which Sir George White had resisted the first fury of the Boers when, with 25,000 men under the Commandant-General [Joubert] himself, they burst into Natal—was relegated to some days of pusillanimous[196] waiting in front of a position held by scarcely 2,500 men.

After breakfast on the morning of the unfought battle [22nd] I climbed to the top of the hill the cavalry had seized two days before, and which the soldiers had christened 'Brab's kopje.' A fifteen hundred yards musketry duel was proceeding, and it was dangerous to put one's head over the stone shelters even for a minute to look at the Boer entrenchments on the green slope opposite.

Troops of the Dublin Fusiliers defending a piquet.[152] Troops of one of the British mounted regiments aiming their rifles at the enemy from behind a rock barricade.[153]

So we passed a pleasant day skirmishing with the Boers, and as soon as evening fell another long column of infantry arrived." Ch. XXVI
PF: I have arranged the MEL narrative to conform to IHM's. Neither are consistent. Both are relevant. Some inconsistencies remain.

192 Winston participates in the attempt to seize the Kopje along with McNeill's Scouts on the 21st.

193 [After their arrival] We now had nearly eleven thousand men and eighteen guns. Ch. XXVI.

194 After Talana Hill, Elanslaagate, Stormberg, Magersfontein, Colenso, and Spion Kop had demonstrated the costliest of this tactic.

195 At daybreak [22nd] when the whole force was drawn up for attack and our yeomanry awaited the signal to ride round the enemy's left flank, suddenly there arrived a staff officer with the news that the battle was again put off for that day at least. This was too much for Brabazon. He rode towards me wagging his head, and with a droll expression emitted suddenly in a loud voice and before everyone the words 'Bob Acres.' Whether the staff officer was so spiteful as to repeat this indiscretion, I cannot tell. Ch. XXVI

196 Showing ignoble cowardice, or contemptible timidity.

But such was not my purpose. I scanned the northern horizon. Far away on a peak of the misty blue hills there flashed a diamond. It was Pole-Carew. Half an hour later another star began to twinkle further to the eastward. French and his cavalry were riding steadily forward, 'fighting, too,' said the heliograph, 'but pushing them back.'

Unit of the Mounted Infantry in advance.[154]

The scale of the operations had grown indeed. No less than five infantry and three cavalry brigades, with more than seventy guns, were involved in the business of dislodging 2,500 Boers from their position in front of Dewetsdorp.

The 23rd passed quietly, except for an intermittent bombardment of our camp by the Dutch guns and a Vickers-Maxim [pom pom] and the usual patter of musketry along the outposts.

ZAR Norden Maxim Mafeking.[155] ZAR Staats Artillery using Maxim Pom Pom. Note belt feed ammunition.[156]

The diamond points on the distant hills seemed nearer and more to the east than before, and in the afternoon Brabazon was sent to reconnoitre towards them. As the Yeomanry emerged from the shelter of the plateau the Boer Creusot gun espied them. Brabazon, with half a dozen officers or orderlies, was riding fifty yards in front of his brigade.

Yeomanry on Patrol.[157]

'See there,' said the Dutch gunners, 'there is the Hoofd[197] Commandant himself; take good aim.' So they did, and from a range of 5,000 yards burst their shell within two yards of the General's horse.

'Wonderful,' said Brabazon; 'why can't our forsaken artillery shoot like that?' and he ordered the brigade to canter by troops across the dangerous ground. I watched the scene that followed from comparative safety, 600 yards nearer the Boer gun. Troop by troop the Yeomanry emerged from shelter. As each did so the men opened out to dispersed order and began to gallop; and for every troop there was one shell. From where I stood the spectacle was most interesting. Between the shrieking of the shell overhead and its explosion among the galloping horsemen there was an appreciable interval, in which one might easily have wagered whether it would hit or miss.

The Yeomanry were very steady, and for the most part ran the gauntlet at a nice, dignified canter, pulling into a walk as soon as the dangerous space was crossed. After all no one was hurt, except three men who broke their crowns through their horses falling on the rocky ground. Indeed, I think, speaking from some experience, that we can always treat these Creusot 9-pounders[198] with contempt. They fling a small shell an immense distance with surprising accuracy, but unless they actually hit someone they hardly ever do any harm. An ordinary bullet is just as dangerous, though it does not make so much noise.

7.5 cm Krupp Field Gun.[199]

Front View of 7.5cm Gun.[158]

At Vaal Krantz, in Natal, Dundonald's Brigade and other troops lived quite comfortably for three days under the fire of a 98-pounder gun, which in all that time only killed one soldier of the Dublin Fusiliers, two natives, and a few beasts.

Staats Artillery at start of conflict. All arms & accoutrements are pristine, as is the Long Tom itself.[159]

[197] Head (of an organisation), chief, boss. Also slang, see: hoofd - Wiktionary

[198] For an example of a British design of the same sized weapon see: RML 9-pounder 8 and 6 cwt guns - Wikipedia See also: List of field guns - Wikipedia

[199] Officially it was denoted as: 7.5cm Krupp: Feldkanone L/27, but during the Boer War it became known as the "Free State Krupp." Courtesy of Neville Constantine: Anglo Boer War - Artillery and Ammunition - Page 26 - Boer War Forum

The wholesale aspect of artillery fire is not obtained unless at least a dozen guns are firing percussion shell or unless shrapnel can be used. At present the Boers often cause us a great deal of trouble with single guns, which, though they do scarcely any material harm, disturb every one, so that camps are shifted and marching columns ordered to make long *détours*; whereas we ought to shrug our shoulders, as Ladysmith did, pay the small necessary toll, and go our ways uninterruptedly. But I am being drawn into detail and discussion, which, if I am ever to catch up the swift march of events, must be rigorously excluded.

The 23rd passed quietly for times of war, and the Boer riflemen and artillerists fired busily till dusk without doing much harm. We wondered how much they knew of the 'increased scale' of the operations. Did they realise the enormous strength of the forces closing round them? Were they going to be caught as Cronje was caught?[200] It was hardly likely.

Yet they were certainly holding all their positions in force at nightfall, and meanwhile the spring of the trap was compressed and the moment for releasing it arrived.

The morning of the 24th was unbroken by a single shot. Rundle, now in touch with Pole-Carew, swung his division to the left, pivoting on Chermside, to whom he entrusted the defence of the plateau. Brabazon with his Mounted Brigade formed the extreme outer flank of this sweeping movement. His orders were to join French, who drove inward from the north, somewhere behind Dewetsdorp on the Modder River. So we started, and, with much caution and the pomp of war, turned the enemy's left, and in solemn silence bore down on the flank and rear of his position.

Meanwhile, Chermside on the plateau was struck by the entire cessation of fire from the Boer lines opposite to him. He sent scouts to reconnoitre. Single men crept up the hill, looked into the trenches, and found—nothing. The Boers had retreated swiftly in the night. They enjoyed good information of all our movements and designs, had foreseen the impossibility of withstanding the great forces operating against them. They delayed us with the appearance of strength until the last minute. On the night of the 22nd they sent off their waggons [North] towards Thabanchu.

On the 23rd they made their effort against Wepener, and attacked the garrison heavily, and on the night of the 24th, having failed at Wepener, they performed a masterly retreat, the assailants of Wepener marching northwards *via* Ladybrand, the covering force at Dewetsdorp moving on Thabanchu.

And so it was that when, as directed, Brabazon circled round the enemy's left flank and struck the Modder River—here only a rocky ditch with occasional pools of mud—and when French, moving from Leeukop round and behind their right flank, met him, they found the Dutch already departed, and Dewetsdorp again under the Union Jack.

Long column of British troops, both mounted and infantry, crossing a river.[160] Cape Mounted Irregulars on a reconnaissance.[161]

The strong jaws of the rat-trap shut together with a snap. I saw them—black across the open plain—two great horns of cavalry and guns; but the rat had walked comfortably away some hours before. Chermside moving over the empty trenches occupied the town. Rundle, reaching it an hour later, owing to his turning movement, hurried

[200] At the battle of Paardeburg: See Frankland's diary infra.

on through it to the Modder, and laid Brabazon's dusty squadrons on the retreating enemy. Indeed, the latter officer was already at the trot towards Thabanchu when French himself arrived—a large and magnificent staff, 'pom-poms,' horse artillery, and two cavalry brigades—and assumed supreme command.

Royal Artillery Column.[162]

He immediately stopped the pursuit, sent Brabazon back to relieve Wepener—which place had by its plucky defence, like Jellalabad,[201] relieved itself—and entered Dewetsdorp, where he remained until the next day.

Such is the story of Dewetsdorp, which cannot be contemplated with feelings of wild enthusiasm. The Wepener situation was cleared up, and the Boers were persuaded to retire from the right-hand bottom corner of the Free State towards Ladybrand and Thabanchu at an exceedingly small price in blood. On the other hand, the enemy might boast that 2,500 Burghers with six guns had contained 13,000 troops with thirty guns for a week, while their brethren worked their wicked will on Wepener, and had only been dislodged by the setting in motion of more than 25,000 men and seventy guns.

The movements which followed the occupation of Dewetsdorp need not take long in the telling. French's occupation of the town instead of pursuing the enemy was not in accordance with the Commander-in-Chief's ideas, and the cavalry leader was forthwith ordered to follow the Boers at his best pace to Thabanchu. He [French] started accordingly at daylight on the 25th, and Rundle with the Eighth Division followed at noon. Chermside remained at Dewetsdorp with part of the Third Division, and was entrusted with the re-establishment of order through the disturbed districts. [Previously assigned to Gatacre]

Brabazon marched on Wepener and collected the garrison. Their defence of seventeen days, under continual rifle and shell fire, in hastily dug trenches, which they were unable to leave even at night; exposed to several fierce attacks; in spite of heavy losses and with uncertain prospects of relief, will deserve careful attention when full accounts are published, and is a very honourable episode in the history of Brabant's Colonial Brigade,[202] and particularly in the records of the Cape Mounted Rifles, who lost nearly a quarter of their strength.[203]

[201] The Battle of Jellalabad in 1842 was an Afghan siege of the isolated British outpost at Jellalabad (now Jalalabad) about 90 miles (140 km) east of Kabul. The siege was lifted after five months when a British counterattack routed the Afghans, driving them back to Kabul. Battle of Jellalabad - Wikipedia

[202] Brabant occupied Jamestown and the Herschel district in February 1900. His units operated round the Queenstown/Dordrecht area and moved north to hold the Jammersburg Drift at Wepener, which they did under appalling rain and cold against a superior Boer force led by Christiaan De Wet. Edward Brabant - Wikipedia See Also: Anglo Boer War - Name search

[203] See: Wepener Account and Medal Roll by David and Alexander Biggins. Anglo Boer War - Wepener Account and Medal Roll Published by Token Publishing: Coin, Banknote and Medal Collector's Magazines. Token Publishing – Numismatic Interest

Brabant's Horse.[163]

Edward Yewd Brabant.[164]

Brabant's Horse Artillery.[165]

Bringing the defenders with him, and having communicated with Hart and Brabant, Brabazon returned to Dewetsdorp, and was ordered to move thence to Thabanchu, which he did in an exceedingly convenient hour, as it turned out, for a certain convoy with an escort of Scots Guards and Yeomanry.

Pole-Carew and the Eleventh Division returned to Bloemfontein to take part in the main advance.

The Boers [De Wet brothers] made good their retreat. They took with them twenty-five prisoners of the Worcester Regiment,[204] who had blundered into their camp before Dewetsdorp, armed only with cooking pots, which they meant to carry to their regiment on 'Brab's kopje,' and great quantities of sheep and oxen. They halted in Ladybrand, and to the north and east of Thabanchu in a most pugnacious mood. Indeed, they had no reason to be discontented with the result of their southern incursion.

They had captured seven guns and nearly 1,000 prisoners.[205] They had arrested and carried off a good many farmers who had laid down their arms and made their peace with the British Government. They had harried all who received the troops kindly, had collected large quantities of supplies which they had sent north, and, lastly, had delayed the main advance by more than five weeks.

Captured British Cannon.[166]

[204] Worcestershire Regiment: https://www.angloboerwar.com/?option=com_content&view=article&id=672

[205] Many from the Sanna's Post engagement.

More British Prisoners Arriving at Pretoria. Note Red Cross on Train.[167]

Owing to the great disproportion of the forces the fighting had not been of a severe nature, and the losses were small. In the skirmishes before Dewetsdorp about forty men were killed and wounded, mostly in Brabazon's Brigade. In the action at Leeukop and the subsequent fighting which attended French's march several officers and fifty men were stricken, and a squadron of the 9th Lancers, which was required to attack a kopje, suffered severely, having nearly twenty casualties, including Captain Stanley,[206] a very brave officer, who died of his wounds, and Victor Brooke[207] (of whom more will be heard in the future) who had his left hand smashed. Captain Brasier-Creagh,[208] 9th Bengal Lancers, commanding Roberts's Horse,[209] was killed at Leeukop, and his many friends along the Indian frontier will not need to be told that by his death Lord Roberts's Army suffered a loss appreciable even among the great forces now in the field.

[206] Herbert Foster Wentworth Stanley Captain. He died April 28th, 1900, of wounds received in action near Dewetsdorp. He was the third son of Sidney Stanley of Longstone Hall, Cambridgeshire, was born July 1863, and educated at Eton. He entered the 9th Lancers in February 1887, being promoted Lieutenant November 1888, and Captain September 1894. Captain Stanley accompanied his regiment to South Africa from India in 1899 and served in Natal and the North of Cape Colony, and subsequently at the relief of Kimberley and the advance on Bloemfontein. He afterwards served in the operations near Dewetsdorp. Source: Donner.

[207] Victor Reginald Brooke Lieutenant, was born at 53, Eaton Square, London, 22 January 1873, son of Sir Victor Alexander Brooke, Baronet, and Alice Sophia, daughter of Sir Alan Bellingham, Baronet. He joined the 9th Lancers 12 December 1894; became Lieutenant 29 April, 1896, and went to South Africa at the beginning of the Boer War (1899-1902). He took part in the Advance on Kimberley, including the actions at Belmont, Enslin, Modder River and Magersfontein; in the Relief of Kimberley; the operations in the Orange Free State, February to May 1900, including operations at Paardeberg; actions at Poplar Grove and Karee Siding. He was one of the first of our men to be wounded, and in a diary picked up off a wounded Boer was set down the enemy's regret that so good a friend and fighter had been, as they thought, killed. He returned to England to recover from his wounds, and had to have a finger removed. He passed the Staff College in the autumn; became Captain 6 May 1901; returned to South Africa in October 1901, as ADC to Lieutenant General Sir Ian Hamilton, Chief of the Staff (9 November 1901, to 11 September 1902). Captain Brooke became ADC to Lord Kitchener, and served with him in that capacity in India (28 November 1902, to 21 December 1905); and as Assistant Military Secretary from 22 December 1905, to 31 May 1907. He was sent to Kabul with Sir Louis Dane's Mission at the end of 1904, when the Amir of Afghanistan bestowed an Afghan decoration on him. . In August 1914, he proceeded to France, forming part of General Headquarters Staff, as Liaison Officer, with General Sardet's Cavalry Corps, during the retreat from Belgium.
Lord Kitchener said of him: "Victor Brooke is one of the best Staff Officers I have ever had, and quite one of the best all-round men to be found." Source: DSO recipients (VC and DSO Book).

[208] George Percy Brasier-Creagh Captain He was severely wounded near Karreefontein on April 23rd, 1900 and died four days later at Eirstelaagte. Born in 1864, he entered the East Surrey Regiment from the 9th Battalion King's Royal Rifle Corps, May 1884, and joined the Indian Staff Corps, May 1886, being promoted Captain May 1895. From 1889 to 1894, he was ADC to the Viceroy of India. Captain Brasier-Creagh acted as Sir William Lockhart's orderly officer in the second Miranzai Expedition (mentioned in despatches). He also took part in the Isazai Expedition of 1892, and the Chitral Expedition of 1895 with the relief force, being awarded the medal with clasp. In the operations on the North-West Frontier in 1897, he served with the Malakand Field Force, and with the Tirah Expeditionary Force, being mentioned in despatches LG June 7th, 1898. At the time of his death Captain Brasier-Creagh was in command of Roberts' Horse, which corps greatly distinguished itself. Source: Donner.

[209] Roberts' Horse Article: https://www.angloboerwar.com/?option=com_content&view=article&id=464

VII. Ian Hamilton's March

Winburg[210]: May 8th 1900.

The unsatisfactory course of the operations in the south-eastern corner of the Free State, and the indecisive results to which they led, were soon to be arrested and reversed by a series of movements of surprising vigour and remarkable success. Of all the demonstrations which had been intended against the enemy to the east of the railway, Hamilton's advance towards the waterworks position, being the most northerly, was to have been the least earnestly pressed. The orders were: 'If you find the waterworks weakly held, which is not likely, you may try to occupy them, and, in the event of success, may call up Smith-Dorrien's Brigade[211] to strengthen you.'

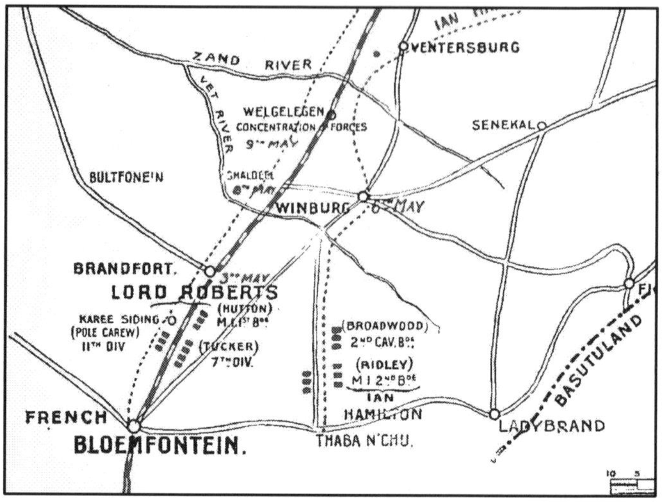

British general Horace Smith-Dorrien[168] Caricature of Gen. Smith-Dorrien Caption "Doreen."[169] Crop of Creswicke Map. Winston is in Winburg.

On this General Ian Hamilton, who now commanded the imposing, but somewhat scattered, Mounted Infantry Division, started from Bloemfontein on the 22nd of April with about 2,000 Light Horse, Australians, and Mounted Infantry, and one battery of Horse Artillery.

On the 23rd he arrived before the waterworks, reconnoitred them, found them weakly held, or, at any rate, thought he could take them, attacked, and before dark made himself master of the waterworks themselves, and of the drift over the river which led to the hills beyond, into which the enemy had retired.

Smith-Dorrien's Brigade was called up at once, arrived after dark, and the next morning the force crossed at the drift, and the whole position was occupied. The enemy offered a slight resistance, which was attributed by some to a deep design on their part to lure the column into a trap further to the east, and by others to the manner in which the attack was delivered.

The news of the capture of this strong and important place, which secures the Bloemfontein water supply, was received with great satisfaction at headquarters.

Meanwhile the operations round Dewetsdorp came to their abortive conclusion, and it became evident that the Boers had evaded the intercepting columns and were making their way northwards by Thabanchu. What was to be done? Had the officer commanding at the waterworks any suggestion to make? Most certainly, and the suggestion was that he should be permitted to advance himself and occupy Thabanchu. This was the answer that was expected and desired. Permission, and with it a field battery, was accordingly given, and, on the 25th of

[210] The town was the site of a concentration camp for women and children captured by the British Army during their scorched earth campaign during the Second Boer War. 355 children and 132 adults died in this camp due to malnutrition and contagious diseases, while kept in tents without any infrastructure or protection during the bitter cold winters of 1899–1901. The famous General Koos de la Rey was born in the district of Winburg on the farm Doornfontein. Winburg had a black armed commando supporting the British soldiers during the war. Winburg - Wikipedia

[211] General Sir Horace Lockwood Smith-Dorrien, (26 May 1858 – 12 August 1930) was a British Army General. One of the few British survivors of the Battle of Isandlwana as a young officer, he also distinguished himself in the Second Boer War. "At Sanna's Post (31 March 1900), Smith-Dorrien ignored inept orders from Colvile to leave wounded largely unprotected and managed an orderly retreat without further casualties." Horace Smith-Dorrien - Wikipedia

April, the column moved out of the waterworks position towards Thabanchu. It consisted of Ridley's [212] Brigade of Mounted Infantry, which included a large proportion of colonials—Australians and New Zealanders—Smith-Dorrien's Infantry Brigade (Gordons, Canadians, Shropshires, and Cornwalls), with twelve guns.

The country to the east of Bloemfontein is at first smooth and open. Great plains of brownish grass stretch almost to the horizon, broken to the eye only by occasional scrub-covered hills. To any one unaccustomed to the South African veldt they appear to offer no obstacle to the free movement of cavalry or artillery; nor is it until one tries to ride in a straight line across them that the treacherous and unimagined donga and the awkward wire fence interpose themselves.

High Veld near Krugersdorp.[170]

South African Vista.[171]

But beyond the Modder River, on which the waterworks are situated, the surface of the ground becomes rocky and hilly, and the features increase in prominence until Thabanchu Mountain is reached, and thereafter the country uprears itself in a succession of ridges to the rugged and lofty peaks of Basutoland.[213]

Drakensberg mountains in Lesotho.[214]

[212] C. P. Ridley Lieutenant Colonel Entered 1873; Brevet Colonel, 1899. Staff service: Station Commandant, South Africa, 1899; AAG (Assistant Inspector General Lines of Communication), South Africa, 1899-1900; Brigade General Mounted Infantry Brigade, South Africa, February 1900. War service: Egyptian Expedition 1882 (medal; bronze star); Miranzai Expedition 1891 (medal with clasp); Boer War, 1899-1900; on Staff. Source: List of CB recipients.

[213] Lesotho; formerly Basutoland - Wikipedia Now Lesotho - Wikipedia

[214] Lesotho, Drakenberg mountains: File:Lesotho, Drakenberg mountains (50182014816).jpg - Wikimedia Commons Author: Dr. Alexey Yakovlev.

Landscape of the Country.[172]

Thabanchu, a small village, as we should regard it in England, a town of comparative commercial importance in the Orange Free State, and of undoubted strategic value during this phase of the operations, stands at the foot of the precipitous feature that bears its name.

Village and Mountain the Thaba N'Chu.[173]

Thabanchu Mountain.[174 215]

[215] Thaba 'Nchu is a town in the Free State, 63 km east of Bloemfontein and 17 km east of Botshabelo. Thaba 'Nchu - Wikipedia

It is approached from the direction of Bloemfontein by a long, broad, flat-bottomed valley, whose walls on either side rise higher and higher by degrees as the road runs eastward. The eastern end of this wide passage is closed by a chain of rocky kopjes, whose situation is so curious and striking that they seem to be devised by nature to resist the advance of an invader. The kopjes, rising abruptly from the flat glacis-like ground, are a strong rampart, and the whole position, resting on apparently secure flanks, creates a most formidable barrier, which is called locally Israel's Poorte.

Along the valley, on the 25th of April, Hamilton proceeded to march with his entire force, Ridley and the Mounted Infantry being a considerable distance in front of the main body. At ten o'clock a heavy fire of musketry and artillery was opened at an extreme range from the hills on the left hand side of the column. Ignoring this, which proved afterwards to be only a Boer demonstration, Ridley continued his march, and Hamilton followed, until, at a little after eleven o'clock, both were brought to a standstill before the Israel's Poorte position, which was found to be occupied by the enemy, estimated at 800 strong, with several guns.

After a personal reconnaissance, and in spite of a most disquieting report that the Boers had just been reinforced by 'two thousand men in four lines,' the General resolved to attack. His plan was simple but effective. It resembled very closely Sir Bindon Blood's forcing of the 'Gate of Swat' at Landakai in 1897.[216] The front was to be masked and contained by a sufficient force of infantry and all the guns. The rest of the troops were to stretch out to the left and swing to the right, the infantry along the left hand wall of the valley, the mounted men actually the other side of the wall.

Accordingly, the Canadian Regiment[217] and the Grahamstown Volunteers (Marshall's Horse)[218] moved forward in extended order—25 yards interval between men—to within about 800 yards of the enemy's position, and here, just out of the range of serious harm, they lay down and opened a continuous musketry fire.

Both batteries came into action forthwith and shelled the crest line with satisfactory energy. Smith-Dorrien, with the remaining three battalions of his brigade, moved to the left, and began working along the ridges. Ridley, breaking out of the valley into the more open ground beyond, began to move against the enemy's line of retreat.

Battery of the Royal Artillery heavy guns in action.[175]

Four hours passed, during which the Boers indulged the hope that the frontal attack would be pushed home, and at the end of which they found their right flank turned and their rear threatened. Immediately, with all the hurry of undisciplined troops who feel a hand on their communications, they evacuated the position, and, running to their horses, galloped away. The Canadians and Grahamstown Volunteers thereupon arose and occupied the line of kopjes, and thus the door was opened and the road to Thabanchu cleared. Our losses in this smart action were about twenty killed and wounded, among whom were no less than five officers of the Grahamstown Volunteers. The Dutch left five corpses on the field, and doubtless carried away a score of wounded.

[216] Siege of Malakand - Wikipedia

[217] Canadian Contingents: https://www.angloboerwar.com/?option=com_content&view=article&id=375

[218] Marshall's Horse was mainly composed of the mounted portions of the First City (Grahamstown) Volunteers and Uitenhage Rifle Volunteers. In the despatches and in unofficial accounts the corps are mixed up, and sometimes are called Grahamstown Volunteers, sometimes Marshall's Horse. https://www.angloboerwar.com/?option=com_content&view=article&id=704 1,050 Records.
And: The First City Volunteers were formed in 1875 in Grahamstown. They fought in the Cape Frontier Wars (9th Frontier War, 1877–1878), the Basuto Gun Wars (1880–1881), the Bechuanaland campaign (1897) and Boer War. https://www.angloboerwar.com/unit-information/south-african-units/2537-grahamstownvols 234 Records.

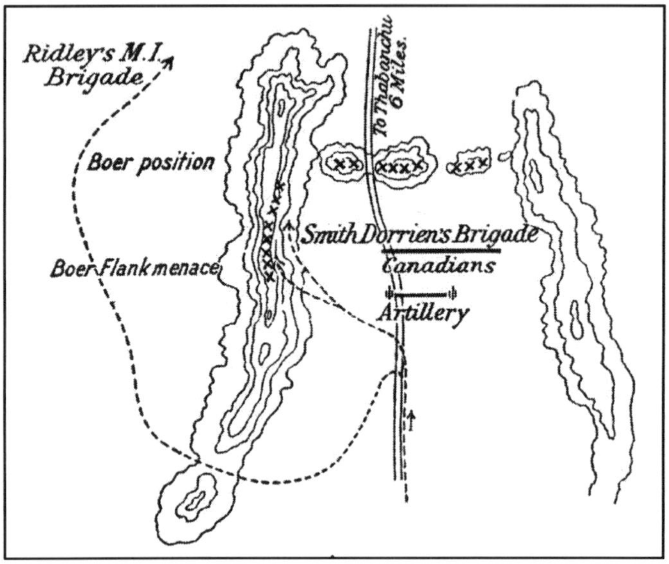

Diagram Explaining Hamilton's Action at Israel's Poorte, The 25th of April, 1900. Crosses indicate Boers.[176]
Taken under fire. The 2nd Queens in action.[177]

General Hamilton, pushing on, entered Thabanchu the same night, and the British flag was again hoisted over the town. The Imperialist section of the community, who had in the interval between the evacuation and reoccupation of the town been subjected to much annoyance at the hands of the Boers, were naturally shy, and afraid to make any sign of welcome. The southern commandos from Dewetsdorp and Wepener had by hard marching already passed behind Thabanchu with their convoys. On the 26th French and his Cavalry, covering the march of Rundle's (Eighth) Division, arrived, and, since he was a lieutenant-general, took the command out of Hamilton's hands for a time.

I had come northwards from Dewetsdorp with the Cavalry Brigades, and was an eyewitness of the operations round Thabanchu which occupied the 26th and 27th. Thabanchu Mountain is a lofty and precipitous feature of considerable extent, and, towards the south, of indefinite shape.

To the north, however, it presents a wide bay, on whose grassy shores rising from the more arid plain the Boer laagers were reported to stand.

A Boer Encampment "Waggons Massed on the Veldt.[178]

A typical Boer camp.[179]

Free State Artillery & Heliograph.[180]

The enemy held the crest of the crescent-shaped mountain with guns and riflemen, and in order that no one should pry behind it they extended on their right a few hundred trustworthy fellows, who, working in the most scattered formation, gave to their position an enormous front of doubtful strength.

Thabanchu

On the afternoon of the 26th, with a view to further operations on the following day, a force of Mounted Infantry, supported by galloping Maxims and a Horse Battery, was sent to reconnoitre, and if possible to hold the hill, henceforward called 'Kitchener's Horse Hill.'

15 pounder guns being transported by a detachment of the Royal Artillery.[181] Colt Machine Gun and Dundonald Galloping carriage with crew. To the left a Red Cross Wagon.[182]

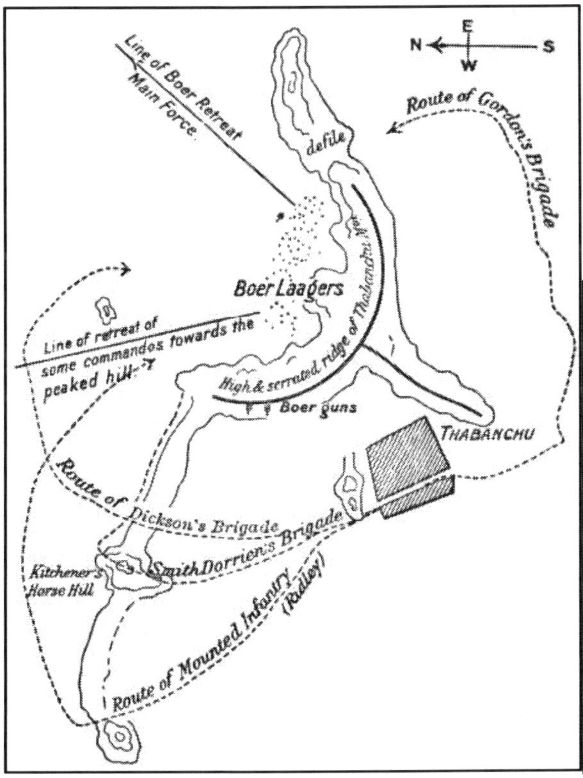

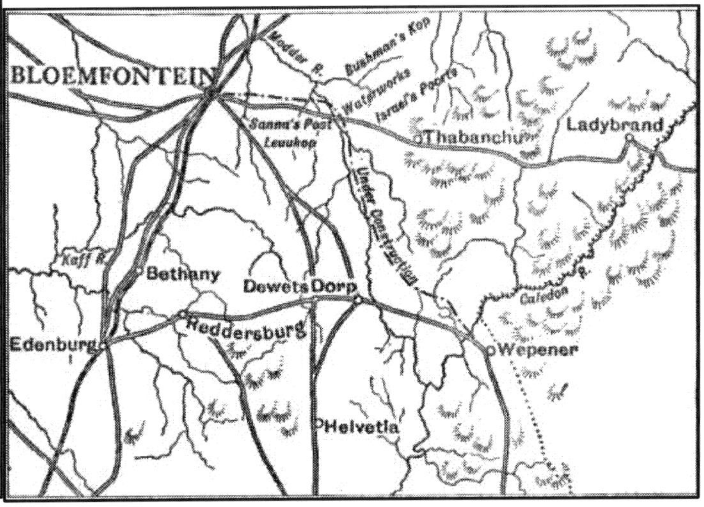

Diagram Explaining French's Operations Round Thabanchu, The 26th And 27th April.[183] In the Orange Free State.[184]

The troops gained possession of the feature without fighting, though a few Boers were seen galloping along the ridges to the right and left, and an intermittent musketry fire began. A garrison to hold the hill was detailed, consisting of Kitchener's Horse,[219] a company of the Lincoln Mounted Infantry,[220] and two Maxim guns; but

[219] Kitchener's Horse Article: https://www.angloboerwar.com/?option=com_content&view=article&id=352 2,082 Records.

[220] Linconshire Regiment https://www.angloboerwar.com/?option=com_content&view=article&id=2053 766 Records. Captain Gamble, Lincoln Regiment.

just as the sun sank this plan was changed by the officer commanding the force, and the whole were ordered to retire into Thabanchu. On the Indian frontier it is a cardinal rule to retire by daylight and sit still when overtaken by night in the best position at hand. In this war experience has shown that it is usually better to remain on the ground, even at a heavy cost, until it is quite dark, and then to retreat, if necessary. The reason of the difference is, that while close contact with an Afridi[221] armed with a four-foot knife, active as a cat and fierce as a tiger, is to be avoided as much as possible, no soldier asks better than the closest contact with a Dutchman. But though the teaching of both wars may seem contradictory on many points, on one point it is in complete agreement: twilight is the worst time of all to retire.

The consequences of this ill-timed change of plan were swift. The Boers saw the retrograde movement, and pressed boldly forward, and Kitchener's Horse, finding themselves closely engaged, were unable to move. A sharp and savage little fight followed in the gloom.

The Boers crept quite close to the soldiers, and one fierce greybeard was shot through the head eight paces from the British firing line, but not until after he had killed his man. The reports which reached the town, that Kitchener's Horse were 'cut off' on a kopje four miles from the camp, induced General French to send the Gordon Highlanders to their relief.

This battalion started at about ten o'clock, and were put on their road to the northward. But in the darkness and the broken ground they lost their way, marched five miles to the south, occupied another hill, and did not rejoin the command until the afternoon of the next day, an absence which, since no inquiries could discover them, caused much anxiety. Kitchener's Horse meanwhile, under Major Fowle,[222] of the 21st Lancers,[223] made a plucky defence, beat off the Boers, and managed at about eleven o'clock to effect their retreat undisturbed. The losses in the affair were twelve or fourteen men killed and wounded, including one officer, who was shot through the head.

Very early the next morning the whole force marched out of the town, and French's operations were this day designed to compel the enemy to retreat from his positions in rear of Thabanchu Mountain, and if possible to surround some part of his force. The information at General French's disposal could not, however, have been very accurate, for in my telegram of the 26th I wrote that 'more than 2,000 Boers' were collected to the north of Thabanchu, and the Press Censor erased this and substituted the words 'small parties.' If this latter view had been correct it is probable that the operations of the following day would have been attended by a greater measure of success.

The plan was clear and vigorous. Gordon's[224] Cavalry Brigade was to move to the right, round the east of Thabanchu Mountain, and force their way into the plains behind it. It was hoped that the Lancers, of which this brigade is entirely composed, would find some opportunity for using their dreaded weapon.

[221] The Afrīdī are a Pashtun tribe present in Pakistan, with substantial numbers in Afghanistan. The British classified the peoples of Colonial India with fixed personality or "racial" traits and regarded the Pashtun Afridi tribesmen "martial" under the martial races theory, and some served with the Khyber Rifles. Afridi - Wikipedia

[222] Fowle John Major QSA (3). Landed at Cape town with Mules from New Orleans, 06 Feb 00. With Kitchener's Horse, 12 Feb 00 - 23 May 00. Source: QSA medal rolls.

[223] 21st Lancers https://www.angloboerwar.com/?option=com_content&view=article&id=513 64 records.

[224] Probably: William Eagleson Gordon Captain He was born on 4 May, 1866, son of W E Gordon, MD, of Homehill, Bridge-of-Allan, Stirlingshire, and joined the 1st Gordon Highlanders, then in Malta, as Second Lieutenant from the Militia, 6 June, 1888, and became Lieutenant 1 September 1891. His first active service was with the Chitral Relief Expedition in 1895, when he was present at the storming of the Malakand Pass, etc. He took part in the operations on the North-West Frontier of India, 1897-8, with the Tirah Expeditionary Force. From 21 January 1899 to 20 January 1903, Captain Gordon was Adjutant to the 1st Battalion Gordon Highlanders, and as such he served throughout the South African War of 1899-1902. He took part in the advance on Kimberley, including the action at Magersfontein (dangerously wounded); operations in the Orange Free State, February to May, 1900, including operations at Paardeberg, 17 to 21 February 1903 (slightly wounded); actions at Poplar Grove, Driefontein, Houtnek (Thoba mount), Vet River (5 and 6 May), and Zand River; operations in the Transvaal in May and June, 1900, including actions near Johannesburg and Pretoria; operations in the Tranvsaal east of Pretoria, July to 29th November 1900, including actions at Belfast (26 and 27 August), and Lydenberg (5 to 8 September); operations in the Transvaal west of Pretoria, July to 29 November 1900; operations in Cape Colony south of Orange River 1899-1900; operations in Cape Colony north of Orange River; operations in the Transvaal 30 November 1900 to 31 May, 1902. Queen's Medal with five clasps; King's Medal with two clasps; Brevet of Lieutenant Colonel on attaining rank of Major.

A scout of the 12th Royal Lancers.[185] 12th Prince of Wales Royal Lancer.[186] Battle Summary.[225]

Hamilton was to push back the weak Boer right, and open the way for Dickson's[226] Cavalry Brigade to pass through and join hands with Gordon. Rundle, whose infantry were tired from their long march from Dewetsdorp, was to demonstrate against the Boers' centre and hold the town.

The action opened with the reoccupation of Kitchener's Horse Hill by Smith-Dorrien's Infantry Brigade, who advanced in determined style, and by a sweeping movement of Ridley's Mounted Infantry. Both these undertakings, which were directed by Hamilton, prospered. The Boer right, which was very thin, was brushed aside, and the road for the cavalry was opened. At, and not until, nine o'clock, French's leading squadrons began to appear on the plain, and by ten the whole of Dickson's Brigade had passed through the gap and were safely extended in the undulating plains beyond.

Wishing to see, for the first time, Cavalry and Horse Artillery working in suitable country, I rode down from my post of observation on Kitchener's Horse Hill and trotted and cantered until I caught up the squadrons. It was evident that the left enveloping arm was making good progress. Already we could almost look into the bay behind Thabanchu Mountain. If Gordon were only getting on as well we might join hands with him, and enclasp a goodly catch of prisoners. So the brigade continued to advance from ridge to ridge, and presently Boers began

He was awarded the Victoria Cross [London Gazette, 28 September 1900]: 'William Eagleson Gordon, Captain, Gordon Highlanders. Date of Act of Bravery: 11 July 1900. On the 11th July, 1903, during the action near Leehoehoek (or Doornboschfontein, near Krugersdorf), a party of men, accompanied by Captains Younger and Allan, having succeeded in dragging an artillery waggon under cover when its horses were unable to do so by reason of the heavy and accurate fire of the enemy, Captain Gordon called for volunteers to go out with him to try and bring in one of the guns. He went out alone to the nearest gun under a heavy fire, and with the greatest coolness fastened the drag-rope to the gun, and then beckoned to the men, who immediately doubled out, to join him, in accordance with his previous instructions. While moving the gun Captain Younger and three men were hit. Seeing that further attempts would only result in further casualties, Captain Gordon ordered the remainder of the party under cover of the kopje again, and, having seen the wounded safely away, himself retired. Captain Gordon's conduct under a particularly heavy and most accurate fire at only 603 yards' range was most admirable, and his manner of handling his men most masterly; his devotion on every occasion that his battalion has been under fire has been remarkable.'

He was Staff Captain, Highland Grouped Regimental District, 6 June 1905, to 31 March 1908; was promoted to Major on 1 January 1907, and gazetted Brevet Lieutenant Colonel 2 January 1907. From April, 1903 to 5 June, 1903, he was DAA and QMG, Highland Division, Scottish Command, and on 9 October 1913, he was appointed ADC to the King, with Brevet of Colonel. Colonel Gordon served in the European War from 1914; he was taken prisoner, and was released by exchange, unconditionally, in 1916. From 15 September 1917, he commanded No 1 (Midland) District, Scottish Command. See also: William Eagleson Gordon - Wikipedia

[225] The 8th King's Royal Irish Hussars along with the 7th Dragoon Guards and the 14th Hussars they formed the 4th Cavalry Brigade under Brigadier General Dickson. On 1 May 1900, the Boers made a stand in a strong position at Houtnek, where the forces of Ian Hamilton faced stiff competition. In a telegram of 2 May Lord Roberts said: "Hamilton speaks in high terms of the services of the 8th Hussars under Colonel Clowes and a made-up regiment of Lancers, which came into Broadwood's brigade and assisted in making the Boers evacuate their position." The 8th then marched from Machadodorp to Heidelberg with the 14th Hussars and M Battery, under the command of Colonel Mahon. On 13 October, Mahon "became heavily engaged near Geluk with a body of 1100 men with four guns." Mahon succeeded in holding his position until the French came to his assistance when the Boers were driven back in a south-easterly direction, having sustained some losses. The 8th lost 2 officers, Lieutenants P A T Jones and F H Wylam and 7 men, with 2 officers and 8 men wounded. 8th King's Royal Irish Hussars - Wikipedia See Also: 7th Dragoon Guards - Wikipedia And: 14th King's Hussars - Wikipedia

[226] Probably: Dickson-Poynder J P (Bart) Captain MID LG: 10 September 1901, page: 5963. Source: Field Marshal Roberts. 4 September 1901. Re: General mentions.

to gallop across the front to escape, as was thought, from the net we were drawing round them. At all of these—the Horse Artillery and the pom-poms—British pom-poms at last—fired industriously. But as the enemy kept a respectful distance and an open formation, only a few were seen to fall. The others did not fly very far, but gathered together in what soon became considerable numbers outside the net, near a peaked hill, which does not appear in my sketch, but which the reader may bear in mind as lying to the left rear of the turning Cavalry.

At last Dickson's advance reached a point between Thabanchu Mountain and the peaked hill, so that no more Boers could escape by that road; and we saw the others, three or four hundred in number, riding about, up and down, or round and round in the bay, like newly-caught rats in a cage.

At this everyone became very excited. 'Gordon must have headed them back,' it was said. 'Only a few more men and we might make a bag.' Where could men be found? Somebody suggested asking Hamilton. The helio[graph] twinkled: 'Come and help us make a bag,' it said, in somewhat more formal language. And Hamilton came forthwith, leaving positions which were of much value; collecting every man he could lay his hands on—weary mounted Infantry, a tired-out battery, and all of Smith-Dorrien's Brigade that could march fast at the end of a long day—he hurried to seize and line the northern spurs of Thabanchu Mountain, prepared to risk much to strike a heavy blow.

British Pom Pom in action.[187] Signaling section of the Royal Artillery detachment passing information to other unit by a heliograph.[188]

The movement of Infantry and guns to support him encouraged Dickson to press still further forward, and the whole brigade advanced nearly another mile. At length we overtopped a smooth ridge, and found ourselves looking right into the bay or horseshoe of mountains. Now at last we must see Gordon. 'There he is,' cried several voices, and looking in the direction shown I saw a majestic body of horse streaming out of the centre of the bay towards the north-west. But was it Gordon? At least 4,000 mounted men were riding across our front, hardly two miles away. Surely no brigade was so numerous. Yet such was the precision of the array that I could not believe them Boers.

Boers their numbers, however, proved them to be; and not their numbers alone, for before we had watched this striking spectacle long, two large puffs of smoke leapt from the tail of the hostile column, and two well-aimed shells burst near our Horse Battery. At the same time patrols from the left rear hurried in with the news that the Boers who had already escaped from our imagined 'trap' were advancing in force, with two more guns, to cut us from the rest of the army.

As for Gordon, there was no longer any doubt about his fortunes. Far away to the eastward the horseshoe wall of mountains dipped to a pass, and on the sides of this gateway little puffs of smoke, dirty brown against the darkening sky, showed that Gordon was still knocking with his artillery at the door, and had never been able to debouch in the plains behind it.

Moreover, the dangerous hour of twilight was not long distant. Dickson determined to retreat while time remained, and did so without any unnecessary delay. Whereat the Boers came down on our rear and flank, opening furious fire at long range, and galloping eagerly forward, so that the brigade and its guns, so far from entrapping the enemy, were all but entrapped themselves; indeed, the brigadier's mess cart, the regimental water carts, and several other little things, which, being able only to trot, could not 'conform to the general movement,' were snapped up by the hungry enemy, who now pressed on exulting.

The Royal Artillery crew in action.[189]

Loading crates of supplies onto a cart.[190]

Meanwhile Hamilton had taken some risks in order to promote the expected entrapping. He had now to think of himself. First, the Boer advance must be stopped, and, secondly, the force which had, in the hopes of grasping the Boers, let go its hold on Kitchener's Horse Hill, must be withdrawn within the Thabanchu picket line. The General, however, was equal to both requirements. Judiciously arranging some force of Infantry and guns, he peppered the advancing Boers heavily, so that at 800 yards they wheeled about and scurried to the shelter of adjacent kopjes. This advantage restored the situation. Hamilton remained on the ground till dark, and then, with the whole of Ridley's and Smith-Dorrien's commands, returned safely into Thabanchu. [On the evening of the 28th]

During the day rifle and artillery fire had been constant; but as the fighting had been conducted at extreme ranges, which neither side showed much anxiety to diminish, the slaughter was small. Indeed, I do not think that a dozen men were stricken in either army. So far as the British were concerned, the result of the day's operations was a qualified success.

The Boers were evidently prepared to retreat from Thabanchu, but they proposed to do so in their own time and at their most excellent discretion, and it was quite evident that we had not succeeded in any way in hindering or preventing them. It was also clear that, far from being 'in small parties', their strength was nearly 6,000, so that on the whole we might congratulate ourselves on having moved in ignorance and taken no great hurt, The only point about the action difficult to understand was the behaviour of the Boers who had ridden about like caged rats. Why should they do so when they knew that their line of retreat to the north-east was perfectly secure? I can only conclude that this particular commando had arranged to retire northwards towards the peaked hill, and were annoyed at being prevented from joining their comrades at the point where their waggons, and, consequently, their dinners, were awaiting them.

Group of Boer leaders. General Koos de la Rey is the man with black beard in right foreground with his rifle near vertical.[191] General Jacob Hercules De La Rey.[192]

On the evening of this instructive, but unsatisfactory, day, Hamilton received orders from Lord Roberts to march north on Winburg[227] in conformity with the general advance of the army. For this purpose his force was to be largely increased, and the operations which followed require the space of another letter. French remained for some days at Thabanchu, but attempted no further serious operations against the enemy.

Only one other incident of interest occurred in the neighbourhood of Thabanchu. After his relief of Wepener, Brabazon was ordered thither *via* Dewetsdorp.

On the 28th, dusty and tired at the end of a long march, he arrived with his Yeomanry at the foot of a pass among the hills. A Kaffir lounged into the bivouac and asked the General whether he would like to see some pretty shelling, for that there was a fine show at the top of the valley. Brabazon, much interested, mounted his horse forthwith, and, guided by the Kaffir through devious paths, reached a point which afforded an extensive view.

There, in the twilight, lay a British convoy, stoutly defended by a company of the kiddies and a few Yeomanry, and shelled—as the Kaffir had said—with great precision by two Boer guns. The General thereupon gave the Kaffir a 'fiver' to carry a letter through the Boer lines to the commander of the convoy, telling that officer to hold out manfully, and promising that with the dawn Brabazon and the Imperial Yeomanry would come to his aid.

The Kaffir succeeded in his mission. The convoy was encouraged, and, good as his word, with the daylight came the General, at whose approach the Boers fled incontinently, so that Brabazon, the Yeomanry, and the convoy came in safety and triumph into Thabanchu together. [193]

Dusty Column toiling up a grade.[194]

A British mule train.[195]

British Army supply column crossing a river, possibly the Molopo River.[196]

[227] The famous Boer General Koos de la Rey was born in the district of Winburg on the farm Doornfontein. Winburg - Wikipedia

VIII. Ian Hamilton

London: August 10th 1900.[228]

Ian Standish Monteith Hamilton was born at Corfu[229] in 1853. His father, the late Colonel Christian Monteith Hamilton—then a captain, but who eventually commanded the 92nd Highlanders—was the eldest son of John George Hamilton and of Christina Cameron Monteith, daughter of Henry Monteith of Carstairs, sometime Member of Parliament for Lanarkshire. His mother, the late Maria Corunia Vereker, was daughter of John, third Viscount Gort, by Maria O'Grady, daughter of Viscount Guillamore.[230]

The Hamilton family is one of the elder branches of the Scottish Hamiltons, and represents the male line of the Hamiltons of Westport. One of his ancestors on his father's side, a Colonel Hamilton, was for several years an aide-de-camp of the first Duke of Marlborough,[231] and it was therefore something in the nature of a coincidence when Ian Hamilton found the present Duke of Marlborough[232] serving in a similar capacity on his staff. It would not be quite correct to call him a pure Celt, but some notice should be taken by those interested in these questions that his blood is mostly Celtic: both of his grandmothers, Monteith and O'Grady, being of Celtic stock, Scottish and Irish respectively.

When Ian Hamilton was born his father was serving with a detachment of the 92nd Highlanders at Corfu. His mother died in 1856, and for the next ten years, the father being constantly on duty with the regiment, he and his younger brother, Vereker Hamilton, who was born in 1856, lived with their grandparents at Hapton, in the Holy Loch in Argyllshire. Such a childhood on moor and loch in a fine wild country was likely to develop and brace nerve and muscle, and stir the keen blood inherited from many generations of warlike ancestors.

He was educated first at Cheam, and as he grew sufficiently old at Wellington College. Here he was very happy, and although he was not especially noted for industry, his success in the examinations at the end of each term excused any neglect in its course. In 1872 he passed the tests for the army, and, according to the system at that time in force, was offered the choice of going to Sandhurst or living for a year abroad to learn a foreign language thoroughly. The cadet chose the latter, and was sent to Germany. Here he had the good luck to make the close friendship of a most distinguished old man. General Dammers was a Hanoverian who had fought against the Prussians at Langesalze, and who, refusing a very high command under the Prussians, lived at Dresden. Although he himself remained aide-de-camp to the ex-King of Hanover, he became the centre of a group of Hanoverian officers who had entered the Saxon service. He was thus in touch with the latest school of military thought, stimulated to its utmost activity by the lessons of the great war which had lately been concluded.[233]

From General Dammers, Ian Hamilton learned the German language, military surveying, something of military history, and something doubtless of strategy and the art of war. The year thus passed very profitably. On his return to England, however, the War Office announced that they had changed their minds and that for the future everybody must go through Sandhurst. Such protests as his father, himself an officer, was entitled to make were overruled by the authorities, and Ian Hamilton embarked upon his military career having lost, through no fault of his own, one year of seniority—a year which Fortune had perhaps even then determined to restore to him manifold.

In 1873 he entered the 12th Foot, and after some months joined his father's old regiment, the 92nd. At first with the 92nd, and after 1881 with the 2nd battalion of the regiment, the Gordon Highlanders, Ian Hamilton followed the drum from garrison to garrison, going through the military routine, and plodding slowly up the first few steps of the long ladder of promotion. From the very first he interested himself in musketry. He became himself a keen and good rifle shot, and not with the military rifle alone. He spent a long leave in Kashmir on the fringe

[228] PF: Note this was penned in August of 1900, after his return to London. I have kept it in its original place in the book.

[229] The Greek island in the Ionian Sea.

[230] WSC Note: Vide Peerage, Gort and Guillamore.

[231] John Churchill, 1st Duke of Marlborough - Wikipedia

[232] Sunny, as we have seen. Charles Spencer-Churchill, 9th Duke of Marlborough - Wikipedia

[233] The Franco-Prussian War in which Germany was decisively the victor and accelerated the creation of the German Empire. Franco-Prussian War - Wikipedia

of the snows, and made a remarkable bag. Indeed, some of his heads attained nearly to the record dimensions, and one big single-horned markhor[234] enjoyed the actual supremacy for several months.

"Ian H. and Record Bag."[197] "Camp, Baramoola, September, 1876. Ian H and Rajvelli."[198] It appears that he is holding a single 'screw horned' Markhor skull.

Then came the [Second] Afghan war. Ian Hamilton, although only an infantry soldier, became aide-de-camp, with Brabazon as Brigade Major, to the unfortunate commander of the British Cavalry Brigade. Early in the campaign he was stricken down with fever, and so avoided being drawn into the controversy which raged for several years in military circles around the actions in the Chardeh valley.[235] It would indeed have been unfortunate if at this early stage in his career he had been led into any antagonism to the great General with whom his fortunes were afterwards so closely associated.

The Boer war of 1881 found Hamilton still a subaltern. He was ordered to South Africa with his regiment, and went full of eager anticipation. The regiment, composed almost entirely of soldiers inured to the hardships and disdainful of the dangers of war, was in the most perfect condition to encounter the enemy, and, as is usual in British expeditions on the outward voyage, they despised him most thoroughly. It was not to be dreamed of that a parcel of ragged Boers should stand against the famous soldiers of Kabul and Kandahar. They discussed beforehand the clasps which would be given upon the medal for the campaign. They were to be Laing's Nek, Relief of Potchefstroom, and Pretoria 1881. No one had then ever heard the name of Majuba Mountain. Yet there was to be the first encounter between Highlanders and Dutchmen.

The dismal story of Majuba is better known than its importance deserves. Had that action been fought in this war it would perhaps have gone down to history as the affair of the 27th of February. Instead, it was accepted as a stricken field, and might, such was the significance that was attached to it, have changed the history of nations. It needs no repetition here save in so far as it is concerned with Ian Hamilton.[236]

Majuba Mountain may in general terms be described as a saucer-topped hill. Sir George Colley and his six hundred soldiers, picked from various units (that all might share the glory), sat themselves down to rest and sleep, and dig a well in the bottom of the saucer. One weak picket of Gordon Highlanders was thrust forward over the rim on to the outer slope of the hill to keep an eye on those silent grey patches which marked the Boer laagers far below.

[234] The markhor is the national animal of Pakistan, where it is also known (in English) as the "screw-horn" or screw-horned goat. Markhor - Wikipedia

[235] During the Second Anglo-Afghan War - Wikipedia On the evening of 5 October 1879, Roberts reached Charasiab, a small town 12 kilometres (7.5 mi) south of Kabul, where he camped overnight, awaiting the arrival of his force's baggage. The next morning, 6 October, saw a large force of regular Afghan soldiers, equipped with artillery and reinforced by local tribesmen, on the ridge of the hills ahead. The fighting lasted several hours, but eventually the Afghan army was pushed back, and by 3:45 pm the British-led force had opened the route to Kabul. After this, Roberts resumed his advance and occupied Kabul on 13 October. Battle of Charasiab - Wikipedia See: 14th Murray's Jat Lancers - Wikipedia The regiment participated in a difficult cavalry charge over extremely difficult ground and routed the Afghan line at the Battle of Charasiab, 15 km from Kabul, on 6 October 1879 for which the regiment was awarded the battle honour "Charasiah" and the theatre honour "Afghanistan 1878–79" See Also: Walter Kitchener - Wikipedia He served in the Second Anglo-Afghan War as a transport officer to the Kabul Field Force and took part in the first Battle of Charasiah and the battle of Karez Meer. Kitchener also saw action in the Chardeh Valley.

[236] Battle of Majuba Hill - Wikipedia

Arrival of a commando at Majuba Hill, 1899.[237] Top of Majuba Hill.[238] Summit of Majuba.

Hamilton was the subaltern in command. As the day gradually broke and the light grew stronger, he saw from the very lifting of the curtain the course of the tragedy. Boers awoke, bustled about their encampments; looked up just as [Wm. Penn] Symons' Brigade looked up on the morning of Talana Hill,[239] and saw the skyline fringes with men. More bustle, long delay, much argument and hesitation below, a little boasting rifle fire from some of the British soldiers: 'Ha, ha! got you this time I think!'—and then, straggle of horsemen riding in tens and twenties towards the foot of the mountain. Hamilton reported accordingly. The action of Majuba Hill had begun. Pause.

There was—so it has been described to me—a long donga that led up the steep slope. Into the lower end of this the Boer horsemen disappeared. Hamilton moved his score of men a little to their right, where they might command this zigzag approach as much as the broken ground would allow, and reported again to the General or whoever was directing affairs—for Colley, wearied with the tremendous exertion of the night climb, was sleeping— 'Enemy advancing to attack.' He also made a few stone shelters. Pause again.

Suddenly, quite close, darting forward here and there among the rocks and bushes of the donga—Boers! Fire on them, then. The Gordons' rifles spluttered accordingly, and back came the answer hot and sharp—a close and accurate musketry fire pinning the little party of Regulars to the earth behind their flimsy shelters. No one could show his head to fire. Soldiers would hold a helmet up above the sheltering stone and bring it down with two and three bullets through it. Could half a company fight a battle by itself? What were others doing? Hamilton felt bound to send another report. He left the half company in charge of the sergeant, got up, ran up the slope, and dropped into safety the other side of the saucer-shaped rim. The distance was scarcely forty yards, yet two bullets passed through his kilt in crossing it. Where was the General?

A staff officer, ignorant and therefore undisturbed, said that the General was sleeping. 'He knew,' said the staff officer, 'what was going on. No need for a subaltern of Highlanders to concern himself.' Hamilton returned, running the gauntlet again, to his men. The fire grew hotter. The Boers began to creep gradually nearer. Their front attack widened and drew around the contours of the hill. Were all the force asleep? One more warning at any rate they should have. Again he darted across the open space with the swish of bullets around him. Again he found the staff. But this time they were annoyed. It is such a bore when young officers are jumpy and alarmist. 'It's all right,' they said: and so it was within the saucer. The bullets piped overhead as the wind howls outside the well-warmed house. But a sudden change impended.

Hamilton rejoined his men just as the Boers attacked at all points. The little picket of Highlanders, utterly unable to withstand the weight of the enemy's advance, ran back to the rim of the saucer intermingled with the Boers, who fired their rifles furiously at them, even putting the muzzles to the men's heads and so destroying them. In Sir William Butler's book,[240] written almost entirely with the view of exonerating Sir George Colley, it is suggested that his advanced picket fell back in a panic. The truth is that they were swept backward by

[237] Hillegas, Boers in War, 1900.

[238] Arthur Mann The Boer in Peace and War, 1900 and War with the Boers, H. Brown, London [1902?] Vol 1.

[239] The Battle of Talana Hill, also known as the Battle of Glencoe, was the first major clash of the Second Boer War. A frontal attack by British infantry supported by artillery drove Boers from a hilltop position, but the British suffered heavy casualties in the process, including their commanding general Sir William Penn Symons.

[240] William Butler (British Army officer) - Wikipedia

overwhelming force after they had three times reported to the General the development of a heavy attack. Of the seventeen men under Ian Hamilton in this advanced position twelve were shot dead.

The survivors of the picket with the pursuing Boers reached the rim together, and became visible to the main force. Astounded by this apparition, the troops who were lying down in the saucer rose up together, and, some accoutred, some with their coats off, Highlanders, sailors, and linesmen, ran forward and fired a ragged volley. The Boers immediately lay down and replied, causing heavy loss. A furious musketry fight followed between the Dutch in cover along the rim and the British among the rocks across the centre of the saucer. This was ended by the appearance of other Boers on the high ground at the northern end of the plateau. Without orders or order, exposed to a terrible fire, ignorant of what was required of them, the soldiers wavered. One last chance presented itself. Hamilton rushed up to the General in the impetuosity of youth: 'I hope you'll forgive my presumption, sir, but will you let the Gordon Highlanders charge with the bayonet?'

'No presumption, young gentleman,' replied Colley, with freezing calmness. 'We'll let them charge us, and then we'll give them a volley and a charge.'

On the word the whole scene broke into splinters. The British troops abandoned their positions and fled from the ground. The Boers, standing up along the rim, shot them down mercilessly—sporting rifles, crack shots, eighty yards' range. Hamilton saw a figure scarcely ten yards away aiming at him, raised the rifle he found himself somehow possessed of to reply. Both fired simultaneously. The British officer went down with his wrist smashed to pieces. He rose again: the rear crest was near. The last of the fugitives were streaming over it. One dash for liberty! The fire was murderous. Before the distance was covered his tunic was cut by one bullet, his knee by another, and finally a splinter of rock striking him behind the head brought him down half stunned to the ground—luckily behind the shelter of a small rock.

The firing stopped. The Boers began to occupy the position. Two discovered the wounded man. The younger, being much excited, would have shot him. The elder restrained him. 'Are you officer, you damned Englishman?' said they.

'Yes.'

'Give your sword.'

Now Hamilton's sword had belonged to his father before him. He replied by offering them money instead.

'Money!' they cried; 'give it up at once,' and were about to snatch it away when a person of authority—it is said Joubert himself—arrived. 'Voorwarts,'[241] he said to the burghers, and in spite of their desire to plunder he drove them on. Hamilton thanked him. 'This is a bad day for us.'

'What can you expect,' was the answer characteristic of the Boer—the privileged of God— from fighting on a Sunday?'

Then they collected the prisoners and helped Hamilton to walk back to the British position. Colley lay dead on the ground. The Boers would not believe it was the General. 'Englishmen are such liars.' Hector Macdonald[242]—grim and sad—hero of the Afghan war, now a prisoner in the enemy's hand, watched the proceedings sullenly. The Boers picked out the surrendered prisoners. They looked at Hamilton. He was covered with blood from head to foot. They said: 'You will probably die. You may go.' So he went; staggered, and crawled back to camp, arrived there delirious the next morning.

The wrist joint is composed of eight separate bones. The bullet, breaking through, had disarranged them sadly, had even carried one or two away. If he had consented to amputation he would soon have been convalescent. But a soldier must preserve all he can. What with fever and shock he nearly died. For six months he was an invalid. But the hand was saved, so that now the General can hold an envelope between his paralysed and withered fingers, and sometimes hold a cigarette. For all other purposes it is useless, and when he rides it flaps about helplessly—a glorious deformity.

After some months of doubt as to whether he should leave the army and throw himself entirely into the literary pursuits which had always possessed for him a keen attraction, Hamilton decided to remain a soldier.

[241] Forwards.

[242] Sir Hector Archibald MacDonald, (4 March 1853 – 25 March 1903), also known as Fighting Mac, was a British Army soldier. MacDonald served as a subaltern in the First Boer War (1880–81), and at the Battle of Majuba Hill, where he was made prisoner, his bravery was so conspicuous that General Joubert gave him back his sword. Hector MacDonald - Wikipedia

He next saw service in the Soudan. He was not intended to make this campaign, for the battalion to which he belonged was serving in India, and there has always been much jealousy between the Indian and the Egyptian British officer. But he happened to be coming home on leave, and when the steamer reached Suez it occurred to him to ask himself why he should not go up the Nile with the columns which were being formed. He got out of the ship accordingly and ran across the sands to the train which was standing in the station. Had he not caught it he would have returned to the ship. But he was in time. Next day he arrived in Cairo, and while waiting there for his luggage he applied for employment. It was refused, officers were not allowed to volunteer.

The Gordon Highlanders, his only hope, had their full complement of officers. They had no vacancy for him. Hamilton did not, however, give up his idea easily. He resolved to travel as far as Wady Halfa and renew his application there. He journeyed south with Colonel Burnaby, and after a week of train and river-boat arrived at the whitewashed mud huts in the midst of a vast circle of sand which marked the base of the British Expeditionary forces, both desert and river columns.

What followed has happened so often that it is well worth the attention of young officers. Be it always remembered that the regulations of the army are formed to make all people quite alike one uniform pattern and on one level of intelligence—not yet the highest. You do not rise by the regulations, but in spite of them.

Therefore in all matters of active service the subaltern must never take 'No' for an answer. He should get to the front at all costs. For every fifty men who will express a desire to go on service in the mess or the club, and will grumble if they are not selected, there is only about one who really means business and will take the trouble and run the risk of going to the front on the chance. The competition is much less keen when you get there. I know something of this myself, and am convinced of its truth.

The subaltern really stands on velvet in the matter.[243] If he succeeds all is well. If he gets rebuked and ordered down, he must try again. What can the authorities do? They cannot very well shoot him. At the worst they can send him back to his regiment, stop his leave for six months, and some choleric old martinet who was a young man once, though he had half forgotten it, will write in some ponderous book in Pall Mall against the offender's name: 'Keen as mustard—takes his own line—to be noted for active service if otherwise qualified.'

Of course everyone was delighted to see Hamilton at Wady Halfa.[244] They appointed him to a vacancy which had meanwhile occurred in the Gordon Highlanders, and gave him a company and a boat in the River Column. Through all the hard campaign that followed he served with credit. The fortunes of the troops who worked their way up the Nile have not been so closely studied as those of the columns which plunged into the desert and fought at Abu Klea[245] and Abu Kru.[246] But it was nevertheless one of the most picturesque enterprises of our military history.

The broad boats toiling forward against the current of the river, making perhaps three miles a day, obstructed by frequent cataracts and menaced continually by the enemy, the scouts on the banks, the lines of men on the tow ropes, the red sand of the desert, the hot steel sky, and the fierce sunlight slanting in between rocks of the Nile gorge, are materials from which a fascinating sketch might be painted. Hamilton's boat became somehow the head of the rear column. At length there came a day when they told of expected opposition, dervish encampments, and a certain rocky ridge said to be lined with riflemen. The leading column of boats was hurried forward. By some mischance Hamilton's boat became the rear boat of the leading column. At any rate, his company alone of the Gordon Highlanders fought in the action of Kirbeckan[247] next day. Nothing succeeds like success. Hamilton received the Distinguished Service Order for his services.

[243] To place one's bets in such a way as not to lose in any event.

[244] A city in Northern Sudan. Wadi Halfa - Wikipedia

[245] Battle of Abu Klea - Wikipedia

[246] Battle of Abu Kru - Wikipedia

[247] The Battle of Kirbekan was a battle in the Mahdist War. It was fought February 10, 1885, when the British Nile Column, about 1,000 strong, under General Earle, stormed the heights of Kirbekan, which were held by a strong Mahdist force, and totally routed them, with heavy loss. The British lost 60, among whom was General William Earle, killed. It was the first appearance of the Egyptian Army Camel Corps under Bimbashi Marriott in action. Battle of Kirbekan - Wikipedia

The Nile Expedition for the relief of Gordon.[199]

After the Nile Expedition of 1885 had reached its sad conclusion,[248] Hamilton returned to India and became an aide-de-camp on the staff of Lord Roberts, who was then commanding the Madras army. The question of musketry training for Infantry was at that time much discussed, and Lord Roberts was determined to do something to improve the shooting of the British army. In his book 'Forty-one Years in India'[249] he tells us how he and his staff formed themselves into a team and had many exciting rifle matches with the regiments in the Madras command. In all this Hamilton's skill with the rifle and the keen interest he had always shown for musketry—his first regimental appointment had been to be Musketry Instructor—stood him in good stead, and when Lord Roberts became Commander-in-Chief in India his aide-de-camp, who had meanwhile served in the Burmah[250] campaign, was made Assistant Adjutant-General for Musketry.

In 1886 he married Jean, daughter of Sir John Muir, Baronet, of Deanston, Perthshire. He had now determined to persevere in the military profession, and devoted himself to it with great assiduity. His literary talents were turned to military subjects. He published a book on musketry in the army entitled 'The Fighting of the Future.' It was strong and well written. The introduction of the magazine rifle has modified many of his conclusions, but at the time the book attracted a great deal of attention. He found time, however, to write on other things, and there are still extant from his pen: 'A Jaunt in a Junk,' an account of a cruise which he made with his brother down the west coast of India; a volume of verses, 'The Ballad of Hadji and the Boar'; and one or two other writings. He preserved and extended his acquaintance with literary men, particularly with Andrew Lang,[251] whom he powerfully impressed, and who inscribed a volume of poems to him in the following compulsive lines:

[248] The Nile Expedition, sometimes called the Gordon Relief Expedition (1884–1885), was a British mission to relieve Major-General Charles George Gordon at Khartoum, Sudan. It was not able to arrive in time to prevent Khartoum from falling to the Mahdist forces. Nile Expedition - Wikipedia

[249] Frederick Sleigh Roberts Forty-one years in India: From Subaltern To Commander-In-Chief 1898, Longman's Green & Co. London. University of Michigan Scan: Forty-one years in India : Roberts, Frederick Sleigh, Earl : Free Download, Borrow, and Streaming : Internet Archive Project Gutenberg Version: Forty-one years in India: From Subaltern To Commander-In-Chief : Roberts, Frederick Sleigh : Free Download, Borrow, and Streaming : Internet Archive

[250] British rule in Burma lasted from 1824 to 1948, from the successive three Anglo-Burmese wars through the creation of Burma as a province of British India to the establishment of an independently administered colony, and finally independence. British rule in Burma - Wikipedia

[251] Andrew Lang FBA (31 March 1844 – 20 July 1912) was a Scottish poet, novelist, literary critic, and contributor to the field of anthropology. He is best known as a collector of folk and fairy tales. The Andrew Lang lectures at the University of St Andrews are named after him. Andrew Lang - Wikipedia

"Jean in her wedding finery"[200]

To Colonel Ian Hamilton

To you, who know the face of war,
You, that for England wander far,
You that have seen the Ghazis fly
From English lads not sworn to die,
You that have lain where, deadly chill,
The mist crept o'er the Shameful Hill,
You that have conquered, mile by mile,
The currents of unfriendly Nile,
And cheered the march, and eased the strain
When Politics made valour vain,
Ian, to you, from banks of Ken,
We send our lays of Englishmen!

After doing much useful work in the Musketry Department he became one of the Assistant Quartermaster-Generals in India. From this office he managed to sally forth to the Chitral Expedition,[252] for his services in which on the lines of communication he was made Commander of the Bath.[253] He next became Deputy Quartermaster-General, and it was evident that if he chose to continue to serve in India he would ultimately become the head of the Department. In 1897 the Great Frontier War broke out. Hamilton was appointed to command one of the brigades of the Tirah Expeditionary Force.[254] He was at the time on leave in England. He returned at speed, assumed command, and led his brigade through the Kohat Pass in the first movement of the general advance. It looked as if his chance in life had come. He had a magnificent force under him. He enjoyed the confidence of the General-in-chief, Sir William Lockhart,[255] and only a few miles away the enemy awaited the advancing army on the heights of Dargai. The next morning his horse shied suddenly. He was thrown to the ground and broke his leg. They carried the brigadier away in a doolie, his brigade passed to another, and the campaign in Tirah was fought without him.

Indians Carrying a Wounded Officer in a Dhoolie.[256]

[252] The Chitral Expedition was a military expedition in 1895 sent by the British authorities to relieve the fort at Chitral, which was under siege after a local coup following the death of the old ruler. An intervening British force of about 400 men was besieged in the fort until it was relieved by two expeditions, a small one from Gilgit and a larger one from Peshawar. Chitral Expedition - Wikipedia

[253] The Most Honourable Order of the Bath is a British order of chivalry founded by King George I on 18 May 1725. Order of the Bath - Wikipedia

[254] The Tirah campaign, often referred to in contemporary British accounts as the Tirah expedition, was an Indian frontier campaign from September 1897 to April 1898. Tirah campaign - Wikipedia

[255] General Sir William Stephen Alexander Lockhart (2 September 1841 – 18 March 1900) was a British General in the British Indian Army. William Lockhart (Indian Army officer) - Wikipedia

[256] With the Flag to Pretoria, H. W. Wilson Vol 1.

Ian Hamilton took this bitter disappointment with philosophical composure. 'Perhaps,' he said to me one day in Calcutta, 'I should have lost my reputation had I held my command.' But it was easy to see how much he felt the lost opportunity and the enforced inaction. At length his leg was mended—after a fashion. He persuaded a medical board to pass him as sound. The campaign continued. There was, however, no vacancy at the front. For several weeks he waited.

Presently Sir Bindon Blood[257]—who was preparing for his invasion of Buner,[258] and who knew Hamilton well—applied for him to command his lines of communication. Obstacles were, however, raised by the Indian War Office, and the proposal fell through. At last, in February, when it seemed certain that a spring campaign must be undertaken against the Afridis,[259] Sir William Lockhart decided to replace General Kempster[260] by some other brigadier, and Ian Hamilton was again sent to the front. The hopes or fears of a further campaign proved unfounded. The Afridis gradually paid their toll of rifles, and their jirgahs made submission. The fighting was practically over. Yet in much skirmishing as occurred while Hamilton's brigade were holding the advanced posts in the Bara Valley[261] his care and eagerness attracted attention, and, small as was his share in the campaign, Sir William Lockhart gave him an honourable mention in the despatches.

On the restoration of order along the North-West Frontier Hamilton was offered the temporary position of Quartermaster-General in India. Anxious, however, for home employment, and fully alive to the importance of not becoming too closely identified with any particular military set, he declined this important office and proceeded to England on a year's leave.

After some delay he was appointed commandant of the School of Musketry at Hythe,[262] and from this post he was twice withdrawn to command brigades at the Manoeuvres. When Sir George White[263] was sent to Natal in September 1899 Hamilton accompanied him as Assistant Adjutant-General. The War Office are therefore entitled to plume themselves upon his successes, for he is one of the few men originally appointed who have increased their reputation.

Ian Hamilton's part in the Boer war is so well known that it will be unnecessary to do more than refer to it here. He displayed a curious facility for handling troops in close contact with the enemy, and practically from the beginning of the fighting he held the command of a brigade. It was Hamilton whose influence went so far to counteract the astounding optimism of the gallant Penn Symons. It was Hamilton who was to have led the bayonet attack by night on the Boer laagers two days before Talana Hill was fought.

It was Hamilton to whom French[264] entrusted the entire disposition of the Infantry and Artillery at Elandslaagte,[265] who arranged the attack, rallied the struggling line, and who led the final charge upon the Boer entrenchment. Again after Lombard's Kop, when the army reeled back in disorder into Ladysmith,[266] it was Hamilton's brigade which, judiciously posted, checked the onset of the victorious enemy. During the defence of Ladysmith Hamilton's section of the defence included Caesar's Camp and Wagon Hill. He has been censured in the Press for not having fortified these positions on their outer crests, and it was said in the army after the 6th of

[257] General Sir Bindon Blood, (7 November 1842 – 16 May 1940) was a British Army commander who served in Egypt, Afghanistan, India, and South Africa. Bindon Blood - Wikipedia

[258] Buner District is a district in the Malakand Division of the Khyber province of Pakistan. Buner District - Wikipedia

[259] The Afridi are a Pashtun tribe present in Pakistan. Afridi - Wikipedia

[260] Francis James Kempster.

[261] The Barah Valley is located on the bank of Shyok river in the Khaplu, (Ghanche District) of Baltistan. Barah Valley - Wikipedia

[262] Small Arms School Corps - Wikipedia

[263] George White (British Army officer) - Wikipedia

[264] John French, 1st Earl of Ypres - Wikipedia

[265] The Battle of Elandslaagte (21 October 1899) was a battle of the Second Boer War, and one of the few clear-cut tactical victories won by the British during the conflict. However, the British force retreated afterwards, throwing away their advantage. Battle of Elandslaagte - Wikipedia

[266] A large British force which had concentrated at the garrison town of Ladysmith launched a sortie on 30 October 1899, against Boer armies which were slowly surrounding the town. The result was a disaster for the British. The main body was driven back into the town, and an isolated detachment of 800 men was forced to surrender to Commandant De Wet. The Boers did not follow up their advantage by proceeding towards the strategically important port of Durban, and instead began a siege of Ladysmith, which was relieved after 118 days. Battle of Ladysmith - Wikipedia

January that this neglect caused unnecessary loss of life.[267] How far this criticism may be just I do not now propose to examine. The arguments against entrenching the outer crest were that heavy works there would draw the enemy's artillery fire, and that the Imperial Light Horse,[268] who were to have defended this section, said they preferred to avail themselves of the natural cover of rocks and stones. The reader would be well advised to defer judgement until some serious and historical work on the campaign in Natal is published. At present all accounts are based on partial and imperfect evidence, nor do I think that the whole true account of a single action has yet been written.

Whatever the rights of this question may be, it is certain that on the 6th of January Ian Hamilton, by his personal gallantry and military conduct, restored the situation on Wagon Hill. Indeed, the Homeric contest, when the British General and Commandant Prinsloo[269] of the Free State fired at each other at five yards' range, the fierce and bloody struggle around the embrasure of the [British] naval gun, and the victorious charge of the Devons, may afterwards be found to be the most striking scene in the whole war.

Devons Firing on the Boers Retreating from Pepworth Hill.[201] Devons taking cover and resting in the trenches, Ladysmith.[202]

After the relief of Ladysmith, Roberts, who knew where to find the men he wanted, sent for Hamilton, much to the disgust of Sir Redvers Buller, who proposed to keep this good officer for the command of one of his own brigades. On reaching Bloemfontein he was entrusted with the organisation of the Mounted Infantry division, a post from which he could conveniently be drawn for any service that might be required. Of the rest some account will be found in these letters.

Ian Hamilton is, as the fine portrait by Sargent, reproduced as the frontispiece of this book, shows him, a man of rather more than middle height, spare, keen eyed, and of commanding aspect. His highly nervous temperament animating what appears a frail body imparts to all his movements a kind of feverish energy.

[267] See infra for battle summary.

[268] Imperial Light Horse Article: https://www.angloboerwar.com/?option=com_content&view=article&id=345

[269] There are at least two general (Commandant) Prinsloo's. I am unable to determine which one was at this engagement.

General Ian Hamilton, C. B., D.SO.[203] Ian Hamilton in camp, 1898, Kohat, Pakistan.[270]

Two qualities of his mind stand forward prominently from the rest. He is a singularly good and rapid judge of character. He takes a very independent view on all subjects, sometimes with a slight bias towards or affection for their radical and democratic aspects, but never or hardly ever influenced by the set of people with whom he lives. To his strong personal charm as a companion, to his temper never ruffled or vexed either by internal irritation or the stir and contrariness of events, his friends and those who have served under him will bear witness.

He has a most happy gift of expression, a fine taste in words, and an acute perception of the curious which he has preserved from his literary days. But it is as a whole that we should judge. His mind is built upon a big scale, being broad and strong, capable of thinking in army corps and if necessary in continents, and working always with serene smoothness undisturbed alike by responsibility or danger. Add to all this a long experience in war, high military renown both for courage and conduct, the entire confidence and affection of the future Commander-in-Chief, the luck that has carried him through so many dangers, and the crowning advantage of being comparatively young, and it is evident that here is a man who in the years that are to come will have much to do with the administration of the British Army in times of peace and its direction in the field.

[270] Listening for the Drums by General Sir Ian Hamilton Faber And Faber Ltd 24 Russell Square London 1944.

IX. The Action of Houtnek 5-1-1900

Winburg: May 8th 1900.

Ian Hamilton's orders were to march north from Thabanchu on Winburg by the Jacobsrust [271] road, and he was expected, if no opposition was encountered, to reach his destination by the 7th of May. The column with which he started from Thabanchu was composed of Smith-Dorrien's 19th Infantry Brigade, Ridley's Mounted Infantry Brigade, and two batteries of artillery; but at Jacobsrust he would receive a strong reinforcement, consisting of Bruce-Hamilton's[272] 21st Brigade of Infantry, Broadwood's Cavalry Brigade,[273] two batteries of field and one of horse artillery, and two 5-in. guns. This accession would raise his force to a total of 7,500 Infantry, 4,000 mounted men, and thirty-two guns—an imposing command for an officer who had not yet had time to take the badges of a colonel off his shoulders.[274] The first thing, however, was to reach Jacobsrust,[275] and effect the junction with Bruce-Hamilton's force.

Sir Bruce Meade Hamilton.[204]

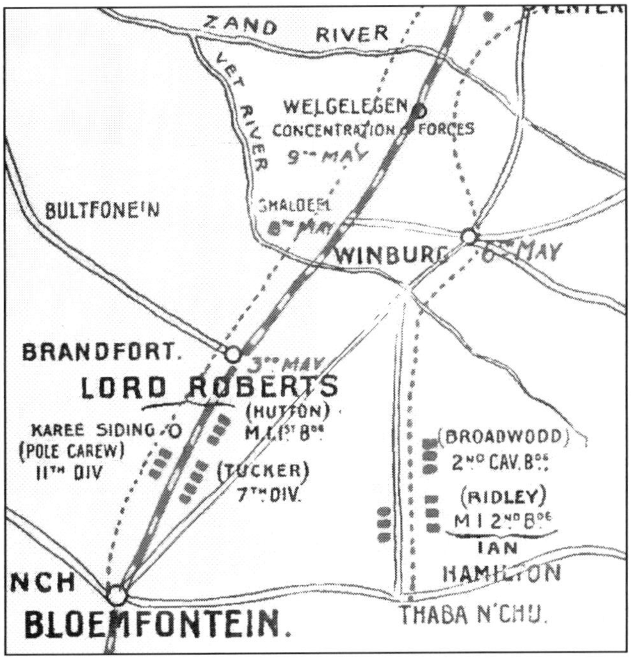

Order of Battle Summary and Map.[276]

The Thabanchu column started at daybreak on the 30th of April, and when it was within three or four miles of Houtnek Poorte the enemy suddenly unmasked field guns and 'pom-poms,' and opened a long-range fire with them from the east on the right flank of the marching troops. Colonel Bainbridge,[277] with the 7th Corps of Mounted Infantry, wheeled up to contain this force of the enemy, and at the same time De Lisle[278] —of polo

[271] Unable to identify this location.

[272] General Sir Bruce Meade Hamilton, (7 December 1857 – 6 July 1936) was a British Army general during the Second Boer War and the First World War. Bruce Hamilton (British Army officer) - Wikipedia

[273] 12th (Prince of Wales's Royal) Lancers Article: https://www.angloboerwar.com/?option=com_content&view=article&id=497 1,665 Records.

[274] He had been promoted to Brigadier General.

[275] See infra: "Jacobsrust, as it is called by the inhabitants; Isabellasfontein" Unable to identify this location.

[276] The 8th King's Royal Irish Hussars along with the 7th Dragoon Guards and the 14th Hussars formed the 4th Cavalry Brigade under Brigadier General Dickson. On 1 May 1900, the Boers made a stand in a strong position at Houtnek, where the forces of Ian Hamilton faced stiff competition. In a telegram of 2 May Lord Roberts said: "Hamilton speaks in high terms of the services of the 8th Hussars under Colonel Clowes and a made-up regiment of Lancers, which came into Broadwood's brigade and assisted in making the Boers evacuate their position." The 8th then marched from Machadodorp to Heidelberg with the 14th Hussars and M Battery, under the command of Colonel Mahon. 8th King's Royal Irish Hussars - Wikipedia See also: 7th Dragoon Guards - Wikipedia And: 14th King's Hussars - Wikipedia

[277] Probably E. G. T. Bainbridge: Lieutenant Colonel 5th Regiment Mounted Infantry. MID LG: 8 February 1901, page: 847. Source: Field Marshal Roberts. 31 March 1900. Re: Orange Free State.

[278] General Henry de Beauvoir de Lisle. (27 July 1864 – 16 July 1955), was a British Army officer and sportsman. He served in both the Second Boer War and the First World War. During the Second Boer War he commanded the Australian Brigade, a mobile column comprising the 6th Battalion Mounted Infantry, the West Australian Mounted Infantry Beauvoir De Lisle - Wikipedia

fame—pushed forward boldly at a gallop with the 6th Corps and the New Zealanders,[279] and seized a commanding position about 2,000 yards south of the actual nek. Colonel Legge,[280] meanwhile advancing on the left front, noticed that Thoba Mountain was weakly held by the enemy, and thereupon ordered Kitchener's Horse to attack it, thus anticipating the order which the General was himself about to send. These dispositions, which were made on their own initiative by the various Mounted Infantry officers, enabled a deliberate view of the situation to be taken.

The pass of Houtnek consists of two parallel grassy ridges separated by a smooth shallow valley a little more than a mile across, and devoid of cover. On the east the pass runs up into sharp rocky kopjes, strengthened by successive lines of stone walls trailing away towards the main laagers of the enemy. Both the centre and the left flank of the Boer position refused all opportunity of attack. The Dutch right was scarcely more encouraging. On the west of the pass rose the great mountain of Thoba, an uneven battlefield, better suited to Boers than to British troops. Yet as it was on Hamilton's safer flank, dominated the rest of the enemy's position, could be turned by mounted troops making a very wide detour, and being, moreover, the only way, the General resolved to attack it.

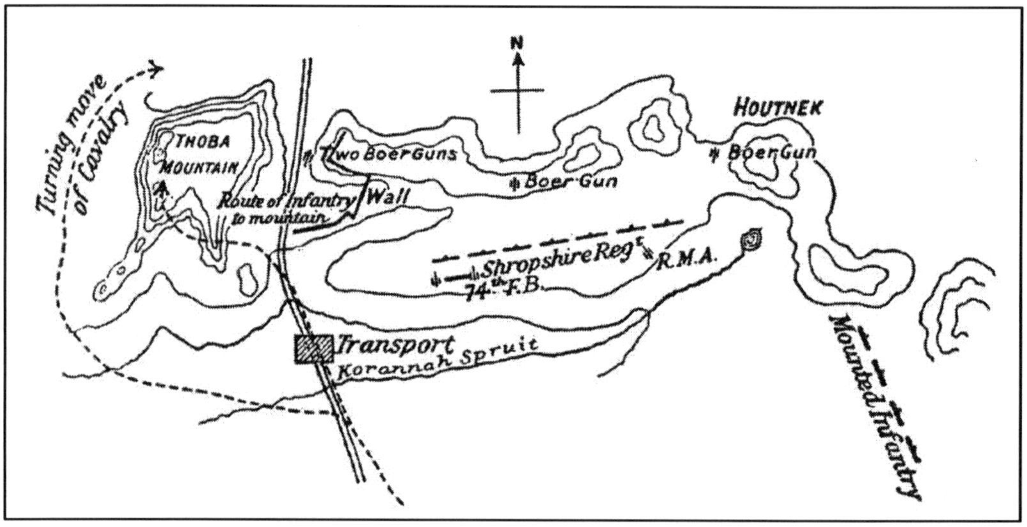

Diagram Explaining the Action of Houtnek. April 30th and May 1st.[205]

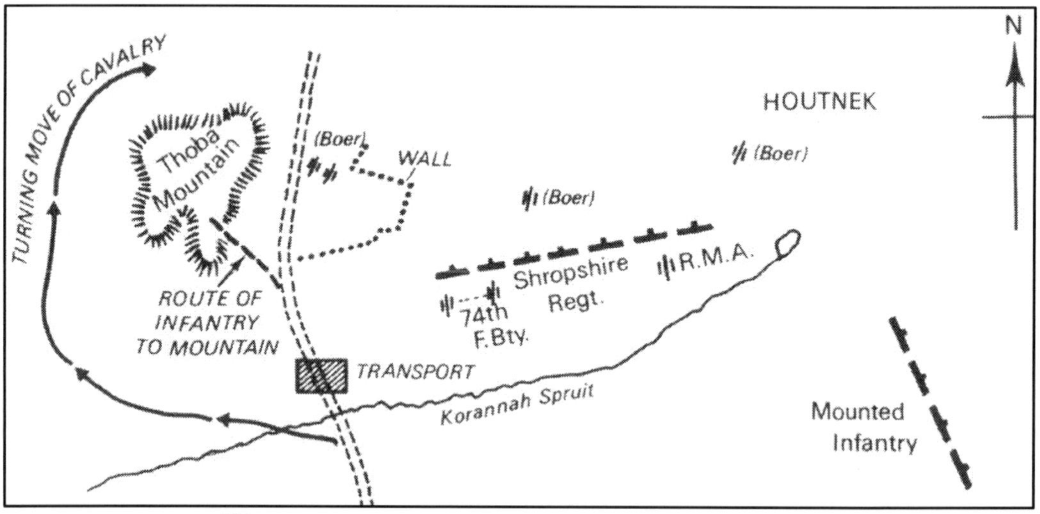

Action of Houtnek.[206]

At 9.30 the Infantry began to come up, and at ten o'clock the approaches to the Boer position were strongly occupied. As soon as Kitchener's Horse were seen to have made good their footing on Thoba Mountain,

[279] New Zealand Contingents Article: https://www.angloboerwar.com/?option=com_content&view=article&id=1733 6,441 Records.
[280] Unable to identify this individual.

Hamilton ordered General Smith-Dorrien to support them with part of his brigade, which was accordingly done, two companies of the Shropshires,[281] the Gordon Highlanders,[282] and four companies of the Canadians being successively worked up on to the hill under a heavy shell fire from the enemy. This practically disposed of the whole force, which was soon engaged all along the line, the Mounted Infantry holding the enemy off the right and right rear, the Cornwalls[283] guarding the baggage, one-half Smith-Dorrien's Brigade containing the front, and the other half with Kitchener's Horse pushing the flank attack on Thoba Mountain. As soon as the Boers understood the designs of the British on Thoba they made a strong effort to regain and hold that important feature. At first the troops made good progress; but as the enemy received continual reinforcements the resistance became more severe, until, presently, far from gaining ground, they began to lose it.

At last, about two o'clock, some one hundred and fifty of the German corps of the Boer force advanced from the northern point of Thoba in four lines across the table top to drive the British off the hill. So regular was their order that it was not until their levelled rifles were seen pointing south that they were recognised as foes, and artillery opened on them. In spite of an accurate shell fire they continued to advance boldly against the highest part of the hill, and, meanwhile, cloaked by a swell of the ground, Captain Towse, of the Gordon Highlanders, with twelve men of his own regiment and ten of Kitchener's Horse, was steadily moving towards them. The scene on the broad stage of the Thoba plateau was intensely dramatic. The whole army were the witnesses.

The two forces, strangely disproportioned, drew near to each other. Neither was visible to the other. The unexpected collision impended. From every point field glasses were turned on the spectacle, and even hardened soldiers held their breath.

The German Corps.[207]

Officers of the German Corps.[284]

A German Commando.[208 285]

[281] Shropshire Light Infantry (King's) Article: https://www.angloboerwar.com/?option=com_content&view=article&id=662

[282] Gordon Highlanders Article: https://www.angloboerwar.com/?option=com_content&view=article&id=557

[283] Duke of Cornwall's Light Infantry article: https://www.angloboerwar.com/?option=com_content&view=article&id=545

[284] Second Row: Dr. W. Vaheutin Lieut. Von Quitzow Lieut. Grothaus (Transvaal Artillery) Front Row: Prince von Jutrzenka Lieut. Von Zelewski Dr. Elsberger (standing) The man leaning against the tree is not identified. Note black puppy at bottom right. Image source: With the Flag to Pretoria, H. W. Wilson Vol 1 1900 and britishbattles.com

British troops watching battle.[209]

At last, with suddenness, both parties came face to face at fifty yards' distance. The Germans, who had already made six prisoners, called loudly on Captain Towse and his little band to surrender. What verbal answer was returned is not recorded; but a furious splutter of musketry broke out at once, and in less than a minute the long lines of the enemy recoiled in confusion, and the top of the hill was secured to the British. Among the foreigners wounded in this encounter, was Colonel Maximoff.[286]

Colonel Maximoff of the Russian Corps.[210 287] "Captain Towse shoots at Lt Col Maximov at point-blank range."[211] Sir Ernest Beachcroft Beckwith Towse, VC.[212 288]

[285] German Volunteers List: https://www.angloboerwar.com/?option=com_content&view=article&id=3243 863 records.

[286] Yevgeny Yakovlevich Maximov (March 1849 – 14 October 1904) was a Russian adventurer, soldier and journalist mostly remembered for his service with the South African Republic during the Second Anglo-Boer War Yevgeny Maximov - Wikipedia See also: The Russians and the Anglo-Boer War: Microsoft Word - Davidson1.doc (narod.ru)

[287] At Taba Nchu, Maximov suffered a splintered shoulder, a badly damaged shoulder blade and a serious wound to his skull, which rendered him unfit for further military service, leading to him being honorably discharged from the Transvaal service. One of the Transvaal leaders, State Secretary Francis William Reitz told the Russian nurse Sophia Izedinova:

Captain Towse, for his conspicuous gallantry, and for the extraordinary results which attended it, has been awarded the Victoria Cross; but, in gaining what is above all things precious to a soldier, he lost what is necessary to a happy life, for in the moment when his military career was assured by a brilliant feat of arms, it was terminated by a bullet which, striking him sideways, blinded him in both eyes. Thus do Misery and Joy walk hand in hand on the field of war.

All this time the rifle and gun fire along the whole front had been continuous, and as the day wore on without the British making good their hold on Thoba Mountain the enemy gathered in a more and more threatening attitude on the right of the column, and by four o'clock at least 1,500 men were collected, with guns and 'pom-poms,' which threw shell into the rear guard and transport.

Village and Mountain the Thaba N'Chu.[289]

Hamilton, however, was determined to fight the matter out. He therefore directed that all troops should post guards on their front, lie down wherever darkness found them, and prepare to renew the action at daybreak. He then telegraphed to General French for some assistance, the need of more mounted troops being painfully felt.

Troops resting.[213]

Two British soldiers, one infantry and another yeomanry, cooking their rations.[214]

"You know, sister...that I am not greatly impressed by all these foreigners who come offering us their services, but concerning your countryman, I am happy to say that we were both mistaken in our caution. All our commandos who have come into contact with Colonel Maximov praise his bravery, his ability to discipline men and the value of his advice. General Louis Botha said that although he is no coward himself and has seen many brave men, he not only never saw, but never imagined anything like the bravery of this Russian colonel. And one of the Dutch who served under him, said that one couldn't but follow him, he is a real leader." Haggman, Bertil. "Haggman on Davidson and Filatova, 'The Russians and the Anglo Boer War.'" H-South Africa. Retrieved 22 February 2021.

[288] Ernest Beachcroft Beckwith Towse, Captain, (April 1864, June 1948) and educated at Wellington College. On the 30th April, 1900, Captain Towse, with twelve men, took up a position on the top of Mount Thaba, far away from support. A force of about 150 Boers attempted to seize the same plateau, neither party appearing to see the other until they were but one hundred yards apart. Some of the Boers then got within forty yards of Captain Towse and his party, and called on him to surrender. He at once caused his men to open fire, and remained firing himself until severely wounded (both eyes shattered), thus succeeding in driving off the Boers. The gallantry of this officer in vigorously attacking the enemy (for he not only fired, but charged forward) saved the situation, notwithstanding the numerical superiority of the Boers'. Captain Towse was decorated by Queen Victoria. Captain Towse became an expert typist, and when the European War broke out he went to the front to type letters for wounded soldiers, and was mentioned in Sir Douglas Haig's Despatch in June 1916. (VC and DSO book) And Beachcroft Towse - Wikipedia

[289] Hillegas, With the Boer forces, 1901 (Photograph by the Author)

At dawn on May-day fighting recommenced, and soon after six o'clock parties of the Gordons and Canadians succeeded in gaining possession of the two peaks of Thoba Mountain. Besides this, half a company of the Shropshires, under Colour-sergeant Sconse, managed to seize the nek between them, and though subjected to a severe cross fire, which caused in this small party ten casualties out of forty, maintained themselves stubbornly for four hours. The points which dominate the flat top of the mountain were thus gained.

Meanwhile reinforcements, consisting of the 8th Hussars,[290] a composite Lancer regiment, the East Yorkshire, and a field battery, had arrived from Thabanchu, and the approach of Bruce-Hamilton's force from the direction of Kranz Kraal was also felt. General Ian Hamilton now ordered Colonel Clowes,[291] commanding the Cavalry, to move right round Thoba Mountain and threaten the Boer line of retreat as a preliminary and accompaniment of the main Infantry assault, which had now become inevitable. Clowes's force was strengthened by the addition of a horse battery. The newly-arrived Infantry and the field battery had to be diverted to support the right and right rear, where the pressure was now very strong.

Brabant's Horse Artillery.[215]

At about eight A.M. General Smith-Dorrien had himself gone up to the top of Thoba Mountain to direct personally the decisive movement when the time should come. A little before one o'clock, the progress of the Cavalry being satisfactory, he determined to settle the matter, so that if successful the force might get its baggage over the pass before dark. He therefore formed a line of Infantry right across the plateau, two companies of the Shropshires in the centre, and one and a half company of the Gordons on either flank. The advance was sounded.

The troops moved forward with alacrity. For a few moments the fire was heavy, but the Boers knew themselves bested, and on the soldiers raising the cheer that precedes the actual assault they rushed to their horses, and the whole of Thoba Mountain was won. The rest of the position now became untenable, and the enemy, to the number of 4,000, promptly evacuated it, galloping swiftly back in the direction of Jacobsrust.

A few troops of the 8th Hussars alone got near enough to charge; half a dozen Dutchmen were sabered, and one was shot dead by an officer, Lieutenant Wylam.[292] The Boers who were making the attack on the right retreated at the same time as their comrades, and the transport, no longer molested, passed safely over the pass and parked for the night on the northern side. No trustworthy estimate can be formed of the enemy's loss; but a score of prisoners were taken, and an equal number of bodies were found on the position.

[290] The 8th King's Royal Irish Hussars along with the 7th Dragoon Guards and the 14th Hussars they formed the 4th Cavalry Brigade under Brigadier General Dickson. On 1 May 1900, the Boers made a stand in a strong position at Houtnek, where the forces of Ian Hamilton faced stiff competition. In a telegram of 2 May Roberts said: "Hamilton speaks in high terms of the services of the 8th Hussars under Colonel Clowes and a made-up regiment of Lancers, which came into Broadwood's brigade and assisted in making the Boers evacuate their position." The 8th then marched from Machadodorp to Heidelberg with the 14th Hussars and M Battery, under the command of Colonel Mahon. 8th King's Royal Irish Hussars - Wikipedia See Also: 7th Dragoon Guards - Wikipedia And: 14th King's Hussars - Wikipedia

[291] Clowes Peter Legh Lieutenant Colonel. QSA (6). Entered 1875 Lieutenant Colonel, 1897. Staff service: ADC to Commander in Chief, Bombay, 1890-91. War Service: Afghan War, 1879-80 (medal); Boer War, 1899-1900.

[292] Wylam Frederick Herbert Lieutenant. He was killed in action between Machadodorp and Heidelberg October 13th, 1900. He was the only son of Edward Wylam of Runnymede Park, Staines, was born in September 1876, and educated at Harrow. He entered the 8th Hussars in February 1897, being promoted Lieutenant March 1898. Lieutenant Wylam accompanied his regiment to South Africa in February 1900, and served in the ORC taking part in the advance on Pretoria, and the subsequent operations in the Transvaal. Source: Donner.

The British casualties were fortunately slight considering the fire and its duration, and did not exceed a hundred officers and men.

The next day the junction between the columns was effected, and Ian Hamilton's force formed, with reference to the main advance, the Army of the Right Flank, and was composed as follows:

Infantry. (19th Brigade) - Smith-Dorrien
(21st Brigade) - Bruce-Hamilton
Mounted Infantry. (1st Mounted Infantry Brigade)- Ridley
Cavalry. (2nd Cavalry Brigade) - Broadwood
Artillery (3 Batteries F.A.)
(2 Batteries H.A.) All under Waldron[293]
(2 5-in. Guns.)[294]

This force was supported by the Highland Brigade and two 4.7 naval guns, under General Colvile,[295] who was directed to follow the leading column at a distance of ten miles. Hamilton proposed to march forward on the 2nd of May, but an order from headquarters enjoined a halt; nor was it until the afternoon of the 3rd that the force reached Jacobsrust, as it is called by the inhabitants; Isabellasfontein, as our maps record. A little cavalry skirmishing in the neighbourhood of the camp resulted in the death of one Lancer.

On the 4th of May the whole army moved forward again, Lord Roberts passing through Brandfort towards Smaldeel, Hamilton continuing his march on Winburg. This day did not pass without fighting, for scarcely had the troops left camp when a patter of musketry warned the General that his Cavalry had become engaged. Riding forward, he was the witness of a very dashing cavalry exploit.

Across the line of advance was drawn up a strong force of the enemy, estimated at 4,000 men and thirteen guns. These, in a good position along a range of wooded bluffs, promised a sufficient task for the troops during the day. But now, suddenly, from the direction of Brandfort, a new army of Boers began to appear, riding swiftly down to join hands with their comrades athwart the road, and fall on the left flank of the column.

The thing was urgent, and perhaps vital. But between the fast converging Boer forces, at the angle where they would meet, ran a long ridge of indefinite extent. General Broadwood at once, without a moment's delay, galloped forward, and with two squadrons of the Guards' Cavalry and two of the 10th Hussars[296] seized it. The Boers were already scrambling up its lower slopes. A sharp fight immediately opened. Kitchener's Horse, hurrying up in support, occupied a further point of the ridge, and the Dutch, after a determined but futile attempt to clear the hill, fell back. The junction of the two Boer columns was prevented. It seems that the whole of their plan for the day was based on this first condition, and in an army where every individual soldier must have the details of any plan explained to him it is not easy to make fresh dispositions on the field.

Indeed, a sort of panic seems to have taken hold of the enemy, for without waiting for the Infantry attack to develop they fled forthwith at great speed, galloping madly across the drift—as the British proprietor of Welcome Farm told me—horsemen and guns, pell-mell, in downright rout, pursued, so swift was their departure, only by the shells of the Horse Artillery.

The losses in this brief affair were not large, and almost entirely among the Cavalry. In those few minutes of firing on the ridge about a dozen troopers had been hit. Lord Airlie[297] was slightly wounded in the arm, and Lieutenant Rose,[298] Royal Horse Guards,[299] was killed. He had been sent forward to see what lay beyond the further crest of the hill, and found that deadly riflemen lay there waiting for a certain victim. He fell pierced by

[293] F. Waldron Lieutenant Colonel Entered 1873; Lieutenant Colonel, 1899. Staff Service: DAAG, Canada, 1890-95. War service: Boer War, 1899-1900. Royal Artillery Article: https://www.angloboerwar.com/?option=com_content&view=article&id=582

[294] For full composition of the Field Force see Appendix.

[295] General Sir Henry Colville, Henry Edward Colvile - Wikipedia [a year later Churchill as MP would demand an inquiry over his dismissal from South Africa].

[296] 10th (Prince of Wales's Own Royal) Hussars Article: https://www.angloboerwar.com/?option=com_content&view=article&id=495 Also simply search Hussars at ABW.com for all Hussar Units. https://www.angloboerwar.com/name-search

[297] Lieutenant colonel David Stanley William Ogilvy, 11th and 6th Earl of Airlie (20 January 1856 – 11 June 1900) was a Scottish peer. David Ogilvy, 11th Earl of Airlie - Wikipedia

[298] Rose Adrian Lieutenant QSA (3). Source: QSA medal rolls Royal Horse Guards.

[299] Royal Horse Guards: https://www.angloboerwar.com/?option=com_content&view=article&id=562 697 Records.

several bullets, and lived only for half an hour. This officer was a most zealous soldier. Though possessed of private means which would have enabled him to lead a life of ease and pleasure, he had for several years devoted himself assiduously to the military profession. He went to India as a volunteer during the Tirah Campaign, and served with distinction on Sir Penn Symons' staff—general and aide-de-camp both vanished now, as the foam fades in the wake of a fast ship! From India he hastened to West Africa, and in that vile and pestilential region won a considerable reputation; indeed, he was to have received the Distinguished Service Order for his part in recent operations there had not another war intervened. He arrived at the Cape, scarcely a month ago, full of hope and energy. This is the end; and while it is one which a soldier must be ready to meet, deep sympathy will be felt for the father, from whom the public necessities have now required two gallant sons.

Though the disorderly and demoralised nature of the Boer flight through Welcome Farm was known throughout the British Army, it was not expected that so strong a position as the bluffs behind the Vet River[300] would be yielded without a shot fired.

This, nevertheless, proved to be the case, for when, on the morning of the 6th, Hamilton resumed his advance, he found that no force of the enemy stood between him and Winburg.

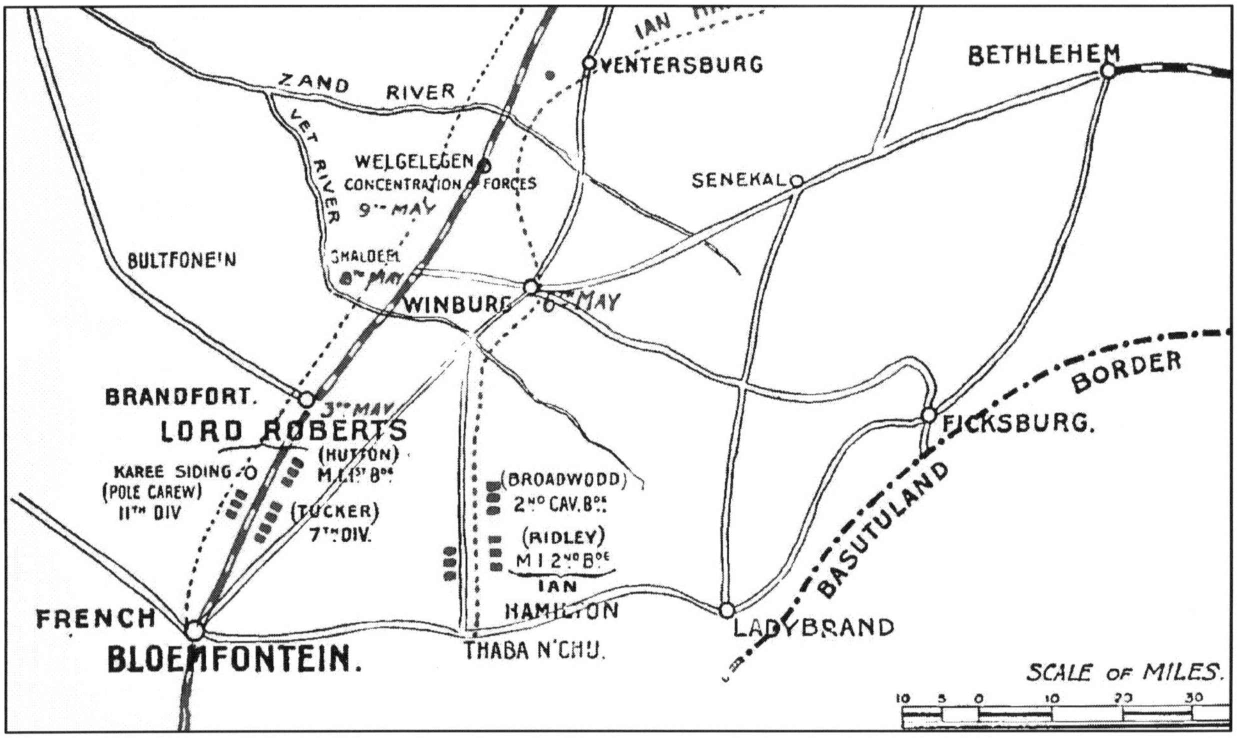

The Vet River passes just south of Winburg, north of Bloemfontein.

He therefore sent, shortly after noon, a staff officer, Captain Balfour[301] to wit, under flag of truce, with a letter to the mayor of the town summoning him forthwith to surrender the town and all stores therein, and promising that

[300] The Vet River (Afrikaans: Vetrivier) is a westward-flowing tributary of the Vaal River in central South Africa. Its sources are between Marquard and Clocolan and the Vet River flows roughly northwestwards to meet the Vaal at the Bloemhof Dam near Hoopstad. Vet River - Wikipedia

[301] Balfour John Edmond Heugh Captain, was born 22 January 1868, son of George Edmond Balfour. He was educated at Eton. He served in the South African War, 1899-1900, as ADC to the GOC, Mounted Infantry Division, and took part in the advance on Kimberley and in the Relief of Kimberley; operations in the Orange Free State, February to May. 1900, including operations at Paardeberg (17 to 26 February); actions at Poplar Grove, Driefontein, Houtnek (Thoba Mountain), Vet River (5 and 6 May) and Zand River; operations in the Transvaal in May and June 1900, including actions near Johannesburg, Pretoria and Diamond Hill (11 and 12 June); operations in Orange River Colony, May to July 1900, including actions at Bethlehem (6 and 7 July) and Wittebergen (1 to 28 July 1900). He was mentioned in Despatches [London Gazette, 10 September 1901]; was promoted to Major, Reserve of Officers; received the Queen's Medal with six clasps, and was created a Companion of the Distinguished Service Order [London Gazette, 27 September 1901]: "John Edmond Heugh Balfour, Captain and Honorary Major (Reserve of Officers), 1st Devon Yeomanry Cavalry. In recognition of services during the operations in South Africa." He served in the European War, 1914-18; was twice mentioned in Despatches, and created a CMG in 1918; He was Colonel, Royal 1st Devon Yeomanry Cavalry; Major, Reserve of Officers, late Captain, 11th Hussars. Colonel Balfour married, in 1910, Evelyn, daughter of Honourable R J Gerrard-Dicconson, and they had one daughter. Source: DSO recipients (VC and DSO Book).

if this were done he would use every effort to protect private property, and that whatever foodstuffs were required by the troops should be paid for. This message, which was duly heralded by the sound of a trumpet, concluded by saying that unless an acceptance was received within two hours the General would understand that his offer had been declined.

Thus accredited, Captain Balfour made his way into the town and was soon the centre of an anxious and excited crowd of burghers and others who filled the market square. The mayor, the landdrost, and other prominent persons—indeed, all the inhabitants—were eager to avail themselves of the good terms, and a satisfactory settlement was almost arranged when, arriving swiftly from the northeast, Philip Botha[302] and a commando of 500 men, mostly Germans and Hollanders, all very truculent since they were as yet unbeaten, entered the town.

A violent and passionate scene ensued. Botha declared he would never surrender Winburg without a fight. Dissatisfied with the attentions paid him by Captain Balfour, he turned furiously on him and rated him soundly. Several of the Free Staters had asked what would be done to them if they laid down their arms. Balfour had replied that they would be permitted to return to their farms, unless actually captured on the field. This Botha held to be a breach of the laws of war, and he thereupon charged the officer with attempting to suborn his burghers. What had he to say that he should not be made a prisoner? 'I ask favours of no Dutchman,' replied Balfour, sternly.

'Arrest that man!' shouted Botha, in a fury; 'I shall begin shooting soon.' At these shameful words a great commotion arose. The women screamed, the mayor and landdrost rushed forward in the hopes of averting bloodshed. The Boers raised their rifles in menace, and the unarmed British envoy flourished his white flag indignantly.

For several minutes it seemed that an actual scuffle, possibly a tragedy, would occur. But the influence of the townsfolk, who knew that their liberty and property lay in the hands of the Imperial General, and that the great siege guns were even then being dragged into effective range, prevailed, and Philip Botha, followed by his men, galloped furiously from the square towards the north.

That afternoon General Ian Hamilton entered Winburg at the head of his troops. Under a shady tree outside the town the mayor and landdrost tendered their submission and two large silver keys. The Union Jack was hoisted in the market-place amid the cheers of the British section of the inhabitants, and, as each battalion marching through the streets saw the famous emblem of pride and power, bright in the rays of the setting sun, these feeble or interested plaudits were drowned in the loud acclamations of the victorious invaders.

Hamilton was expected to arrive on the 7th, if no opposition was encountered, He had fought nearly every day, and reached the town on the evening of the 5th.

[302] Botha Philip Rudolf General He was born in 1851. He was noted for his conduct at Paardeburg. While in command of the Harrismith burghers, he made a breakthrough at Sprinkaan's Nek. He was killed in action while attacking the British position near Ventersburg in 1901.

X. The Army of the Right Flank

Kroonstadt: May 16, 1900.

On the same day that Ian Hamilton's force won their fight at Houtnek, to wit, the 1st of May, the advance of the main army towards Pretoria, long expected, long prepared, long delayed, began, and the Eleventh Division marched north from Bloemfontein to join the Seventh, which was entrenched at Karree Siding. On the 3rd both Infantry divisions moved forward along the railway, their left protected by Gordon's Cavalry Brigade and Hutton's[303] Mounted Infantry, and after a sharp cannonade drove the Boers from their positions covering Brandfort and entered the town.

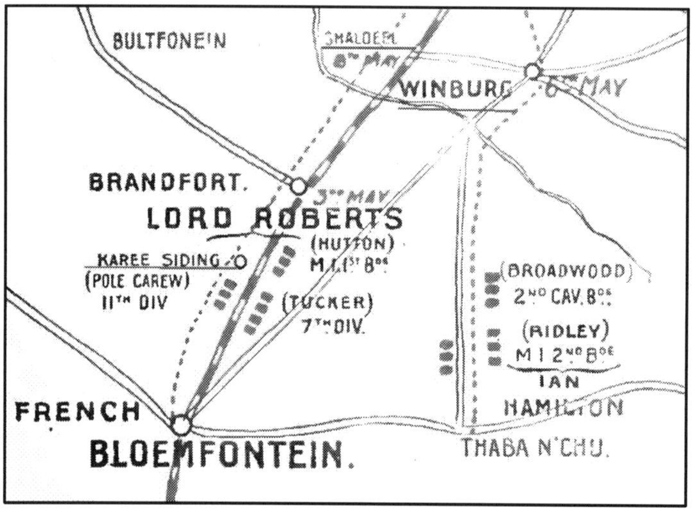

Brandfort area. Note Karee Siding and Smaldel at top. The Vet River flows just south of Winburg.

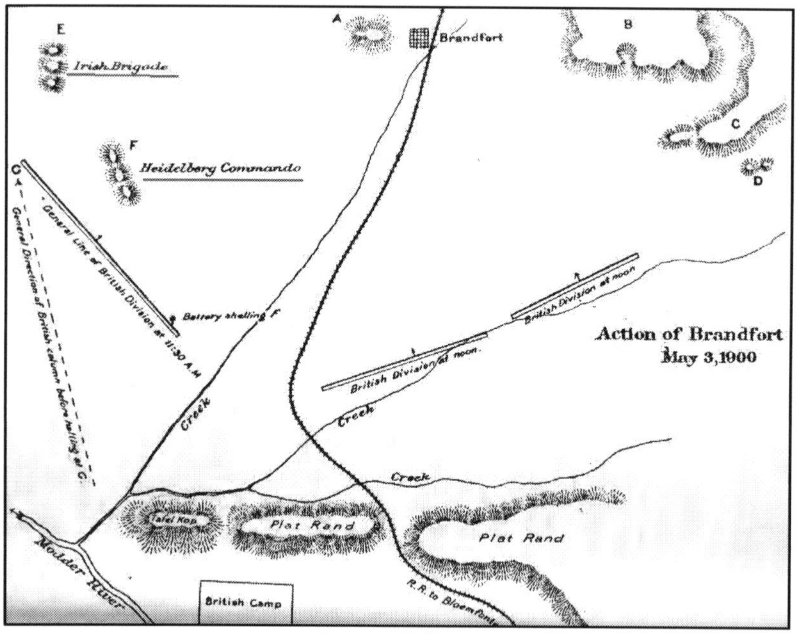

Action at Brandfort. May 3rd, 1900.[216] Note Irish Brigade and Heidelberg Commando participating upper left.[304]

[303] E. T. H. Hutton Colonel Entered 1867; Colonel, March 1900. Staff service: ADC to Major General Expedition Force, Egypt, 1882; Assistant Military Secretary to GOC, Egypt, 1882-83; Brigade Major, Aldershot, 1883-84; DAA and QMG, Egypt, 1884-85; DAAG Aldershot, 1887-89, 1689-92; ADC to the Queen, 1892; Commandant Colonel Forces, NSW, 1893-96; AAG, Dublin, 1896-97; Curragh, 1897-98; GOC Mila., Dominion of Canada, 1898-1900; Special Service, South Africa, 1900; Major General Infantry Brigade, South Africa, March 1900. War Commanding 1st Mounted Infantry Bde. Service: South African War, 1879-81 (Despatches; medal with clasp); Egyptian Expedition 1882 (Despatches; medal with clasp; bronze star; Brevet of Major; 4th class Medjidie); Sudan Expedition 1885 (clasp); Boer War, 1899-1900; on Staff. This gallant officer, who has energetically interested himself in the Colonial patriotic movement from its inception, was the son of Mr. E T Hutton of Beverley, Yorkshire. He was born in 1848, and married in 1889 the daughter of Lord Charles Paulet.

[304] Heidelberg Commando: https://www.angloboerwar.com/?option=com_content&view=article&id=3050

Col. John Blake[305] of the Irish Brigade.[217] The Chicago Irish-American Volunteers, (Col. Blake marked by X).[218]

The advance was resumed on the 5th, and the enemy were again met with, this time holding the line of the Vet River. Another artillery action ensued, in which the British 5-inch [Howitzers] and naval 4.7 guns were very effective, and at the end of which the West Australians and other parts of Hutton's Mounted Infantry force, pushed across the river in gallant style and captured an important kopje. The Dutchmen then retreated, and the Field-Marshal's headquarters on the 6th were fixed in Smaldeel. His losses since leaving Bloemfontein had not amounted to twenty-five men.

President Steijn (sic) [On Train] Exhorting His Burghers at Smaldeel.[219]

Marthinus Theunis Steyn. President of the Orange Free State 1898-1900.[220] President Steyn [Seated] with his Burghers.[221]

[305] John Young Filmore Blake, also known as John Y.F. Blake and J.Y.F. Blake was an Irish-American soldier and writer. He was born October 6, 1856, in Bolivar, Missouri in the United States, and died January 24, 1907, in New York City. Blake served as a foreign volunteer for the Boers of the South African Republic during the Second Boer War. John Blake (soldier) - Wikipedia

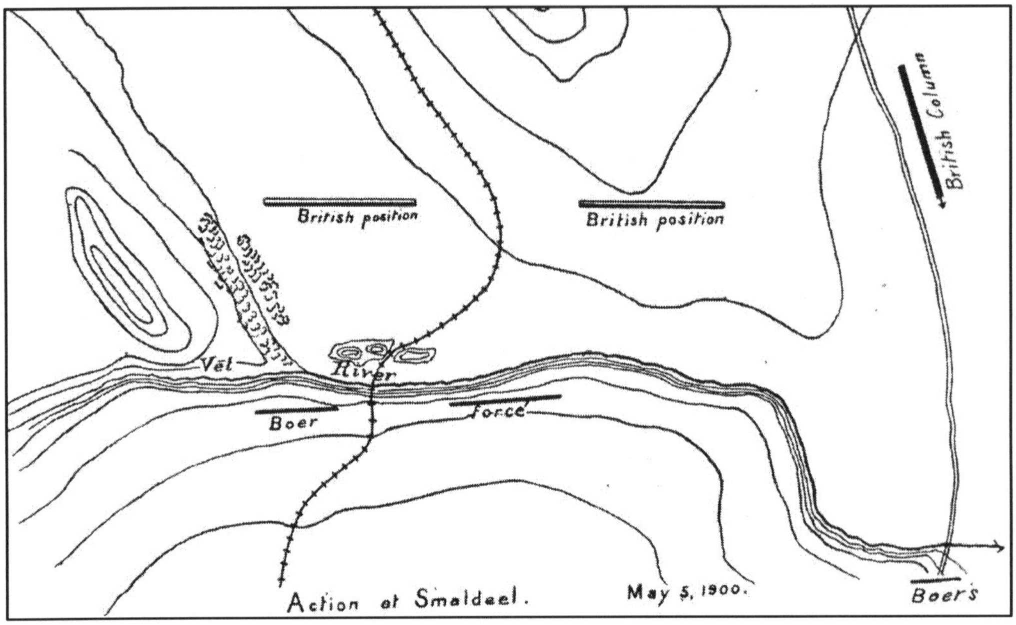

Action on Vet River at Smaldeel. May 5th, 1900.[222]

Five Inch Howitzers (Siege Guns).[223]

4.7 Inch Gun on Display.[224] Probably Captain Percy Scott in the dress whites.

We were just congratulating ourselves upon the success of these curious operations—curious because the drill books do not contemplate both sides fighting rear-guard actions at the same time—when half a dozen riderless horses galloped in from somewhere miles away on the right flank. Evidently sharp fighting was proceeding there; the flow of Boers had meant mischief. The peaceful landscape told no tale. No sound of musketry, nor sign of action could be distinguished. Indeed, in this scattered warfare one part of a force may easily be destroyed without the rest even knowing that a shot has been fired. 'Why scatter them?' asks the armchair strategist. 'Because if you don't scatter, and haven't got soldiers who are good enough to act when scattered, you will all get destroyed in a lump together.'

The General sent directions to the rear guard to communicate with the flank guard; kept another corps of Mounted Infantry handy to support either if necessary, and turned his attention to getting his brigades across the Rhenoster River.[sic] While this was proceeding the head of Smith-Dorrien's column, which had marched prosperously from their bivouac near Kaalfontein,[225] came into view, and the Army of the Right Flank stood again united, a fact which suggests some consideration of its functions in the general scheme of Lord Roberts's advance.

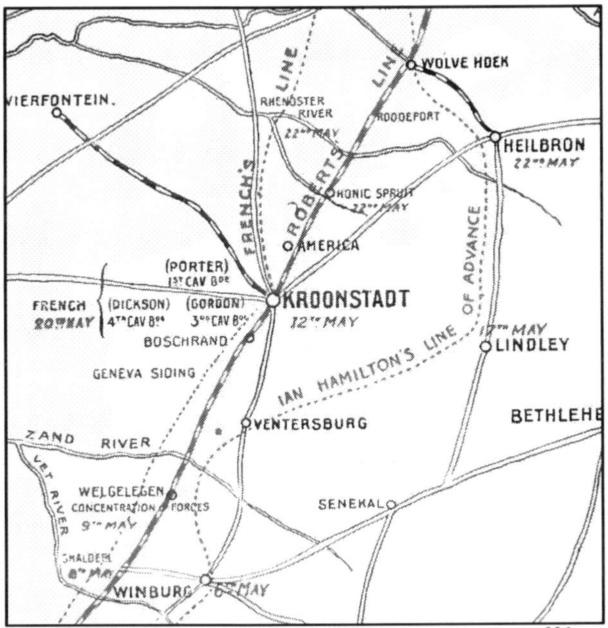

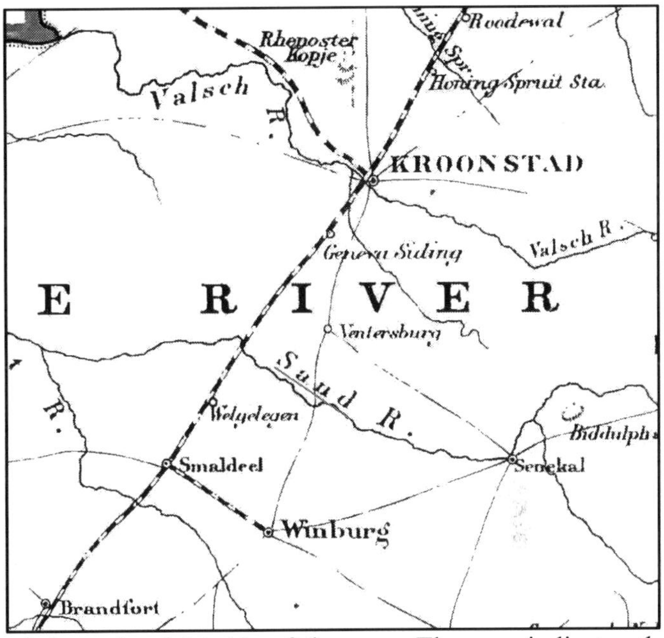

Map of March from Bloemfontein to Pretoria.[226] North is to the right side of the page. The map indicates that Kaalfonstein was reached on May 17th, Lindley on May 18th, and actions at the upper reaches of the Rhenoster River on May 20th and May 21st. His Despatch is datelined May 16th from Kroonstad. Second Map: The Rhenoster River is North of Kroonstadt. He must mean the Zand (Sand) River or the Valsch River.

Rhenoster Area.[227] Note Rhenoster Kopje on Valsch River. WSC has reason to think he is near this river.

After Kroonstadt had been captured the republican forces on the railway retreated to the line of the Rhenoster. Half a mile to the north of this river there rises abruptly from the smooth plain a long line of rocky hills, and in this strong position the Boers had determined to make a stubborn stand. Any force advancing along the railway would indeed have found it a difficult and costly business to cross the river and dislodge an enemy so posted.

Other low hills trending away to either flank would have made any turning movement an exceedingly extended and probably a useless operation, for the enemy being on the inside of the circle would have been able to confront the attack wherever it might fall. But the Rhenoster River, as the reader will see by a glance at the map, rises considerably south of the point where it intersects the railway; and so soon as Ian Hamilton's force was across it, the Boers holding the kopjes position were in considerable danger of being cut off.

The effect of our crossing the Rhenoster between Lindley and Heilbron should therefore be to clear the march of the main army. All fell out as Lord Roberts had expected; although the Boers had made great preparations to defend Rhenoster, had constructed strong entrenchments and made sidings to detrain their heavy guns, they evacuated the whole position without a shot being fired, compelled by the movement of a column forty miles away to their left flank.

Ian Hamilton, in spite of the long marches his troops had made, was impatient to push on from Winburg without delay, and, following the track to Ventersburg,[306] to seize the drifts across the Sand River,[307] twenty miles to the north. The great speed of his last movement had outpaced the Boers, and their convoys were struggling along abreast of, and even behind, the British column, trying vainly to slip across our front, and join the burgher forces accumulating for the defence of Kroonstadt. By marching forthwith—great though the strain might be—the General hoped to secure the bloodless passage of the river, and perhaps cut up some of these same toiling convoys. Accordingly, having collected from the town about three days' stores—Sir Henry Colvile helping him unselfishly with mule waggons—he set his brigades in motion on the afternoon of the 6th, and marched nine miles towards the Sand. [River]

Column of the Royal Horse Artillery.[228]

Column of the Royal Artillery on the move.[229]

But Lord Roberts had decided to remain at Smaldeel until his temporary bridge over the Vet River was made and the trains running, and he did not choose to run the risk of the Boers concentrating all their forces upon any single division of his army, such as would be incurred if Hamilton pushed forward alone.

Vet River Bridge before repairs.[230]

[306] Ventersburg is a small town in the Lejweleputswa District Municipality, 148 km north of Bloemfontein. Ventersburg - Wikipedia
[307] Sand River (Free State) - Wikipedia

The principle was indisputable; but, of course, in practice it resolved itself into another instance of balancing drawbacks, for delay gave the enemy time to get his breath, and meant that the Sand River passage would be opposed. Besides, if the Boers had flung all their strength upon Hamilton, we were 7,000 bayonets, 3,000 horse, and nearly forty guns, and would have beat them off with a shocking slaughter. To us it seemed a great pity to wait; but to the Chief, in whose eyes the Army of the Right Flank was but one column of that far-flung line which stretched from Rundle near Senekal,[308] along the front of the main army to Methuen near Boshof, Hunter at Warrenton, and Mahon far away on the fringe of the Kalahari desert, it must have been a very small matter, and certainly not one justifying any loss of cohesion in the general scheme. So I have no doubt that it was right to make us halt on the 7th and 8th.

President Steyn [center] on his Way to Sand River Battle. PF: That appears to be Christiaan De Wet on the left side. President Steyn of the Orange Free State.[309]

On the former of these two days of rest Lord Roberts sent for General Hamilton to meet him at a point on the branch railway line mid-way between Winburg and Smaldeel, and they had a long private conference together. On the 9th, the whole army marched forward again towards the Sand River. I rode with the General, who managed somehow to find himself among the cavalry patrols of the right flank guard, and we watched with telescopes three long lines of dust in the eastward, which, under examination, developed into horsemen and waggons marching swiftly north and turning more and more across our front. [Northwest] It was clear that if we had pushed on without halting, all these commandos would have been prevented from reaching Kroonstadt. The General contemplated them hungrily for some time, but they were too far off to attack, bearing in mind the great combination of which we were a part. The flanking patrols, however, exchanged a few shots.

The march was not a long one, and by mid-day we reached the halting-place, a mile south of the [Sand] river. The headquarters were fixed in a large farm which stood close to the waggon-track we followed.

This farmhouse was certainly the best purely Dutch homestead I have ever seen in the 500 miles I have ridden about the Free State. It was a large square building, with a deep verandah, and a pretty flower-garden in front, and half a dozen barns and stables around it. The construction of a dam across the neighbouring spruit had formed a wide and pleasant pool, in which many good fat ducks and geese were taking refuge from the wandering soldier. At the back, indeed, on all sides but the front of the farm, rose a thick belt of fir trees. Within the house the ground-floor was divided into three excellent bedrooms, with old-fashioned feather-beds and quaint wooden bedsteads, a prim but spacious parlour, a kitchen, pantry, and storeroom. The parlour deserved the greatest attention.

[308] Senekal is a town situated on the banks of the Sand River in the eastern part of the Free State province of South Africa. It was named after Commandant FP Senekal. Senekal - Wikipedia

[309] Richard Harding Davis With Both Armies in South Africa. Charles Scribner's Sons, New York 1900.

The furniture was dark and massive. The boards of the floor were deeply stained. In the middle was a good carpet upon which an ample oval table stood. The walls were hung with curious prints or coloured plates, and several texts in Dutch. One pair of plates I remember represented the ten stages of man's life and woman's life, and showed both in every period from the cradle to the grave, which latter was not reached until the comfortable age of one hundred. The woman's fortunes were especially prosperous. At birth she sprawled contentedly in a cradle, while loving parents bent over her in rapture, and dutiful angels hung attendant in the sky. At ten she scampered after a hoop. At twenty she reclined on the stalwart shoulder of an exemplary lover. At thirty she was engaged in teaching *seven* children their letters. At forty, she celebrated a silver wedding. At fifty, still young and blooming, she attended the christening of a grandchild. At sixty, it was a great-grandchild. At seventy she enjoyed a golden wedding. [anniversary] At eighty she was smilingly engaged in knitting. Even at ninety she was well preserved, nor could she with reason complain of her lot in life when, at a hundred, the inevitable hour arrived. 'Be fruitful and multiply,' was the meaning of a Dutch text on the opposite wall, and a dozen children black and white (little Kaffirs, the offspring of the servants, playing with the sons and daughters of the house) showed that the spirit of the injunction was observed; and these are things with which the statesman will have to reckon.

Boer farm in the neighborhood of the Modder River.[231] Note trees to shade house on probably the North side in the Southern Hemisphere.

Prosperous Boer Farm.[232]

The inmates of the farm consisted of the old man, a venerable gentleman of about sixty years, his dame, a few years younger, three grown-up daughters, a rather ill-favoured spinster sister, and seven or eight children or grandchildren of varying ages. There were in all seven sons or grandsons—two were married and had farms of their own; but all, including even one of fourteen, were 'on commando' at the wars, some, perhaps, looking at us and their home from the heights across the river.

The General politely requested shelter for the night, and a bedroom and the parlour were placed at his disposal; not very enthusiastically, indeed, but that was only natural. The staff settled down in the verandah so as not to disturb the family.

Ian Hamilton, keenly interested in everything, began at once to ask the old lady questions through an interpreter. She gave her answers with no good grace, and when the General inquired about her youngest fighting son—he of fourteen—her sour face showed signs of emotion, and the conversation ended for the day. On the morrow, however, just before he crossed the river, he had to come back to the telegraph-tent pitched near the farm, and found time to see her again.

'Tell her,' he said to the interpreter, 'that we have won the battle to-day.'

They told her, and she bowed her head with some dignity.

'Tell her that the Dutch will now certainly be beaten in the war.'

No response.

'Perhaps her sons will be taken prisoners.'

No answer.

'Now tell her to write down on a piece of paper the name of the youngest, and give it to my aide-de-camp; and then when he is captured she must write to me or to the Hoofd-General, and we will send him back to her, and not keep him a prisoner.'

She thawed a little at this, and expressed a hope that he had been comfortable while beneath her roof, and then—for the guns were still firing—he had to hurry away. But the aide-de-camp remained behind for the paper.

During the time we spent in this homely place I made a thorough inspection of the farm, especially the parlour, where I found one very curious book. It was a collection of national songs and ballads, compiled, and in part written, by Mr. Reitz.

I afterwards succeeded in buying another copy in Ventersburg; indeed, it has been widely disseminated. The first part consists of patriotic Boer poems—the Volkslied, the Battle of Majuba, the Battle of Laings Nek, and other similar themes. The second half of the book is filled with Reitz's translations of English songs and well-known ditties into the *taal*.[310] John Gilpin,[311] besides being a burgher of credit and renown, was eke [not a misprint] a Field-Cornet of famous Bloemfontein. Young Lochinvar had come from out of the Boshof district. The Landdrost's daughter of Winburg found a lover no less faithful than a famous swain of Islington. The pictures were mightily diverting. The old Field-Cornet Gilpin— 'Jan Jurgens,' as he called himself now—was shown galloping wildly along, on a pulling Basuto pony, through the straggling streets of, let us say, Ventersburg, his slouch hat crammed over his eyes, his white beard flapping in the wind, while a stately *vrouw*,[312] four children, and a Kaffir, flung up their hands in mingled wonder and derision.

Francis William Reitz, 1890.[233]

A Similar English Illustration for The Diverting History of John Gilpin.[234 313]

[310] From the earlier tale.

[311] John Gilpin was featured as the subject in a well-known comic ballad of 1782 by William Cowper, entitled The Diverting History of John Gilpin. Cowper had heard the story from his friend Lady Austen. John Gilpin - Wikipedia

[312] His spouse.

[313] The Diverting History of John Gilpin Shewing how he went Farther than he intended, and came safe Home again is a comic ballad by William Cowper written in 1782. The ballad concerns a draper called John Gilpin who rides a runaway horse. Cowper heard the story from Lady Anna Austen at a time of severe depression, and it cheered him up so much that he put it into verse. The Diverting History of John Gilpin - Wikipedia

One piece began:

Engels! Engels! alles Engels! Engels wat jij siet en hoor.

Ins ons skole, in ons kerke, word ons modertaal vermoor.[314]

I cannot read Dutch, but the meaning and object of the book were sufficiently clear without that knowledge.

F. W. Reitz, sometime President of the Free State, now State Secretary of the Transvaal, looked far ahead, and worked hard. This, the foundation-stone of a vernacular literature, was but one act in the long scheme of policy, pursued, year in year out, with tireless energy, and indomitable perseverance, to manufacture a new Dutch nation in South Africa—the policy which, in the end, had brought a conquering army to this quiet farm, and scattered the schemers far and wide. But what a game it must have been to play! Only a little more patience, a little less pride and over-confidence, concessions here, concessions there, anything to gain time, and then, some day—a mighty Dutch Republic, 'the exchange of a wealthier Amsterdam, the schools of a more learned Leyden,' and, above all—no cursed Engels.

I was considering these matters, only suggested here, when messengers and the sound of firing came in from the eastward. The news that small parties of Boers were engaging our right flank guard did not prevent Hamilton riding over to meet the Chief, [Roberts] nor tempt us to quit the cool verandah of the farm; but when, suddenly, at about three o'clock, fifty shots rang out in quick succession, scarcely 500 yards away, every one got up in a hurry, and, snatching pistols and belts, ran out to see what mischance had occurred. The scene that met our eyes was unusual. Down the side of the hill there poured a regular cascade of antelope—certainly not less than 700 or 800 in number—maddened with fear at finding themselves in the midst of the camp, and seeking frantically for a refuge. This spectacle, combined with the hope of venison, was too much for the soldiers, and forthwith a wild and very dangerous fire broke out, which was not stopped until fifteen or twenty antelopes were killed, and one Australian Mounted Infantryman wounded in the stomach. The injury of the latter was at first thought to be serious, and the rumour ran that he was dead; but, luckily, the bullet only cut the skin.

Thus disturbed, I thought it might be worth while to walk up to the outpost line and see what was passing there. When I reached the two guns which were posted on the near ridge, the officers were in consultation. Away across the Sand River, near two little kopjes, was a goodly Boer commando. They had just arrived from the east of our line of march, and having skirted round our pickets had set themselves down to rest and refresh. Spread as they were on the smooth grass, the telescope showed every detail. There were about 150 horsemen, with five ox waggons and two guns. The horses were grazing, but not off-saddled.

Danie Theron's Scouts: at Dinner.[235]

The men were lying or sitting on the ground. Evidently they thought themselves out of range. The subaltern commanding the guns was not quite sure that he agreed with them. Some Colonial Mounted Infantry officers standing near were almost indignant that the guns should let such a chance slip. The subaltern was very anxious

[314] English! English! all English! English that you see and hear. In our schools, in our churches, we are being murdered in modern language. Google Translation to Afrikaans or Dutch returns the same result.

to fire—'really think I could reach the brutes'; but he was afraid he would get into trouble if he fired his guns at any range greater than artillery custom approves.

His range finders said '6,000.' Making allowances for the clear atmosphere, I should have thought it was more. At last he decided to have a shot. 'Sight for 5,600, and let's see how much we fall short.' The gun cocked its nose high in the air and flung its shell accordingly. To our astonishment the projectile passed far over the Boer commando, and burst nearly 500 yards beyond them: to our astonishment and to theirs. The burghers lost no time in changing their position.

The men ran to their horses, and, mounting, galloped away in a dispersing cloud. Their guns whipped up and made for the further hills. The ox waggons sought the shelter of a neighbouring donga. Meanwhile, the artillery subaltern, delighted at the success of his venture, pursued all these objects with his fire, and using both his guns threw at least a dozen shells among them. Material result: one horse killed. This sort of artillery fire is what we call waste of ammunition when we do it to others, and a confounded nuisance when they do it to us.

After all, who is there who enjoys being disturbed by shells just as he is settling himself comfortably to rest, after a long march? And who fights the better next day for having to scurry a mile and a half to cover with iron pursuers at his heels?

Even as it was an opportunity was lost. We ought to have sneaked up six guns, a dozen if there were a dozen handy, all along the ridge, and let fly with the whole lot, at ranges varying from 5,000 to 6,000 yards with time [fused] shrapnel. Then there would have been a material as well as a moral effect. 'Pooh,' says the scientific artillerist, 'you would have used fifty shells, tired your men, and disturbed your horses, to hit a dozen scallywags and stampede 150. That is not the function of artillery.' Nevertheless, function or no function, it is war, and the way to win war. Harass, bait, and worry your enemy until you establish a funk. Once he is more frightened of you than you are of him, all your enterprises will prosper; and if fifty shells can in any way accelerate that happy condition, be sure they are not wasted.

The afternoon passed uneventfully away, though the outposts were gradually drawn into a rifle duel with the Dutch sharpshooters in the scrub across the river.

Representative of the landscape. This from the Modder River Battlefield.[236]

In the evening the General returned from his conference with Lord Roberts, and told us the passage was to be forced on the morrow all along the line. The Army of the Right Flank would cross by the nearest drift in our present front. The Seventh Division inclining to its right would come into line on our left. The Field-Marshal, with the Guards and the rest of Pole-Carew's Division, would strike north along the line of the railway. French, with two Cavalry brigades and Hutton's Mounted Infantry brigade, was to swing around the enemy's right and push hard for Ventersburg siding. Broadwood from our flank, with the Second Cavalry Brigade, and such of the Second Mounted Infantry Brigade as could be spared, was to be thrust through as soon as the Boer front was broken, and try to join hands with French, thus, perhaps, cutting off and encircling the Boer right. The diagram—it is not a map—[on page 172 (of original publication)] will help to explain the scheme.

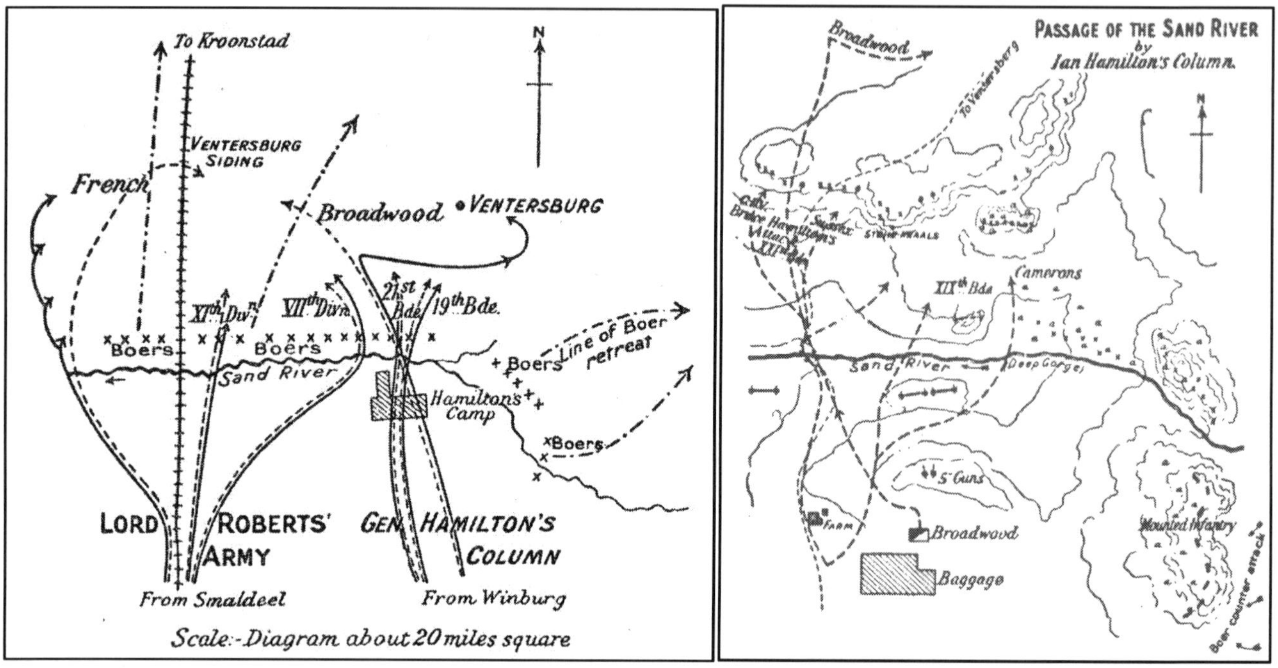

Diagram to Explain the Passage of the Sand River,[315] May 10th, 1900. The dotted lines show what was proposed, the continuous lines show what was done. The crosses indicate the Boers. Scale is 4.47* 4.47 Miles for 20 Square Miles.[237] Ian Hamilton's Action at the Sand River, May 10, 1900.[238] The crosses indicate the Boers.

The operation of the next day was one of the largest and most extended movements of the war, although, probably from this cause, it was attended by very little loss of life. Upon the British side six Infantry and six Mounted brigades, with rather more than 100 guns, were brought into action along a front of over twenty-five miles. The Boers, however, still preserved their flanks. Upon the west they succeeded in holding up [General] French, and on the east they curled round Hamilton's right and rear so that his action here, which in its early stages resembled that afterwards fought at Diamond Hill, was of a piercing rather than a turning nature. But in thus amazingly extending their scanty forces, which, altogether, did not number more than 9,000 men, with twenty-five guns, the enemy became so weak all along their front that the attacking divisions broke through everywhere, as an iron bar might smash thin ice, with scarcely any shock.

On the evening of the 10th, the British forces, in their extended line, lay spread along the south bank of the river, just out of cannon shot of the Boer positions on the further side. French, indeed, did not rest content with securing his ford twelve miles to the west of the railway, but pushed his two brigades across before dark. The wisdom of this movement is disputed. On the one hand, it is contended that by crossing he revealed the intention of the Commander-in-Chief, and drew more opposition against himself the next day. On the other, it is urged that he was right to get across unopposed while he could, and that his purpose was equally revealed, no matter which side of the river he stayed.

During the night Ian Hamilton, at the other end of the line, seized the drift in his front with a battalion, which promptly entrenched itself. Tucker, [316] who proposed to cross near the same point, despatched the Cheshire regiment[317] for a similar purpose. The single battalion was sufficient; but the importance and wisdom of the

[315] Sand River (Free State) - Wikipedia

[316] Tucker C Major General Entered 1855; Major General, 1893. Staff service: Colonel on Staff, Natal, 1891-93; Brigadier General, Natal, 1893-95; Major General, India, 1895-99; Lieutenant General Infantry Division, South Africa, 1899. War service: Bhootan Expedition 1865-66 (medal with clasp); Boer War, 1878-79 (Despatches, April and August 1879; medal with clasp; CB); Boer War, 1899-1900; on Staff. General Tucker, born in 1838, was a son of Mr. Tucker of Ashburton and a daughter of Mr. Hayter, Painter-in-Ordinary to Queen Victoria. As a practical, resourceful, and rough-and-ready soldier, he has no equal. He was rewarded in 1896 for "distinguished and meritorious service with a good-service pension. He was Colonel of the South Staffordshire Regiment from 1911.

[317] Cheshire Regiment Article: https://www.angloboerwar.com/?option=com_content&view=article&id=541 7,687 Records

At the Zand River the battalion was sent forward on the evening of the 9th May to seize and hold a drift. This they successfully accomplished, and the brigade crossed early on the 10th, clearing the enemy from strong positions north of the river.

movement was proved by the fact that the enemy during the night sent 400 men to occupy the river bank and hold the passage, and found themselves forestalled.

At daybreak the engagement was begun along the whole front. I am only concerned with Ian Hamilton's operations; but, in order that these may be understood, some mention must be made of the other forces. French advanced as soon as it was light, and almost immediately became engaged with a strong force of Boers, who barred his path, and prevented his closing on the railway as intended. A sharp Cavalry action followed, in which the Boers fought with much stubbornness; and the Afrikander Horse, a corps of formidable mercenaries, even came to close quarters with Dickson's brigade, and were charged.

French persevered throughout the day, making very little progress towards the railway, but gaining ground gradually to the north. Although his casualties numbered more than a hundred, he was still some distance from Ventersburg siding at nightfall. The centre attack properly awaited the progress of the flanking movements, and was, during the early part of the day, contented with an artillery bombardment, chiefly conducted by its heavy guns. Tucker and Hamilton, however, fell on with much determination, and were soon briskly engaged.

Ian Hamilton began his action at half-past five, with his heavy guns, which shelled the opposite heights leisurely, while the Infantry and Cavalry were moving off. The Boer position before us ran along a line of grassy ridges, with occasional kopjes, which sloped up gradually and reached their summits about a mile from the river. But besides this position, which was the objective of the force, the Boers, who held all the country to the east, began a disquieting attack along our right and right rear, and although the Mounted Infantry, and principally Kitchener's Horse, under Major[John] Fowle,[318] held them at arm's length throughout the day, the firing in this quarter caused the General some concern, and occupied the greater part of his attention.

At six o'clock the Twenty-first Brigade began to cross the river, and Bruce-Hamilton, stretching out to his left, soon developed a wide front. The Boers now opened fire with two or three field guns and a 'pom-pom,' which latter was quickly silenced by our heavy pieces. At the same time, the Nineteenth Brigade, who were containing the enemy's left, became engaged with their skirmishers in the scrub by the river. The four batteries of Field Artillery also came into action, and were pushed forward across the drift as soon as sufficient space was gained by the Infantry. At a little after seven the head of General Tucker's Division appeared on the plain to our left, and that determined officer thrust his men over the river in most vigorous style. Moreover, seeing Bruce-Hamilton committed to an assault, he swung two of his own batteries round to the eastward, and so rendered us material assistance.

Both Smith-Dorrien, who directed the two Infantry brigades, and Ian Hamilton were fully alive to the grave dangers of crowding too many troops on to a narrow front, and the Infantry attack was very sparingly fed with supports, until it became completely extended. This condition was attained about eleven o'clock, when the Camerons were sent across the river to clear the scrub and prolong the line to the right.

Bruce-Hamilton now had his deployment completed, and with an admirable simultaneity the whole of the assaulting Infantry rose up and advanced together upon the enemy's position, covered by the heavy fire of twenty-six guns. The panorama was now very extensive. Far away to the left the smoke of lyddite shells, and the curious speck of the war balloon high in the clear air, showed that the centre was engaged. The whole of the Seventh Division had crossed the Sand, and were now curving to the north-west amid a crackle of fire. Before us the slopes were sprinkled with brown dots moving swiftly upwards. The crest of the ridge was fringed with exploding shells. For a few minutes the Boers fired steadily, and the dust jumped amid the Sussex Regiment[319] and the City Imperial Volunteers.[320] But both Infantry and Artillery attacks were far beyond the capacity of the defence to resist, and by noon the whole of the heights beyond the Sand were in the British possession.

Ian Hamilton had meanwhile ordered baggage and Cavalry to cross. Broadwood was over the enemy's position almost as soon as the Infantry. He proceeded to move in the direction of Ventersburg siding. The enemy, however, had covered themselves with a strong rear guard, and the Cavalry were soon opposed by three guns and a force of riflemen of considerable numbers. Whether Broadwood would have thought it worth while to

[318] John Fowle Major Landed at Cape town with Mules from New Orleans, 06 Feb 00. With Kitchener's Horse, 12 Feb 00 - 23 May 00. Source: QSA medal rolls. 21st Lancers.

[319] Royal Sussex Regiment Article: https://www.angloboerwar.com/?option=com_content&view=article&id=668 486 Records.

[320] City Imperial Volunteers Article: https://www.angloboerwar.com/?option=com_content&view=article&id=542 2,084 Records.

make here the effort which he afterwards made in the action of Diamond Hill, and order a charge, is uncertain; for at this moment a misunderstanding arose which induced him to change his plans altogether.

The Boer pressure on our right rear had been growing stronger and stronger all the morning, and at length Hamilton, wishing to check the enemy sharply, so as to draw his rear guard over the river after his baggage, told his chief of artillery to find him a battery. Now it happened that only one of the two horse batteries, 'P,' had been able to go with the Cavalry, the other, 'Q,' being too tired to keep up. The chief of artillery therefore proposed to send for the tired battery. Unfortunately, by some mistake, either in giving or taking the order, the orderly was sent for 'P' instead of 'Q.' The man, a sergeant-major, galloped across the river, and, understanding that the matter was urgent, hurried after Broadwood, overtook him just as he was becoming engaged, and demanded the battery. Broadwood, who knew that Hamilton would never deprive him of his guns except for some very urgent reason, sent them at once, abandoned his movement to the north-west, which indeed was now impracticable without artillery, and concluding that the rear guard was seriously involved, turned sharply to the east to assist them. Explanations arrived too late to make it worth while to revert to the original plan, and, perhaps, seeing that French was unable to make Ventersburg siding, it was just as well that Broadwood did not try alone.

Broadwood's latest movement, or the action of the artillery, or the knowledge that the British had successfully forced the passage of the river at all points, induced the Boers who were assailing the rear guard to desist, and the musketry in that quarter gradually died away. Meanwhile, by the exertions of Lieutenant-Colonel Maxse,[321] the baggage had mostly been dragged across the river, and Ian Hamilton made haste to overtake his victorious Infantry, who had already disappeared into the valley beyond the enemy's position. By the time that we reached the top of the high ground, Bruce-Hamilton's leading battalions were nearly a mile further on, and the tail of Broadwood's brigade was vanishing in a high cloud of dust to the eastward.

Troops of one of the British mounted regiments on the move.[239]

The City Imperial Volunteers, who had lost a few men in the attack, were resting on the hill after their advance, and eating their biscuits. Several dead Boers had been found lying among the rocks, and a burial party was at work digging a grave for these and for four of our own men who had fallen close by. There were also a few prisoners—Transvaalers for the most part—who had surrendered when the troops fixed bayonets. Four miles away to the north-east the trees and houses of Ventersburg rose from a grassy hollow.

The General decided to bivouac in the valley beyond the enemy's position, and to set his pickets upon the hills to the northward. He also sent an officer with a flag of truce into Ventersburg to demand the surrender of the town, and directed Broadwood to detach a regiment and some Mounted Infantry to occupy it, should the enemy comply. In case they should desire to hold the town the 5-inch guns were brought into position on the captured heights.

[321] General Sir Frederick Ivor Maxse, KCB CVO DSO DL (22 December 1862 – 28 January 1958) was a senior British Army officer who fought during the First World War, best known for his innovative and effective training methods. Ivor Maxse - Wikipedia

Flag of Truce by Clive M. Dixon.[240]

Boer prisoners awaiting transportation to a POW camp.[241]

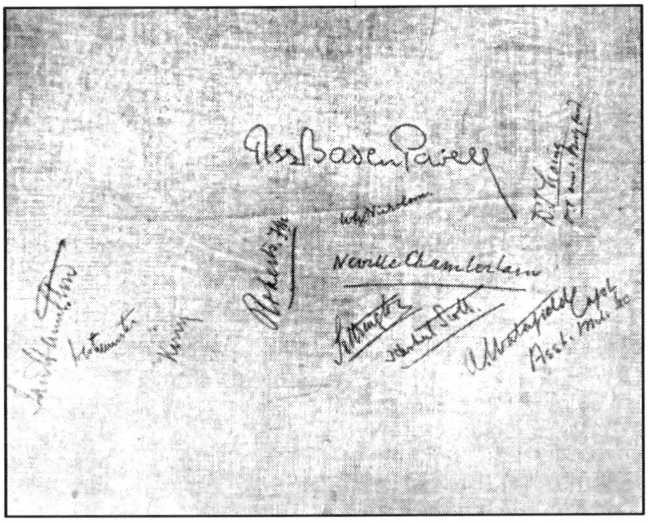

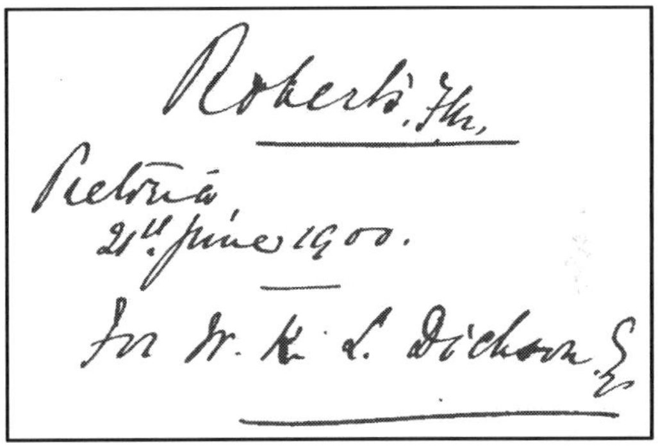

Flag of Truce:[242] What is most instructive is how the flag of truce was authenticated. It is inscribed with the following signatures for authenticity: Roberts (Vertical in Center), Baden Powell, Roberts' Secretary Neville Chamberlain. Kerry, Nicholson, Scott, and others. . Roberts' Autograph for comparison.[243]

Hoping to secure some supplies, particularly bottled beer, before everything should be requisitioned by the army, I rode forward after the flag of truce had gone in and waited where I could see what followed. When, about an hour later, a cavalry force began to advance from the direction of Broadwood upon the town, I knew that all was well, and trotted on to join them. My road led me within a few hundred yards of the town, but, luckily for me, I did not enter it alone, and hurried to join the troops. All of a sudden the ominous patter of rifle shots broke the stillness of the evening, and, turning to whence the sound came, I saw a score of Boers standing on the skyline about a mile away and firing at the advancing Cavalry, or, perhaps, for I was much nearer, at me.

The next minute there galloped out of the town about a score of Dutchmen, who fled in the direction of their friends on the western skyline. Had I ridden straight into the town I should have run into these people's jaws. I lost no time in joining the Cavalry, and entered the streets with the squadron of Blues.[322] It was a miserable little place, not to be compared with Winburg. There were a few good stores and a small hotel, where I found what I sought; but the whole town was very dirty and squalid. Thirty or forty troopers of Roberts's Horse were firing at the fugitive burghers from the edge of the buildings and gardens, while a score of reckless fellows were galloping after them in excited pursuit. The Boers on the hill kept up a brisk fire to help their comrades in, and not a few of the bullets kicked up the dust in the village streets, without in the least disturbing the women and children who crowded together to look at the war, in blissful ignorance of their danger. When some of these people were told that they would perhaps be killed if they came out of their houses while the fighting was going on, they clutched their children and sought shelter with an energy at which, since, after all, nobody was hurt, it was pardonable to laugh.

[322] The Blues and Royals (Royal Horse Guards and 1st Dragoons) (RHG/D) is a cavalry regiment of the British Army, part of the Household Cavalry. Blues and Royals - Wikipedia

Night put an end to all skirmishing, and under its cover the Boers retreated—the greater part to Kroonstadt, which, be it remembered, they meant to hold to the death; but a considerable proportion to the east, where they collected with the commandos under Christian de Wet.

Broadwood's brigade had captured about a dozen waggons and thirty prisoners. In all there were fifty-two unwounded and seven wounded Boers in our hands at the end of the day. The casualties in Hamilton's force were under fifty. Tucker and Pole-Carew may have lost the same number between them. French, who encountered the most stubborn resistance, had a little over 120. But, in any case, the passage of the Sand River in this long straggling action was cheaply won at a cost of under 250 officers and men.

All our beasts were so exhausted by the labour of dragging the waggons through the steep and rocky drift of the Sand, and by the long pull up the hills on the opposite side, that few of the regiments got their baggage that night, and hence it was impossible to make an early start next morning. But it was known that the Field-Marshal meant to reach Kroonstadt on the next day, and as all the information at our disposal indicated that the Boers were entrenching a strong position along a line of wooded bluffs called the Boschrand, just south of the town, every minute of halt was grudged.

We moved at eleven o'clock, heading direct for Kroonstadt, and persevered for two hours after the sun had set, making in all nearly seventeen miles. The country to our left was flat and open, and as we converged upon the main army we could see, like red clouds with the sunset behind them, the long parallel lines of dust, which marked the marches of the Seventh and Eleventh Divisions; and we knew besides, that, beyond both columns and west of the railroad, French was driving his weary squadrons forward upon another wide swoop. The army drew together in the expectation of a great action. But for all our marching we could never make up the extra distance we had to cover in coming diagonally from the flank, and as darkness fell we realised that the Seventh Division was drawing across our front, and that Pole-Carew with the guard was striding along ahead of us all. That night Lord Roberts slept at America Siding, [Geneva Siding] scarcely six miles from the Boschrand position.

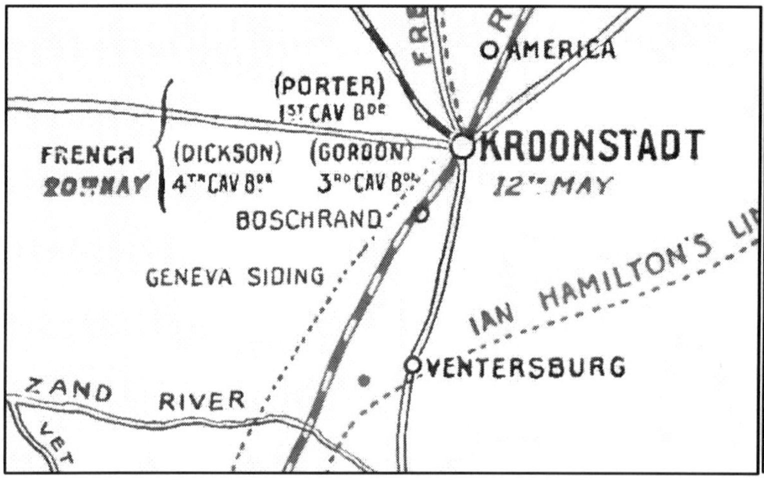

Geneva Siding is just south of the Boschrand station. America (siding) is north of Kroonstadt. Lord Roberts' infantry crossing the Zand River - balloon watching ahead for the Boers.[323]

Ian Hamilton marched on again at dawn, [May 11th ?] transport and convoys struggling along miles behind, and the fine-drawn yet eager Infantry close upon the heels of the Cavalry screen. At times we listened for the sound of guns, for if the enemy stood, the Field-Marshal must come into contact with them by eight o'clock. And when, after nine o'clock, no cannonade was heard, the rumour ran through the army that the Boers had fled without giving battle, the pace slacked off, and the Infantry began to feel the effects of their exertions.

[323] Lord Roberts' infantry crossing the Zand River - balloon watching ahead for the Boers, S.A. Digital ID: (b&w film copy neg.) cph 3c07542 http://hdl.loc.gov/loc.pnp/cph.3c07542 Reproduction Number: LC-USZ62-107542 (b&w film copy neg.)

'President Kruger and President Steijn Addressing the Boers in the Market Hall at Kroonstad after the occupation of Bloemfontein.'[244]

At eleven a message from Lord Roberts reached General Broadwood to say that it did not matter by which road Hamilton's column marched in, as the enemy was not holding his positions. Thereupon I determined, since there was to be no battle, to see the capture of Kroonstadt, and being mounted on a fresh pony I had bought at Winburg, a beautiful and tireless little beast, by an English blood sire out of a Basuto mare, [324]

A Basuto-Pony.[245]

Basuto Pony & Rider, Howard C. Hillegas.[246]

I soon left the Cavalry behind, caught up the rear of Tucker's transport, pushed on four or five miles along the line of march of his division, struck the tail of the Eleventh Division, and finally overtook the head of the Infantry columns about three miles from the town.

Lord Roberts entered Kroonstadt at about mid-day with all his staff. The Eleventh Division, including the Guards' Brigade, marched past him in the market square, and then, passing through the town, went into bivouac on the northern side.

[324] The Basuto is a pony breed from Lesotho and South Africa. The Basuto is considered a small horse, since it possesses horse-like characteristics, such as an exceptionally long stride. Basutos have, long necks and long, straight backs, and straight shoulders. They have very tough legs and sound, very hard hooves. They can be up to 14.2 hands high (56.8 inches or ~142 cm). Basutos can be chestnut, brown, bay, gray or black. They are sure-footed, fast, fearless and are known for their stamina. Basuto pony - Wikipedia

Guards Marching from Sand River to Ventersburg.[247]

Field Marshal Frederick Roberts riding with his escort through the valley of the Zand River.[248] Reporters and photographers await the arrival of Roberts in Kroonstad on Saturday, May 12, 1900.[249]

Kroonstadt Camp.[250]

Arrival of the English Mail at Kroonstadt.[251]

The rest of the army halted south of Kroonstadt. Gordon's Cavalry Brigade a mile from the town; the Seventh Division and Ian Hamilton's force three miles away, in a wide valley among the scrub-covered, trench-rimmed hills the Boers had not dared defend. French, whose turning movement had again been obstinately opposed, reached the railway line north of the town too late to intercept any rolling stock. Indeed, Major Hunter Weston,[325] a daring and enterprising engineer,[326] arrived at the bridge he had hoped to blow up only to find that it had been blown up by the enemy.

[325] Lieutenant General Sir Aylmer Gould Hunter-Weston, KCB, DSO (23 September 1864 – 18 March 1940) In 1900 he was promoted brevet lieutenant-colonel and awarded the Distinguished Service Order. He was described as having "reckless courage combined with technical skill and great coolness in emergency," was mentioned in despatches (including 31 March 1900) and received the Queen's South Africa Medal. Later he was a British Army officer who served in the First World War at Gallipoli in 1915 and in the very early stages of the Somme Offensive in 1916. He was also a Scottish Unionist MP. Aylmer Hunter-Weston - Wikipedia

[326] Royal Engineers Article: https://www.angloboerwar.com/?option=com_content&view=article&id=586

Taaibosch Spruit Bridge, from the north bank.[252] Bridge across the Valsch River at Kroonstadt, blown up by the Boers.[253] Convoy of the 11th Division "outspanned" on the march to Kroonstadt.[254]

Thus, by one long spring from Bloemfontein, Kroonstadt, the new capital of the Free State, was captured. It has the reputation of being one of the prettiest places in the Republic, but even when allowances are made for the circumstances under which we saw it, it does not seem that its fame is just. The town looked a little larger than Winburg, though not nearly so clean and well-kept, and the whole place was smothered in reddish dust, and dried up by the sun. The Boers retreated northward along the railway, in spite of all President Steyn's exhortations, which included the public sjambokking[327] of several unwilling burghers, and did not stop except to wreck the permanent way until they reached Rhenoster kopjes.

The President, with the members of the Executive Council and the seat of Government—which needs to have a good pair of legs beneath it in times like these—withdrew to Lindley,[328] whither, for various reasons, it soon became desirable to follow them.

Ambulance Train blown up at Kroonstad.[255]

The Red Cross hospital train wrecked.[256]

Kroonstad. Station building.[257]

The Lace Diamond Mines near Kroonstad.[258]

[327] Sjambok: A stout whip, especially made of rhinoceros or hippopotamus hide.

[328] Lindley, Free State - Wikipedia

XI. Lindley

Heilbron: May 22, 1900.

Having arrived thus prosperously at Kroonstadt, Lord Roberts determined to halt until his supplies were replenished and the railway line from Bloemfontein in working order. Moreover, in the expectation of a general action outside the town, he had concentrated all his troops and had drawn the Army of the Right Flank close in to the main force. Before he advanced again towards the enemy's position on the Rhenoster River, [Renoster][329] he wished to extend his front widely, as he had done in the previous operation. The scheme of advance by converging columns required a pause after each concentration before the movement could be repeated; so that while the Field-Marshal himself remained stationary his energetic Lieutenant was again on the move.

General Ian Hamilton, with the same troops as before and an addition of four 'pom-poms,' started from his camp outside Kroonstadt on the 15th, and after a short march encamped on the eastern side of the town preparatory to moving on Lindley, whither President Steyn had withdrawn. The question of supplies was a very troublesome one, and it was no light matter to thrust out fifty miles into a hostile country with only three and a half days' food and forage in hand. Suppose anything should happen to the convoys which were to follow. Meat in plenty could be found everywhere, but the stores of flour and other farinaceous goods which the farm-houses might contain were insufficient and precarious. Even the benefits of the abundant meat supply were to some extent discounted by the scarcity of wood, for it is not much satisfaction to a soldier to be provided with a leg of mutton if he has no means of cooking. The deficiencies were hardly made good by the arrival of a small convoy, the greater part of which consisted of disinfectants for standing camps, and the rest—so valuable in a grass country—of compressed hay.

Camp butchery a pleasant change from Bully Beef.[259]

Field Kitchen.[260]

Handiwork of the retreating Boers - destroyed Railway Bridge between Kroonstadt and Pretoria, South Africa.[261]

Engineers repairing a Railway destroyed by the Boers near Kroonstadt.[262]

[329] The Renoster River is a tributary of the Vaal River in South Africa. The river originates in Petrus Steyn in the Free State. From its origin, the river flows north west, passing Heilbron into the Koppies Dam. From there, it flows beyond Koppies as it flows under the N1 national road. Near Viljoenskroon it flows northwest, flows under the R59 and R501 and joins the Vaal River at Renovaal, about 25 km east of Orkney. https://en.wikipedia.org/wiki/Petrus_Steyn#The_origin_of_the_Renoster_river

Highlanders Repairing the Line. [In kilts][263]

Wrecked railway water tank, Taaibosch Spruit.[264]

Rhenoster bridge midway through reconstruction.[265]

Nevertheless, being determined, and trusting, not without reason, in his supply officer, Captain Atcherley,[330] Hamilton started on the 16th, and the Infantry bivouacked eighteen miles from Kroonstadt on the Lindley road—it would perhaps be less misleading to write track. The Cavalry brigade with one corps of Mounted Infantry under Broadwood were pushed ten miles further on, and seized a fine iron bridge, not marked on any map, which spans an important spruit at Kaalfontein. Here trustworthy information was received that a large force of Boers with guns was retreating before Rundle's column (Eighth Division) northwards upon Lindley, and deeming it important to occupy the town before they arrived, Hamilton ordered the Cavalry to hurry on and take possession of the heights to the north of it.

It was a double march when ordinary marches were long. The result, however, justified the effort. Broadwood 'surprised'—the word is taken from the Boer accounts—Lindley on the 17th. Scarcely fifty Boers were at hand to defend it. A waggon with £60,000 in specie barely escaped from the clutches of the Cavalry. After a brief skirmish the town surrendered. The British loss was three men wounded. Broadwood then retired as directed by his chief to the commanding hill to the north to bivouac. This hill may for convenience be called 'Lindley Hill' in the subsequent narrative.

The Infantry and baggage also made a long march on the 17th, but as the road was obstructed by several bad spruits or *dongas*, they were still fourteen miles from Lindley when night closed in. Even then the transport was toiling on the road, and a large part of it did not come in, and then in an exhausted condition, until after midnight. I wonder how many people in England realise what a *spruit* is, and how it affects military operations.

[330] L. W. Atcherley Brevet Major MID LG: 29 July 1902, page: 4852. Source: General Kitchener. 23 June 1902. Re: Final despatch & mentions. Army Service Corps Article: https://www.angloboerwar.com/?option=com_content&view=article&id=531 2,647 Records.

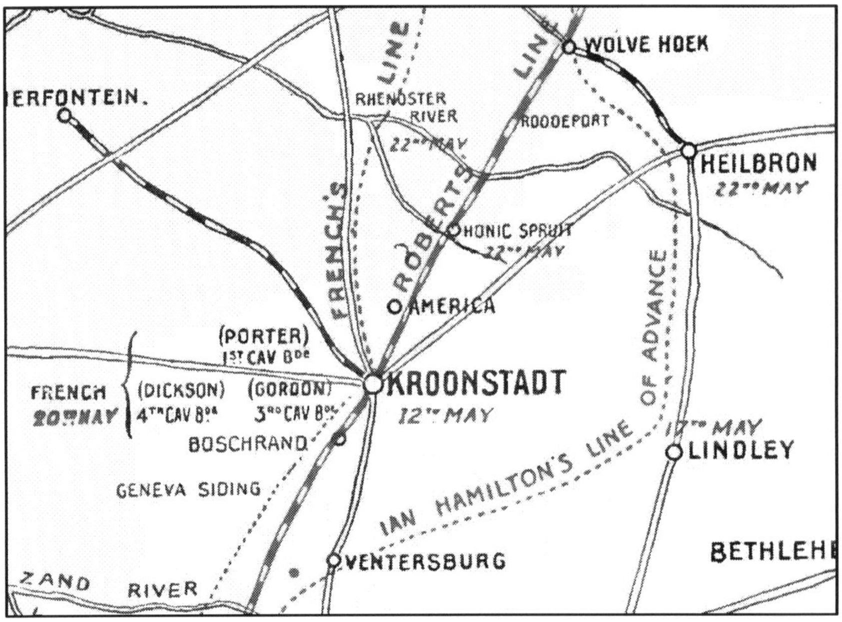

Those who live in highly developed countries, where the surface of the earth has been shaped to our convenience by the patient labour of many years, are accustomed to find the road running serenely forward across the valleys, and they scarcely notice the bridges and culverts over which it passes. All is different in South Africa. The long column of transport trails across the plain. The veldt in front looks smooth and easy going. Presently, however, there is a block. What is the matter? Let us ride forward to see: and so onward to where the single string of waggons merges in a vast crowd of transport, twenty rows abreast, mule carts, Cape carts, ox waggons, ambulances, and artillery, all waiting impatiently, jostling each other, while drivers and conductors swear and squabble.

Supply wagon being pulled out mud by troops and oxen.[266]

Supply column of the British Army on the move.[267]

An Army on the March: Guns and Ambulance Carts crossing a Donga.[268]

Horses and [Cape]Cart.[269] [331] Horse cart, possibly an ambulance, attached to one of the British Army units.[270]

Here is the spruit—a great chasm in the ground, fifty feet deep, a hundred yards from side to side. The banks are precipitous and impassable at all points except where the narrow single track winds steeply and unevenly down. The bottom is a quagmire, and though the engineers are doing their best to level and improve the roadway, it is still a combination of the Earl's Court water chute and the Slough of Despond.[332] One by one, after a hot dispute for precedence, the waggons advance. The brakes must be screwed up to their tightest grip lest the ponderous vehicles rush forward down the slope and overwhelm their oxen.

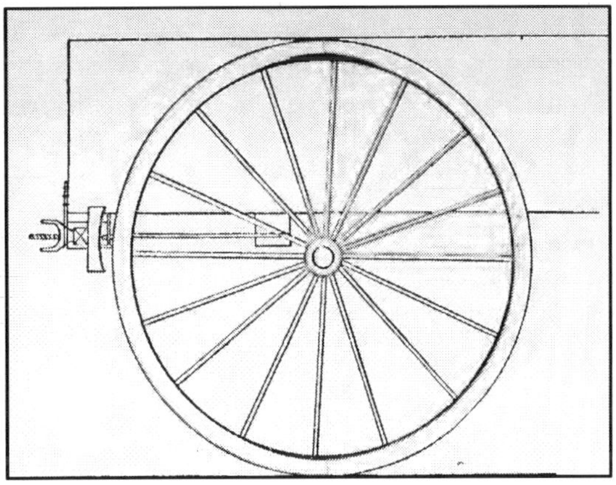

General service wagon showing wagon brakes at rear.[271] Wagon Brake used by the British Army.[272]

Even with this precaution the descent of each is a crash, a scramble, and a bump. At the bottom like a feather-bed lies the quagmire. Here one waggon in every three sticks. The mules give in after one effort—unworthy hybrids. The oxen strain with greater perseverance. But in the end it is the man who has to do the hauling. Forthwith come fatigue parties of weary men—it has been a long march already to soldiers fully equipped. Drag ropes are affixed, and so with sweat, blood, and stretching sinew, long whips cracking and whistling, white men heaving and natives yelping encouragement, another waggon comes safely through. And there are seven miles of transport!

British supply column on the move.[273] Wagons on train.[274]

331 A cape cart is a two-wheeled four-seater carriage drawn by two horses and formerly used in South Africa. Cape cart - Wikipedia

332 The Slough of Despond (swamp of despair) is a fictional, deep bog in John Bunyan's allegory The Pilgrim's Progress, into which the protagonist Christian sinks under the weight of his sins and his sense of guilt for them. Slough of Despond - Wikipedia

An Army in Being.[275] Miles of Bullock Waggons.[276]

On the morning of the 18th the Infantry were about to move off, when a patter of rifle shots to the north of the road reminded us of the presence of the enemy. A foraging party of Major Rimington's[333] Guides[334] had ridden up to a farm, which stood in full view of the camp and flew (or was it hoisted afterwards?) a white flag. Arrived there, they were received by a volley from five Boers in hiding near. Conceive the impudence of these people: five Boers, within a mile of eight thousand British and a powerful Cavalry force, fire on a foraging party! Luckily no harm done; Cavalry gallop out angrily; Boers vanish among remoter kopjes. 'But,' said the General, 'what about my convoys?'

Burning of a farmhouse near Dewetsdorp in May, 1900.[277] "This was one of the houses burnt under Lord Roberts' proclamation; the offence In this case was the treacherous firing on British troops from the building while a white flag flew from its gable." Farms set alight.[278]

So it was arranged that Smith-Dorrien should be left where he was (twelve miles west of Lindley) with his own brigade, one battery, and a corps of Mounted Infantry to help in the expected convoy, and should cut off the corner and rejoin the column at the end of its first march towards Heilbron. Ian Hamilton with the rest of the troops then moved on to Lindley. The march lay through the same class of country hitherto traversed—a pleasant grassy upland which, if not abundantly supplied with water by nature, promised a rich reward to man, should he take the trouble to construct even the simplest irrigation works. Spruits ran in all directions, and only required an ordinary dam, like the *bunds* the peasants build in India, to jewel each valley with a gleaming vivifying lake.

[333] M.F. Rimington Lieutenant Colonel Born in Wales, entering the Inniskilling Dragoons in 1881. Colonel, September 1900. After service in the Bechuanaland Expedition in 1884 and 1885, as well as in Zululand in 1888, he saw active service in the Boer War when as commander of Rimington's Scouts/Guides, he won distinction at Magersfontein, at Thaba Nchu, the Brandwater Basin and elsewhere. In his later career he was Inspector-General of Cavalry in India.
Source: List of CB recipients. Various sources.

[334] Rimington's Guides were raised in South Africa at the outbreak of the war by Major M F Rimington, 6th (Inniskilling) Dragoons, who had been on special service in South Africa prior to the commencement of hostilities. Article: https://www.angloboerwar.com/?option=com_content&view=article&id=463 1,548 Records Listed under Damant's Horse.

The husbanding of water would repair the scarcity of wood, and the tenth year might see the naked grass clothed and adorned with foliage. But at present the country-side is so sparsely populated that the energies of its inhabitants could not produce much effect upon the landscape. The unamiable characteristic of the Boer, to shun the sight of his neighbour's barn, has scattered the farms so widely that little patches of tillage are only here and there to be seen, and the intervening miles lie neglected, often not more than twenty acres of a six thousand acre property being brought into cultivation, which seems rather a pity.

The fair face of the land under its smiling sky was not unmarked by the footprints of war. In the dry weather the careless habits of the soldiers were the constant cause of grass fires. The half-burnt match, tossed idly aside after a pipe was lighted, or an unguarded spark from a cooking fire, kindled at once an extensive conflagration. The strong winds drove the devouring blaze swiftly forward across the veldt, clouding the landscape by day with dense fumes of smoke and scarring the scene by night with vivid streaks of flame.

A Domiciliary Visit Firing the outhouse [outbuilding] of a Boer who has been shooting natives.[279] A useless haul. A captured wagon for which no oxen are available is burnt.[280]

So frequent were these grass fires that they became a serious nuisance, wasting in an hour many acres of grazing, proclaiming the movement and marking the track of the army, stifling the marching columns with pungent odours, destroying the field telegraph, and only extinguished by the heavy dews of the early morning. But in spite of repeated injunctions in the daily orders, the accidents—for which, indeed, there was every excuse—continued, and the plains of brownish grass were everywhere disfigured with ugly patches of black ashes which, as the fires burnt outwards, would spread and spread, like stains of blood soaking through khaki.

Grass fire on the side of Bryn Glas.[281]

Grass fire near Potlatch.[282]

At length the track, which had been winding among the smooth undulations, rounded an unusually steep hillock of kopje character, and we saw before us at the distance of a mile the pretty little town of Lindley. The Cavalry bivouacs covered the nearer slopes of the high hill to the northward. The houses—white walls and blue-grey roofs of iron—were tucked away at the bottom of a regular cup, and partly hidden by the dark green Australian trees.[335] We rode first of all to Broadwood's headquarters, following the ground wire which led thither.

[335] Eucalyptus trees had been imported to South Africa since the 1820s.

British Mounted Infantry Camp.[283]

A cyclist telegraph engineer laying 'ground wire.'[284]

Making camp in the field.[285]

Arrived there we learned the news. Boer laagers and Boer patrols had been found scattered about the country to the south-east and north-east. There was occasional firing along the picket line. The town had upon most searching requisition yielded nearly two days' supply, and, most important of all, Piet De Wet, brother of the famous Christian, had sent in a message offering to surrender with such of his men as would follow his example, if he were permitted to return to his farm. Broadwood had at once given the required assurance, and Hamilton on his arrival had wired to Lord Roberts fully endorsing the views of his subordinate, and requesting that the agreement might be confirmed.

The answer came back with the utmost despatch, and was to the effect that surrender must be unconditional. De Wet, it was remarked, was excluded from the favourable terms of the Proclamation to the Burghers of the Orange Free State, by the fact that he had commanded part of the Republican forces. He could not therefore be permitted to return to his farm.

I need not say with what astonishment this decision was received. The messenger carrying the favourable answer was luckily overtaken before he had passed through our picket line and the official letter was substituted. Piet De Wet, who awaited the reply at a farmhouse some ten miles from Lindley, found himself presented with the alternative of continuing the war or going to St. Helena, or perhaps Ceylon; and as events have shown he preferred the former course to our loss in life, honour, and money.

Diyatalawa Camp in Ceylon, situated at an altitude of 1,499 m (4,918 ft.) [336] Ceylon Camps.[286] Over 5,000 Boers were interned in Ceylon (Sri Lanka) during the Boer War. The Ceylon camps were the second set of overseas POW camps after those in St. Helena had filled. See Infra for St. Helena camp material.

In the afternoon I rode into Lindley to buy various stores in which my waggon was deficient. It is a typical South African town, with a large central market square and four or five broad unpaved streets radiating therefrom. There is a small clean-looking hotel, a substantial gaol, a church and a schoolhouse. But the two largest buildings are the general stores. These places are the depôts whence the farmers for many miles around draw all their necessaries and comforts. Owned and kept by Englishmen or Scotchmen, they are built on the most approved style. Each is divided into five or six large well-stocked departments. The variety of their goods is remarkable. You may buy a piano, a kitchen range, a slouch hat, a bottle of hair wash, or a box of sardines over the same counter. The two stores are the rival Whiteley's[337] of the country-side; and the diverse tastes to which they cater prove at once the number of their customers, and the wealth which even the indolent Boer may win easily from his fertile soil.

Personally I sought potatoes, and after patient inquiry I was directed to a man who had by general repute twelve sacks. He was an Englishman, and delighted to see the British bayonets at last. 'You can't think,' he said, 'how we have looked forward to this day.'

I asked him whether the Dutch had ill-used him during the war.

'No, not really ill-used us; but when we refused to go out and fight they began commandeering our property, horses and carts at first and latterly food and clothing. Besides, it has been dreadful to have to listen to all their lies and, of course, we had to keep our tongues between our teeth.'

It was evident that he hated the Boers among whom his lot had been cast with great earnestness. This instinctive dislike which the British settler so often displays for his Dutch neighbour is a perplexing and not a very hopeful feature of the South African problem. Presently we reached his house (where the potatoes were stored). Above the doorway hung a Union Jack. I said—

'I advise you to take that down.'

'Why?' he asked, full of astonishment.

'The British are going to keep the country, aren't they?'

'This column is not going to stay here for ever.'

'But,' with an anxious look, 'surely they will leave some soldiers behind to protect us, to hold the town.'

I told him I thought it unlikely. Ours was a fighting column. Other troops would come up presently for garrison duty. But there would probably be an interval of at least a week. Little did I foresee the rough fighting which would rage round Lindley for the next three months. He looked very much disconcerted; not altogether without reason.[338]

[336]Diyatalawa meaning 'the watered plain') is a former garrison town in the central highlands of Sri Lanka. It is situated at an altitude of 1,281 m (4,203 ft.) and has become a popular destination for local holiday makers. Diyatalawa - Wikipedia

[337] Whiteleys was a shopping centre in Bayswater, London. The original Whiteleys department store was created by William Whiteley, who started a drapery shop at 31 Westbourne Grove in 1863. By 1867 it had expanded to a row of shops containing 17 separate departments. Whiteleys - Wikipedia

[338] Thursday 5/31/1900 Battalion of Irish Yeomanry captured at Lindley. Gen. Piet de Wet defeated the 13th Battalion Imperial Yeomanry, killing and wounding 80 soldiers and capturing 530 British soldiers. Lindley, Free State - Wikipedia

'It's very hard on us,' he said after a pause. 'What will happen when the Boers come back? They're just over the hill now.'

'That's why I should take the flag down if I were you. If you don't fight, keep your politics till the war is over!' He looked very disappointed, and I think was asking himself how much his enthusiasm had compromised him. After we had settled the potato question to his satisfaction and I had sent the sack away upon my pack pony, he perked up. 'Come and see my garden,' he said, and nothing loth I went. It was not above a hundred yards square, but its contents proclaimed his energy and the possibilities of the soil. He explained how he had dammed a marshy sluit in the side of the hills to the eastward. 'Plenty of water at all seasons: this pipe you see, only a question of piping: as much water as ever I want :twenty gardens: grow anything you like, potatoes mostly, cabbages (they were beauties), tomatoes and onions, a vine of sweet white grapes, a bed of strawberries over there—anything: it only wants water, and there's plenty of that if you take the trouble to get it.'

The signs of industry impressed me. 'How long,' I asked, 'have you been here?'

'Eight years last February,' he replied; 'see those trees?'

He pointed to a long row of leafy trees about twenty feet high, which gave a cool shade and whose green colour pleased the eye after looking at so much brown grass. I nodded.

'I planted those myself when I came: they grow quickly, don't they? Only a question of water, and that is only a question of work.'

Then I left him and returned to the camp with my potatoes and some information thrown in.

A Homestead.[287] Note the tall eucalyptus trees on presumably the North side of the house. This is the southern hemisphere. Also the campaign is extending into the Winter season at this latitude with the winter solstice occurring on June 21st.

The next morning before breakfast-time there was firing in the picket line south of Lindley. The patter of shots sounded across the valley, and upon the opposite slopes the British patrols could be seen galloping about like agitated ants. I was at the moment with General Hamilton. He watched the distant skirmish from his tent door for a little while in silence. Then he said:

'The scouts and the Kaffirs report laagers of the enemy over there, and over there, and over there' (he pointed to the different quarters). 'Now either I must attack them today or they will attack me tomorrow. If I attack them today, I weary my troops; and if I don't we shall have to fight an awkward rearguard action to get out of this place to-morrow.'

He did not say at the time which course he meant to follow, but I felt quite sure he would not take his troops back very far to the south or south-east to chastise impalpable[339] laagers. We were running on schedule time and had to make our connections with the main army, to securing whose smooth and undisturbed march all our efforts must be directed. So I was not surprised when the day passed without any movement on our part.

[339] Impalpable: Not able to be perceived by the senses (especially by touch); intangible or insubstantial; not easily grasped or understood.

British troops at their camp.[288]

"Feet Inspection at Estcourt."[289] Shaving in Camp.[290] PF: Why is a man getting a shave while another two are aiming a machine gun right behind him?

Morning wash at Orange River.[291]

Very early on the 20th the brigades were astir, and as soon as the light was strong Broadwood's Cavalry began to stream away over the northern ridges. The guns and the greater part of the Infantry followed them without delay, so that by seven o'clock the great column of transport was winding round the corner of Lindley Hill on the road to Heilbron. The fact that parties of the enemy had been observed on all sides except the west, made the operation of disentangling the force from Lindley difficult and dangerous. Broadwood's duty was to clear the way in front. Legge's corps of Mounted Infantry guarded the right flank: and Ian Hamilton himself watched the movement of the rear guard, which consisted of the Derbyshire Regiment,[340] Bainbridge's corps of Mounted Infantry and, as a special precaution, the 82nd Field Battery.

[340] Derbyshire Regiment (Sherwood Foresters) Article: https://www.angloboerwar.com/?option=com_content&view=article&id=546

The full light of day had no sooner revealed the march of the troops than the watching Boers began to feel and press the picket line: and an intermittent musketry spread gradually along the whole three quarter circle round Lindley. At eight o'clock our troops evacuated the town itself, at nine, the convoy being nearly round Lindley Hill, the pickets commenced to draw in. This was a signal for decided increase in the firing. No sooner were the outposts clear of the town than the Boers in twos and threes galloped into it and began to fire from the houses.

All kinds of worthy old gentlemen, moreover, who had received us civilly enough the day before, produced rifles from various hiding-places and shot at us from off their verandahs.

A Free State burgher ready for the fray.[292]

An Old Takhaar.[293]

Indeed, so quickly did the town revert to the enemy's hands that Somers Somerset,[341] the despatch rider of the 'Times,' was within an ace of being caught. He had arrived late the night before, and having found a comfortable bed at the hotel went to sleep without asking questions. The next thing he remembers is the landlord rushing into his room and crying in great excitement that the Boers were in the town. He scrambled into his clothes and, jumping on his horse galloped through the streets and was not fired at till he was more than a quarter of a mile away. History does not record whether among such disturbing events he retained his presence of mind sufficiently to settle his hotel bill.

The General and his staff had watched the beginnings of the action from the now deserted camping ground, a dirty waste, littered with rubbish and dotted with the melancholy figures of derelict horses and mules. So soon as the retiring pickets drew north of the town, he mounted and made his way to the top of Lindley Hill. From this commanding table top the whole scene of action, indeed the whole surrounding country, was visible. At our feet beyond the abandoned bivouac lay the houses of Lindley giving forth a regular rattle of musketry. On either side, east and west, rose two prominent kopjes held by companies of Mounted Infantry briskly engaged. The tail of the transport serpent was twisting away into safety round the base of our hill.

Far away on the broad expanse of down parties of Dutch horsemen cantered swiftly forward; and along a road beyond the eastern kopje rose a steady trickle of mounted men. They moved in true Boer fashion—little independent groups of four and five, now and then a troop of ten or a dozen, here and there a solitary horseman riding back against the general flow. At no particular moment were more than thirty to be seen on the mile of dusty road. Yet to an experienced eye the movement seemed full of dangerous significance. One became conscious of a growing accumulation of force somewhere among the hills to the eastward.

The General, who had served on the Indian frontier, understood rear-guard actions, and his face was grave, as I had not seen it when larger operations were toward; and at this moment the boom of a heavy gun told us that the advanced troops were also engaged. The Boers knew what they wanted. There was an air of decision about their movements which boded no good to rear or right flank guard. Gallopers were sent off, one to warn the right corps of Mounted Infantry, another to bid the main body of the force go dead slow, another to the threatened eastern kopje to learn the state of affairs there. The rear-guard battery was brought up on to the table top, and

[341] H C S A Somerset: Correspondent for The Times. QSA issued: 20 Feb 03 Source: QSA Medal roll.

came into action. This was, I think, the key of the situation. The battery planted on Lindley Hill, and casting its shells now in one direction, now in another, compelled the assailants to keep their distance, and helped the pickets into safety and new positions further back. It called to mind some famous knight of history or romance holding an angry rabble back beyond the sweep of his long sword, while his comrades made good their retreat.

Under this good protection the pickets, having dutifully held their positions until the convoy was well on its road, scampered in, and the battery itself began to think about retiring. But the trickle of Boers along the eastern roadway had not stopped. Seven or eight hundred men must have passed already; and those that now came galloped as if they had some very tangible objective. 'Look out, the right flank!'

But now, the rear guard having disengaged itself from Lindley town, the General's place was with his main body, and we set off to trot and gallop the seven miles that intervened between the head and tail of our force. The firing in front had ceased before we came up. Indeed, the affair had not been of any importance. About seven hundred Boers with three or four guns had obstructed the advance near the Rhenoster River; had even checked the Cavalry screen; Tenth Hussars had two officers wounded; a dozen other casualties in the Brigade; Infantry and guns wanted to clear the way. A Cavalry brigade is not a kopje-smashing machine. 'Never mind, here come the cow guns. Now we shall see.' Indeed, as soon as the head of the 21st Brigade began to deploy, the five-inch guns and a field battery opened on the enemy, who thereupon fled incontinently across the river, pursued by the fire of the guns and of the Cavalry 'pom-poms.'

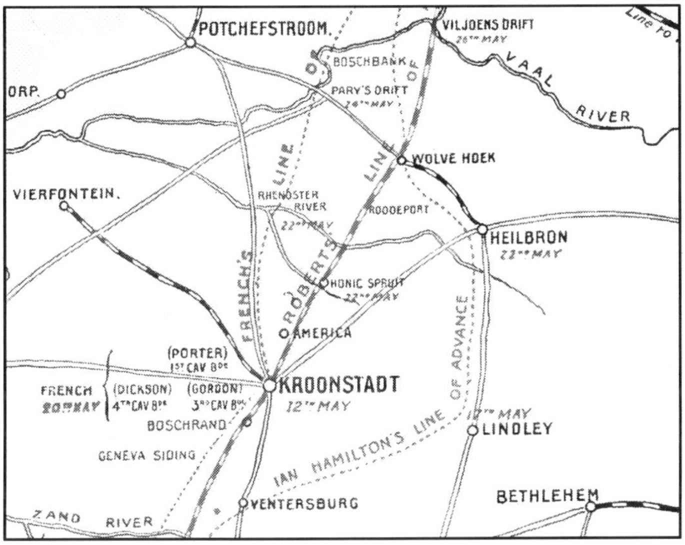

Note Rhenoster River in center of map below Heilbron.

All who understood the scope and cohesion of the operations were delighted at the prospect of getting across the Rhenoster River. The General was determined, rear and flank guard actions notwithstanding, to have his army and transport over that night: and two practicable crossings having been found, Infantry, Cavalry, guns and baggage began to push across. The last was now increased by the arrival of Smith-Dorrien, who brought with him a much-needed convoy with sufficient supplies to carry us on to Heilbron and a march beyond. It was midnight before all the waggons were across; but though this cruel day of march and sun tore the hearts out of the transport animals, and the flocks of sheep were so weary they could scarcely be driven along, we knew that the exertions had not been made in vain.

Herd of cattle, probably livestock belonging to the British Army, being moved to another location by native African workers.[294] Native African workers tending cattle. They are employed by the British Army to look after livestock.[295]

Family of Boer refugees passing by a flock of sheep.[296]

Late in the evening came the news from the right flank guard. They had waited, fearing to expose the rear guard to a flank attack. The rear guard had made good its retreat. A gap had sprung up between the two bodies. The vigilant Boers had pounced in and stampeded the horses of one Mounted Infantry company.

A sharp, fierce fight followed; rear guard hearing the fusillade swung in to help. Ultimately the Boers were checked sufficiently to enable rear and flank guards to cut inwards together and draw off: but it was by general agreement of participants a very unpleasant affair. The officer commanding the company whose horses were stampeded had particularly interesting experiences. The Boers galloped right in among his men, and a confused scrimmage followed: officer was running towards stampeded horses; on the way he passed a burgher; 'Surrender,' cried the Dutchman. 'No,' retorted the officer—an Irishman—(with suitable emphasis) and ran on, whereupon burgher dismounted and began shooting; had four shots and missed every one. Meanwhile officer reached shelter of a convenient rock, turned in just indignation, fitted his Mauser pistol together and fired back.

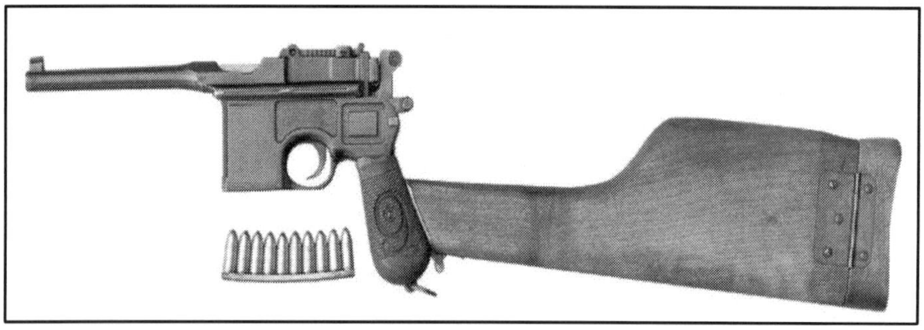

Mauser C96 in its carbine form.[297 342]

The burgher, finding his enemy behind cover, and himself in the open—by no means the situation for a patriot—jumped on his horse, and would have galloped away but that the officer managed to hit him in the leg with his pistol, and so he dropped, according to the account of an eyewitness, 'like a shot rook.'[343]

The local advantage, however, rested with the Boers, who hit or captured the greater part of the squadron, including twenty wounded. Concerning these latter, Piet De Wet sent in a flag of truce during the night offering to hand them over if ambulances were sent, and several wounded Boers whom we had taken were given up. This was accordingly done. Our total losses during the 20th were about sixty, some of whom were officers. The Boers admitted a loss of twenty killed and wounded, and it may easily have been more. The army bivouacked on the north bank of the Rhenoster within two marches of the town of Heilbron, upon which it was now designed to move.

[342] Mauser C96 - Wikipedia The Mauser C96 (Construktion 96) Also known as the 'Broomhandle Mauser' for the shape of the grip; is a semi-automatic pistol originally produced by German arms manufacturer Mauser from 1896 to 1937. Mauser supplied the C96 to Westley Richards in the UK for resale. A ten round magazine was accepted as a standard after 1905, earlier models also used 6 or 20 round magazines. It's loaded with single rounds or via stripper-clips. It was very popular with British officers at the time, and many purchased it privately. Unlicensed copies of the gun were also manufactured in Spain and China in the first half of the 20th century. Over 1,000,000 were produced.

[343] The Rook is a large, gregarious, black-feathered bird, with a whitish featherless area on the face. Rooks nest collectively in the tops of tall trees, often close to farms or villages, the groups of nests being known as rookeries. Rook (bird) - Wikipedia

Two Boer fighters, one of them holding a white flag, coming up to a British soldier, probably with intention to negotiate.[298] [299]

Netherlands Ambulance at Modderspruit.[300]

"Doctor under the protection of the Red Cross. moving across the outskirts of Frere Camp"[301]

An Ambulance Wagon Receiving its Freight.[302]

XII. Concerning a Boer Convoy

Heilbron: May 22nd 1900.

Heilbron[344] lies in a deep valley. About it on every side rolls the grassy upland country of the Free State, one smooth grey-green surge beyond another, like the after-swell of a great gale at sea; and here in the trough of the waves, hidden almost entirely from view, is the town itself, white stone houses amid dark trees, all clustering at the foot of a tall church spire. It is a quiet, sleepy little place, with a few good buildings and pretty rose gardens, half a dozen large stores, a hotel, and a branch line of its own.

For a few days it had been capital of the Free State. The President, his secretaries, and his councillors arrived one morning from Lindley, bringing the 'seat of government' with them in a Cape cart. For nearly a week Heilbron remained the chief town. Then, as suddenly as it had come, the will-o'-the-wisp dignity departed, and Steyn, secretaries, councillors, and Cape cart, hurried away to the eastward, leaving behind them rumours of advancing hosts—and (to this I can testify) three bottles of excellent champagne. That was on Sunday night. [20th] The inhabitants watched and wondered all the next day.

On the Tuesday morning, [22nd] shortly after the sun had risen, Christian De Wet appeared with sixty waggons, five guns, and a thousand burghers, very weary, having trekked all night from the direction of Kroonstadt, and glad to find a place of rest and refreshment. 'What of the English?' inquired the new-comers, and the Heilbron folk replied that the English were coming, and so was Christmas, and that the country to the southward was all clear for ten miles. Thereat the war-worn commando outspanned their oxen and settled themselves to coffee.

A burgher and his breakfast.[303]

Outspanned at breakfast.[345]

Forty minutes later the leading patrols of Broadwood's Brigade began to appear on the hills to the south of the town.

Looked at from any point of view, the British force was a formidable array: Household Cavalry,[346] 12th Lancers[347] and 10th Hussars, with P and Q Batteries Royal Horse Artillery (you must mind your P's and Q's with them), two 'pom-poms,' and two galloping Maxims; and, hurrying up behind them, Light Horse, Mounted Infantry, Nineteenth and Twenty-first Brigades, thirty field guns, more 'pom-poms,' two great 5-in. ox-drawn siege pieces ('cow guns' as the army calls them), and Ian Hamilton. It was an army formidable to any foe; but to those who now stared upwards from the little town and saw the dark, swift-moving masses on the hills—an avalanche of armed men and destructive engines about to fall on them—terrible beyond words.

344 Heilbron - Wikipedia

345 South Africa; its history, heroes and wars by Mackenzie, William Douglas, 1859-1936; assisted by Alfred Stead, 1877-1933, joint authors. Published 1899 by The Co-Operative Publishing Company, Chicago, Ill.

346 The Household Cavalry (HCav) is made up of the two most senior regiments of the British Army, The Life Guards and The Blues and Royals (Royal Horse Guards and 1st Dragoons). Household Cavalry - Wikipedia

347 Lancer's Article: https://www.angloboerwar.com/?option=com_content&view=article&id=497

Oxen drawn column of the British Army carrying an artillery gun (5-inch gun?) across a river, possibly the Modder River.[304]

The 5-inch Howitzer or Siege Gun.[305]

37th R.A. Howitzer battery.[306]

'And then,' as the poet observes, 'there was mounting in hot haste,' saddling up of weary ponies, frantic inspanning of hungry oxen cheated of their well-earned rest and feed, cracking of long whips, kicking of frightened Kaffirs; and so pell-mell out of the town and away to the northward hurried the commando of Christian De Wet.

The Cavalry halted on the hills for a while, the General being desirous of obtaining the formal surrender of Heilbron, and so preventing street-fighting or bombardment. An officer—Lieutenant M. Spender-Clay,[348] of the 2nd Life Guards[349]—was despatched with a flag of truce and a trumpeter; message most urgent, answer to be given within twenty minutes, or Heaven knows what would happen; but all these things take time. Flags of truce (prescribe the customs of war) must approach the enemy's picket line at a walk; a mile and a half at a walk—twenty minutes; add twenty for the answer, ten for the return journey, and nearly an hour is gone. So we wait impatiently watching the two solitary figures with a white speck above them draw nearer and nearer to the Boer lines; 'and,' says the brigadier, 'bring two guns up and have the ranges taken.'

[348] Unable to identify this individual.

[349] Life Guards Article: https://www.angloboerwar.com/?option=com_content&view=article&id=561 1,358 Records. And Life Guards (United Kingdom) - Wikipedia

There was just a chance that while all were thus intent on the town, the convoy and commando might have escaped unharmed, for it happened that the northern road runs for some distance eastward along the bottom of the valley, concealed from view. But the clouds of dust betrayed them.

'Hullo! what the deuce is that?' cried an officer.

'What?' said everyone else.

'Why, that! Look at the dust. There they go. It's a Boer convoy. Gone away.'

And with this holloa the chase began. Never have I seen anything in war so like a fox hunt. At first the scent was uncertain, and the pace was slow with many checks.

Before us rose a long smooth slope of grass, and along the crest the figures of horsemen could be plainly seen. The tail of the waggon train was just disappearing. But who should say how many rifles lined that ridge? Besides, there were several barbed-wire fences, which, as anyone knows, will spoil the best country.

Broadwood began giving all kinds of orders—Household Cavalry[350] to advance slowly in the centre; 12th Lancers to slip forward on the right, skirting the town, and try to look behind the ridge, and with them a battery of horse guns; 10th Hussars, to make a cast to the left, and the rest of the guns to walk forward steadily.

Slowly at first, and silently besides; but soon the hounds gave tongue. Pop, pop, pop—the advanced squadron—Blues—had found something to fire at, and something that fired back, too; pip-pop, pip-pop came the double reports of the Boer rifles. Bang—the artillery opened on the crest line with shrapnel, and at the first few shells it was evident that the enemy would not abide the attack. The horsemen vanished over the skyline.

The leading squadron pushed cautiously forward—every movement at a walk, so far. Infantry brigadiers and others, inclined to impatience, ground their teeth, and thinking there would be no sport that day, went home criticising the master. The leading squadron reached the crest, and we could see them dismount and begin to fire.

Column of troops of one of the British mounted regiments on the move.[307] Cape Mounted Irregulars on a reconnaissance.[308]

We were over the first big fence, and now the scent improved. Beyond the first ridge was another, and behind this, much nearer now, dust clouds high and thick. The General [Hamilton] galloped forward himself to the newly-captured position and took a comprehensive view. 'Tell the brigade to come here at once—sharp.'

A galloper shot away to the rear. Behind arose the rattle of trotting batteries. The excitement grew. Already the patrols were skirting the second ridge. The Boer musketry, fitful for a few minutes, died away. They were abandoning their second position. 'Forward, then.' And forward we went accordingly at a healthy trot.

In front of the jingling squadrons two little galloping Maxims darted out, and almost before the ridge was ours they were spluttering angrily at the retreating enemy, so that four burghers, as I saw myself, departed amid a perfect hail of bullets, which peppered the ground on all sides.

[350] Household Cavalry - Wikipedia

"Captain Arthur Hill MP Colt battery in action."[351]

With The 3rd "Queen's."[309] A Section with a Maxim Gun in Action.

Carriage-mounted .303-inch Vickers Sons & Maxim machine gun, shortly after landing at Durban with the 2nd Bn. Queen's (Royal West Surrey) Regiment.[310]

But now the whole hunt swung northward towards a line of rather ugly-looking heights. Broadwood looked at them sourly. 'Four guns to watch those hills, in case they bring artillery against us from them.' Scarcely were the words spoken, when there was a flash and a brown blur on the side of one of the hills, and with a rasping snarl a shell passed overhead and burst among the advancing Cavalry. The four guns were on the target without a moment's delay.

Loading Drill.[311]

Firing Drill.[312]

The Boer artillerists managed to fire five shots, and then the place grew too hot for them—indeed, after Natal, I may write, even for them. They had to expose themselves a great deal to remove their gun, and the limber and its six horses showed very plainly on the hillside, so that we all hoped to smash a wheel or kill a horse, and thus capture a real prize. But at the critical moment our 'pom-poms' disgraced themselves. They knew the range,

351 Dundonald had a Colt Company of four .303 calibre machine guns, mounted on 'galloping carriages'" I believe the Colts were commanded by volunteer officers, three of whom may have been: Lt. Clutterbuck, Capt Hill, MP, and Maj Villiers. Photos courtesy MC Heunis; the second shows Capt. Hill MP's Colt. Anglo Boer War - Colt machine guns - calibre? - Boer War Forum

they saw the target. They fired four shots; the aim was not bad. But four shots—four miserable shots! Just pom-pom, pom-pom. That was all. Whereas, if the Boers had had such a chance, they would have rattled through the whole belt, and sent eighteen or twenty shells in a regular shower. So we all saw with pain how a weapon, which is so terrible in the hands of the enemy, may become feeble and ineffective when used on our side by our own gunners.

Pom Pom in action.[313]

British Pom Pom Unit.[314] Note shell belt over barrel.

After the menace of the Boer artillery was removed from our right flank, the advance became still more rapid. Batteries and squadrons were urged into a gallop. Broadwood himself hurried forward. We topped a final rise.

Then at last we viewed the vermin. There, crawling up the opposite slope, clear cut on a white roadway, was a long line of waggons—ox waggons and mule waggons—and behind everything a small cart drawn by two horses. All were struggling with frantic energy to escape from their pursuers. But in vain.

The batteries spun round and unlimbered. Eager gunners ran forward with ammunition, and some with belts for the 'pom-poms.' There was a momentary pause while ranges were taken and sights aligned, and then----! Shell after shell crashed among the convoys. Some exploded on the ground, others, bursting in the air, whipped up the dust all round mules and men. The 'pom-poms,' roused at last from their apathy by this delicious target and some pointed observations of the General, thudded out strings of little bombs. For a few minutes the waggons persevered manfully. Then one by one they came to a standstill. The drivers fled to the nearest shelter, and the animals strayed off the road or stood quiet in stolid ignorance of their danger.

And now at this culminating moment I must, with all apologies to 'Brooksby,' change the metaphor, because the end of the chase was scarcely like a fox hunt. The guns had killed the quarry, and the Cavalry dashed forward to secure it. It was a fine bag—to wit, fifteen laden waggons and seventeen prisoners.

Such was the affair of Heilbron, and it was none the less joyous and exciting because, so far as we could learn, no man on either side was killed, and only one trooper and five horses wounded. Then we turned homewards.[352]

On the way back to the town I found, near a fine farmhouse with deep verandahs and a pretty garden, Boer ambulance waggons, two German doctors, and a dozen bearded men. They inquired the issue of the pursuit; how many prisoners had we taken? We replied by other questions. 'How much longer will the war last?'

'It is not a war any more,' said one of the Red Cross men. 'The poor devils haven't got a chance against your numbers.'

[352] Back towards Heilbron from where this installment was penned.

Boer Doctors.[315]

Red Cross Wagon With Driver and attendant.[316]

Boer Prisoners of War.[317]

Boer prisoners awaiting to be transported to one of the POW camps.[318]

'Nevertheless,' interposed another, 'they will fight to the end.'

I looked towards the last speaker. He was evidently of a different class to the rest.

'Are you,' I asked, 'connected with the ambulance?'

'No, I am the military chaplain to the Dutch forces.'

'And you think the Free State will continue to resist?'

'We will go down fighting. What else is there to do? History and Europe will do us justice.'

'It is easy for you to say that, who do not fight; but what of the poor farmers and peasants you have dragged into this war? They do not tell us that they wish to fight. They think they have been made a catspaw for the Transvaal.'

'Ah,' he rejoined, warmly, 'they have no business to say that now. They did not say so before the war. They wanted to fight. It was a solemn pledge. We were bound to help the Transvaalers; what would have happened to us after they were conquered?'

'But, surely you, and men like you, knew the strength of the antagonist you challenged. Why did you urge these simple people to their ruin?'

'We had had enough of English methods here. We knew our independence was threatened. It had to come. We did not deceive them. We told them. I told my flock often that it would not be child's play.'

'Didn't you tell them it was hopeless?'

'It was not hopeless,' he said. 'There were many chances.'

'All gone now.'

'Not quite all. Besides, chances or no chances, we must go down fighting.'

'You preach a strange gospel of peace!'

'And you English,' he rejoined, 'have strange ideas of liberty.'

So we parted, without more words; and I rode on my way into the town. Heilbron had one memory for me, and it was one which was now to be revived. In the hotel—a regular country inn—I found various British subjects who had been assisting the Boer ambulances—possibly with rifles. It is not my purpose to discuss here the propriety of their conduct. They had been placed in situations which do not come to men in quiet times, and for the rest they were mean-spirited creatures.

While the Republican cause seemed triumphant they had worked for the Dutch, had doubtless spoken of 'damned rooineks,' and used other similar phrases; so soon as the Imperial arms predominated they had changed their note; had refused to go on commando in any capacity, proclaimed that Britons never should be slaves, and dared the crumbling organism of Federal government to do its worst.

We talked about the fighting in Natal which they had seen from the other side. The Acton Homes affair cropped up.[353] You will remember that we of the irregular brigade plumed ourselves immensely on this ambuscading of the Boers—the one undoubted score we ever made against them on the Tugela.

'Yes,' purred my renegades, 'you caught the damned Dutchmen fairly then. We were delighted, but of course we dared not show it.' (Pause.) 'That was where De Mentz[354] was killed.'

De Mentz! The name recalled a vivid scene—the old field-cornet lying forward, grey and grim, in a pool of blood and a litter of empty cartridge cases, with his wife's letter clasped firmly in his stiffening fingers. He had 'gone down fighting;' had had no doubts what course to steer. I knew when I saw his face that he had thought the whole thing out. Now they told me that there had been no man in all Heilbron more bitterly intent on the war, and that his letter in the 'Volksstem,' calling on the Afrikanders to drive the English scum from the land, had produced a deep impression.

'Let them,' thus it ran, 'bring 50,000 men, or 80,000 men, or even'—it was a wild possibility—'100,000, yet we will overcome them.' But they brought more than 200,000, so all his calculations were disproved, and he himself was killed with the responsibility on his shoulders of leading his men into an ambush which, with ordinary precautions, might have been avoided. Such are war's revenges. His widow, a very poor woman, lived next door to the hotel, nursing her son who had been shot through the lungs during the same action. Let us hope he will recover, for he had a gallant sire.

[353] In Natal where WSC & Co. captured a portion of a commando. See London to Ladysmith Ch. XVI Trichardt's Drift and the Affair of Acton Homes in despatch from Venter's Spruit: Sunday, January 22, 1900.

[354] Unable to identify this individual. About 60 'Mentz' possibilities: https://www.angloboerwar.com/name-search

XIII. Action of Johannesburg

Johannesburg: June 1st, 1900.

On the 24th of May, Ian Hamilton's force, marching west from Heilbron, struck the railway and joined Lord Roberts's main column. The long marches, unbroken by a day's rest, the short rations to which the troops had been restricted, and the increasing exhaustion of horses and transport animals seemed to demand a halt. But a more imperious voice cried 'Forward!' and at daylight the travel-stained brigades set forth, boots worn to tatters, gun horses dying at the wheel, and convoys struggling after in vain pursuit—'Forward to the Vaal.'

And now the Army of the Right Flank became the Army of the Left; for Hamilton was directed to move across the railway line and march on the drift of the [Vaal] river near Boschbank. Thus, for the first time it was possible to see the greater part of the invading force at once.[355]

French, indeed, was already at Parys,[356] but the Seventh and Eleventh Divisions, the Lancer brigade, the corps troops, the heavy artillery, and Hamilton's four brigades were all spread about the spacious plain, and made a strange picture; long brown columns of Infantry, black squares of batteries, sprays of Cavalry flung out far to the front and flanks, 30,000 fighting men together, behind them interminable streams of waggons, and, in their midst, like the pillar of cloud that led the hosts of Israel, the war balloon, full blown, on its travelling car.

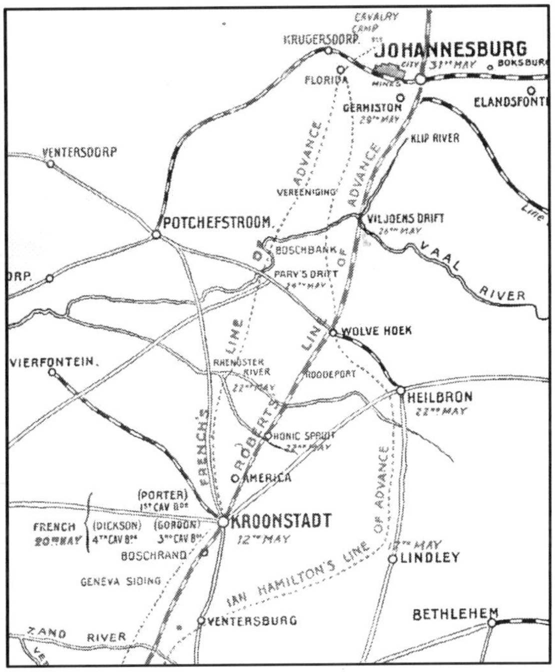

Military reconnaissance balloon with an escort. It took four carts to carry the gas tubes and one to carry the equipment, each cart was drawn by four horses. Two extra carts carried the baggage.[357] Creswicke's map shows Hamilton crossing at Wolve Hoek. America Siding is just north of Kroonstadt. Churchill was just in Heilbron. SA War Vol 5 Map page VIII.[319]

[355] WSC Quote: "Lord Roberts, against the advice of his Intelligence Officer, believed that the enemy would retreat into the Western rather than the Eastern Transvaal. Accordingly, as we approached the frontiers of the Transvaal, Sir Ian Hamilton's column was shifted from the right of the main army to the left. We crossed the central line of railway at America siding [See: Creswicke SA War Vol 5 Map page VIII and reproduced herein for location.] and marched to the fords of the Vaal River. In this disposition we were so placed as to turn the western flank of the Johannesburg district and so compel its evacuation by the enemy without requiring the main army to deliver a costly frontal attack. The Boers were alive to the purpose of this manoeuvre, and although ready to evacuate Johannesburg, they sent a strong force to oppose the advance of Hamilton's column at a point called Florida on the Johannesburg-Potchefstroom route." WSC My Early Life Ch. XXVI.

[356] The name of the city is the Afrikaans translation of Paris. The origin of the name 'Parys' is attributed to German surveyor Schilbach, who named it after the city of Paris because of the similarity between its location next to the Vaal River and that of Paris on the River Seine. Parys - Wikipedia

[357] Roberts' army advancing toward Johannesburg - 84th Battery and Balloon Corps, S.A. Digital ID: (b&w film copy neg.) cph 3c07541 http://hdl.loc.gov/loc.pnp/cph.3c07541 Reproduction Number: LC-USZ62-107541 (b&w film copy neg.) https://www.loc.gov/pictures/item/93505023/resource/

We crossed the Vaal on the 26th prosperously and peacefully. Broadwood, with his Cavalry, had secured the passage during the previous night, and the Infantry arriving found the opposite slopes in British hands. Moreover, the engineers, under the indefatigable Boileau, assisted by the strong arms of the Blues and Life Guards, had cut a fine broad road up and down the steep river banks.

Once across we looked again for the halt. Twenty-four hours' rest meant convoys with full rations and forage for the horses.

Mounted Infantry Crossing the Vaal River.[320] Thursday, May 24, 1900 British Army entered the Transvaal, crossing the Vaal near Parys, unopposed.

First camp inside the Transvaal.[321]

But in the morning there came a swift messenger from the Field-Marshal: main army crossing at Vereeniging,[358] demoralisation of the enemy increasing, only one span of the railway bridge blown up, perhaps Johannesburg within three days—at any rate, 'try,' never mind the strain of nerve and muscle or the scarcity of food.

Forward again. [27th] That day Hamilton marched his men eighteen miles—('ten miles,' say the text-books on war, 'is a good march for a division with baggage,' and our force, carrying its own supplies, had ten times the baggage of a European division!)—and succeeded besides in dragging his weary transport with him. By good fortune the Cavalry discovered a little forage—small stacks of curious fluffy grass called manna, and certainly heaven-sent—on which the horses subsisted and did not actually starve. All day the soldiers pressed on, and the

[358] Vereeniging is a city located where the Klip River empties into the northern loop of the Vaal River. The city is the location where the Treaty of Vereeniging ending the Second Boer War (1899–1902) was negotiated by the delegates of the South African Republic, Orange Free State and the British Empire. Vereeniging - Wikipedia

sun was low before the bivouac was reached. Nothing untoward disturbed the march, and only a splutter of musketry along the western flank guard relieved its dullness.

Troops of one of the British mounted regiments collecting forage for their horses.[322]

At first, after we had crossed the Vaal, the surface of the country was smooth and grassy, like the Orange River Colony, but as the column advanced northwards the ground became broken—at once more dangerous and more picturesque. Dim blue hills rose up on the horizon, the rolling swells of pasture grew sharper and less even, patches of wood or scrub interrupted the level lines of the plain, and polished rocks of conglomerate or auriferous quartz showed through the grass, like the bones beneath the skin of the cavalry horses. We were approaching the Rand.[359]

On the evening of the 27th, Hamilton's advance guard came in touch with French, who, with one Mounted Infantry and two Cavalry brigades, was moving echeloned forward on our left in the same relation to us as were we to the main army.

The information about the enemy was that, encouraged by the defensive promise of the ground, he was holding a strong position either on the Klip Riviersburg, or along the line of the gold mines crowning the main Rand reef. On the 28th, in expectation of an action next day, Hamilton made but a short march. French, on the other hand, pushed on to reconnoitre, and if possible—for the Cavalry were very ambitious—to pierce the lines that lay ahead.

I rode with General Broadwood, whose brigade covered the advance of Hamilton's column. The troops had now entered a region of hills which on every side threatened the march and limited the view.

At nine o'clock we reached a regular pass between two steep rocky ridges. From the summit of one of these ridges a wide landscape was revealed. Northwards across our path lay the black line of the Klip Riviersburg,[360] stretching to the east as far as I could see, and presenting everywhere formidable positions to the advancing force. To the west these frowning features fell away in more grassy slopes, from among which, its approach obstructed by several rugged under features, rose the long smooth ridge of the Witwatersrand reef.

[359] The Witwatersrand, locally the Rand or, less commonly, the Reef is a 56-kilometre-long (35 mi), north-facing scarp in South Africa. It consists of a hard, erosion-resistant quartzite metamorphic rock, over which several north-flowing rivers form waterfalls, which account for the name Witwatersrand, meaning "white water ridge" in Afrikaans Witwatersrand - Wikipedia

[360] Klipriviersberg Nature Reserve - Wikipedia

Witwatersrand National Botanical Gardens.[323] Klipriviersberg Nature Reserve.[324]

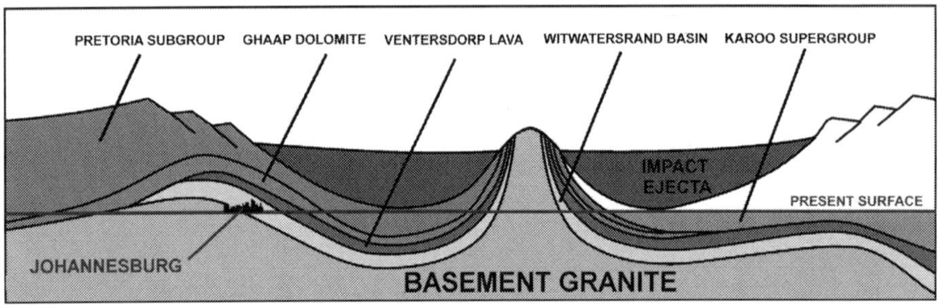

Vredefort crater[361] Cross section[325] Witwatersrand.[362]

The numerous grass fires which attend the march of an army in dry weather—the results of our carelessness, or, perhaps, of the enemy's design—veiled the whole prospect with smoke, and made the air glitter and deceive like the mirages in the Soudan. But one thing showed with sufficient distinctness to attract and astonish all eyes. The whole crest of the Rand ridge was fringed with factory chimneys.

Gold Mines at Johannesburg 1899.[326]

We had marched nearly 500 miles through a country which, though full of promise, seemed to European eyes desolate and wild, and now we turned a corner suddenly, and there before us sprang the evidences of wealth, manufacture, and bustling civilisation. I might have been looking from a distance at Oldham.[363]

[361] Vredefort crater - Wikipedia In 1886 a large deposit of gold was discovered in the Witwatersrand range of hills, which forms part of the central rebound peak of the Vredefort crater. The asteroid that created this structure struck the Earth two billion years ago and exposed rocks that had been laid down almost a billion years earlier when Archean rivers from the north and west formed fan deltas, with many braided channels, before flowing into the "Witwatersrand Sea" to the south.

[362] The Witwatersrand; locally the Rand or, less commonly, the Reef) is a 56-kilometre-long (35 mi), north-facing scarp in South Africa. It consists of a hard, erosion-resistant quartzite metamorphic rock, over which several north-flowing rivers form waterfalls, which account for the name Witwatersrand, meaning "white water ridge" in Afrikaans. This east-west-running scarp can be traced with only one short gap, from Bedfordview (about 10 km [6 mi] west of O.R. Tambo International Airport) in the east, through Johannesburg and Roodepoort, to Krugersdorp in the west. Witwatersrand - Wikipedia

The impression was destroyed by the booming of shotted guns, unheard, by God's grace, these many years in peaceful Lancashire. French was at work. The haze and the distance prevented us from watching closely the operations of the Cavalry. The dark patches of British horsemen and the white smoke of the Dutch artillery were the beginning and the end of our observations. But, even so, it was easy to see that French was not making much progress.

As the afternoon wore on the loud reverberations of heavy cannon told that the Boers had disclosed their real position, and we knew that something more substantial than Cavalry would be required to drive them from it. In the evening French's brigades were seen to be retiring across the Klip River, and the night closed in amid the rapid drumming of the Vickers-Maxims covering his movement, bringing with it the certainty of an Infantry action on the morrow.

Royal Dublin Fusiliers Maxim Gun Team.[327]

Navy Bluejackets with Maxim Gun.[328]

At twelve o'clock a despatch from the Cavalry division reached Hamilton. French's messenger said that the cavalry were having a hot fight and were confronted by several 40-pounder guns, but the stout-hearted commander himself merely acquainted Hamilton with his orders from headquarters, to march *via* Florida[364] to Driefontein,[365] and made no allusion to his fortunes nor asked for assistance. Indeed, as we found out later, his operations on the 28th had been practically confined to an artillery duel, in which, though the expenditure of ammunition was very great and the noise alarming, the casualties—one officer and eight men—were fortunately small.

But the Boers, seeing the Cavalry retire at dusk, claimed that they had repulsed the first attack; their confidence in the strength of the Rand position was increased; their resistance on the next day was consequently more stubborn; and the 'Standard and Diggers' News' was enabled to terminate a long career of exaggeration and falsehood by describing one more 'bloody British defeat with appalling slaughter.'

The event of the next day admitted of no such misinterpretation.

The orders from headquarters for the 29th were such as to involve certain fighting should the enemy stand. French, with the Cavalry Division, was to march around Johannesburg to Driefontein; Ian Hamilton was directed on Florida; the main army, under the Field-Marshal, would occupy Germiston[366] and seize the junctions of the Natal, Cape Colony, and Potchefstroom lines. These movements, which the chief had indicated by flags on the map, were now to be executed—so far as possible—by soldiers on the actual field.

The operations of the main army are not my concern in this letter; but it is necessary to state the result, lest the reader fail to grasp the general idea, and, while studying the detail, forget their scale and meaning.

Advancing with great speed and suddenness through Elandsfontein,[367] Lord Roberts surprised the Boers in Germiston, and after a brief skirmish drove them in disorder from the town, which he then occupied. So precipitate was the flight of the enemy, or so rapid the British advance, that **nine locomotives** and much other rolling stock were captured, and the line from Germiston southward to Vereeniging was found to be undamaged.

[363] Oldham is a town in Greater Manchester, England. Winston will be elected MP from here when he returns. Oldham - Wikipedia

[364] Florida is located about 16 kilometres west of Johannesburg. Florida, Gauteng - Wikipedia

[365] Driefontein is the Driefontein Mine in the West Witwatersrand Basin (West Wits) mining field. Driefontein, Gauteng - Wikipedia

[366] Germiston, also known as kwaDukathole,is a city in the East Rand. Germiston - Wikipedia

[367] Alberton is a town situated on the southern part of the East Rand close to Johannesburg. Alberton, South Africa - Wikipedia

Refugee engines at Viljoen's Drift.[368] At the time the British troops were closing in on Bloemfontein, the Boers shifted all rolling stock further north. Before the troops arrives at Viljoen's Drift, these engines were sent across the border to the Transvaal. It will be seen that all the connecting rods gave been removed. A Goods Shed at Viljoen's Drift.[329]

The importance of these advantages on the success of the operations can scarcely be over-estimated. The problem of supply was at once modified, and though the troops still suffered privations from scarcity of food, the anxieties of their commanders as to the immediate future were removed.

French had camped for the night south of the Klip River, just out of cannon shot of the enemy's position, and at eight o'clock on the morning of the 29th he moved off westward, intending to try to penetrate, or, better still, circumvent, the barrier that lay before him.

Such ground as he had won on the previous day he held with Mounted Infantry, and thus masking the enemy's front he attempted to pierce if he could not turn his right. For these purposes the force at his disposal—three horse batteries, four 'pom-poms,' and about 3,000 mounted men—was inadequate and unsuited. But he knew that Ian Hamilton, with siege guns, field guns, and two Infantry brigades, was close behind him, and on this he reckoned.

Firing began about seven o'clock, when the Boers attacked the Mounted Infantry Corps holding the positions captured on the 28th, and who were practically covering the flank movement of the rest of the Cavalry Division and the march of Hamilton's column. The Mounted Infantry, who were very weak, were gradually compelled to fall back, being at one time enfiladed by two Vickers-Maxims and heavily pressed in front.

But their resistance was sufficiently prolonged to secure the transference of force from right to left. By ten o'clock French had gone far enough west to please him, and passing round the edge of a deep swamp turned the heads of his regiments sharply to their right (north), and moved towards the Rand ridge and its under features.

By the vigorous use of his Horse Artillery he [French] cleared several of the advanced kopjes, and had made nearly two miles progress north of the drainage line of the Klip River, when he was abruptly checked. A squadron sent forward against a low fringe of rocks, clumping up at the end of a long grass glacis, encountered a sudden burst of musketry fire, and returned, pursued by shell, with the information that mounted men could work no further northwards.

Meanwhile Hamilton, who had determined to lay his line of march across the Doornkop ridges[369] [370](of inglorious memory),[371] and whose Infantry, baggage, and guns were spread all along the flat plain south of the Klip, was drawing near.

[368] These engines were sent across the border to the Transvaal. Note that all the connecting rods have been removed so they can be towed by a front engine. Anglo Boer War - Images from 'On the Transvaal border' published by W H Gill, 1900 - Page 3 - Boer War Forum Credit Dr. David Biggins.

[369] Doornkop (literally "thorn hill") is a ridge and locality on the western outskirts of Johannesburg. Doornkop - Wikipedia

[370] The Doornkop area is also the site of the third day (29 May 1900) of the Battle of Johannesburg that took place nearby from 27 to 29 May 1900 during the Second Boer War. The ridge was taken by cavalry under command of Lt Gen John French while a nearby ridge, now covered by several Soweto suburbs, was taken by seven infantry battalions, including the following units on the left and right flank.
Left Flank (French)
21st Brigade;
the City Imperial Volunteers
The Derbyshire Regiment
The Cameron Highlanders
The 1st Battalion The Royal Sussex Regiment.

English tourists with horses and a cart at the battlefield of Doornkop, 1896.[330]

French halted his brigades and awaited him. The instructions from headquarters defined very carefully the relations which were to be observed between the two Generals. They were to co-operate, yet their commands were entirely separate. Should they attack the same hill at once, French, as a lieutenant-general and long senior to Hamilton, would automatically assume command. But this contingency was not likely to arise from the military situation, and the good feeling and mutual confidence which existed between these two able soldiers, and which had already produced golden results at Elandslaagte, made the possibility of any misunderstanding still more remote.

French was joined by Hamilton at one o'clock, and they discussed the situation together. French explained the difficulty of further direct advance. He must move still more to the west. On the other hand, Hamilton, whose force was eating its last day's rations, could make no longer *détour*, and must break through there and then—frontal attack, if necessary. So all fitted in happily. The Cavalry division moved to the left to co-operate with the Infantry attack by threatening the Boer right, and, in order that this pressure might be effective, Hamilton lent Broadwood's Brigade and two corps of Mounted Infantry to French for the day. He himself prepared to attack what stood before him with his whole remaining force.

By two o'clock the Cavalry in brown swarms had disappeared to the westward, both Infantry brigades were massed under cover on the approaches of the Rand ridge, and the transport of the army lay accumulated in a vast pool near the passage of the Klip—here only a swamp, but further east a river. The artillery duel of the morning had died away. The firing on the right, where the Mounted Infantry still maintained themselves, was intermittent. The reconnaissance was over.

The action was about to begin, and in the interval there was a short, quiet lull—the calm before the storm. The soldiers munched their biscuits silently under the sun blaze. The officers and staff ate a frugal luncheon. Ian Hamilton with his aide-de-camp, the Duke of Marlborough,[372] shared the contents of my wallets.[373] I watched the General closely. He knew better than the sanguine people who declared the Boers had run away already. No one understood better than he what a terrible foe is the rock-sheltered Mauser-armed Dutchman. In spite of its cavalry turning movement, and other embellishments, the impending attack must be practically frontal. Supply did not allow a wider circle: to stop was to starve; and the position before us—half a dozen clusters of rock, breaking from the smooth grass upward slopes, except in colour like foam on the crest of waves, natural parapet and glacis combined, and, beyond all, the long bare ridge of the Rand lined with who should say what

Right Flank of Lt Gen Ian Hamilton
19 Brigade;
The Gordon Highlanders.
The 2nd Battalion The Royal Canadian Regiment.
The Kings Shropshire Light Infantry.
The Duke of Cornwall's Light Infantry.

[371] Doornkop is where Dr. Leander Starr Jameson was defeated on January 2nd 1896 at the conclusion of the Jameson Raid. Jameson Raid - Wikipedia

[372] Who has accompanied his first cousin Winston during this entire campaign.

[373] A bag or pouch. In this case saddlebags.

entrenchments or how many defenders—a prospect which filled all men who knew with the most solemn thoughts.

For my part, having seen the Infantry come reeling back in bloody ruin two or three times from such a place and such a foe, though I risked no repute on the event—scarcely my life—I confess to a beating heart. But the man who bore all the responsibility, and to whom the result meant everything, appeared utterly unmoved. Indeed, I could almost imagine myself the General and the General the Press Correspondent, though perhaps this arrangement would scarcely have worked so well.

At three o'clock precisely the Infantry advanced to the attack. Major-General Bruce-Hamilton directed the left attack with the Twenty-first Brigade, and Colonel Spens the right with the Nineteenth Brigade. The whole division was commanded by General Smith-Dorrien. The lateness of the hour gave scarcely any time for the artillery preparation, and the artillery came into action only a few minutes before the infantry were exposed to fire.

It must be noticed that the combination of the batteries and the support which they afforded to the attack was scarcely so effective as might have been expected from the number of guns available. But the General commanding a mixed force is bound to trust the various specialists under him, at least until experience has shown them to be deficient in energy or ability.

The Infantry advance was developed on the most modern principles. Each brigade occupied a front of more than a mile and three-quarters, and the files of the first line of skirmishers were extended no less than thirty paces. Bruce-Hamilton, with the left attack, started a little earlier than the right brigade, and, with the City Imperial Volunteers in the first line, soon had his whole command extended on the open grass.

A few minutes after three, French's guns were heard on the extreme left, and about the same time the firing on the right swelled up again, so that by the half-hour the action was general along the whole front of battle—an extent of a little over six miles.

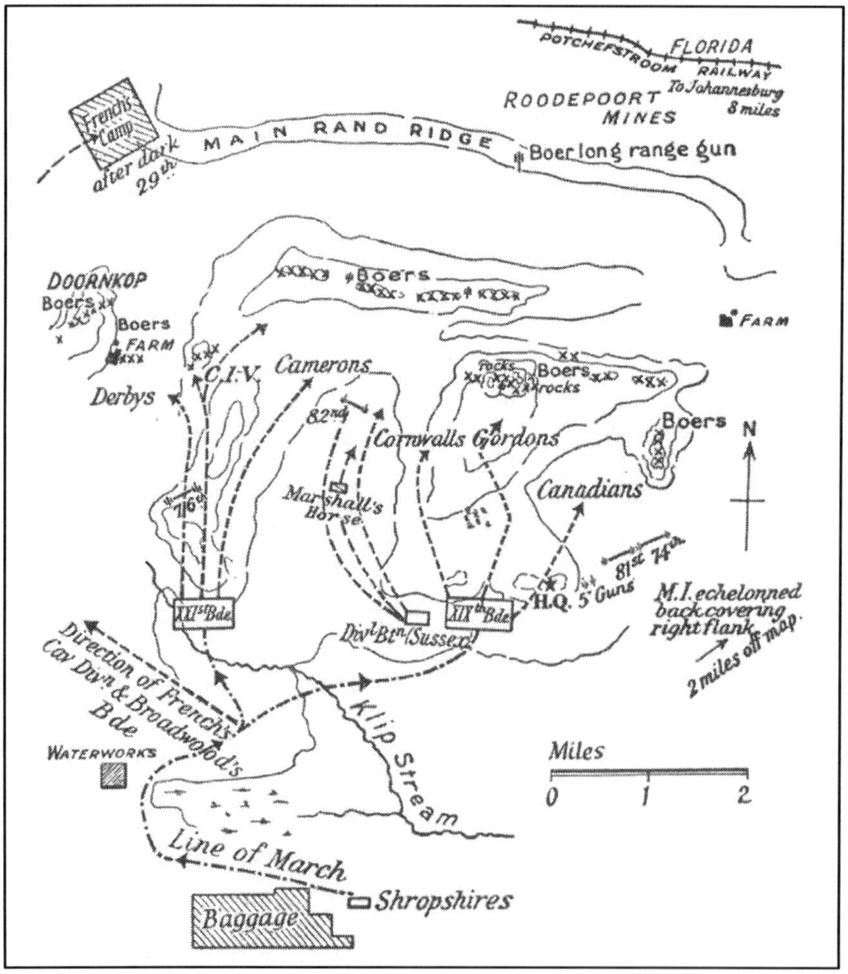

Actions before Johannesburg. May 29th 1900. [331] Left (21st Brigade) - Bruce-Hamilton Center (19th Brigade) - Smith-Dorrien.

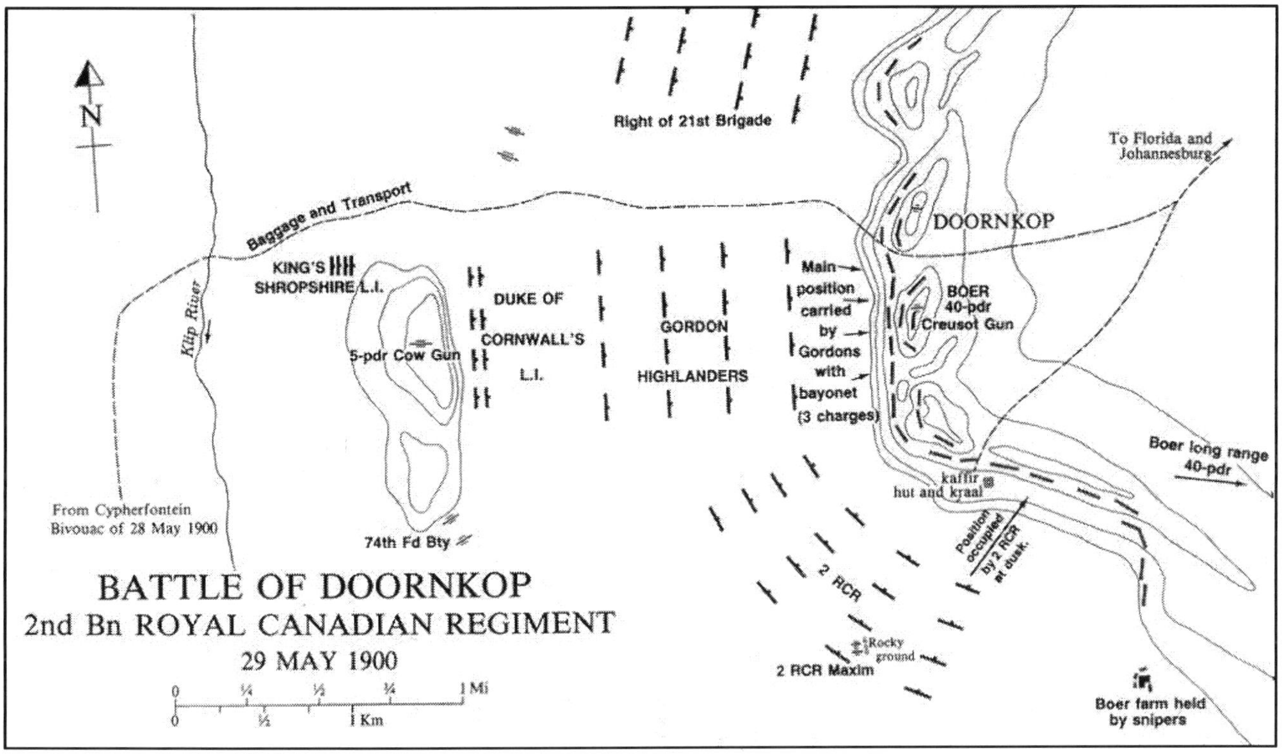

Battle of Doornkop.[332] [374]

The left attack, pressed with vigour, and directed with skill by General Bruce-Hamilton, led along a low spur, and was designed to be a kind of inside turning movement to assist the right in conformity with the Cavalry action now in full swing. The City Imperial Volunteers moved forward with great dash and spirit, and in spite of a worrying fire from their left rear, which increased in proportion as they moved inwards towards the right, drove the Boers from position after position. While there is no doubt that French's pressure beyond them materially assisted their advance, the rapid progress of this Twenty-first Brigade entitled them and their leader to the highest credit. The Cameron Highlanders and the Sherwood Foresters supported the attack. The Boers resisted well with artillery, and their shells caused several casualties among the advancing lines; but it was on the right that the fighting was most severe.

The leading battalion of the Nineteenth Brigade chanced—for there was no selection—to be the Gordon Highlanders; nor was it without a thrill that I watched this famous regiment move against the enemy. Their extension and advance were conducted with machine-like regularity. The officers explained what was required to the men. They were to advance rapidly until under rifle fire, and then to push on or not as they might be instructed.

With impassive unconcern the veterans of Chitral,[375] Dargai,[376] the Bara Valley,[377] Magersfontein, Paardeburg, and Houtnek walked leisurely forward, and the only comment recorded was the observation of a private: 'Bill, this looks like being a kopje day.' Gradually the whole battalion drew out clear of the covering ridge, and long dotted lines of brown figures filled the plain. At this moment two batteries and the two 5-in. guns opened from the right of the line, and what with the artillery of French and Bruce-Hamilton there was soon a loud cannonade.

The Dutch replied at once with three or four guns, one of which seemed a very heavy piece of ordnance on the main Rand ridge, and another fired from the kopje against which the Gordons were marching.

[374] Battle of Doornkop - Wikipedia

[375] Chitral is situated on the Chitral River in northern Khyber Pakhtunkhwa, Pakistan. Chitral - Wikipedia

[376] Dargai is one of the tehsils of Malakand District in Pakistan's northwestern Khyber Pakhtunkhwa province. Dargai - Wikipedia

[377] See: Bara River - Wikipedia In Pakistan.

Long Tom showing breech and powder bag used along with ramrod/swab and water bucket.[333] Creusot Gun at Mafeking.[334]

But the Boer riflemen, crouching among the rocks, reserved their fire for a near target. While the troops were thus approaching the enemy's position, the two brigades began unconsciously to draw apart. Colonel Spens'[378] battalions had extended further to the right than either Ian Hamilton or Smith-Dorrien had intended. Bruce-Hamilton, pressing forward on the left, found himself more and more tempted to face the harassing attack on his left rear. Both these tendencies had to be corrected. The Gordons were deflected to their left by an officer, Captain Higginson,[379] who galloped most pluckily into the firing line in spite of a hail of bullets. Bruce-Hamilton was ordered to bear in to his right and disregard the growing pressure behind his left shoulder. Nevertheless a wide gap remained. But by this mischance Ian Hamilton contrived to profit. Smith-Dorrien had already directed the only remaining battalion—the Sussex—to fill up the interval, and the General-in-Chief now thrust a battery forward through the gap, almost flush with the skirmish line of the Infantry on its left and right.

The fire of these guns, combined with the increasing pressure from the turning movements both of Bruce-Hamilton and French, who was now working very far forward in the west, weakened the enemy's position on the [Doornkop] kopje which the Gordons were attacking. Yet, when every allowance has been made for skilful direction and bold leading, the honours, equally with the cost of the victory, belong more to the Gordon Highlanders than to all the other troops put together.

Officers of the Gordon Highlanders.[335]

[378] J. Spens Lieutenant Colonel Source: Field Marshal Roberts. 31 March 1900. Shropshire Light Infantry (King's).

[379] Cecil Pickford Higginson Captain Shropshire Light Infantry (King's) Extensive Bio: Anglo Boer War - Name search

NCO's of the Gordon Highlanders.[336]

The rocks against which they advanced proved in the event to be the very heart of the enemy's position. The grass in front of them was burnt and burning, and against this dark background the khaki figures showed distinctly. The Dutch held their fire until the attack was within 800 yards, and then, louder than the cannonade, the ominous rattle of concentrated rifle fire burst forth. The black slope was spotted as thickly with grey puffs of dust where the bullets struck as with advancing soldiers, and tiny figures falling by the way told of heavy loss. But the advance neither checked nor quickened.

With remorseless stride, undisturbed by peril or enthusiasm, the Gordons swept steadily onward, changed direction half left to avoid, as far as possible, an enfilade fire, changed again to the right to effect a lodgment on the end of the ridge most suitable to attack, and at last rose up together to charge. The black slope twinkled like jet with the unexpected glitter of bayonets. The rugged skyline bristled with kilted figures, as, in perfect discipline and disdainful silence, those splendid soldiers closed on their foe.

The Boers shrank from the contact. Discharging their magazines furiously, and firing their guns at point-blank range, they fled in confusion to the main ridge, and the issue of the action was no longer undecided.

Still the fight continued. Along the whole Infantry front a tremendous rifle fire blazed. Far away to the left French's artillery pursued the retreating Boers with shells. The advanced batteries of Hamilton's force fired incessantly. The action did not cease with the daylight. The long lines of burning grass cast a strange, baleful glare on the field, and by this light the stubborn adversaries maintained their debate for nearly an hour.

At length, however, the cannonade slackened and ceased, and the rifles soon imitated the merciful example of the guns. The chill and silence of the night succeeded the hot tumult of the day. Regiments assembled and reformed their ranks, ambulances and baggage waggons crowded forward from the rear, the burning veldt was beaten out, and hundreds of cooking fires gleamed with more kindly meaning through the darkness.

After the Battle. The Irish Brigade assembles.[337]

Grass Fire at dusk.[338]

The General [Hamilton] rode forward, to find the Gordons massed among the rocks they had won. The gallant Burney,[380] who commanded the firing line, was severely wounded. St. John Meyrick[381] was killed. Nine officers

[380] Herbert H. Burney Major MID LG: 16 April 1901, Re: General mentions List of CB recipients.

[381] Unable to identify this individual. Possibly see: Meyrick baronets - Wikipedia

and eighty-eight soldiers had fallen in the attack; but those that remained were proud and happy in the knowledge that they had added to the many feats of arms which adorn the annals of the regiment—one that was at least the equal of Elandslaagte or Dargai; and, besides all this, they may have reflected that by their devotion they had carried forward the British cause a long stride to victory, and, better than victory, to honourable peace. Ian Hamilton spoke a few brief words of thanks and praise to them—'the regiment my father commanded and I was born in'—and told them that in a few hours all Scotland would ring with the tale of their deeds. And well Scotland may, for no men of any race could have shown more soldier-like behaviour.

Then we rode back to our bivouac, while the lanterns of searching parties moved hither and thither among the rocks, and voices cried 'Bearer party this way!' 'Are there any more wounded here?' with occasional feeble responses.

Searching for wounded at night.[339]

Bearer party and ambulances.[340]

Owing to the skilful conduct of the attack, the losses, except among the Gordons, were not severe—in all about 150 killed and wounded. The result of the fight—the action of Johannesburg, as we called it—was the general retreat of all the enemy west of the town under Delarey and Viljoen northwards towards Pretoria, and, in conjunction with the Field-Marshal's movements, the surrender of the whole of the Witwatersrand.

French, continuing his march at dawn to Driefontein, captured one gun and several prisoners. Ian Hamilton entered Florida, and found there and at Maraisburg[382] sufficient stores to enable him to subsist until his convoys arrived.

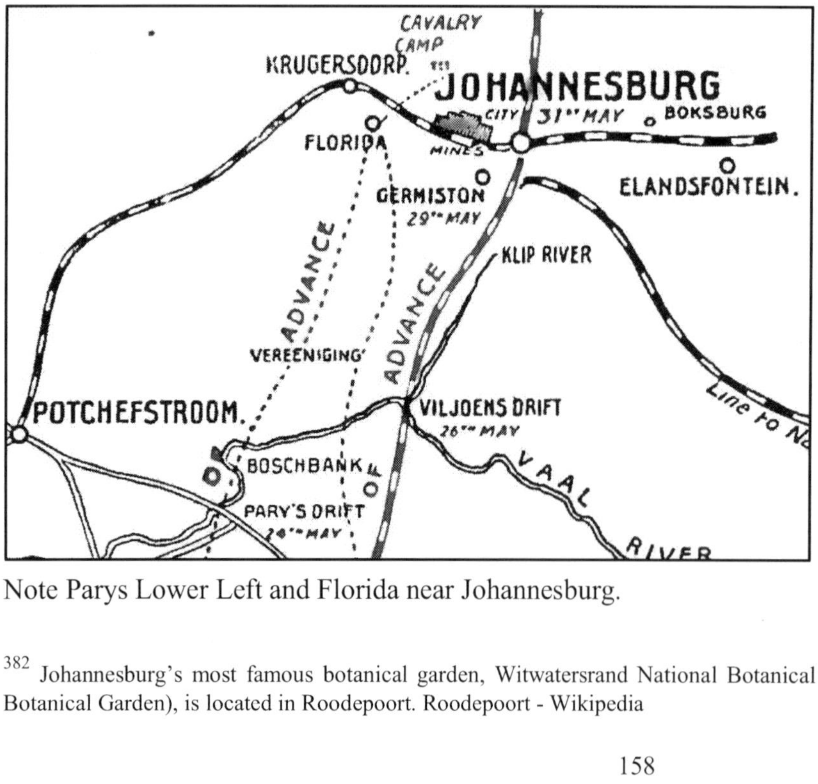

Note Parys Lower Left and Florida near Johannesburg.

[382] Johannesburg's most famous botanical garden, Witwatersrand National Botanical Gardens (now renamed Walter Sisulu National Botanical Garden), is located in Roodepoort. Roodepoort - Wikipedia

Quote From My Early Life

It is difficult to contextualize this quote with respect to the events described above. [383] It is so dramatic an episode that I need to recite Winston's first person account in order to assemble a complete record of the days' events. This episode is not reported anywhere else; nor is it the only time that Winston participates in an action but does not report it in his despatches. As we shall see, Ian Hamilton credits his actions at the Battle of Diamond Hill on June 12th; but Winston does not record himself participating so directly in the engagement.

Here on June 1, 1900,[384] on the very ground where the Jameson raiders had surrendered four years before, was fought what in those days was considered a sharp action.

The Boers, buried amid the jagged outcropping rocks of the ridges, defied bombardment and had to be dislodged by the bayonet. [This is related supra.]

The Gordon Highlanders, with a loss of nearly a hundred killed and wounded, performed this arduous task, while at the same time French's mounted forces tried rather feebly to turn the enemy's right flank and rear.

[This to some degree contradicts the narrative and the map since he noted that French advanced two miles.]

I had myself a fortunate escape in this fight. After the ridge had been taken by the Highlanders, General Smith-Dorrien, who commanded one of Sir Ian Hamilton's brigades,[385] wished to bring his artillery immediately on to the captured position, and as time was short, determined to choose the place himself.

Inviting me to follow him, he cantered forward alone across the rolling slopes. The Boers had, according to their usual custom, lighted the dry grass, and long lines of smoke blotted out the landscape in various directions. In these baffling veils we missed the left flank of the Gordon Highlanders on the ridge, and coming through the smoke curtain with its line of flame, found ourselves only a few score yards distant from the enemy. There was an immediate explosion of rifle fire. The air all around us cracked with a whip-lash sound of close-range bullets. We tugged our horses' heads round and plunged back into our smoke curtain. One of the horses was grazed by a bullet, but otherwise we were uninjured.

Johannesburg Defence, May 28th, 1900' "Group of Boer Commandants on the Klip River, Near Johannesburg. This picture was taken just before the great battle on 28 May 1900, when all the forces united under Lord Roberts were needed to capture the positions held by 2,000 Boers. Lord Roberts succeeded, but only after he had encountered the most determined resistance. His losses were very heavy. Johannesburg was occupied by him on the 31st of May."[341]

383 My Early Life Ch. XXVII.

384 This is the May 29th Battle of Doornkop.

385 Major-General Smith-Dorrien, 19th Brigade. See Appendix.

XIV. The Fall of Johannesburg

Johannesburg: June 2nd 1900

[Wednesday, May 30th] Morning broke and the army arose ready, if necessary, to renew the fight. But the enemy had fled. The main Rand ridge still stretched across our path. Its defenders had abandoned all their positions under the cover of darkness. Already French's squadrons were climbing the slopes to the eastward and pricking their horses forward to Elandsfontein (North). So Hamilton's force, having but six miles to march to Florida, did not hurry its departure, and we had leisure to examine the scene of yesterday's engagement. Riding by daylight over the ground of the Gordon's attack, we were still more impressed by the difficulties they had overcome. From where I had watched the action the Boers had seemed to be holding a long black kopje, some forty feet high, which rose abruptly from the grass plain. It now turned out that the aspect of steepness was produced by the foreshortening effects of the burnt grass area; that in reality the ground scarcely rose at all, and that what we had thought was the enemy's position was only a stony outcrop separated from the real line of defence by a bare space of about 200 yards.

Looking around I found a Highlander, a broad-shouldered, kind-faced man, with the Frontier ribbon, which means on a Gordon tunic much hard fighting; and judging with reason that he would know something of war, I asked him to explain the ground and its effect.

'Well, you see, sir,' he said, in quick spoken phrases, 'we was regularly tricked. We began to lose men so soon as we got on the burnt grass. Then we made our charge up to this first line of little rocks, thinking the Boers were there. Of course they weren't here at all, but back over there, where you see those big rocks. We were all out of breath, and in no order whatever, so we had to sit tight here and wait.'

'Heavy fire?' I asked. He cocked his head like an expert.

'I've seen heavier; but there was enough. We dropped more than forty men here. 'Tis here poor Mr. ---- was wounded; just behind this stone. You can see the blood here yet, sir—this mud's it.'

I looked as required, and he proceeded:

'We knew we was for it then; it didn't look like getting on, and we couldn't get back—never a man would ha' lived to cross the black ground again with the fire what it was, and no attack to fright them off their aim. There was such a noise of the bullets striking the rocks that the officers couldn't make themselves heard, and such confusion too! But two or three of them managed to get together after a while, and they told us what they wanted done ... and then, of course, it was done all right.'

'What was done? What did you do?'

'Why, go on, sir, and take that other line—the big rocks—soon as we'd got our breath. It had to be done.'

He did not seem the least impressed with his feat of arms. He regarded it as a piece of hard work he had been set to do, and which—this as a matter of course—he had done accordingly. What an intrepid conquering machine to depend on in the hour of need!—machine and much more, for this was a proud and intelligent man, who had thought deeply upon the craft of war, and had learned many things in a severe school.

I had not ridden a hundred yards further, my mind full of admiration for him and his type, when a melancholy spectacle broke upon the view. Near a clump of rocks eighteen Gordon Highlanders—men as good as the one I had just talked with—lay dead in a row. Their faces were covered with blankets, but their grey stockinged feet—for the boots had been removed—looked very pitiful. There they lay stiff and cold on the surface of the great Banket Reef. I knew how much more precious their lives had been to their countrymen than all the gold mines the lying foreigners say this war was fought to win. And yet, in view of the dead and the ground they lay on, neither I nor the officer who rode with me could control an emotion of illogical anger, and we scowled at the tall chimneys of the Rand.

Commando with Johannesburg and smokestacks in the background.[342]

Gold mine near Johannesburg.[343]

Johannesburg in 1900.[344]

General Ian Hamilton, General Smith-Dorrien, all their staffs, and everyone who wished to pay a last tribute of respect to brave men, attended the funerals. The veteran regiment stood around the grave, forming three sides of a hollow square—Generals and staff filled the other. The mourning party rested on their arms, reversed; the Chaplain read the Burial Service, the bodies were lowered into the trench, and the pipes began the lament. The wild, barbaric music filled the air, stirring the soldiers, hitherto quite unmoved, with a strange and very apparent force. Sad and mournful was the dirge wailing of battles ended, of friendships broken, and ambitions lost; and yet there were mingled with its sadness many notes of triumph, and through all its mourning rang the cry of hope.

Military funeral.[345]

Gordon Highlander.[346]

The whole of Hamilton's force had marched by ten o'clock, but even before that hour the advanced guard had passed through Florida and picketed the hills beyond. Florida is the Kew Gardens of Johannesburg. A well-built dam across a broad valley has formed a deep and beautiful lake. Carefully planted woods of Australian pines offer a welcome shade on every side. The black and white pointed chimneys of the mine buildings rise conspicuous above the dark foliage. There is a small but comfortable hotel, called 'The Retreat,' to which on Sundays, in times of peace, the weary speculators whose minds were shattered by the fluctuations of the Exchange were wont to resort for rest or diversion.

Auckland Park, Johannesburg.[347]

Stock exchange in Johannesburg.[348]

Everywhere along the reef the signs of industry and commerce were to be seen. Good macadamised roads crossed each other in all directions; flashy advertisements caught the eye. A network of telegraphs and telephones ran overhead. The ground was accurately marked out with little obelisks of stone into 'Deeps' and 'Concessions,' and labelled with all the queer names which fill the market columns of the newspapers. In a word, it seemed—to us dirty, tattered wanderers—that we had dropped out of Africa and War, and come safely back to Peace and Civilisation.

Street in Johannesburg, 1899.[349]

Trolley Cars in Johannesburg.[350]

The Johannesburg Hospital.[351]

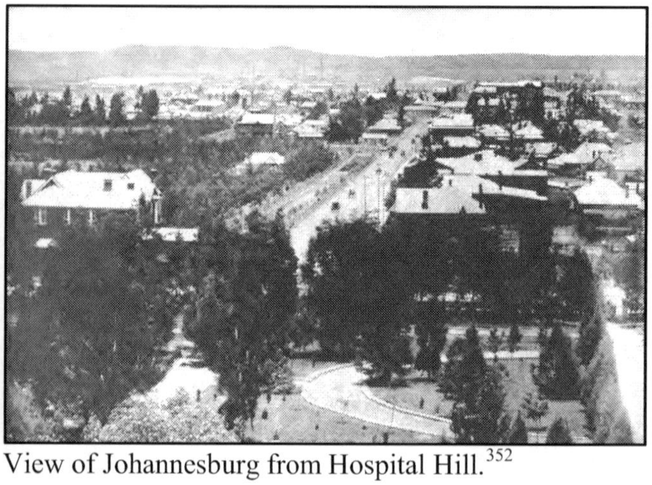

View of Johannesburg from Hospital Hill.[352]

Morning market in Johannesburg.[353]

Since the soldiers had eaten their last day's rations, and the only food they had had that morning came from any odds and ends the regiments might have saved, it was imperative to find some supplies. The Field-Marshal had ordered that no troops should enter Johannesburg until he should specially direct; but, finding little to eat in Florida, Hamilton sent his supply officer and a squadron as far as Maraisburg; whence they presently returned with a quantity of tinned rabbit and sardines, and with the news that the Boers were said to be occupying a position near Langlaagte[386] mine.

During the morning we caught a train and some prisoners. The train was returning from Potchefstroom, guarded by six armed burghers, and on rifles being pointed, it stopped obediently and surrendered. The other prisoners were brought in by the Cavalry and Mounted Infantry, who had caught them wandering about without their horses. Among them was Commandant Botha—not Louis or Philip—but Botha[387] of the Zoutspansburg[388] commando, a brave and honest fellow, who had fought all through the war from Talana Hill until the last action; but who was quite content that Fate had decided he should fight no more. Hearing of him under guard, and near headquarters, I went to see him. He displayed no bitterness whatever, and seemed quite prepared to accept the decision of war. He inquired anxiously whether he would be sent to St. Helena,[354] and evinced a childish horror of the sea.

Boer Prisoners of War.[355]

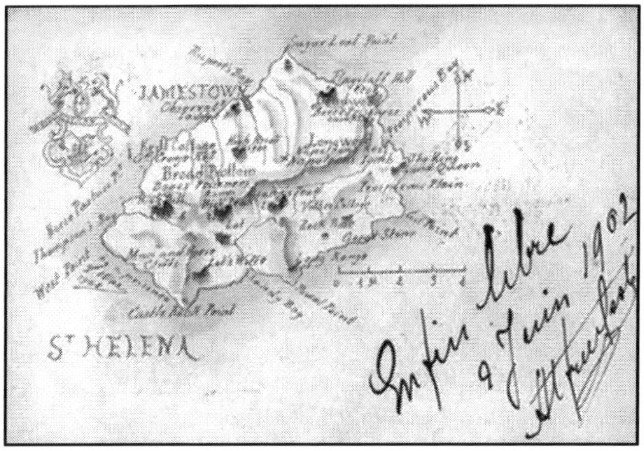

St. Helena postcard depicting the island.[356]

[386] Langlaagte North is a suburb of Johannesburg, adjacent to Mayfair West. Langlaagte North - Wikipedia

[387] Too many Botha possibilities amongst 'Boer Forces.' No Article about the unit itself.

[388] The Soutpansberg, (formerly Zoutpansberg) meaning "Salt Pan Mountain" in Afrikaans, is a range of mountains in far northern South Africa. Soutpansberg - Wikipedia

Arriving at the wharf on St. Helena.[357]

Marching from the wharf to the town.[358]

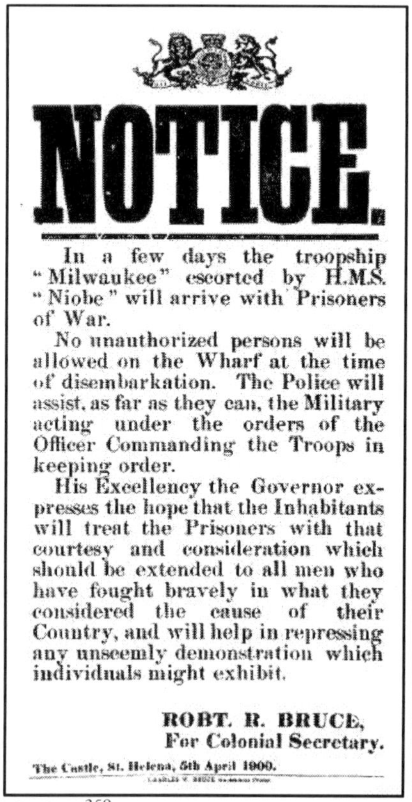

NOTICE.

In a few days the troopship "Milwaukee" escorted by H.M.S. "Niobe" will arrive with Prisoners of War.

No unauthorized persons will be allowed on the Wharf at the time of disembarkation. The Police will assist, as far as they can, the Military acting under the orders of the Officer Commanding the Troops in keeping order.

His Excellency the Governor expresses the hope that the Inhabitants will treat the Prisoners with that courtesy and consideration which should be extended to all men who have fought bravely in what they considered the cause of their Country, and will help in repressing any unseemly demonstration which individuals might exhibit.

ROBT. R. BRUCE,
For Colonial Secretary.

The Castle, St. Helena, 5th April 1900.

Notice.[359]

Prisoners on main street in Jamestown.[360]

Deadwood Camp from Flagstaff Direction.[361]

General Cronje [center] and his officers.[389]

Prisoners visiting Napoleon's Tomb.[362]

While we were chatting, one of the other Boer prisoners, who had been looking hard at us, said, suddenly, in very good English:

'The last time I saw you, you were in my position and I in yours.'

Winston Churchill and other officers upon arrival in Pretoria, November 18th 1899.[390]

He then went on to tell me that he had been in the commando that destroyed the armoured train. 'I felt very sorry for you that day,' he said.

I remarked that it was much worse to be taken prisoner at the beginning of a war than near the end, as he was.

'Do you think this is the end?' asked the Commandant quickly.

'I should ask you that.'

'No, no—not yet the end. They will fight a little more. Perhaps they will defend Pretoria—perhaps you will have to go to Lydenburg;[391] but it will not be very long now.'

[389] The first shipment of 514 Pow's arrived on the 10th April 1900, including General Cronjé and his wife, Colonel Schiel and 21 other officers. Image Source: Exiles | Saint Helena Island Info: All about St Helena, in the South Atlantic Ocean

[390] Winston Churchill and apparently Captain Haldane with his face obscured by his helmet. The figure is wearing an officer's sash but no Sam Browne belt. Haldane is listed as 6 foot 1 inch in his wanted poster. It is likely that Adam Brockie would also be with the 'officers' at the front of the column (Far left, although according to Haldane he went to prison for a day of interrogation. PF Note: Churchill is wearing the same "Irish Fusilier" (Dublin Fusilier) cap referred to supra, that was tossed to him by a Boer during his march through the rain on November 15th. Also, the consensus is that the young 'officer' standing at the front next to Churchill is not Thomas Frankland. It may be Chudleigh, or Lieutenant Grimshaw or Lieutenant Southey. In the background are other soldiers of 1st Royal Dublin Fusiliers all arriving in Pretoria On November 18th, 1899. Image Credit Rob Davidson. Anglo Boer War - Armored Train Ambush 11-15-99 Chuchill & Haldane - Page 5 - Boer War Forum at Anglo Boer War.

[391] Lydenburg, officially known as Mashishing, is a town on the Mpumalanga highveld, South Africa. It is situated on the Sterkspruit/Dorps River tributary of the Lepelle River at the summit of the Long Tom Pass. Lydenburg - Wikipedia

And then, since both he and his companion had been through the Natal campaign, we fell to discussing the various actions. Ian Hamilton came up while we were talking. I had just told the Commandant that we considered the Boers had made a fatal strategic mistake in throwing their main strength into Natal, instead of merely holding the passes, masking Mafeking and Kimberley, and marching south into the [Cape] colony with every man and gun they could scrape together. He admitted that perhaps that might be so; 'but,' said he, 'our great mistake in Natal was not assaulting Ladysmith—the Platrand position, you know—the day after our victory at Lombard's Kop.[392]

Note directions to locations around Ladysmith including Lombard's Kop and Umbulwana hill.[363]

We blame Joubert[393] for that. Many of us wanted to go on then. There were no fortifications; the soldiers were demoralised. If once we had taken the Platrand (Cesar's Camp) you could not have held the town. How many men had you on top of it?'

'Only a picket for the first week,' said the General.[394]

'Ah! I knew we could have done it. What would have happened then?'

'We should have had to turn you out.'

The Commandant smiled a superior smile. The General continued: 'Yes—with the bayonet—at night; or else, as you say, the town could not have been held.'

'Presently,' said Botha, 'you pulled yourselves together, but for three days after Nicholson's Nek there was no fear of bayonets. If we had stormed you then—(then we had all our men and no Buller to think about)—you would not have been able to turn us out.'

Hamilton reflected. 'Perhaps not,' he said, after a pause. 'Why didn't Joubert try it?'

'Too old,' said Botha, with complete disdain; 'you must have young men for fighting.'[395]

That was, so far as I remember, the end of the conversation; but, a fortnight later, I met Botha a free man in the streets of Pretoria. He told me he had been released on parole, so that evidently his frank manliness had not been lost upon the General.

[392] The Battle of Ladysmith was one of the early engagements of the Second Boer War. A large British force which had concentrated at the garrison town of Ladysmith launched a sortie on 30 October 1899, against Boer armies which were slowly surrounding the town. The result was a disaster for the British. The main body was driven back into the town, and an isolated detachment of 800 men was forced to surrender to Commandant De Wet. The Boers did not follow up their advantage by proceeding towards the strategically important port of Durban, and instead began a siege of Ladysmith, which was relieved after 118 days. Battle of Ladysmith - Wikipedia

[393] Piet Joubert - Wikipedia

[394] Ian Hamilton who was there.

[395] General Piet J. Joubert, the Boer commander, came under heavy criticism from his men for permitting the British troops under Hamilton and Grimwood to withdraw into Ladysmith without a more vigorous pursuit. When urged to order a general assault on the retreating British, Joubert is reputed to have said, "When God holds out a finger don't take the whole hand."

Boers Surrendering at the government buildings, Pretoria.[364]

After lunch I became very anxious to go into, and, if possible, through, Johannesburg. An important action had been fought, witnessed by only two or three correspondents; and since the enemy lay between the force and the telegraph wire no news could have been sent home. Hamilton, indeed, had sent off two of Rimington's Guides early in the morning with despatches; but they were to make a wide sweep to the south, and it was not likely, if they got through at all, that they would reach Lord Roberts until late. The shortest, perhaps the safest, road lay through Johannesburg itself. But was the venture worth the risk?

While I was revolving the matter in my mind on the verandah of the temporary headquarters, there arrived two cyclists from the direction of the town. I got into conversation with one of them, a Frenchman, Monsieur Lautré by name. He had come from the Langlaagte mine, with which undertaking he was connected. There were no Boers there, according to him. There might or might not be Boers in the town.

Could a stranger get through? Certainly, he thought, unless he were stopped and questioned. He undertook there and then to be my guide if I wished to go; and it being of considerable importance to get the telegrams through to London, I decided, after a good many misgivings, to accept his offer. The General, who wanted to send a more detailed account of his action, and to report his arrival at Florida, was glad to avail himself even of this precarious channel. So the matter was immediately settled. Lautré's friend, a most accommodating person, got off his bicycle without demur and placed it at my disposal. I doffed my khaki, and put on a suit of plain clothes which I had in my valise, and exchanged my slouch hat for a soft cap. Lautré put the despatches in his pocket, and we started without more ado.

The tracks were bad, winding up and down hill, and frequently deep in sand; but the machine was a good one, and we made fair progress. Lautré, who knew every inch of the ground, avoided all highways, and led me by devious paths from one mine to another, around huge heaps of tailings, across little private tram lines, through thick copses of fir trees, or between vast sheds of machinery, now silent and idle.

Johannesburg, 24 April 1900. Explosion accident at the Begbies munition works.[365]

Head-Gear of the Van Rhyn Gold Mine destroyed by the Boers, January 29th, 1901.[366] PF: The battery house is visible in the background. Crushing Battery at a Johannesburg mine.[367]

In three-quarters of an hour we reached Langlaagte, and here we found one of Rimington's scouts pushing cautiously forward towards the town. We held a brief parley with him, behind a house, for he was armed and in uniform. He was very doubtful of the situation ahead; only knew for certain that the troops had not yet entered Johannesburg. 'But,' said he, 'the Correspondent of the Times[396] passed me more than two hours ago.'

'Riding?' I asked.

'Yes,' he said, 'a horse.'

'Ah,' said my Frenchman, "that is no good. He will not get through on a horse. They will arrest him." And then, being quite fired with the adventure: "Besides, we will beat him, even if, unhappily, he escape the Boers."

So we hurried on. The road now ran for the most part down hill, and the houses became more numerous. The day was nearly done, and the sun drew close to the horizon, throwing our long shadows on the white track before us. At length we turned into a regular street.

'If they stop us,' said my guide, 'speak French. Les François sont en bonne odeur ici.[397] You speak French, eh?'

I thought my accent might be good enough to deceive a Dutchman, so I said yes; and thereafter our conversation was conducted in French.

We avoided the main thoroughfares, bicycling steadily on through the poorer quarters. Johannesburg stretched about me on every side, silent, almost deserted. Groups of moody-looking people chatted at the street corners, and eyed us suspiciously. All the shops were shut. Most of the houses had their windows boarded up. The night was falling swiftly, and its shades intensified the gloom which seemed to hang over the town, on this the last day of its Republican existence.[398] [399]

Suddenly, as we crossed a side lane, I saw in the street parallel to that we followed, three mounted men with slouch hats, bandoliers, and that peculiar irregular appearance which I have learned to associate with Boers. But to stop or turn back was now fatal. After all, with the enemy at their gates, they had probably concerns of their

[396] Leo Amery (22 November 1873 – 16 September 1955), was a British Conservative politician and journalist. In 1887, Amery went to Harrow School, where he was a contemporary of Winston Churchill. Amery represented Harrow at gymnastics and held the top position in examinations for a number of years. Leo Amery - Wikipedia

[397] Loosely translated: The French are well thought of here.

[398] PF: This journey is courageous but also dangerous to the point of foolhardiness. As an escaped prisoner of war he is a wanted man in the Transvaal. He is out of uniform while carrying military despatches. His face is known. If recognized it is entirely possible that he will be summarily executed as a spy. WSC notes this himself 30 years later.

[399] As we passed our farthest outpost lines I experienced a distinct sensation of adventure. We were soon in the streets of Johannesburg. Darkness was already falling. But numbers of people were about, and at once I saw among them armed and mounted Boers. They were still in possession of the city, and we were inside their lines. According to all the laws of war my situation, if arrested, would have been disagreeable. I was an officer holding a commission in the South African Light Horse, disguised in plain clothes and secretly within the enemy's lines. No court-martial that ever sat in Europe would have had much difficulty in disposing of such a case. On all these matters I was quite well informed. WSC My Early Life Ch. XXVII.

own to occupy them. We skimmed along unhindered into the central square, and my companion, whose coolness was admirable, pointed me out the post-office and other public buildings, speaking all the time in French. The slope now rose against us so steeply that we dismounted to push our machines. While thus circumstanced I was alarmed to hear the noise of an approaching horse behind me. With an effort I controlled my impulse to look back.

'*Encore un Boer*,'[400] said Lautré lightly.

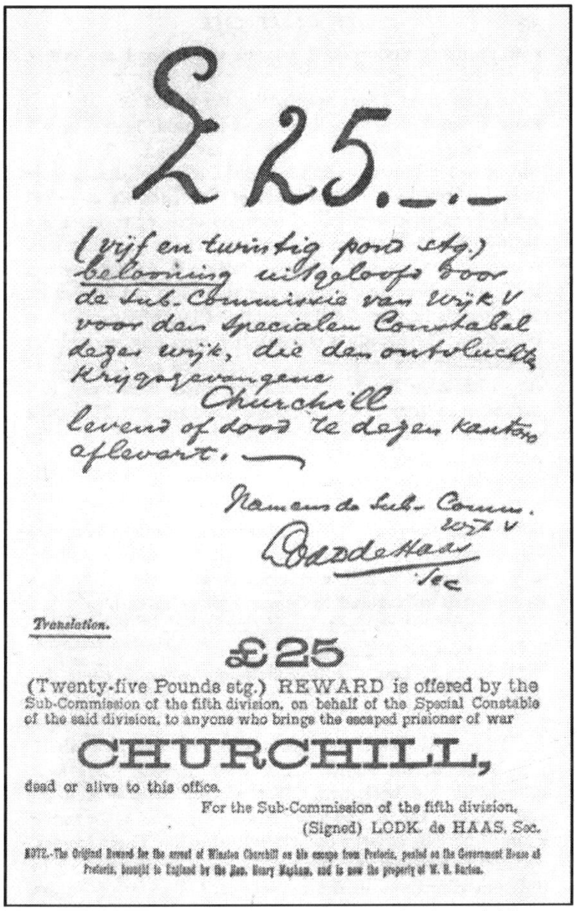

Reward Poster for the Recapture of Churchill Dead or Alive.[401] Illustrated Account of Churchill's Escape.[368]

I was speechless. The man drew nearer, overtook and pulled his horse into a walk beside us. I could not help—perhaps it was the natural, and, if so, the wise, thing to do—having a look at him. He was a Boer sure enough, and I think he must have been a foreigner. He was armed *cap-à-pie*.[402] The horse he rode carried a full campaigning kit on an English military saddle. Wallets, saddle-bags, drinking-cup, holsters—all were there. His rifle was slung across his back, he wore two full bandoliers over his shoulders and a third round his waist—evidently a dangerous customer. I looked at his face and our eyes met. The light was dim, or he might have seen me change colour. He had a pale, almost ghastly visage, peering ill-favoured and cruel from beneath a slouch hat with a large white feather. Then he turned away carelessly. After all, I suppose he thought it natural a poor devil of a townsman should wish to look at so fine a cavalier of fortune. Presently he set spurs to his horse and cantered on. I breathed again freely. Lautré laughed.

'There are plenty of cyclists in Johannesburg,' he said. 'We do not look extraordinary. No one will stop us.'

We now began to approach the south-eastern outskirts of the town. If the original scheme of advance had been carried out, Lord Roberts's leading brigade should be close at hand. Lautré said, 'Shall we inquire?' But I

[400] Another Boer.

[401] The Dutch text tracks the English text closely, if not exactly. Issued in the name of "Lok de Hass, a volunteer from Europe, in the Hollander Corps who was secretary of the Commission of Peace and Order." Churchill later corresponded with him on the occasion of his (Churchill's) marriage and chided him that "'I think you might have gone as high as £50 without an overestimate of the value of the prize—if living.'" Sandys p. 103. Published in 'My Early Life.'

[402] Head to foot.

thought it better to wait. As we progressed the streets became still more deserted, and at last we found ourselves quite alone. For more than half a mile I did not see a single person. Then we met a shabby-looking man, and now, no one else being in sight, the night dark, and the man old and feeble, we decided to ask him.

'The English,' he said with a grin, 'why, their sentinels are just at the top of the hill.'

'How far?'

'Five minutes—even less.'

Two hundred yards further on three British soldiers came in sight. They were quite unarmed, and walking casually forward into the town. I stopped them and asked what brigade they belonged to. They replied Maxwell's.[403]

'Where is the picket line?'

'We haven't seen no pickets,' said one of them.

'What are you doing?'

'Looking for something to eat. We've had enough of 'arf rations.'

I said, 'You'll get taken prisoners or shot if you go on into the town.'

'Wot's that, guvnor?' said one of them, deeply interested in this extraordinary possibility.

I repeated, and added that the Boers were still riding about the streets.

'Well, then, I ain't for it,' he said with decision. 'Let's go back and try some of them 'ouses near the camp.'

So we all proceeded together.

I discovered no picket line at the edge of the town. Maxwell must have had one somewhere, but it certainly did not prevent anyone from passing freely; for we were never challenged, and, walking on, soon found ourselves in the middle of a large bivouac. I now became of some use to my companion, for if he knew the roads I knew the army. I soon found some officers of my acquaintance, and from them we learned that Lord Roberts's headquarters were not at Elandsfontein (South), but back at Germiston, nearly seven miles away. It was now pitch dark, and all signs of a road had vanished; but Lautré declared he knew his way, and, in any case, the messages—press and official—had to go through.

We left the camp of Maxwell's Brigade and struck across country in order to cut into the main southern road. A bicycle now became a great incumbrance, as the paths wound through dense fir woods, obstructed by frequent wire fences, ditches, holes, and high grass. Lautré, however, persisted that all was well, and, as it turned out, he was right. After about an hour of this slow progress we reached the railway, and, seeing more camp fires away to the left, turned along it. Half a mile in this direction brought us to another bivouac, which we likewise entered unchallenged. I asked a soldier whose brigade he belonged to, but he did not know, which was painfully stupid of him.

A group of officers were gathered round an enormous fire a few yards away, and we went up to them to ask. Chance had led me to General Tucker's[404] mess. I had known the commander of the Seventh Division in India, when he was stationed at Secunderabad,[405] and he welcomed me with his usual breezy courtesy.

[403] General John Maxwell John Maxwell (British Army officer) - Wikipedia

[404] C. Tucker Major General Entered 1855; Staff service: Colonel on Staff, Natal, 1891-93; Brigadier General, Natal, 1893-95; Major General, India, 1895-99; Lieutenant General Infantry Division, South Africa, 1899. War service: Bhootan Expedition 1865-66 (medal with clasp); Boer War, 1878-79 (Despatches, April and August 1879; medal with clasp; CB); Boer War, 1899-1900; on Staff. General Tucker, born in 1838, was a son of Mr. Tucker of Ashburton and a daughter of Mr. Hayter, Painter-in-Ordinary to Queen Victoria. As a practical, resourceful, and rough-and-ready soldier, he has no equal. He was rewarded in 1896 for "distinguished and meritorious service with a good-service pension. He was Colonel of the South Staffordshire Regiment from 1911.

[405] Secunderabad, is a twin city of Hyderabad and one of the six zones of the Greater Hyderabad Municipal Corp.in the Indian state of Telangana. Secunderabad - Wikipedia

James street circa 1880, an important shopping district in Secunderabad.[369]

He had been sent off with his leading brigade late in the afternoon to try to join hands with French, and so complete the circle round Johannesburg; but darkness had curtailed his march. Besides this, no communications having yet come through from the Cavalry, he was uncertain where French was. Naturally he was interested to hear what had passed on the west of the town, and about the stirring action of the previous day. From him I got some whisky and water, and clear directions to the Field-Marshal's headquarters.

They were, it appeared, two miles beyond Germiston, a mile and a half west of the road, in a solitary house on a small hill which stood beyond a large tank. And in case these indications might have been of little avail in the dark, he led us a few feet up the slope, and there we saw that, on the blackness of the night, flamed a regular oblong of glittering lights. It was the camp of the Eleventh Division. Somewhere near that were the Chief's headquarters. Thus instructed, we resumed our journey.

Another half-hour of walking brought us, as Lautré had promised, to a good firm road, and the bicycles quickly made amends for their previous uselessness. The air was cold, and we were glad to spin along at a fair ten miles an hour. At this rate twenty minutes brought us into Germiston. Not knowing where I should be likely to find dinner, or a bed, I dismounted opposite the hotel, and, seeing lights and signs of occupation, went inside. Here I found Mr. Lionel James,[406] the principal Correspondent of the *Times*. I asked him if his subordinate [Leo Amery] had arrived from Hamilton's force. He said 'No'; and when I told him he had started two hours in front of me, looked much concerned; whereat the Frenchman could not conceal a heartless grimace.

I offered to give him some account of the action for his own use (for what is more detestable than a jealous journalist?), but he said that I had had the good luck to come through, and that he would not think of depriving me of my advantage. Alas! the days of newspaper enterprise in war are over. What can one do with a censor, a forty-eight hours' delay, and a fifty-word limit on the wire? Besides, who can compete with Lord Roberts as a special correspondent? None against the interest of his daily messages; very few against their style and simple grace. Never mind. It is all for the best.

[406] James Lionel Correspondent for The Times. QSA issued: 20 Feb 03. Colonel Lionel James was born in 1871 and educated at Cranleigh. Employed as a journalist, he was Reuter's special correspondent in the Chitral, Mohmand, Malakand and Tirah Expeditions and for the Sudan in 1898. On the staff of The Times from 1899 until he retired in 1913. Amongst his many assignments he was special correspondent in South Africa 1899-1901, Manchuria 1904, with the Spanish Army in Morocco in 1909, with the Turkish Army in Albania in 1910, with the Turks in Thrace in 1912 and with the Bulgarians in Thrace in 1913. With the onset of the Great War he commanded King Edward's Horse in the BEF, France and Italy, 1915-18. He was awarded the DSO (London Gazette 16.9.1918) and mentioned in despatches twice. He was awarded the CBE in 1924. CBE (1st, Mil) Order of the British Empire, 1st type, Military Division, silver-gilt and enamel; DSO GV, IGS (4) Relief of Chitral 1895, Punjab Frontier 1897-8, Malakand 1897, Tirah 1897-8, (QSA (6) CC Eland DofL OFS Joh DH, IGS (1) NWF 1908, 1914 Star; British War and Victory Medals; Italy, Order of the Crown, Officer's, Khedive's Sudan (1) Khartoum, Japan, Russo-Japanese War Medal. Miniatures only, DNW Sep 03 £550. Source: QSA Medal roll.

We dined hastily and not too well, secured the reversion of half the billiard table, should all other couches fail, and set out again, this time tired and footsore. After two miles of dusty track the camp was reached. I found more officers who knew where Army Headquarters were, and at last, at about half-past ten, we reached the solitary house. We sent the despatches in by an orderly, and after a few minutes Lord Kerry[407] came out and said that the Chief wanted to see the messengers.

Now, for the first time in this war, I found myself face to face with our illustrious leader. The room was small and meanly furnished, and he and his staff, who had just finished dinner, sat round a large table which occupied the greater part of the floor. With him were Sir William Nicholson[408] (who arranges all the transport of the army, a work the credit of which is usually given to Lord Kitchener) and Colonel Neville Chamberlayne,[409] his private secretary, both of them soldiers of the practical Indian school, where you have real fighting, both of them serving once more under their commander of Afghan days. There, too, was Sir Henry Rawlinson,[410] whom I had last seen round Sir George White's table, the night Dundonald broke into Ladysmith; and Sir James Hills-Johnes,[411] who won the Victoria Cross in the Indian Mutiny, and aides-de-camp and others whom I cannot remember.

The Field-Marshal rose from his place, shook hands, and bade us, in most ceremonious fashion, to be seated. He had read half of Hamilton's despatch. 'The first part of this,' he said, 'we knew already. Two guides—Rimington's, I think—got in here about an hour ago. They had a dangerous ride, and were chased a long way, but escaped safely. I am glad to hear Hamilton is at Florida. How did you get through?'

I told him briefly. His eye twinkled. I have never seen a man before with such extraordinary eyes. I remember to have been struck with them on several occasions. The face remains perfectly motionless, but the eyes convey the strongest emotions. Sometimes they blaze with anger, and you see hot yellow fire behind them. Then it is best to speak up straight and clear, and make an end quickly. At others there is a steel grey glitter—quite cold and uncompromising—which has a most sobering effect on anyone who sees it. But now the eyes twinkled brightly with pleasure or amusement or approbation, or, at any rate, something friendly.

'Tell me about the action,' he said.

So I told him all I knew, much as it is set down in these pages, though not nearly at such length; but I don't think the tale lost in the telling. From time to time he asked questions about the Artillery concentration, or the length of front of the Infantry attack, and other technical matters, on which I was luckily well-informed. The fact that the troops had no rations seemed to disturb him very much.

He was particularly interested to hear of Hamilton's novel attack 'at thirty paces extension'; of the manner in which the batteries had been rammed almost into the firing line; but most of all he wanted to hear about the Gordons' charge. When I had done he said: 'The Gordons' always do well.' Then he asked what we proposed to do. Lautré said he would go back forthwith; but the Chief said, 'Much better stay here for the night; we will find you beds'; so of course we stayed.

[407] Henry Petty-Fitzmaurice, 6th Marquess of Lansdowne styled Earl of Kerry until 1927. (14 January 1872 – 5 March 1936) Henry Petty-Fitzmaurice, 6th Marquess of Lansdowne - Wikipedia

[408] Field Marshal William Gustavus Nicholson, 1st Baron Nicholson, (2 March 1845 – 13 September 1918) was a British Army officer who served in the Second Anglo-Afghan War, the Mahdist War, the Third Anglo-Burmese War, the Second Boer War and the First World War. He became Chief of the Imperial General Staff and was closely involved in the reorganisation of the British Army in the early years of the 20th century. William Nicholson, 1st Baron Nicholson - Wikipedia

[409] Neville F.F. Chamberlain Colonel Entered 11th Foot 1873; Colonel, 1899. Staff service: ADC to Lieutenant General, Afghan Campaign, 1878; ADC to Commander in Chief, Madras, 1881-85 Persian Interpreter to Commander in Chief in India, 1885-89 (DAA and QMG, Burmese Expedition 1886-87); Colonel on Staff, India, 1899; **Private Secretary to Field-Marshal Roberts**, 1899-1900. War Service: Afghan War, 1878-80, wounded (Despatches, February, January, May, 1880; medal with 4 clasps; bronze star); Burmese Expedition 1886-87 (Despatches; medal with clasp; Brevet of Lieutenant Colonel); Boer War, 1899-1900; on Staff. Source: List of CB recipients. Various sources.

[410] Spencer Richard Rawlinson Major, was born in 1848, son of Reverend George Rawlinson. He entered the 45th Regiment in 1867; was transferred to the MSC, 1871. He became Captain in 1879; served with the Burma Field Force in 1885-87 (Medal with clasp); mentioned in Despatches 2 September 1887; created a Companion of the Distinguished Service Order [London Gazette, 25 November 1887]: "Spencer Richard Rawlinson, Major, Madras Staff Corps." His DSO was awarded for operations in Burma. He became Major in 1887; Lieutenant Colonel 20 July 1893. Colonel Rawlinson died on 15 April 1903. He married, in 1877, Mary (who died in 1882), daughter of H Mann, of Hedenham Hall, Norfolk. Source: DSO recipients (VC and DSO Book).

[411] Lieutenant General Sir James Hills-Johnes, VC, GCB (20 August 1833 – 3 January 1919) James Hills-Johnes - Wikipedia

He asked me whether I meant to go back next morning. I said that as I had got my messages to the telegraph office I thought, upon the whole, that I would not run any more risks, but wait and see the British occupation of the town. He laughed at this, and said that I was quite right, and would be very ill-advised to be caught again. Then he said that he would send a letter to Hamilton in the morning, bade us all 'good-night,' and retired to his waggon. I, too, found a comfortable bed—the first for a month—and being thoroughly worn out soon fell asleep. [Winston relates this conversation again 30 years later in My Early Life with a few additional personal details.[412]]

Lord Roberts – Head Study by Menpes.[370]

Lord Roberts by Menpes.[371]

Part of Lord Roberts's letter that he wrote to Ian Hamilton next day was published in the orders of the flanking column. In some way it explains why the private soldier will march further for 'Bobs Bahadur' than for any one else in the world.

[412] It was quite dark when at last we reached them. An aide-de-camp whom I knew came to the door.
'Where do you spring from?'
'We have come from Ian Hamilton. I have brought a dispatch for the Commander-in-Chief.'
'Splendid!' he said. 'We have been longing for news.'
He disappeared. My business was with the Press Censor, for whom I had a heavy sheaf of telegrams full of earliest and exclusive information. But before I could find this official the aide-de-camp reappeared.
'Lord Roberts wants you to come in at once.'
The Commander-in-Chief was at dinner with about a dozen officers of his Headquarters Staff. He jumped up from his chair as I entered, and with a most cordial air advanced towards me holding out his hand.
'How did you come?' he asked.
'We came along the main road through the city, sir.'
'Through Johannesburg? Our reports are that it is still occupied by the enemy.'
'There are a few, sir,' I said, 'but they are clearing out.'
'Did you see any of them?'
'Yes, we saw several, sir.'
His eye twinkled. Lord Roberts had very remarkable eyes, full of light. I remember being struck by this at the moment.
'Did you see Hamilton's action yesterday?' was his next question.
'Yes, sir.'
'Tell me all about it.'
Then, while being most hospitably entertained, I gave a full account of the doings of General Hamilton's force to my father's old friend and now once again my own. WSC My Early Life Ch. XXVII.
Please see the biographical sketch Winston penned about Roberts' career published in the US and then UK in 1900, infra. It is similar to the sketch that he wrote on behalf on Hamilton as a chapter supra.

'I am delighted at your repeated successes, and grieve beyond measure at your poor fellows being without their proper rations. A trainfull shall go to you to-day. I expect to get the notice that Johannesburg surrenders this morning, and we shall then march into the town. I wish your column, which has done so much to gain possession of it, could be with us.

'Tell the Gordons that I am proud to think I have a Highlander as one of the supporters on my coat-of-arms.'

XV. The Capture of Pretoria

Pretoria: June 8th 1900.

The Commander-in-Chief had good reasons—how good we little knew—for wishing to push on at once to the enemy's capital, without waiting at Johannesburg. But the fatigue of the troops and the necessities of supply imposed a two days' halt.

"Dorset Yeomanry Entering Johannesburg"[372] "The City [London] Volunteers Entering Johannesburg."[373]
Note: Same picture with different captions. One must be careful about these things.

On the 3rd of June the advance was resumed. The army marched in three columns. The left, thrown forward in echelon, consisted of the Cavalry Division under French; the centre was formed by Ian Hamilton's force; and the right or main column nearest the railway comprised the Seventh and Eleventh Divisions (less one brigade left to hold Johannesburg), Gordon's Cavalry Brigade, and the Corps Troops all under the personal command of the Field-Marshal.

The long forward stride of the 3rd was, except for a small action against French, unchecked or unopposed by the Boers, and all the information which the Intelligence Department could collect seemed to promise a bloodless entry into the capital. So strong was the evidence that at dawn on the 4th of June Hamilton's column was diverted from its prescribed line of march on Elandsfontein (1)[413] and drawn in towards the main army, with orders to bivouac on Pretoria Green, west of the town. French, whom the change of orders did not reach, pursued his wide turning movement, and encountered further opposition in a bad country for cavalry.

At ten o'clock it was reported that Colonel Henry,[414] with the corps of Mounted Infantry in advance of the main column, was actually in the suburbs of Pretoria without opposition. The force continued to converge, and Ian Hamilton had almost joined Lord Roberts's force when the booming of guns warned us that our anticipations were too sanguine. The army had just crossed a difficult spruit, and Colonel Henry with the Mounted Infantry had obtained a lodgment on the heights beyond. But here they were sharply checked. The Boers, apparently in some force, were holding a wooded ridge and several high hills along the general line of the southern Pretoria forts.

Determined to hold what he had obtained, Lord Roberts thrust his artillery well forward, and ordered Ian Hamilton to support Colonel Henry immediately with all mounted troops. This was speedily done. The horsemen galloped forward, and, scrambling up the steep hillsides, reinforced the thin firing line along the ridge. The artillery of the Seventh Division came into action in front of the British centre. The Boers replied with a brisk rifle fire, which reached all three batteries, and drew from them a very vigorous cannonade.

Meanwhile the Infantry deployment was proceeding. The 14th Brigade extended for attack. Half an hour later Pole-Carew's batteries prolonged the line of guns to the right, and about half-past two the corps and heavy

[413] Original Footnote: (1) Yet another Elandsfontein, situated to the west of Pretoria.

[414] G. C. Henry Major Entered 1880; Brevet Colonel, March 1900. Staff service: Employed with Egyptian Army. War service: Expedition to Dongola, 1896 (Despatches; Egyptian medal with 2 clasps); Nile Expedition 1897 (clasp to Egyptian medal); Nile Expedition 1898 (Despatches, September and December 1898; Brevet of Lieutenant Colonel; clasp to Egyptian medal; medal); Nile Expedition 1899 (Despatches; Brevet of Colonel); Boer War, 1899-1900; commanded 4th Corps Mounted Infantry.

artillery opened in further prolongation. By three o'clock fifty guns were in action in front of the main army, and both the Seventh and Eleventh Divisions had assumed preparatory formations. The balloon ascended and remained hanging in the air for an hour—a storm signal.

Observation Balloon.[374]

Military Balloon.[375]

Observation Balloon.[376]

During this time Hamilton was pushing swiftly forward, and Smith-Dorrien's 19th Infantry Brigade occupied the line of heights, and thus set free the mounted troops for a turning movement. The 21st Brigade supported. The heights were so steep in front of Hamilton that his artillery could not come into action, and only one gun and one 'pom-pom' could, by great exertion, be dragged and man-handled into position. The fire of these pieces, however, caught the Boers holding the weeded ridge in enfilade, and was by no means ineffective.

"All together boys."[377]

Dragging a Maxim up.[378]

So soon as Hamilton had collected the mounted troops he sent them to reinforce Broadwood, whom he directed to move round the enemy's right flank. The ground favoured the movement, and by half-past four the Cavalry were seen debouching into the plain beyond the Boer position, enveloping their flank and compromising their retreat.

Colonel de Lisle's corps of Mounted Infantry, composed mainly of Australians, made a much shorter circuit, and reaching the level ground before the Cavalry espied a Boer Maxim retreating towards the town. To this they immediately gave chase, and the strong Waler horses[415] were urged to their utmost speed.

[415] The Waler is an Australian breed of horse developed from horses that were brought to the Australian colonies in the 19th century. The name comes from their breeding origins in New South Wales; they were originally known as "New South Walers." Waler - Wikipedia

Australian light horsemen riding Waler horses.[379] Australian Mounted Rifles.[380]

The appearance of this clattering swarm of horsemen, must have been formidable to those below. But we who watched from the heights saw what Ian Hamilton, who was in high spirits, described as 'a charge of infuriated mice' streaming across the brown veldt; so great are the distances in modern war.

Plan of Pretoria.[381]

Towards four o'clock the cannonade all along the front had died away, and only the heavy artillery on the right of Pole-Carew's Division continued to fire, shelling the forts, whose profile showed plainly on the skyline, and even hurling their projectiles right over the hills into Pretoria itself. So heavy had the artillery been that the Boers did not endure, and alarmed as well by the flank movement they retreated in haste through the town; so that before dusk their whole position was occupied by the Infantry without much loss. Night, which falls at this season and in this part of the world as early as half-past five,[416] then shut down on the scene, and the action—in which practically the whole Army Corps had been engaged—ended.

The fact that the forts had not replied to the British batteries showed that their guns had been removed, and that the Boers had no serious intention of defending their capital. The Field-Marshal's orders for the morrow were,

[416] June is Winter in the Southern Hemisphere.

therefore, that the army should advance at daybreak on Pretoria, which it was believed would then be formally surrendered. Meanwhile, however, Colonel de Lisle, with the infuriated mice—in other words, the Australians[417]—was pressing hotly on, and at about six o'clock, having captured the flying Maxim, he seized a position within rifle shot of the town. From here he could see the Boers galloping in disorder through the streets, and, encouraged by the confusion that apparently prevailed, he sent an officer under flag of truce to demand the surrender. This the panic-stricken civil authorities, with the consent of Commandant Botha, obeyed, and though no British troops entered the town until the next day, Pretoria actually fell before midnight on the 4th of June.

Boer Retreat from Pretoria.[382]

As soon as the light allowed the army moved forward. The Guards were directed on the railway station. Ian Hamilton's force swept round the western side. Wishing to enter among the first of the victorious troops the town I had crept away from as a fugitive six months before, I hurried forward, and, with the Duke of Marlborough, soon overtook General Pole-Carew, who, with his staff, was advancing towards the railway station. We passed through a narrow cleft in the southern wall of mountains, and Pretoria lay before us—a picturesque little town with red or blue roofs peeping out among masses of trees, and here and there an occasional spire or factory chimney. Behind us, on the hills we had taken, the brown forts were crowded with British soldiers. Scarcely two hundred yards away stood the railway station.

Pretoria from the East.[383]

[417] New South Wales Contingents Article: https://www.angloboerwar.com/?option=com_content&view=article&id=378

Pretoria.[384]

Arrived at this point, General Pole-Carew was compelled to wait to let his Infantry catch him up; and while we were delayed a locomotive whistle sounded loudly, and, to our astonishment—for had not the town surrendered?—a train drawn by two engines steamed out of the station on the Delagoa Bay line. For a moment we stared at this insolent breach of the customs of war, and a dozen staff officers, aides-de-camp, and orderlies (no mounted troops being at hand) started off at a furious gallop in the hopes of compelling the train to stop, or at least of shooting the engine-driver, and so sending it to its destruction. But wire fences and the gardens of the houses impeded the pursuers, and, in spite of all their efforts, the train escaped, carrying with it ten trucks of horses, which might have been very useful, and one truck-load of Hollanders.

Three engines with steam up and several trains, however, remained in the station, and the leading company of Grenadiers, doubling forward, captured them and their occupants. These Boers attempted to resist the troops with pistols, but surrendered after two volleys had been fired, no one, fortunately, being hurt in the scrimmage.

Part of Pretoria railway station. (One of the Forts on the hill in the distance.)[385]

After a further delay, the Guards, fixing bayonets, began to enter the town, marching through the main street, which was crowded with people, towards the central square, and posting sentries and pickets as they went.

The Guards Enter Pretoria.[386]

First Troops to Enter Pretoria.[387]

We were naturally very anxious to know what had befallen our comrades held prisoners all these long months. Rumour said they had been removed during the night to Waterfall Boven, 200 miles down the Delagoa Bay line. But nothing definite was known.

The Duke of Marlborough, however, found a mounted Dutchman who said he knew where all the officers were confined, and who undertook to guide us, and without waiting for the troops, who were advancing with all due precautions, we set off at a gallop.

The Birdcage, British Officers' Prison, at Daspoort, Pretoria.[388]

The distance was scarcely three-quarters of a mile, and in a few minutes, turning a corner and crossing a little brook, we saw before us a long tin building surrounded by a dense wire entanglement. Seeing this, and knowing its meaning too well, I raised my hat and cheered. The cry was instantly answered from within. What followed resembled the end of an Adelphi melodrama.[418]

The Duke of Marlborough called on the commandant to surrender forthwith.[419] [420] The prisoners rushed out of the house into the yard, some in uniform, some in flannels, hatless or coatless, but all violently excited. The sentries threw down their rifles. The gates were flung open, and while the rest of the guards—they numbered fifty-two in all—stood uncertain what to do, the long-penned-up officers surrounded them and seized their weapons. Some one—Grimshaw[421] of the Dublin Fusiliers—produced a Union Jack (made during imprisonment out of a Vierkleur). The Transvaal emblem was torn down, and, amid wild cheers, the first British flag was hoisted over Pretoria. Time 8.47, June 5.[422] [423]

[418] The Adelphi Theatre is a West End theatre, located on the Strand in the City of Westminster, central London. In its early years, the theatre was known for melodrama, called Adelphi Screamers. Many stories by Charles Dickens were also adapted for the stage here, Adelphi Theatre - Wikipedia

[419] "Marlborough, resplendent in the red tabs of the staff (officer), called on the Commandant to surrender forthwith, adding by a happy thought that he would give a receipt for the rifles." WSC My Early Life.

[420] Per Frankland on March 25: This evening the new Commandant had roll-call. We call him "Pyjamas," because he wears a suit of clothes for all the world like a pair of pyjamas. His real name is Westernant.

[421] Cecil Thomas Wrigley Grimshaw, Lieutenant, 2nd Btn. Royal Dublin Fusiliers, Prisoner - released at Chieveley. 6 May 1900. Source: Natal Field Force Casualty Roll, page 195 line 47 He was born 22 October 1875, son of Thomas Wrigley Grimshaw, CB, Registrar-General for Ireland. He was educated at Eastman's Royal Naval Academy and at Dublin University, where he took his BA degree, and joined the Royal Dublin Fusiliers, as a Second Lieutenant, 15 May 1897, being promoted Lieutenant 28 December 1898. He served in the South African War, 1899-1902 (employed with Mounted Infantry); took part in the operations in Natal, 1899, including action at Talana; in the Transvaal, east of Pretoria, July to 29 November 1900; in Orange River Colony, May to 29 November 1900; again in the Transvaal 30 November 1900 to 31 May 1902. Was promoted Captain 14 July 1904; became Adjutant, Royal Dublin Fusiliers, 28 December 1911. He served in the European War, and was killed in action at the Dardanelles in 1915. He had married, in 1906, Agnes Violet, youngest daughter of George B Alderson, of Alexandria, and left two sons. Source: DSO recipients (VC and DSO Book).

[422] Per Frankland: To close the scene we hoisted the Union Jack which Burrows (one of the prisoners) had made by cutting up a Vierkleur, on the staff whence the Transvaal colours had so long reminded us of our condition.

Waterval after the liberation of the [British] prisoners.[389]

The commandant then made formal surrender to the Duke of Marlborough of 129 officers and thirty-nine soldiers whom he had in his custody as prisoners of war, and surrendered, besides himself, four corporals and forty-eight Dutchmen. These latter were at once confined within the wire cage, and guarded by their late prisoners; but, since they had treated the captives well, they have now been permitted to take the oath of neutrality and return to their homes.

"Boer Prisoners in the Prison Prepared for the British Officers Just Outside Pretoria."[390]

The anxieties which the prisoners had suffered during the last few hours of their confinement were terrible, nor did I wonder, when I heard the account, why their faces were so white and their manner so excited. But the reader shall learn the tale from one of their number, nor will I anticipate.

[423] WSC, 'My Early Life, Chapter XXVII. Early on the morning of the 5th Marlborough and I rode out together and soon reached the head of an infantry column already in the outskirts of the town. There were no military precautions, and we arrived, a large group of officers, at the closed gates of the railway level crossing. Quite slowly there now steamed past before our eyes a long train drawn by two engines and crammed with armed Boers whose rifles bristled from every window. We gazed at each other dumbfounded at three yards distance. A single shot would have precipitated a horrible carnage on both sides. Although sorry that the train should escape, it was with unfeigned relief that we saw the last carriage glide slowly past our noses.
Then Marlborough and I cantered into the town. We knew that the officer prisoners had been removed from the State Model Schools, and we asked our way to the new cage where it was hoped they were still confined. We feared they had been carried off—perhaps in the very last train. But as we rounded a corner, there was the prison camp, a long tin building surrounded by a dense wire entanglement. I raised my hat and cheered. The cry was instantly answered from within. What followed resembled the end of an Adelphi melodrama. We were only two, and before us stood the armed Boer guard with their rifles at the 'ready.' Marlborough, resplendent in the red tabs of the staff, called on the Commandant to surrender forthwith, adding by a happy thought that he would give a receipt for the rifles. The prisoners rushed out of the house into the yard, some in uniform, some in flannels, hatless or coatless, but all violently excited. The sentries threw down their rifles, the gates were flung open, and while the last of the guard (they numbered 52 in all) stood uncertain what to do, the long-penned-up officers surrounded them and seized their weapons. Someone produced a Union Jack, the Transvaal emblem was torn down, and amidst wild cheers from our captive friends the first British flag was hoisted over Pretoria. Time: 8.47, June 5. Tableau!

At two o'clock Lord Roberts, the staff, and the foreign attachés entered the town, and proceeded to the central square, wherein the Town Hall, the Parliament House, and other public buildings are situated. The British flag was hoisted over the Parliament House amid some cheers. The victorious army then began to parade past it, Pole-Carew's Division, with the Guards leading, coming from the south, and Ian Hamilton's force from the west.

THE ENTRY INTO PRETORIA

This picture shows the arrival of the Duke of Marlborough (on the grey horse), his cousin, Mr. Winston Churchill, and a few other officers in Pretoria, shortly before the formal entry of Lord Roberts

This is the caption that appeared for a reprint of the photograph in The Sphere on July 21st, 1900.[424]

THE FIRST TROOPS ENTERING PRETORIA, 5TH JUNE.

Duke of Marlborough. Major Maude.

Mr Winston Churchill.

This above is the caption appearing in Rosslyn's Book. Twice Captured. Photo by Rosslyn.[391]

Frederick Roberts and Staff Entering the Outer Lines at Pretoria.[425] [Roberts leading on dark horse. His Indian attendants in the background.]

[424] Lord Rosslyn, stationed at a window with his camera, managed to get some capital snapshots. At the head of one body of troops rode the Duke of Marlborough and Mr. Winston Churchill, and they recognised Lord Rosslyn's head at an upper storey window with "Halloo, Harry!" The Sphere - Saturday 21 July 1900 P.13. https://www.britishnewspaperarchive.co.uk/viewer/bl/0001861/19000721/014/0013

Roberts riding in to Pretoria.[392] Roberts is riding the white horse. The entry of Roberts into Pretoria.[393] Drawn by Sidney Paget from a photograph taken by the Earl of Rosslyn.

The surrender of the Golden City was an accomplished fact.[426]

For three hours the broad river of steel and khaki flowed unceasingly, and the townsfolk gazed in awe and wonder at those majestic soldiers whose discipline neither perils nor hardships had disturbed, whose relentless march no obstacles could prevent.

Regimental band of the Gordon Highlanders parading through Church Square in Pretoria.[394] Gordon Highlanders parading through Church Square in Pretoria.[395] Note Kilts.

[425] Richard Harding Davis With Both Armies in South Africa. Charles Scribner's Sons, New York 1900.

[426] Johanna Brandt, The Petticoat Commando Boer Women in Secret Service by, Mills & Boon, London 1913 Contributor University of California Libraries: https://archive.org/details/petticoatcommand00branrich

The scene in the square on June 5th, [1900] when the flag was raised on the Raadzaal.[396]

Colonials "Marching Past" at the formal Occupation of Pretoria, June 5th 1900.[397] "Lord Roberts and staff cheering the Queen--Raising Union Jack on occupation of Pretoria, June 5th, S.A."[398]

British troops just before raising the Union Jack.[399]

Raising the flag at Pretoria.[400]

Pretoria Square occupied by the British, June 5th 1900.[401] Hoisting the British flag on the Raadzaal, Pretoria.[402]

With such pomp and the rolling of drums the new order of things was ushered in. The former Government had ended without dignity. One thought to find the President—stolid old Dutchman—seated on his stoep reading his Bible and smoking a sullen pipe. But he chose a different course. On the Friday preceding the British occupation he left the capital and withdrew along the Delagoa Bay Railway, taking with him a million pounds in gold, and leaving behind him a crowd of officials clamouring for pay, and far from satisfied with the worthless cheques they had received, and Mrs. Kruger, concerning whose health the British people need not further concern themselves.

President and Mrs. Kruger in their cottage in Pretoria, April, 1900.[403] President Kruger's private car used as the capital after evacuation of Pretoria.[404]

I cannot end this letter without recalling for one moment the grave risks Lord Roberts bravely faced in order to strike the decisive blow and seize Pretoria. When he decided to advance from Vereeniging without waiting for more supplies, and so profit by the enemy's disorder, he played for a great stake. He won, and it is very easy now to forget the adverse chances. But the facts stand out in glaring outline: that if the Boers had defended Pretoria with their forts and guns they could have checked us for several weeks; and if, while we were trying to push our investment, the line had been cut behind us, as it has since been cut, nothing would have remained but starvation or an immediate retreat on Johannesburg, perhaps on the Vaal.

Even now our position is not thoroughly secure, and the difficulties of subjugating a vast country, though sparsely populated, are such that the troops in South Africa are scarcely sufficient. But the question of supplies is for the present solved. The stores of Johannesburg, and still more of Pretoria, will feed the army for something over a fortnight, and in the meanwhile we can re-open our communications, and perhaps do much more. But what a lucky nation we are to have found, at a time of sore need and trouble, a General great enough to take all risks and overcome all dangers.

The Model School, Pretoria, Now an Hospital for British Soldiers, Recently the Prison of Our Officers.[405] Three Thousand British Soldiers Released from Prison at Pretoria Being Re-Armed in the Public Square.[406]

Lord Roberts and Staff — First Day in Pretoria at Headquarters. Roberts moved his staff outside so that the moment could be captured by Biograph moving pictures.[407] Arrival of Despatch Rider, Capt. Somers, at Headquarters.[408]

Boers Encrusting the Public Buildings Giving up Their Arms.[409] Natives procuring passes in Pretoria.[410]

Battle of Roodewal Station

What Winston is referring to when he says the army's positon is not secure is the result of the Boers efforts to deprive Roberts of his supplies by attacking the railroad lines. The Battle of Roodewal Station took place on June 7th, two days after the occupation of Pretoria.[427] Roodewal Station is three kilometres south of the Rhino River[428] in the Free State. A force of Gen. Christiaan de Wet, together with Commandant Piet Fourie,[429] attacked the British garrison, which was tending to the fully loaded trains and a supply depot. Once De Wet had cut off the telegraph's wires, he sent a message to the British force asking them to surrender. They refused, to which De Wet began his attack. The battle lasted for five hours and swung in De Wet's favour when General Froneman[430] arrived with two cannons. The British immediately surrendered as the bombs began to fall between them. The Boers looted more than 500 crates full of ammunition. Later that night, the Boers blew up the remaining ammunition, thousands of lignite bombs, and two [sic, two thousand][431] bags of mail. The explosion leaves a gigantic crater. The British losses were 45 killed, 123 wounded and 645 captured. The Boers lost 8 civilians and 24 were wounded."

Destruction of a train of supplies near Greylingstad..[411] Damage to rolling stock after a train was captured by Boer forces.[412]

Derailment on the Johannesburg - Volksrust line.[413]

[427] Battle of Roodewal Station - Wikipedia

[428] The Rhino River is a major tributary of the Vaal River in the Free State Province of South Africa.- Rhino River (Vaal) - Wikipedia

[429] Adriaan Paulus Fourie, Asst Veldkornet: Commando: General Liebenberg. Awarded the DTD. Source: Roll of DTD recipients. Also see: List of administrators of former South African provinces - Wikipedia: Adriaan Paulus Johannes Fourie (1853–1932) 31 May 1910 December 1925 15 years, 6 months 2 Adriaan Paulus Johannes Fourie (1882–1941) January 1926 August 1929 3 years, 7 months

[430] Froneman Christoffel Cornelius Veldkornet Awarded the DTD and received the Wound Ribbon. Senekal Commando Christoffel Cornelis Froneman - Wikipedia

[431] Unfortunately many of the letters written by correspondents dealing with the entry of Lord Roberts into Pretoria seem to have been seized by the Boer; during the attack on the post office train, when 2,000 bags of mails were captured and destroyed Morning Post War Letter 30 July p. 5 Col 7. https://www.britishnewspaperarchive.co.uk/viewer/bl/0000174/19000730/042/0005

XVI. Action of Diamond Hill

Pretoria: June 14th 1900.[432]

The feeble resistance which the Boers offered to our advance from Bloemfontein favoured the hope that with the fall of Pretoria they would sue for peace, and after the almost bloodless capture of the town there was a very general tendency to regard the war as practically over. The troops who had been marching for so many days with Pretoria as their goal, not unnaturally hoped that when that goal was achieved a period of rest and refreshment would be given them. But the imperious necessities of war demanded fresh efforts.

The successes gained in the Free State by the redoubtable Christian De Wet, and the cutting of the communications near Rhenoster, awoke everyone to the fact that further exertions were required.

Though the Boers under Botha had made but a poor resistance in front of their capital, they were encouraged by the news from the Free State to adopt a more defiant attitude, and to make what we hope has been almost a final effort. As to that I will not be sanguine; but it is certain that, whereas on the 7th and 8th of June the Boer leaders in the Transvaal were contemplating surrender, on the 9th and 10th they were making all kinds of bold schemes to harass and even entrap the British army.

On the 7th the news ran through the camp that Mrs. Botha had come through the lines with some mission on her husband's behalf, and General Schoeman[433] had himself made very decided overtures. On the 8th, therefore, an armistice was observed by both sides, and a conference on Zwartskop, where Lord Roberts was to meet the Republican generals, was arranged for the 9th; but when the 9th came circumstances had changed.

The Field-Marshal had actually his foot in the stirrup ready to ride to the meeting-place, when a messenger arrived from Botha declining, unless Lord Roberts had some new proposal to make, to enter into any negotiations. The consequence of this was an immediate resumption of active operations.

The military situation was, briefly, that Lord Roberts's army was spread around and in Pretoria in various convenient camping grounds, with the greater part of its force displayed on the east and north-east sides of the town; and that the Boers, under Botha and Delarey, to the number of about 7,000, with twenty-five guns, held a strong position some fifteen miles to the east astride the Delagoa Bay Railway.

Mrs. Commandant-General Louis Botha.[414]

It was evident that on any grounds, whether moral or material, it was not possible for the conquering army to allow the capital to be perpetually threatened by the enemy in organised force, and, indeed, to be in a state of semi-siege.

With the intention, therefore, of driving the enemy from the neighbourhood, and in the hope of capturing guns and prisoners, a large series of combined operations was begun. Practically all the available troops were to be employed. But the army which had marched from Bloemfontein had dwindled seriously from sickness, from casualties, and, above all, from the necessity of dropping brigades and battalions behind it to maintain the communications. We have already seen how it was necessary to leave the Fourteenth Brigade to hold Johannesburg, and now the Eighteenth Brigade became perforce the garrison of Pretoria, thus leaving only the Eleventh Division, the corps troops, and Ian Hamilton's force free for field operations.

[432] This despatch was originally located after Lt. Frankland's account. For continuity it has been moved to its chronological location and Lt. Frankland's account is reproduced as subsequent chapters.

[433] Probably: David Johannes Schoeman Commandant Awarded the DTD (Dekoratie voor Trouwe Dienst). Lydenburg Commando.

Camp Kitchen, E.E.O.V.R. at Victoria West Road.[415]

Mounted infantry resting in their camp.[416]

Dinner out.[417]

Post Office.[418]

Shaving in camp.[419]

Smiling soldier at table in camp.[420]

General Louis Botha.[421]

General J. H. De La Rey.[422]

The Eleventh Division numbered, perhaps, 6,000 bayonets with twenty guns. Ian Hamilton's force had lost Smith-Dorrien's Brigade, which was disposed along the line between Kroonstadt and Pretoria, and though strengthened by the addition of Gordon's Cavalry Brigade did not number more than 3,000 bayonets, 1,000 sabres, and 2,000 rifle-armed Cavalry, with thirty guns. But the shrinkage had been greatest among the mounted troops. French's command of a Cavalry Division, which should have been some 6,000 mounted men, was scarcely, even with part of Hutton's Brigade of Mounted Infantry, 2,000. The two Cavalry Brigades with Ian Hamilton mustered together only 1,100 men, and Ridley's Mounted Infantry, whose nominal strength was at least 4,000, was scarcely half that number in actuality. Brigades, therefore, were scarcely as strong as regiments, regiments only a little stronger than squadrons, and the pitiful—absurd if it had not been so serious—spectacle of troops of eight and ten men was everywhere to be seen. It must, therefore, be remembered that though the imposing names of divisions and brigades might seem to indicate a great and powerful force, the army at Lord Roberts's disposal was really a very small one.

The enemy's position ran along a high line of steep and often precipitous hills, which extend north and south athwart the Delagoa Bay line about fifteen miles east from Pretoria, and stretch away indefinitely on either side. The plan of the Field-Marshal was to turn both flanks with Cavalry forces, and to endeavour to cut the line behind the Boers, so that, threatened by the attack of the Infantry in front, and their retreat compromised, they would have to fall back, probably without being able to save some, at least, of their heavy guns.

French was directed to make a wide sweep round the enemy's right flank north of the railway. Pole-Carew, with the Eighteenth Brigade and the Guards, was to advance frontally along the railway; Ian Hamilton to move parallel to him about six miles further south; and Broadwood, who, with the rest of the mounted troops, formed part of Hamilton's force, was to endeavour to turn the enemy's left. It was felt that, important as were the objects to be gained, they scarcely justified a very large sacrifice of life. But though the Field-Marshal would be content with the retreat of the enemy, both Cavalry forces were intended to press hard inward.

On the 11th, the whole army was in motion. French on the extreme left of the British front, which was extended from flank to flank about sixteen miles, soon came in contact with the Boers, occupying strong defensive positions, and he became sharply engaged. During the day he continued to persevere, but it was not until nightfall that he was able to make any progress. Pole-Carew, with the Eleventh Division, moved eastward along the railway, extended in battle formation, and engaged the enemy with his long-range guns, to which the Boers replied with corresponding pieces, including a 6-in. gun mounted on a railway truck. Though an intermittent bombardment continued throughout the day, the operations in the centre were confined to a demonstration.

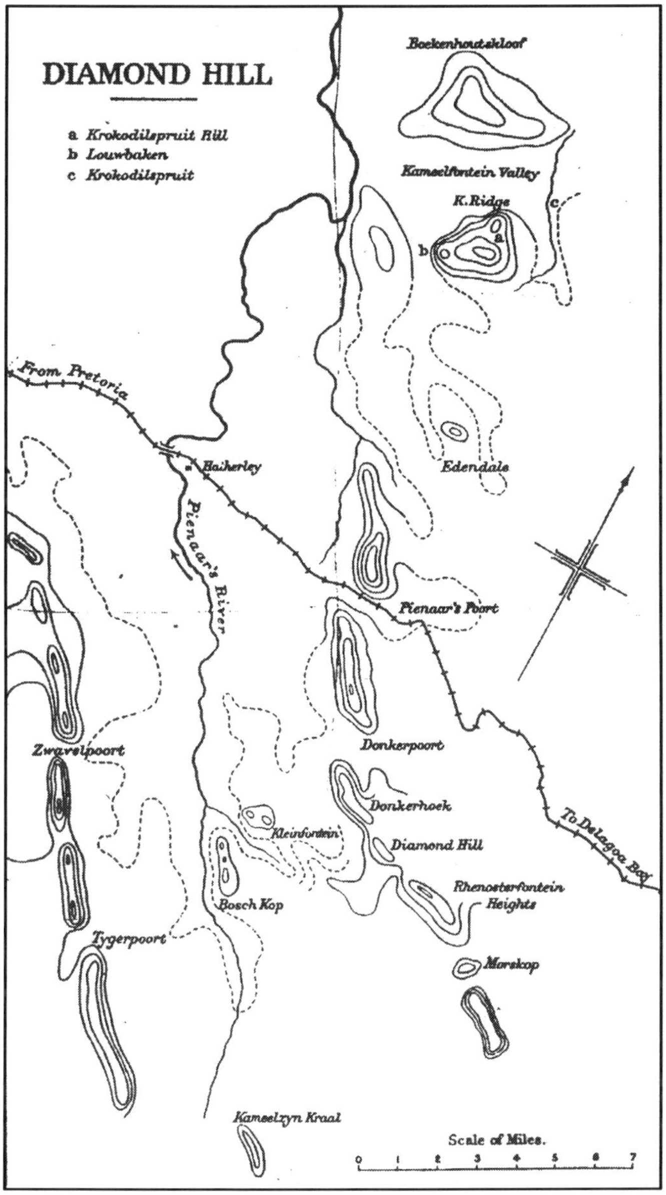

Diamond Hill Area.[423]

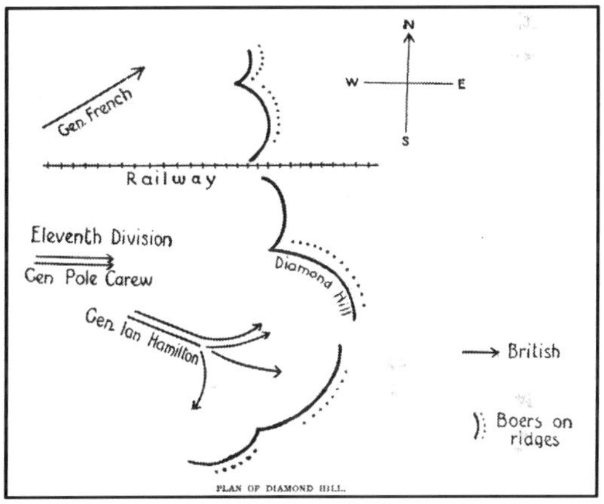

Plan of Diamond Hill.[424]

This is a British Naval 6-Inch gun. The Boer gun would have been on a similar railroad truck.[425]

Meanwhile Broadwood and Ian Hamilton, advancing on the right, found that the Boers, besides occupying the whole line of the Diamond Hill plateau, had also extended their left flank, which was composed of the Heidelburg commando and other South Transvaal burghers, far beyond the reach of any turning movement, and for this reason the operations to the British right and right centre became of a piercing rather than of an enveloping nature. Hamilton endeavoured to hold off the enemy's unduly extended left by detaching a battalion, two field guns, and Gordon's Cavalry Brigade with its horse battery, in the direction of the Tigerspoorte ridges. Ridley's Brigade of Mounted Infantry curved inwards towards the railway, and while these two forces struck out, like the arms of a swimmer, Broadwood's Brigade was intended to push through the gap thus made.

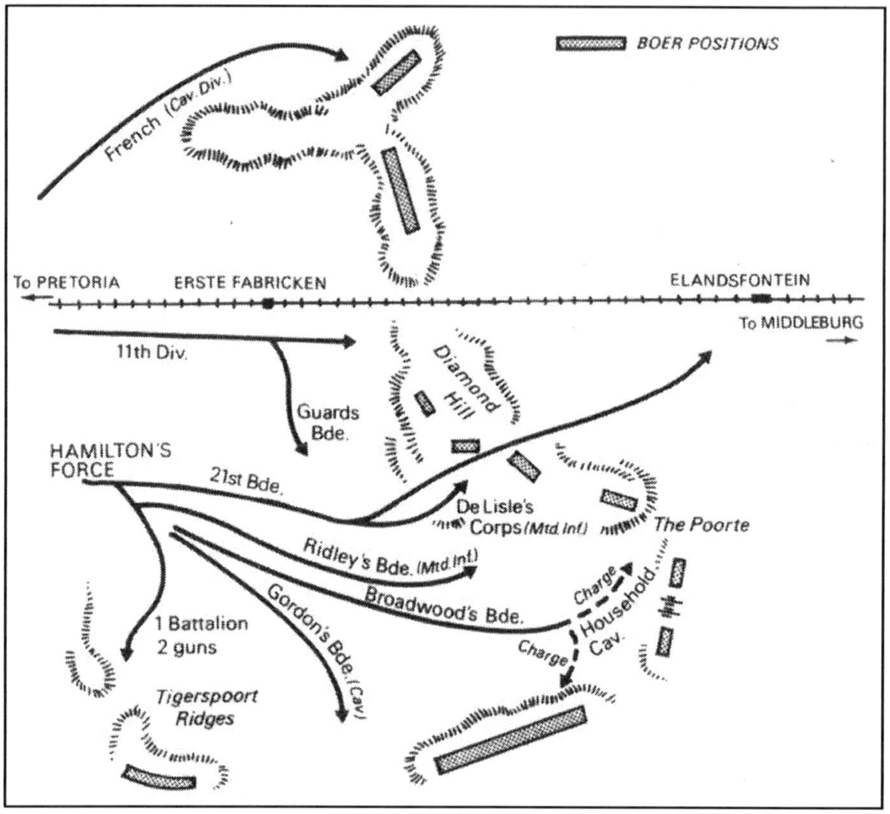

Plan of the Operations for action of Diamond Hill 11-12 June, 1900.[426] Note Tigerspoorte Ridges lower left.

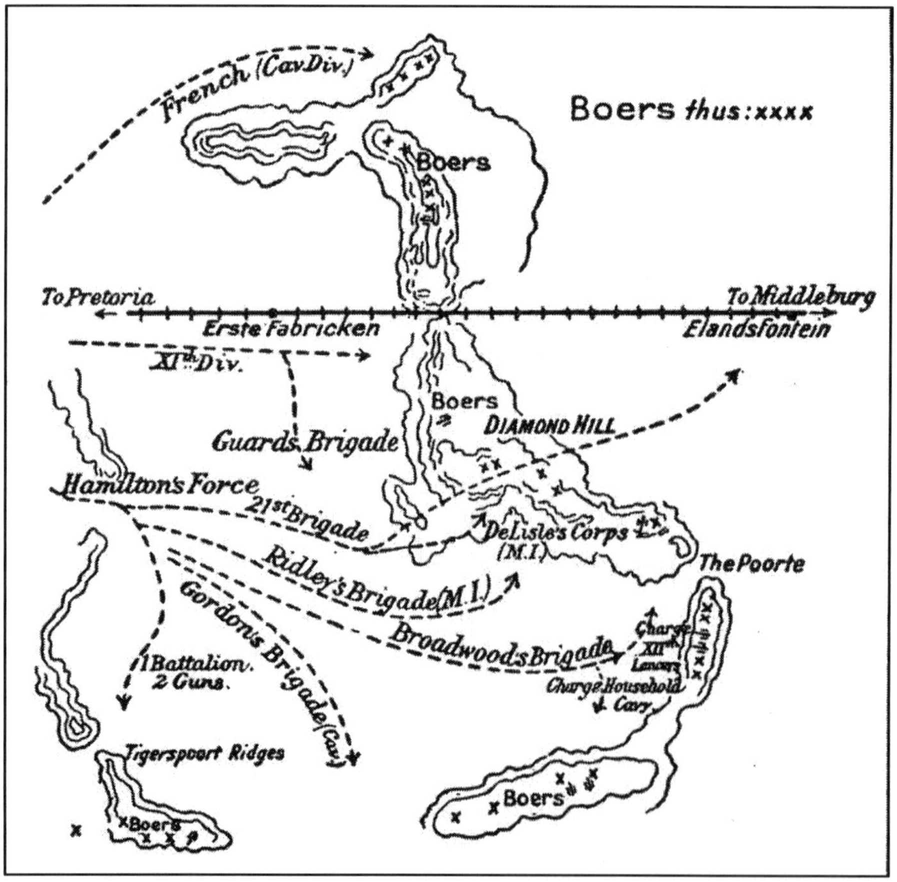

Plan of the Operations of 11th And 12th June, 1900. Scale, Diagram is about 16 miles square. (4*4).[427]

A dropping musketry and artillery fire began shortly after eight o'clock along the front of the force engaged in containing the Boers near Tigerspoorte, and half an hour later Ridley's Brigade was engaged along the southern slopes of Diamond Hill. Meanwhile, Broadwood was advancing steadily to the eastward, and crossing a difficult spruit debouched into a wide, smooth, grass plain, surrounded by hills of varying height, at the eastern end of which was a narrow gap. Through this the line of march to the railway lay. He became immediately engaged with the Boers round the whole three-quarters of the circle, and a scattered action, presenting to a distant observer no picturesque features, and yet abounding in striking incidents, began. The Boers brought seven guns, so far as we could observe, against him, and since the fire of these pieces was of a converging nature, the Cavalry was soon exposed to a heavy bombardment.

In spite of this, Broadwood continued to push on. The country was well suited for Cavalry action, and the gap, or 'poorte,' as it is called in this country, plainly visible among the hills to the eastward, encouraged him to try to break through. Accordingly, at about eleven o'clock, he brought two horse guns, under Lieutenant Conolly, ([1])[434] [435] into a very forward position, with the design of clearing his road by their fire. The Boers, however, fought with a stubbornness and dash which had long been absent from their tactics. They were in this part of the field largely composed of Germans and other foreigners, of colonial rebels, and of various types of irreconcilables.

The Transvaal German Corps.[428]

No sooner had these two guns come into action than a very ugly attack was made on them. The ridge from which they were firing was one of those gentle swells of ground which, curving everywhere, nowhere allows a very extended view; and the Boers, about 200 strong, dashed forward with the greatest boldness in the hope of bringing a close musketry fire to bear on the gunners and of capturing their pieces. So sudden was the attack that their heads were seen appearing over the grass scarcely 300 yards away. In these circumstances the guns fired case shot, but though they prevented the Boers from coming nearer, it was evident that the position was still critical. Broadwood was compelled, therefore, to ask the 12th Lancers to charge.

The continual shrapnel fire of the last few hours had, in spite of their dispersed formation, caused a good deal of loss among the horses of the brigade. The Earl of Airlie, who was riding with the brigadier, had had his horse shot under him, and had gone away to find another. He returned to place himself at the head of his regiment just as it was moving forward to the attack, and, perhaps unacquainted with the latest development of the action, he gave a direction to the charge which was slightly more northerly than that which Broadwood intended; so that, in advancing, the regiment gradually came under the fire of the enemy holding the lower slopes of Diamond Hill, instead of falling on those who were directly threatening the guns. But it was a fine, gallant manoeuvre, executed with a spring and an elasticity wonderful and admirable in any troops, still more in troops who have been engaged for eight months in continual fighting with an elusive enemy, and who must have regarded any action, subsequent to the capture of Pretoria, rather in the nature of an anti-climax.

Its effect was instantaneous. Though the regiment scarcely numbered 150 men, the Boers fled before them—those who were threatening the guns towards the south, and those immediately in the line of the charge eastward

[434] Original footnote (1 (per page numbering)) A younger brother of that brilliant officer of the Scots Greys, whose death at Nitral Nek a few weeks later was so great a loss to his friends, his regiment, and his country.

[435] Probably, Thomas Conolly Lieutenant, He was killed in action at Kaalboschfontein, (Silikats Nek) July 11th, 1900. He was the eldest son of Thomas Conolly, MP, of Castletown, Ireland, and was born in September 1870. Lieutenant Conolly was educated at Harrow and Trinity College, Cambridge, and entered the 2nd Dragoons from the 3rd Battalion Royal Dublin Fusiliers, June 1893, and was promoted Lieutenant December 1894. He served in the Nile Expedition in 1898, being present at the battle of Khartoum, receiving the medal and the Egyptian medal, with clasp. Source: Donner.

and northward, towards Diamond Hill. Had the horses been fresh and strong a very severe punishment would have been administered to the enemy; but with weary and jaded animals—many of them miserable Argentines, and all worn out with hard work and scanty food—they were unable to overtake the mass of fugitives who continued to fly before them.

A few, however, stood boldly, and one man remained firing his rifle until the charge was close on him, when he shot Lieutenant Wright[436] dead at only a few yards distance, and then, holding up his hands, claimed quarter. This was, however, most properly refused. Altogether ten Boers perished by the lance, and the moral effect on those who escaped must certainly have been considerable. But now in pursuit the regiment gradually came nearer to the enemy's main position, and drew a heavy fire on their left flank.

12th Prince of Wales Royal Lancer.[429] Lancer Charge by R. Caton Woodville.[430]

Seeing this, and having obtained the object with which he had charged—the immediate relief of the guns—Lord Airlie gave the order 'files about,' and withdrew his regiment before it became too seriously involved. As he issued this command he was struck by a heavy bullet through the body, and died almost immediately. So fell, while directing his regiment in successful action, an officer of high and noble qualities, trusted by his superiors, beloved by his friends, and honoured by the men he led. The scanty squadrons returned in excellent order to the positions they had won, having lost in the charge, and mostly in the retirement, two officers, seventeen men, including a private of the 10th Hussars, who managed to join in, and about thirty horses.

David Ogilvy, 11th Earl of Airlie.[431] [432] Troops of the Mounted Infantry unit on the move.[433]

[436] Godfrey Charles de Cardonnell Wright Lieutenant He was killed in action at Diamond Hill, near Pretoria, June 11th, 1900. He was the eldest son of Charles Booth E Wright of Bolton Hall, Yorkshire, by his marriage with Edith de Cardonnel, second daughter of R W M Nesfield of Castle Hill, Bakewell, Derbyshire. Lieutenant Wright was born in December 1873, educated at Eton and entered the 12th Lancers, March 1893, being promoted Lieutenant October 1895. He proceeded to South Africa with his regiment in October 1899, and served in Cape Colony. He was afterwards present at the battles of Paardeberg and Driefontien, and took part in the advance on Bloemfontein and Pretoria. Source: Donner.

Meanwhile the pressure on Broadwood's right had become very severe. A large force of Boers who were already engaging the 17th Lancers and the rest of Gordon's Brigade, but who were apparently doubtful of attacking, seeing the advance checked, now swooped down and occupied a kraal and some grassy ridges whence they could bring a heavy enfilading fire to bear. Broadwood, who throughout these emergencies preserved his usual impassive composure, and whose second horse had been shot under him, ordered the Household Cavalry to 'Clear them out.'

The troopers began immediately to dismount with their carbines, and the General had to send a second message to them, saying that it was no good firing now, and that they must charge with the sword. Whereon, delighted at this unlooked-for, unhoped-for opportunity, the Life Guardsmen scrambled back into their saddles, thrust their hated carbines into the buckets, and drawing their long swords, galloped straight at the enemy. The Boers, who in this part of the field considerably outnumbered the Cavalry, might very easily have inflicted severe loss on them. But so formidable was the aspect of these tall horsemen, cheering and flogging their gaunt horses with the flat of their swords, that they did not abide, and running to their mounts fled in cowardly haste, so that, though eighteen horses were shot, the Household Cavalry sustained no loss in men.

Unit of the Mounted Infantry in advance.[434]

These two charges, and the earnest fashion in which they were delivered, completely restored the situation; but though Broadwood maintained all the ground he had won, he did not feel himself strong enough, in face of the severe opposition evidently to be encountered, to force his way through the poorte.

At about noon the Field-Marshal, who was with the Eleventh Division, observing an apparent movement of the enemy in his front, concluded that they were about to retreat, and not wishing to sacrifice precious lives if the strategic object were attained without, sent Ian Hamilton a message not, unless the resistance of the enemy was severe, to weary his men and horses by going too far. Hamilton, however, had seen how closely Broadwood was engaged, and fearing that if he stood idle the enemy would concentrate their whole strength on his Cavalry commander, he felt bound to make an attack on the enemy on the lower slopes of Diamond Hill, and so hold out a hand to Broadwood.

He therefore directed Bruce-Hamilton to advance with the Twenty-first Brigade. This officer, bold both as a man and as a general, immediately set his battalions in motion. The enemy occupied a long scrub-covered rocky ridge below the main line of hills, and were in considerable force. Both batteries of artillery and the two 5-in. guns came into action about two o'clock. The Sussex Regiment, moving forward, established themselves on the northern end of the ridge, which was well prepared by shelling, and while the City Imperial Volunteers and some parts of the Mounted Infantry, including the Corps of Gillies,[437] held them in front, gradually pressed them out of it by rolling up their right.

[437] The Lovat Scouts, Lovat Scouts - Wikipedia a Scottish Highland regiment formed by Simon Fraser, 14th Lord Lovat during the Boer War, is the first known military unit to use ghillie suits. In 1916 It went on to become the British Army's first sniper unit. The Scouts were initially recruited from Scottish Highland estate workers, especially professional stalkers and gamekeepers. Ghillie suit - Wikipedia

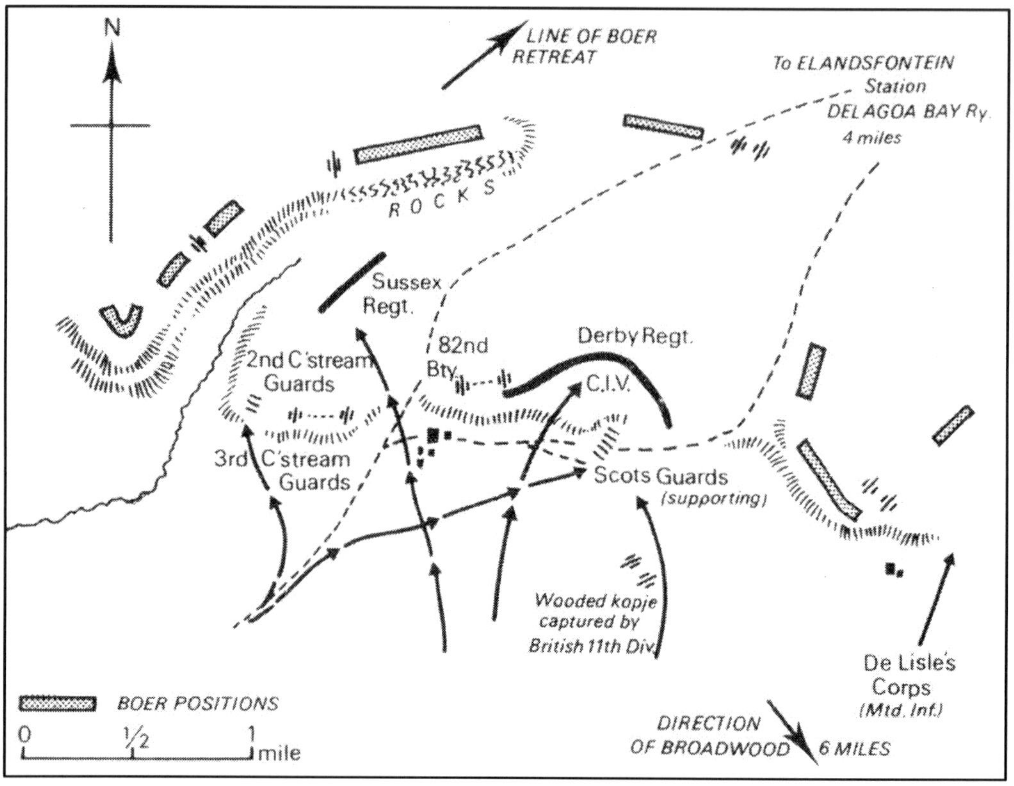

Action of Diamond Hill 11-12 June, 1900.[435]

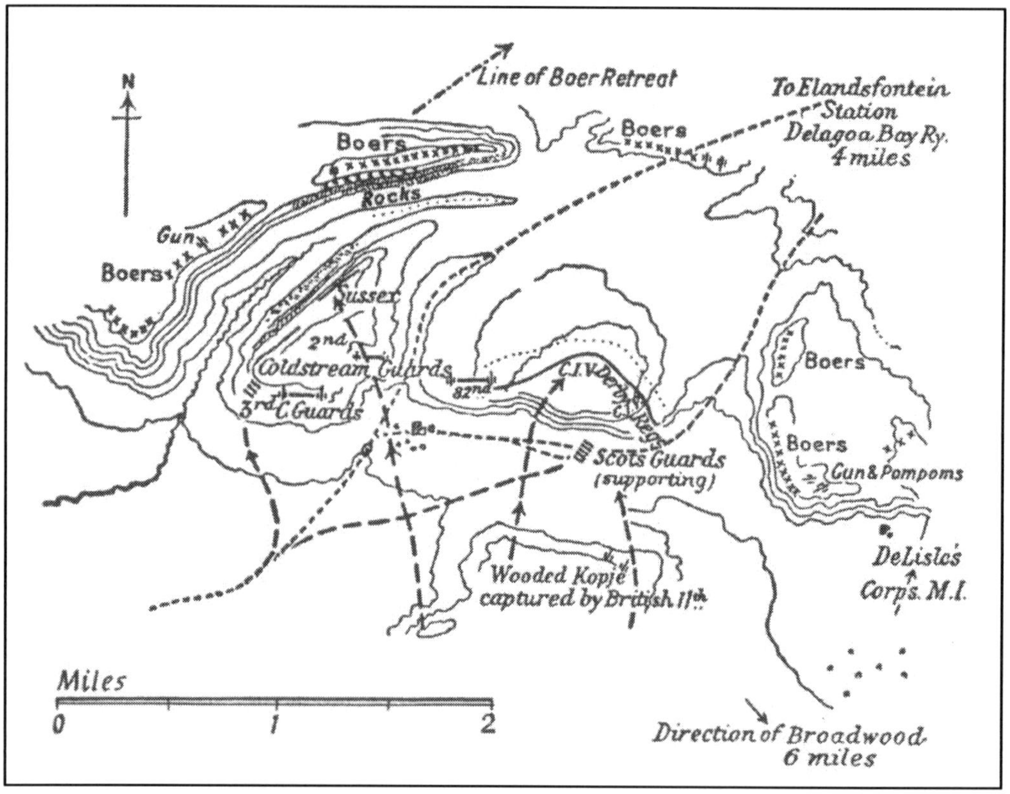

Diagram Explaining the Action of Diamond Hill.[436] One can just make out "CIV' and "Derby's Reg." on the center plateau. The sides look really steep from where they are attacking. Note 'Wooded Kopje Captured by British 11th.

There is no doubt that our Infantry have profited by the lessons of this war. The widely extended lines of skirmishers moving forward, almost invisible against the brown grass of the plain, and taking advantage of every scrap of cover, presented no target to the Boer fire. And once they had gained the right of the ridge it was very difficult for the enemy to remain.

A yeoman of the Lothian and Border Horse Yeomanry.[437]

Taken under fire. The 2nd Queens in action.[438]

Accordingly at 3.30 the Boers in twenties and thirties began to abandon their position. Before they could reach the main hill, however, they had to cross a patch of open ground, and in so doing they were exposed to a heavy rifle fire at 1,200 yards from the troops who were holding the front.

From where I lay, on the left of the Gillies' firing line, I could see the bullets knocking up the dust all round the retreating horsemen, while figures clinging to saddles or supported by their comrades, and riderless horses, showed that some at least of the bullets had struck better things than earth. So soon as they reached fresh cover, the Dutchmen immediately reopened fire, and two of the Gillies were wounded about this time.

The City Imperial Volunteers then occupied the whole of the wooded ridge. One poor little boy, scarcely fourteen years old, was found shot through the head, but still living, and his father, a very respectable-looking man, who, in spite of his orders from the field-cornet, had refused to leave his son, was captured; but with these exceptions the Boers had removed their wounded and made good their retreat to the main position. It being now nearly dark the action was broken off, and having strongly picketed the ground they had won, the Infantry returned to their waggons for the night.

It was now imperative to carry the matter through, and in view of the unexpected obstinacy of the enemy, the Field-Marshal directed Pole-Carew to support Hamilton with the brigade of Guards in his attack the next day.

Three Generations under arms.[438]

Soldiers of one of the British mounted regiments eating their rations.[439]

[438] (L to R) P.J. Lemmer (65) J.D.I. Botha (15) and C.J. Pretorius (43) Hillegas, Boers in War, 1900.

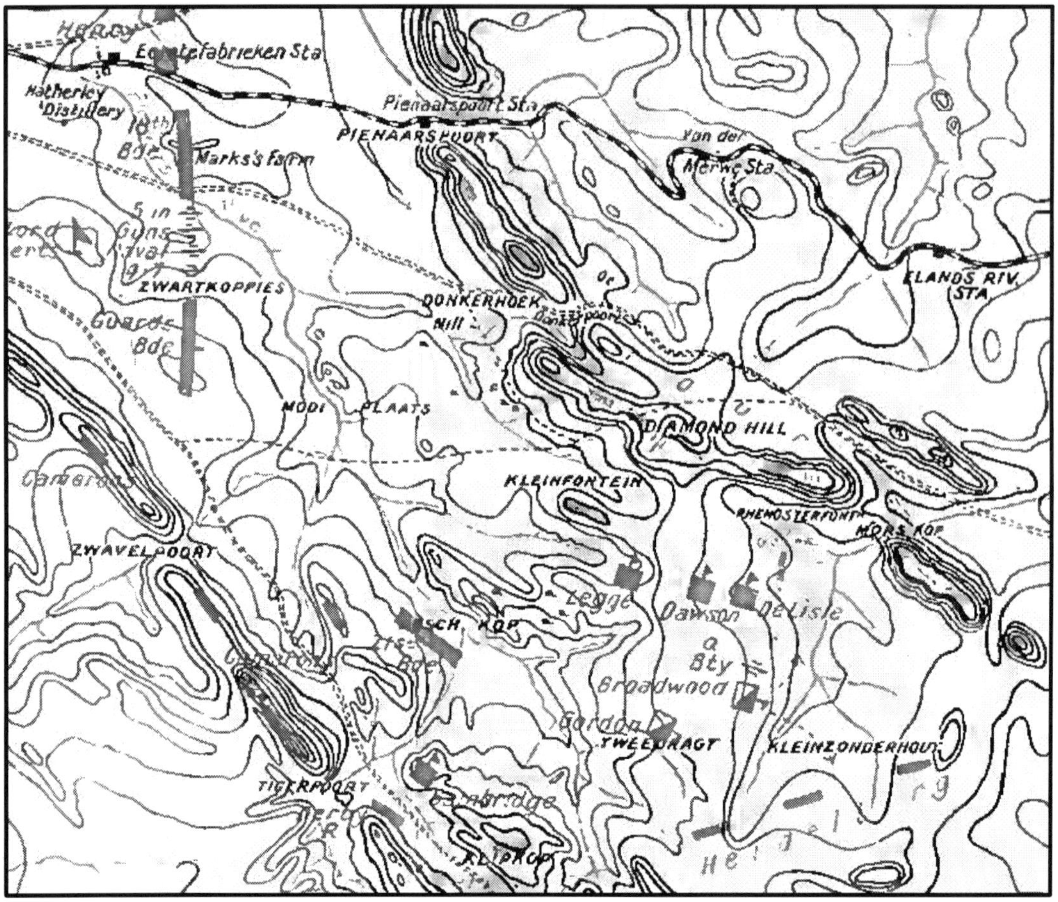

June 11th actions.[440] Note Kleinfontein (not Kleitnzohderout) just southwest of Diamond Hill and Boschkop southwest of it.[439]

Quote from Times History.[440]

> "Ian Hamilton determined to relieve the pressure on [Broadwood]….by sending Bruce Hamilton to attack the Boers on Kleinfontein, where Legge's M.I. had been checked after securing a footing on the edge of the ridge. Bruce Hamilton sent forward Lovat's Scouts [Gillies] to explore the ground between Boschkop and Kleinfontein, and behind them in open order the Sussex on the left and the C.I.V. on the right. The attack was supported by the two field batteries on each flank and by the two 5 -inch guns in the rear. The Boers, who were not numerous on Kleinfontein, did not seriously contest the possession of the ridge, but on the approach of the infantry, retired to their main position on Diamond Hill."

June 12th

Early the next morning Hamilton's Infantry moved forward and reoccupied the whole of the ground picketed the previous night. On the right De Lisle's corps of Mounted Infantry prepared to attack; the Cavalry maintained their wedge-like position, and exchanged shots all along their front with the Boers; but no serious operations were begun during the morning, it being thought better to await the arrival, or, at least, the approach, of the brigade which had been promised.

During this interval the Boers shelled our batteries heavily with their long-range 30-pounder guns, and General Ian Hamilton, who was sitting on the ground with his staff near the 82nd Field Battery, was struck by a shrapnel bullet on the left shoulder. Fortunately, the missile did not penetrate, but only caused a severe bruise with numbness and pain, which did not, however, make it necessary for him to leave the field. The case of this shell,

[439] PF: This is as close as I get to confirming Ian Hamilton's recollection that Winston signaled him with a handkerchief from a forward position, pointing the way for an attack. Either Boschkop or Kleinfontein would have been the location. He was with the scouts and present shortly after the hill was taken. Hamilton inflated the story just a bit by stating it was on the 12th but other facts in his recitation point to the 11th. See the excerpt from Hamilton's autobiography: Listening for the Drums, at the conclusion of this chapter.

[440] Times History Vol. 4. P. 284.

which struck close by, ran twirling along the ground like a rabbit—a very peculiar sight, the like of which I have never seen before.[441]

At one o'clock the leading battalion of the Guards was observed to be about four miles off, and Bruce-Hamilton's brigade was therefore directed to attack. The Derbyshire Regiment, which had been briskly engaged during the morning, advanced up a flat tongue of land on the right. The City Imperial Volunteers moved forward in the centre, and the Sussex on the British left.

Though this advance was exposed to a disagreeable enfilade fire from the Boer 'pom-pom,' the dispersed formations minimised the losses, and lodgments were effected all along the rim of the plateau. But once the troops had arrived here the fight assumed a very different complexion.

The top of the Diamond Hill plateau was swept by fire from a long rocky kopje about 1,800 yards distant from the edge, and was, moreover, partially enfiladed from the enemy's position on the right. The musketry immediately became loud and the fighting severe. The City Imperial Volunteers in the centre began to suffer loss, and had not the surface of the ground been strewn with stones, which afforded good cover, many would have been killed and wounded.

Group of the British Army troops on look-out in the rocky terrain.[441]

Though it was not humanly possible to know from below what the ground on top of the hill was like—we were now being drawn into a regular rat-trap. It was quite evident that to press the attack to an assault at this point would involve very heavy loss of life, and, as the reader will see by looking at the rough plan I have made, the troops would become more and more exposed to enfilade and cross fire in proportion as they advanced.

June 12th actions.[442]

After what I had seen in Natal the idea of bringing guns up on to the plateau[442] to support the Infantry attack when at so close a range from the enemy's position seemed a very unpleasant one. But General Bruce-Hamilton did not hesitate, and at half-past three the 82nd Field Battery, having been dragged to the summit, came into action against the Boers on the rocky ridge at a distance of only 1,700 yards.

This thrusting forward of the guns undoubtedly settled the action. The result of their fire was immediately apparent. The bullets, which had hitherto been whistling through the air at the rate of perhaps fifteen or twenty to the minute, and which had compelled us all to lie close behind protecting stones, now greatly diminished, and it was possible to walk about with comparative immunity. But the battery which had reduced the fire, by keeping the enemy's heads down, drew most of what was left on themselves. Ten horses were shot in the moment of unlimbering, and during the two hours they remained in action, in spite of the protection afforded by the guns and waggons, a quarter of the gunners were hit. Nevertheless, the remainder continued to serve their pieces with machine-like precision, and displayed a composure and devotion which won them the unstinted admiration of all who saw the action.

[441] Per this quote, Winston is with Ian on the 12th and not at the front signaling. He did that yesterday (11th) when he was with the Gillies on the left side of their firing line. See map below with 'Wooded Kopje Captured by British on 11th' identified.
[442] Wooded Kopje captured by British on the 11th on the map.

15 pounder guns being transported by a detachment of the Royal Artillery.[443]

English 15-Pounder Gun.[444]

Getting a 15-Pounder to the top.[445]

15-Pounder on Top.[446]

About four o'clock General Ian Hamilton came himself to the top of the plateau, and orders were then given for the Coldstream Guards[443] to prolong the line to the left, and for the Scots Guards to come into action in support of the right. Two more batteries were also brought forward, and the British musketry and artillery being now in

[443] The Coldstream Guards is the oldest continuously serving regiment in the British Army. Coldstream Guards - Wikipedia

great volume, the Boer fire was brought under control. Ian Hamilton did not choose to make the great sacrifices which would accompany an assault, however, nor did his brigadier suggest that one should be delivered, and the combatants therefore remained facing each other at the distance of about a mile, both sides firing heavily with musketry and artillery, until the sun sank and darkness set in.

General Pole-Carew, who with the Eighteenth Brigade was still responsible for containing the Boer centre across the railway, now rode over to Hamilton's force, and plans were made for the next day. It must have been a strange experience for these two young commanders, who, fifteen years ago, had served together as aides-de-camp on Lord Roberts's staff, to find themselves now under the same chief designing a great action as lieutenant-generals. It was decided that Hamilton's force should move further to the right and attack on the front, which, on the 12th, had been occupied by De Lisle's corps of Mounted Infantry, that the brigade of Guards should take over the ground which the Twenty-first Brigade had won and were picketing, and that the Eighteenth Brigade, which was now to be brought up, should prolong the line to the left. But these expectations of a general action on the morrow were fortunately disappointed. Worsted in the fire fight, with three parts of their position already captured, and with the lodgment effected by Colonel de Lisle's corps on the left threatening their line of retreat, the Boers shrank from renewing the conflict.

During the night they retreated in good order from the whole length of the position which they occupied, and marched eastward along the railway in four long columns. When morning broke and the silence proclaimed the British the victors, Hamilton, in order to carry out his original orders, marched northward and struck the railway at Elandsfontein station, where he halted. The Mounted Infantry and Cavalry were hurried on in pursuit, but so exhausted were their horses that they did not overtake the enemy.

Such were the operations of the 11th, 12th, and 13th of June, by which, at a cost of about 200 officers and men, the country round Pretoria for forty miles was cleared of the Boers, and a heavy blow dealt to the most powerful force that still keeps the field in the Transvaal.

After the action of Diamond Hill the whole army returned to Pretoria, leaving only a Mounted Infantry corps to hold the positions they had won to the eastward. French and Pole-Carew, whose troops had marched far and fought often, were given a much-needed rest. Ian Hamilton, whose force had marched further and fought more than either, was soon sent off on his travels again.

The military exigencies forbade all relaxation, and only three days' breathing space was given to the lean infantry and the exhausted horses. By the unbroken success of his strategy Lord Roberts had laid the Boer Republics low. We had taken possession of the Rand, the bowels whence the hostile Government drew nourishment in gold and munitions of war. We had seized the heart at Bloemfontein, the brain at Pretoria. The greater part of the railways, the veins and nerves, that is to say, was in our hands. Yet, though mortally injured, the trunk still quivered convulsively, particularly the left leg, which, being heavily booted, had already struck us several painful and unexpected blows.

To make an end two operations were necessary: first, to secure the dangerous limb, and, secondly, to place a strangling grip on the windpipe somewhere near Komati Poorte. The second will, perhaps, be the business of Sir Redvers Buller and the glorious Army of Natal. The first set Hamilton's Brigades in motion as part of an intricate and comprehensive scheme, which arranged for the permanent garrisoning of Frankfort, Heilbron, Lindley, and Senekal, and directed a simultaneous movement against Christian De Wet by four strong flying columns.

I had determined to return to England; but it was with mixed feelings that I watched the departure of the gallant column in whose good company I had marched so many miles and seen such successful fights. Their road led them past Lord Roberts's headquarters, and the old Field-Marshal came out himself to see them off. First the two Cavalry Brigades marched past. They were brigades no longer; the Household Cavalry Regiment was scarcely fifty strong; in all there were not a thousand sabres. Then Ridley's 1,400 Mounted Infantry, the remnants of what on paper was a brigade of nearly 5,000; thirty guns dragged by skinny horses; the two trusty 5-inch 'cow guns' behind their teams of toiling oxen; Bruce-Hamilton's Infantry Brigade, with the City Imperial Volunteers, striding along—weary of war, but cheered by the hopes of peace, and quite determined to see the matter out; lastly, miles of transport: all streamed by, grew faint in the choking red dust, and vanished through the gap in the southern line of hills. May they all come safely home.

[End of published Ian Hamilton's March.]

Roberts (left on white horse most likely) and Kitchener (Center with moustache) reviewing troops.[447]

Although Roberts had removed the Boer threat to his eastern flank, [by his victory Diamond Hill] the Boers were unbowed despite their retreat. Jan Smuts wrote that the battle had "an inspiriting effect which could scarcely have been improved by a real victory."[444]

The war would continue for another two years as a guerrilla campaign of unparalleled vehemence and suffering on both sides until ended by the Treaty of Vereeniging on May 31st, 1902.[445]

[444] Pakenham 1992, p. 160.
[445] Treaty of Vereeniging - Wikipedia

Listening for the Drums

Forty-four years after this battle, in 1944, Ian Hamilton wrote this summary of Winston's actions during the Battle of Diamond Hill in his book Listening for the Drums.[446]

> During the South African War Winston gave the embattled hosts at Diamond Hill an exhibition of conspicuous gallantry (the phrase often used in recommendations for the V.C.) for which he has never received full credit. Here is the story:—My Column, including Broadwood's Cavalry and a lot of guns, lay opposite and below a grassy mountain, bare of rocks or trees, not unlike our own South Downs where they meet the sea.
>
> The crest line was held by the Boer left. The key to the battlefield lay on the summit but nobody knew it until Winston, who had been attached to my Column by the High Command, somehow managed to give me the slip and to climb this mountain, most of it being dead ground to the Boers lining the crestline as they had to keep their heads down owing to our heavy gun-fire. He climbed this mountain as our Scouts were trained to climb on the Indian Frontier and ensconced himself in a niche not much more than a pistol shot directly below the Boer Commandos—no mean feat of arms in broad daylight and one showing a fine trust in the accuracy of our own guns. Had even half a dozen of the Burghers run twenty yards over the brow they could have knocked him off his perch with a volley of stones.
>
> Thus it was that from his lofty perch Winston had the nerve to signal me, if I remember right, with his handkerchief on a stick, that if I could only manage to gallop up at the head of my Mounted Infantry we ought to be able to rush the summit.
>
> At that moment another message was handed in from Sir John French saying that our extreme left [on the north side of railroad] was falling back.[447] A strong counter-stroke seemed therefore as if it might come in the nick of time. Lord Airlie's charge (and death)[448] at the head of his Lancers had for the moment completely cleared the face of the mountain—there was nothing moving but a field ambulance—so I called up my men and jumped on to my little black waler.[449]
>
> Before I had got my right foot into the stirrup (luckily) a blow like a punch from the fist of a giant struck my right shoulder and I was flung out of the saddle. A shrapnel bullet had struck me, but the range being extreme had only decorated my back with a big black and blue bruise, nothing more, no blood, no wound pension, no nothing, so on we galloped. [450]]
>
> Pole Carew and the Guards followed suit[451] and we were right in among the Boers firing from horseback, a most glorious scrimmage, before they knew where they were; but they were in the soup all right and lo and behold! the left flank of the Boers was turned. Looks quite simple, doesn't it—on paper? Louis Botha has himself told us in his memoirs how he had been on the point of delivering a crushing blow at French and his Cavalry on our extreme left which would have sent him (so Botha was persuaded) scampering back to Pretoria when, his own left having been turned, he had to forgo his coup and fall back with the whole of his force.
>
> The capture of Diamond Hill[452] meant the winning of the battle, ending as it did in a general retirement by the Boers; also it meant the turning point of the war. The capture of Pretoria had not been the true

[446] Listening for the Drums by General Sir Ian Hamilton Faber And Faber Ltd 24 Russell Square London 1944 p. 248-249 Listening for the Drums : Hamilton : Free Download, Borrow, and Streaming : Internet Archive

[447] This occurs on the 11th See: Times History, Vol 4. P. 285.

[448] As noted above, Airlie's charge and demise took place on the 11th.

[449] The Waler is an Australian breed of horse developed from horses that were brought to the Australian colonies in the 19th century. The name comes from their breeding origins in New South Wales; they were originally known as "New South Walers." Waler - Wikipedia

[450] Winston relates this event seemingly with his personal observation on the 12th. 'General Ian Hamilton, who was sitting on the ground with his staff near the 82nd Field Battery, was struck by a shrapnel bullet on the left shoulder. Fortunately, the missile did not penetrate, but only caused a severe bruise with numbness and pain, which did not, however, make it necessary for him to leave the field. The case of this shell, which struck close by, ran twirling along the ground like a rabbit—a very peculiar sight, the like of which I have never seen before.'

[451] Pole-Carew was with the Coldstream Guards, noted on the left of the map, supra. The Scotts Guards are on the right.

[452] Ian Hamilton Original Footnote: 'So named by me in my Dispatch after they had told me that a large diamond had not very long ago been found there.'

turning point but rather this battle of Diamond Hill which proved that, humanly speaking, Pretoria would not be retaken.

Persistent efforts were made by me to get some mention made or notice taken of Winston's initiative and daring and of how he had grasped the whole lay-out of the battlefield; but he had two big dislikes against him,—those of Bobs and K. And he had only been a Press Correspondent—they declared—so nothing happened.

As it was under me at Gudda Kalai that he had enjoyed a brief but very strenuous course of study in the art of using ground to the best advantage either for attack or defence, this made me furious with impotent rage and I would like the numbers of penmen who are making good copy out of Winston every day to bear this fact more constantly in mind: that he had his full share of bad luck as well as of good before he reached his present high perch on the political Diamond Hill, where now everyone shouts bravo! each time he opens his mouth.

Ian Hamilton, Winston Churchill and Roger Keyes[453] at a Reunion of the Royal Naval Division, Crystal Palace, 1938.[448] Ian Hamilton returning from a Levee in full dress uniform of the Gordon Highlanders.[449]

[453] Admiral of the Fleet Roger John Brownlow Keyes, 1st Baron Keyes, October 1872 – 26 December 1945) was a British naval officer. During the First World War Keyes was heavily involved in the organisation of the Dardanelles Campaign. Roger Keyes, 1st Baron Keyes - Wikipedia

Protecting the Railway.

Protecting the Railway. Cape Town, June 28.[450] Published June 29th.

Railway travelling north of Kroonstad is still exciting owing to the extraordinary activity of De Wet.

A long turn of rails and crossings over frequent dongas, which require broad culverts and even bridges, are extremely vulnerable points.

Railroad Culvert Blown up by the Boers Near the Ruenoster(sic) River. [451]

The iron structures have been destroyed, and the temporary bridges, composed of great baulks of timbers on cribs of sleepers, need only a man with a match and a can of paraffin. Once burnt repairing involves at least forty-eight hours interruption of traffic, twenty-four hours being required to bring a reconstruction train to the spot and an equal period to mend the bridge.

In these circumstances it is necessary to guard every span of 24 feet. and upwards.

The Railway Bridge Over The Sunday River (Natal),Repaired By The Transvaalers. .[452] Bethulie. Bridge over Orange River.[453]

Method of Guarding Bridges.

When this was originally decided a very curious principle was followed.

Bridges were apparently protected according to their size.

A 10-span bridge was given a battalion, a five span bridge four companies, and a culvert half a company.

Now that the whole maintenance of the line from Kroonstad to Pretoria has been entrusted to Smith-Dorrien a more sensible and practical system has been adopted. There is nothing less than two companies anywhere, and with every post there are one, two, or more guns.

Game of Fox and Geese.

The general, who is a man of tremendous energy, lives in a train rushing to and fro along the line entrenching, reinforcing, repairing, mounting guns, and despatching patrols, but even his skill and industry cannot prevent frequent breakdown.

It is a kind of game of fox and geese.

The Boer is the fox, and is darting here and there inflicting continual loss, in spite of the number of his adversaries.

The practice of burning down farms near where the bridges were burnt, though naturally repugnant to all good soldiers, is thoroughly justified by the precedents of 1880.

But the difficulty is that the Boers in the field are desperate men, who have got far beyond burning farms, and who are now careless of the miseries of their country if only they can preserve their military pride and honour.

Burning farms near the railway may very likely enlist the inhabitants of the farms unhurt on our side.

Not only from information I have had, but from what I have seen coming down, I am inclined to think that we shall soon have killed all our hostages, and that there will be no more farms to burn, at any rate near those sections of railway which are exposed to Boer attack.

Travelling South.

I travelled south with the Duke of Westminster,[454] who was carrying Lord Robert's despatches for mail.

All went well until we reached Rhenoster. There signs of the recent conflicts are to be seen on all sides.

Rhenoster kopjes are strongly defiled by a battalion with three guns, including one 5-in. gun under Colonel Spens.[455]

Temporary bridge over Sundays River.[454] Armoured train on temporary trestle bridge.[455] Note the rope cladding covering the locomotive.

Locomotive And Cars Being Armoured In A New Way.[456] A protected armoured train.[457] PF: known as 'Hairy Marys', this heavy rope braiding helped protect the locomotives from shrapnel or bullets.

[454] Hugh Grosvenor, 2nd Duke of Westminster - Wikipedia

[455] This is de Wet's July and August 1900 raid he is running into. See: Times History War in S.A Vol 5 (1907) for map.

Railway Wars June 28 Published July 25th 1900

War Letter - Railway War. By Winston Spencer Churchill. Cape town, June 28.[456 457]

After the action of Diamond Hill the whole Army returned to Pretoria, leaving only a Mounted Infantry corps to hold the positions they had won to the eastward. French and Pole-Carew, whose troops had marched far and fought often, were given a much-needed rest. Ian Hamilton, whose force had marched further and fought more than either—perhaps more than both—was soon sent off on his travels again. The military exigencies forbade all relaxation, and only three days breathing space was given to the lean Infantry and the exhausted horses. By the unbroken success of his strategy Lord Roberts had laid the Boer Republics low. We had taken possession of the Rand, the bowels whence the hostile Government drew nourishment in gold and munitions of war. We had seized the heart at Bloemfontein, the brain at Pretoria. The greater part of the railways, the veins and nerves, that is-to say, was in our hands. Yet, though mortally injured, the trunk still quivered convulsively, particularly the left leg, which, being heavily booted, had already struck us several painful and unexpected blows.

Two Necessary Operations.

To make an end two operations were necessary: first to secure the dangerous limb, and secondly, to place a strangling grip on the windpipe somewhere near Komati Poort. The second will perhaps be the business of Sir Redvers Buller and the glorious Army of Natal. The first set Hamilton's Brigades in motion as part of an intricate and comprehensive scheme, which arranged for the permanent garrisoning of Frankfort, Heilbron, Lindley, and Senekal, and directed a simultaneous movement against Christian de Wet by four strong flying columns.

I had determined to return to England; but it was with mixed feelings that I watched the departure of the gallant column in whose good company I had marched so many miles and seen such successful fights. Their road led them past Lord Roberts's headquarters, and the old field-marshal came out himself to see them off. First the two Cavalry Brigades marched past. They were brigades no longer; the Household Cavalry Regiment was scarcely fifty strong: in all there were not a thousand sabres. Then Ridley's fourteen hundred Mounted Infantry, the remnants of what on paper was a brigade of nearly five thousand; thirty guns dragged by skinny horses; the two trusty 5-inch "cow guns" behind their teams of toiling oxen; Bruce Hamilton's Infantry Brigade, with the City Imperial Volunteers, striding along—weary of war, but cheered by the hopes of peace and quite determined to see the matter out—lastly, miles of transport: all stream by, grew faint in the choking red dust and vanished through the gap in the southern line of hills. May they all come safely home.

Pretoria to Cape town.

Two days later I left Pretoria for Cape town, travelling with the Duke of Westminster, who was going south in charge of Lord Roberts's mails, and who therefore enjoyed many special advantages in the way of carriages, engines, and escorts, of which I was very glad, at his invitation, to avail myself. The first day's journey was without incident, nor was it until the next morning that we entered on the dangerous section of the line. It was still dark when we crossed the Vaal, but day had broken before the train passed Vredefort.

South of the station the advanced guard of great convoy hove in sight, and we presently ran along the whole column. It was preceded and indeed surrounded by a perfect network of Yeomanry and Mounted Infantry, who dotted the rolling plains almost to their horizons.

Within this circle came Infantry—Highlanders and Linesmen, drafts probably for regiments at the front—all extended to nearly fifty yards interval between each man and his neighbour. Nearer to the railway were several companies in close order, a couple of Maxims, and some Artillery.

Through the midst of the troops flowed the line of waggons, perhaps three or four hundred in all, each with its long span of mules or oxen and its little group of Kaffirs and escorting soldiers. It may have carried eight days' supplies for two divisions. Lord Roberts was not relying entirely on the railway.

[456] Morning Post, Railway War p.5 & 6, 25 July https://www.britishnewspaperarchive.co.uk/viewer/BL/0000174/19000725/049/0005

[457] This 'War Letter' was not included in the original publication of Ian's March. It appeared in the Morning Post on July 25th.

The Debatable Ground.

We rattled on. Convoy and escort disappeared behind us, and soon it was evident that the debatable ground was reached. Kopies Station takes its name from the rocky hillocks that rise all round it, and make a regular upward step from the Rhenoster River. Its garrison, since it was retaken by Lord Methuen, consists of a half battalion of the Shropshire Regiment, two field guns (one damaged, but can fire another 100 rounds), and a formidable 5 in. gun, all under Colonel Spens. Both Infantry and guns are strongly sangared among the rocks and stones, of the kopjes, and they are provided with seven days supplies of food at the very least.

The train descended the step by a watercourse between the kopjes, and slowed down to cross the Rhenoster Bridge. The line here had been destroyed with a thoroughness which excited a mournful admiration. Not only had the bridge over the deep chasm through which the river flows been blown up, but the ordinary permanent way had been scientifically wrecked for more than three miles. A slap of gun cotton had been exploded under every alternate joining place of rails, making the strong iron bars curl into the air like paper before it catches light, and excavating a deep hole in the formation beneath.

The new bridge—the second new bridge, that is to say, for the first had been burned by Christian de Wet—a good piece of timber trestle work, was approached by a long deviation, over which the train moved with much deliberation. Gangs of coolies under the direction of men of the Railway Pioneer Regiment—a corps of brave and skilful Volunteers—were still at work on the foundations of the temporary piers. Their cartridge belts and rifles and the old line with its curling rails leading to the tangled ruins of the bridge filled the mind with the realisation of war. Other reminders were forthcoming.

Scene of Woeful Desolation.

Two miles from Rhenoster Bridge is Roodeval Station. When we had marched north a month before this had been quite a pretty little oasis in the great plain. There were trees, a few iron buildings, a well, a water tank, and the stationmaster's house—all standing up amid green grass. Now the eye surveyed a scene of woeful devastation.

For ten days two thousand mailbags, containing the longed-for letters of the Army, thousands of warm coats, and all the reserve of heavy ammunition had been confided to the charge of the garrison of Roodeval—eight men strong. On the evening of the 6th of June a hundred and fifty more arrived to share their fate. The storm broke with the daylight, heralded by a Dutchman with a white flag.

"Have fourteen hundred men and four guns," said the Boer summons. "Give you ten minutes to surrender."

But they would not surrender until more than a quarter had been shot and all hope of relief had gone. Then the Boers got to work on their prey. What they needed they took. The rest, together with the buildings already sorely battered by the shells, they gave to the flames. The enormous lyddite shells for the 9-2 and 5 in. siege guns formed the foundation of the pyre.

Terrific and Continuous Explosions.

An officer who had listened ten miles away told me he thought that a great battle was being fought, so terrific and continuous were the explosions. Sheds, buildings, trees, and water tank were blown to the winds. Nothing-absolutely nothing but one bent and tattered piece of corrugated iron remained, hanging insecurely to the ruins of a railway truck.

The ground was strewn with splinters of the mighty bombs. On the siding a complete train of trucks stood in skeleton. All the woodwork was consumed, and only the iron frames remained—with iron stanchions, bolting air, and rows of funny little nails sticking into nothing. Many of the monstrous projectiles, though curiously stained by the fire and battered by the explosions, were still live and lay in heaps.

For nearly a quarter of mile the veldt was thickly carpeted with the ashes of the warm coats, a pile of half-charred scraps of paper marked the destruction of the two thousand mail bags. And Christian de Wet cannot be complimented on his chivalry thereon, for it seems a poor and spiteful thing to burn up soldiers' letters, nor can I see how this can benefit his cause or his campaign.

"Little Less than an Insult"

So much for Roodeval but I do not desire to leave the letters yet. It is necessary to state soberly and solemnly that the administration of the military post offices during the war in South Africa has not been good. Letters have usually been delayed; many have been sent to the wrong destinations; a considerable proportion has been lost.

Finally mails which, if dealt with week by week, would have imposed no strain on the transport, have been allowed to accumulate in vast heaps, such as this one at the Roodeval. Once they attain these unmanageable dimensions no straining transport can find room for them, and the result is that they lie neglected at the caprice of the enemy.

It must be remembered that the private soldier, and the officer for that matter, regard the deprivation of his letters as little less than a personal insult. In times of war, when life hangs on an uncertain thread, the soldier longs for and clings to the letter from home, which fills his mind with the memory of other scenes and refreshes his love for the country in whose cause he fights.

The German Army, in which these things are studied, was distinguished during the France-Prussian war for the excellence of its postal arrangements in the field. On the Indian frontier, in victory or defeat, letters and newspapers arrive with strict punctuality. But here in South Africa delay and confusion culminate in hopeless irretrievable loss, which has caused, and is causing, intense annoyance and distress throughout the Army, and perhaps some dissatisfaction at home.

The Burning Bridge.

After leaving Roodeval, it being now nearly seven o'clock, we made haste to breakfast, and were soon at a suitable meal of sardines, pickles, and whisky. Suddenly the train came to a standstill, and the servant came into the compartment with the news that the bridge in front of us was burning. Without much thought we jumped out on to the line to see what was the matter.

It was even as the man had said. A culvert about thirty feet broad, temporarily repaired with cribs of railway sleepers and massive longitudinal baulks, was blaring away merrily. My companion with the engine-driver very properly ran on towards the bridge to learn the extent of the damage. They were followed sheeplike by most of the passengers.

Without any precautions whatever, without discipline or order, and quite unarmed, nearly a hundred soldiers and twenty or thirty Kaffirs came scuttling along the track laughing and shouting at the fine joke of a burning bridge, and just as if they had never heard of the word "war."

Meanwhile, I could see two dark horsemen cantering away a quarter of a mile to the left and sixty or seventy more approaching along the railway to the south. The dongas on either side of the line afforded excellent cover to any number of riflemen who might happen to be there. The crowd near the culvert provided them with the best of targets.

Boer Attack on the Train.

I was in the act of climbing back into the Carriage to get my Mauser pistol, from which I have never been separated since that miserable morning at Frere, when a shell skimmed over the engine and exploded in the ditch by the line— some four or five yards away.

Its effect was electrical. Back in comical hurry, with surprise writ, large on their faces, scampered the foolish sightseers, and jumped into their trucks. The engine driver re-entered his cab, and without more ado the engine whistling loudly began to retire. The second shell fell wide

Before the third could come we were out of range. Only the two Boers who had been cantering away to a safe distance dismounted and let fly until their magazines were empty. We replied with a vicious spluttering from pistols and carbines, the noise of which, though it did these gentlemen no more harm than the bullets, had the effect of making their advancing comrades pause and consider. .[458]

Moral of the Tale.

Now the hardened soldier and even the peaceful Correspondent may view with composure the spectacle of a burning bridge so long as it be in front of them. But a burning bridge between you and home is quite a different thing. I could not help thinking of armoured trains—and this train was not even armoured. It was therefore a distinct pleasure to see the remains of Roodeval station once more. Here we learned that the telegraph signal had failed before we started, and that there was no longer any communication with the south. Inevitable incident of war—no doubt.

But—and this is the moral which justifies the tale—in such cases it might be as well to inform the senior officer in the train, so that he can make some sort of preparations for defence. Surely that is a mild criticism. Let me extend it and make it stronger. When we go to war our conceit costs us much in blood and honour. We are always absolutely sure that the enemy will not dare to attack us, and so certain that he will fly at our approach.

I have scarcely seen an action in this war which officers of a certain class have not begun with an incredulous sniff. Never have I shared in a defeat without hearing afterwards that the enemy had actually taken to their heels when we retreated.

Who's Afraid?

It is all due to lack of imaginative power. Our troops are visible. The enemy is unseen. Therefore, it is argued, he cannot be very numerous. Our troops are strong. Therefore, he must be very frightened. Therefore, he will run away, has run away already. Therefore, there is nothing to fear—never was (for who's afraid?)—but nothing to guard against.

So we go on. While there is firing all is well. But let there be three days of peace. The Lancers stick their lances into the ground point foremost to preserve the sharpness. The whole camp wears a bored expression. The Infantry officers carry out the precautions which the Drill Book prescribes with a smile of pitying contempt for the nerves of its writer. Everyone concludes that the war is over—forgets that there has ever been a war, and then suddenly—Bang! and another glorious page is added to British military history.

Thinking of this, and with plenty of time to spare, now that the Boers had cut the line, I strolled back to the spot where the Derby Militia had met their fate. There was the usual litter of biscuit tins and rubbish which mark a war-time camp. Here and there a few splinters of shell and empty twisted shrapnel cases caught the eye. Round a cluster of graves a colony of tiny brown African foxes—dachshunds with brushes—were playing in the sunlight.

[458] Quote from "My Early Life" Chapter XXII. All went well till we reached the neighbourhood beyond Kopjes Station, about 100 miles south of Johannesburg. In the first light of morning I was breakfasting with Westminster, who was travelling on some commission for Lord Roberts, when suddenly the train stopped with a jerk. We got out on to the line, and at the same moment there arrived almost at our feet a shell from a small Boer gun. It burst with a startling bang, throwing up clods from the embankment. A hundred yards ahead of us a temporary wooden bridge was in flames. The train was enormously long, and crowded with soldiers from a score of regiments, who for one reason or another were being sent south or home. No one was in command. The soldiers began to get out of the carriages in confusion. I saw no officers. Kopjes Station, where there was a fortified camp surmounted by two 5-inch guns, was three miles back. My memories of the armoured train made me extremely sensitive about our line of retreat. I had no wish to repeat the experiences of November 15; I therefore ran along the railway line to the engine, climbed into the cab, and ordered the engine-driver to blow his whistle to make the men re-entrain, and steam back instantly to Kopjes Station. He obeyed. While I was standing on the foot-plate to make sure the soldiers had got back into the train, I saw, less than a hundred yards away in the dry watercourse under the burning bridge, a cluster of dark figures. These were the last Boers I was to see as enemies. I fitted the wooden stock to the Mauser pistol and fired six or seven times at them. They scattered without firing back. Then the engine started, and we were soon all safely within the entrenchment at Kopjes Station. Here we learned that a fierce action was proceeding at Honing Spruit, a station farther down the line. The train before ours had been held up, and was at that moment being attacked by a considerable Boer force with artillery. The line had been broken in front of our train, no doubt to prevent reinforcements coming to their aid. However, with a loss of 60 or 70 men our friends at Honing Spruit managed to hold out till the next day when help arrived from the south and the Boers retreated. As it would take several days to repair the line, we borrowed horses and marched all night from Kopjes Station with a troop of Australian Lancers, coming through without misadventure. I thought for many years that the 2-inch Creusot shell which had burst so near us on the embankment was the last projectile I should ever see fired in anger. This expectation however proved unfounded.

Scene of a Record Defence.

The unfortunate regiment had arrived late in the afternoon, and had bivouacked forthwith. Either on the rocky kopjes or near the scarped bed of the Rhenoster River they would have been safe, would at least have had a run for their money, for both were strong defensive positions. River and kopjes were scarcely a mile apart, and if the hour was late why not stop at the first? The river bed offered water and security. The kopjes security only. They hailed midway between the two.

Now here imagination, the torturer of cowards, the protector of brave men, should have intervened; should have peopled the dark rocks with crafty Dutchmen creeping stealthily to the attack; should have flashed on the mental screen the inspiring picture of a small force making a record defence among the deep dongas of a river; should have flashed in vivid colours the appalling spectacle of such a force crowded together in the open, and swiftly destroyed in spite of its gallantry.

"What awful rot. There isn't a Boer within fifty miles."

"Still," argues imagination, "if I were a Boer just at this time I might very possibly attempt something against the communications; and if I did, and got on those kopjes in the dark, and in the morning found you at the bottom, I would smash you to pieces with rifles and artillery without losing a man myself."

"Good Lord my dear fellow, you have got the jumps. Come and let's have some dinner."

Boers Astride the Line.

Near Rhenoster Bridge we found a force of seventy Westralians and two field guns whose orders were to march to Honing Spruit, but who now had to wait for reinforcements. The Mounted Infantry detachments were taken out of the train and arranged in companies under such officers as were available. The ammunition was unloaded.

The Westralians threw out patrols in all directions. A message was sent to Colonel Spens on the kopje to explain the situation. The telegraph, moreover, began to click, telling the Army two hundred miles away that its communications with the south were cut and urgently requiring guns, men, and reconstruction trains.

Presently the mounted patrols came galloping in with their reports: "Four or five hundred Boers with several guns astride of the railway; frequent explosions along the line; heavy firing from the direction of Honing Spruit station." And at this news we backed our train under the protection of the fortified kopjes and the 5 inch gun and waited.

We waited all day long, while contradictory orders succeeded one another, and rumours ran amuck through the bridge-builders' camp. At about four o'clock in the afternoon definite news arrived. Headquarters had received the telegrams from Roodeval, had also received telegrams from Honing Spruit.

How, since the wire was cut? Quite simply, "via Kroonstad, Bloemfontein, Norval's Pont, De Aar, Kimberley, Mafeking, Potchefstroom, and Pretoria: clear the line."

Honing Spruit Attacked.

Telegrams said that Honing Spruit was being sharply attacked by a large force of Boers, but that Kroonstad was marching north to help it. In reply to which headquarters ordered all troops at Rhenoster to halt pending reinforcements or the arrival of the general commanding the communications north of Kroonstad: Smith=Dorrien.

The night was bitterly cold, and we retreated thankfully to the shelter of our railway carriage. Provisions were getting rather scanty. Luckily, I had guarded against the chances of delay. The African winter, if it whets the appetite, chills the blood. We shivered under three blankets. The train was shunted several times during the night, and morning revealed us in a wide basin surrounded by the fortified kopjes. I peered about me. The ground gleamed with hoar frost. The air was biting. The sun had not yet risen, and his advancing glow shot a queer, unreal light on the scene, which might, as far as its natural features were concerned, have been laid in Antarctic regions.

The figures of the sentries among the rocks to the east moved black and active against the dull yellow sky. An early cooking fire crackled briskly in the foreground in pleasing contrast. The effects of cutting the railway were apparent. Three great trains had arrived during the night and stood blocked one behind another on the line, their locomotives blowing off immense white clouds of steam. The hindmost train was laden with huge baulks of timber and crowded with men. The reconstruction gangs had arrived.

Arrival of General Smith-Dorrien.

But it was not till evening that they got to work. All the day passed in shunting the other trains out of the way—quite a puzzle on a single line with one siding too few—and in reconnaissance. In the evening General Smith-Dorrien arrived full of spirits and energy, but disgusted at the cutting of his line. It appeared that troops had been moved from the railway on the order of the chief of the staff, without timely notification to the officer responsible for its safety.

There should have been nearly six hundred mounted men at Rhenoster, in which case what would five hundred Boers be doing across the line near Roodeval. There should have been two field guns at Honing Spruit, in which case—but I shall come to Honing Spruit presently. Lord Kitchener had sent the mounted men to join Methuen, and had taken the two guns to reinforce Hamilton, had meant indeed to replace them with another two guns, the same we had seen at Roodeval, but in the interval Christian de Wet struck his blow. Fortunately no great harm was done.

It was arranged that as the Honing Spruit garrison had given a good account of themselves, and the country between that place and Roodeval was now more or less clear, Lord Roberts's mails should be sent through with a small mounted escort at daybreak next morning. (Observe the advantage of travelling with the mails.)

Rebuilding the Bridge.

Everything and everybody else would wait till the line was repaired, and for aught I knew they are waiting still. Nevertheless, no one could grudge one night with the reconstruction gangs.

Night had fallen before they began. I had known the officer in charge in Egypt. He was one of the subalterns who built the Desert Railway; had written a book about it besides—a plain record of sterling work told by one of the hardest workers. Now he was a major. I hope he may become a general. He gave us an excellent hot dinner in his little travelling saloon at the end of the train, and then we all proceeded to the bridge.

Two powerful gas flares and a huge fire made from the half-burned sleepers illumined the scene. Round, dotted about in the velvet blackness, twinkled the picket fires. Above, the frosty sky was bright with stars. The remains of the bridge swarmed with men—swarthy Basutos directed by British engineers—and there arose a clamour of voices and hammering.

The speed at which they worked took hold of one. Already they had pulled away the charred cribs and built new ones of fresh sleepers. Now they were unbolting the rails to pull away the damaged baulks. Such a crowd of weird figures struggling in the firelight to an accompaniment of crashes and creaks!

The Weird Scene.

Presently on the railroad there arose a loud monotonous chant. My companion dragged me back to look. Walking along the line past the engine which waited under steam ready to fly at a moment's notice were some seventy or eighty men—Basutos—carrying an enormous log of wood. Perhaps it was forty feet long and two feet six in thickness, and may have weighed, Heaven knows what, but it was very heavy at any rate. The men moving it slowly forward on their shoulders looked as if they were the legs of a colossal black caterpillar; and as they moved a dusky fugleman at the rear sang solemnly:

"Awa chowachy: chooka chowee," or sounds to that effect; to which the others chorused with the most extraordinary earnestness and conviction,

"Azziz

Awa chowachy: chooka chowee:

'Azziz,"

and so on, again and again, remorselessly, until at length they reached the edge of the chasm which their log was to span. Then they set it down carefully—woe to unlucky fingers and feet—and went back for another.

I looked back to the workers under the flare lights. They had stripped the damaged rails from the track and had cleared the half-burnt baulks away. Though it was freezing hard all were wet with perspiration.

"Will they Guard it when Finished?"

"How much longer? "I asked the officer.

"Only three hours, with any luck," he answered and then a little bitterly: "This is the third time we've built this blasted bridge. I wonder if they'll take the trouble to guard it when it's finished?"

The escort—twenty Canadians and fifteen Westralians—waited in the morning. The officer commanding the latter lent me horses. The Duke fastened his precious despatches to his saddle; we started at a trot in a formation which soon spread out more than a mile in every direction, and in two hours arrived without adventure at Honing Spruit Station. Here a train had been ordered from Kroonstad to meet the mails; but as it had not yet arrived we were compelled to wait, and passed the time, not unprofitably, in learning the story of the fight two days before.

General Smith-Dorrien, who after the capture of Pretoria had been entrusted with the charge of the railway between Kroonstad and Pretoria, was dissatisfied with the security of the section south of Rhenoster River, and particularly with the garrison of Honing Spruit station. The two guns which should have formed part of the force at this point had been removed by Lord Kitchener's order—because they properly belonged to one of Ian Hamilton's batteries.

A Curious Fight.

Smith-Dorrien, therefore, so soon as he was informed, ordered two other guns from Rhenoster to Honing Spruit and railed thither part of the Provisional Battalion formed from the British prisoners released when Pretoria fell. The guns, as I have described, were late, but the men, four hundred strong, under Colonel Bullock of Colenso repute, arrived in the nick of time. The train pulled up at the station, the soldiers got out and proceeded to take stock of their new abode.

While they were doing this someone observed Boers riding slowly along the grassy ridges on both flanks and other Boers tampering with the line a mile to the northward. The next event was the arrival of a flag of truce with a formal demand for surrender.

"No," said Colonel Bullock with decision, and the poor released prisoners, long starved in Waterfall Cage, added, "Never again."

So the Boer envoy departed, and both sides settled down to their work: the Dutch to surround and bombard the station; the British to scratch up what shelters they could. It was a curious fight. The ground was of that undulating character which gives no marked advantage to either, but if anything it favoured the Boers.

Colonel Bullock's Gallant Defence.

Moreover, their enveloping position enabled them to bring a converging fire to bear on the water tank and three tin houses round which the defenders lay. But the greatest disparity was in respect of arms. The Boers had three modern guns and were all armed with Mauser rifles.

The Provisional Battalion were only armed with Martinis, no other rifles being available in Pretoria. In one-third of these obsolete weapons the carbine ammunition supplied jammed or failed to explode owing to weak strikers, and such as would fire revealed by their puff of smoke every man's position. Moreover, their range was lamentably short compared to the Mauser, and the Boers walking about carelessly beyond reach of their enemies struck them down in safety. The British had no guns, and could therefore make no reply to the enemy's artillery.

By good luck the officers, of whom there were nearly twenty, had armed themselves with Mauser carbines, and it was on their fire that the defence mainly relied, keeping their tell-tale black powder for a supreme moment at close quarters. The unequal fight lasted seven hours, and in spite of the widely-extended formation which Colonel Bullock adopted, the fire caused a considerable proportion of casualties.

Timely Succour.

In the absence of any medical officers or appliances, the wounded were dressed under fire by Mr. Cheatle, a London consulting surgeon, who was assisted in his task by the Hon. Somers Somerset, the despatch rider of the *Times* newspaper.

This gentleman's carbolic tooth powder was the only antiseptic dressing available, and was therefore used in all cases, with what results I recommend the *Lancet* to inquire.

The train standing at the siding was much knocked about by the artillery. The tin houses were pierced from various directions. The enemy, being unused to firing at such short range—two thousand yards—burst most of their shrapnel too close to the target, so that the showers of bullets passed overhead and caused little loss. But to lie exposed hour after hour, armed with useless weapons and unable to reply, was a trying ordeal to men picked haphazard from various regiments and united by no honourable association.

With an exception all behaved with the greatest discipline and fortitude. One mixed company, however, being exposed to the cross shell fire, lost heart, and when two more shells burst among them, though these did little harm, they broke, despite the entreaties of the young officer who commanded, and ran southward along the railway in an attempt to escape.

They succeeded desperately in driving away some Boers who tried to intercept them, and meeting the relieving force from Kroonstad, told a wild tale of disaster. Indeed, five men did not stop until they reached Kroonstad itself (twenty miles) where they were promptly arrested, and, it is said, are like to be shot for their conduct. The rest of the battalion held their ground stubbornly.

Towards evening the Boers, apprised of the approach of succour, drew off, and were pursued by Yeomanry and guns, which had arrived on the scene. Their losses in this unequal combat were certainly very small. The Provisional Battalion lost eight killed and seventeen wounded, including among the former Major Hobbs, of the West Yorkshire Regiment, a bravo and unfortunate officer, who was taken prisoner at the end of his first engagement, and killed at the beginning of his second.

Final Words.

Our train arrived about midday, and without further incidents worthy of record we reached Kroonstad, and so to Cape Town. Here, then, I make my bow to the reader, and bring these letters to their conclusion. They have chronicled as faithfully as lay in my power nearly every exciting adventure or military affair it has been my fortune to share or witness during the last nine months. They have been written amid many difficulties and often adverse conditions, with pen or pencil as the fates decreed, under a waggon, in the shadow of a rock, by the uncertain light of a lantern, or even sitting on the ground in the rain. They had to hurry home without waiting for second thoughts, and spring into print guided by a strange hand; and for these reasons perhaps I may claim some indulgence for their crudities and imperfections. So far as my memory serves me, and subject to the demands of truth, they have maligned no one, certainly no one who was unfortunate; and if there be anybody who feels himself assailed unjustly here and now I crave his pardon.

Such as it is, the work at least represents the changing moods and forecasts of the camp, and if in this fashion it has helped to bring those who have borne the long anxiety at home into closer sympathy with their soldiers at the wars, I shall be content, and even be bold to ask the reader to dismiss me with a smile.

Buller & Botha Peace Negotiations June 1900

Buller - Laing's Nek

From the General Officer Commanding in Natal to the Secretary for War.[459]

(Through the Field-Marshal Commanding the Forces in South Africa.)

Newcastle, Natal, 4th June, 1900.

Sir,

In my letter of the 24th ultimo, I reported that on reaching Laing's Nek on the 19th. [May] I found it too strongly occupied for me to force it with the troops then at my disposal.

2. I have since ascertained that the Wakkerstrom Commando, which had left the force opposing Lord Roberts' advance, reached Laing's Nek just in time to rally sufficient of the fugitives I was pursuing to stop our pursuit.

3. For the next few days the enemy were able to concentrate on and about Laing's Nek quicker than I could bring my force forward, and on the 28th I calculated that they had about 8,000 men distributed in very strong positions in Keteni Nek, Laing's Nek, Molls Nek, and the defiles of the Utrecht—Wakkerstrom road, with some 13 to 15 guns in well-selected positions, while some 6,000 Free Staters were in the north-east corner of the Free State, of whom at least 3,000 were holding Botha's Pass and the tracks adjoining it on my left.

4. On the 28th May I sent a force under General Hildyard across the Buffalo [River] at Wools Drift with instructions to march on Utrecht, and another under General Lyttelton at Ichanga Drift, with orders to turn south and clear the Dornberg, which was at the time held by the enemy in some force.

5. The enemy evacuated the Dornberg on General Lyttelton crossing the Buffalo River, and he then turned north and came up on the left flank of General Hildyard, who, on the 29th, had occupied the Belelas Berg, south and in front of the town of Utrecht.

Commandant Grobelaar, who was commanding the 1,500 men holding the Wakkerstrom—Utrecht road, withdrew from the town, giving a promise that no armed men should enter the town while we were there. On the following day, the Landrost surrendered the town, had several copies of attached Proclamation posted, and asked for more for distribution.

Buller Proclamation:

> The troops of Queen Victoria are now passing through the Transvaal. Her Majesty does not make war on individuals, but is on the contrary anxious to spare them as far as may be possible the horrors of war. The quarrel England has is with the Government, not with the people of the Transvaal.
>
> Provided they remain neutral no attempt will be made to interfere with persons living near the line of march, every possible protection will be given them, and any of their property that it may be necessary to take will be paid for.
>
> But on the other hand those who are thus allowed to remain near the line of march must respect and maintain their neutrality, and the residents of any locality will be held responsible, both in their persons and their property, if any damage is done to railway or telegraph, or any violence done to any member of the British Forces in the vicinity of their homes.

6. On the 30th May, I heard that Lord Roberts had occupied Elandsfontein Junction, and I at once sent a flag of truce to the Commandant at Laing's Nek, asking him, now his communications were cut, whether it was worth his while to continue the struggle. He replied that he had referred the question to his Commandant-General.

7. On the 2nd June, I met Assistant Commandant-General Chris. Botha between our outposts. I attach notes of the conversation that ensued. I may add that my Staff elicited from his Staff that at least half the burghers are

[459] South Africa Despatches: Presented to Parliament by Command of His Majesty, London: Printed for H. M. Stationery Office, by Harrison and Sons 1901. Vol I 286 pages. https://archive.org/details/southafricadespa00lond/page/n3/mode/2up
See Page 146- Start of Natal Field Army Despatches: (Buller & Botha page 228-231) Also: HathiTrust Page 298: Pages 84-87 of the online Volume I and II document: https://babel.hathitrust.org/cgi/pt?id=hvd.32044089026009&seq=298&view=1up&q1=Laing%27s+
And: Proclamations Issued In South Africa. Robert, Frederick Sleigh Roberts, 1st earl. Proclamations Issued in South Africa. London, 1900. 820 pages. Vol I & II 1900. https://babel.hathitrust.org/cgi/pt?id=hvd.32044089026009&seq=15
Start Vol II Page 355. https://babel.hathitrust.org/cgi/pt?id=hvd.32044089026009&seq=355&view=1up

anxious for peace, that they have telegraphic communication with Lydenburg through Standerton and Middelburg, and that their supplies also now come by wagons by the same route.

8. With regard to the terms I offered the troops on Laing's Nek, viz., that their homes should be unmolested, provided they now dispersed, taking their small-arms but leaving their artillery, I may mention that even if I attack and take the position, I cannot hope to get any result better than these terms involve. The position admits of such facilities for escape that I cannot hope to make any prisoners, or indeed capture many guns, and I reserved the right of further disarmament.

In these circumstances, should my terms be accepted (though this is I fear unlikely), I shall have obtained the occupation of Laing's Nek without loss of life and without prejudice to our position.

I have the honour to be,
Sir,
Your obedient Servant,
Redvers Buller, General

The Boer commando guarding Van Reenan's Pass.[458]

Drakensberg Railway Switchbacks.[459]

Tunnel Near Laing's Nek Wrecked by the Boers. North View.[460] Tunnel Near Laing's Nek Wrecked by the Boers. South View.[461] Laing's nek Tunnel reopened.[462]

Buller's Interview Notes

General Christian Botha.[463] Botha. General Sir Redvers Henry Buller.[464] Cape Cart, Requisitioned by Roberts for Peace Negotiations with Louis Botha, Driven by the Military Secretary, Captain Waterfield, to the Boer Lines.[465]

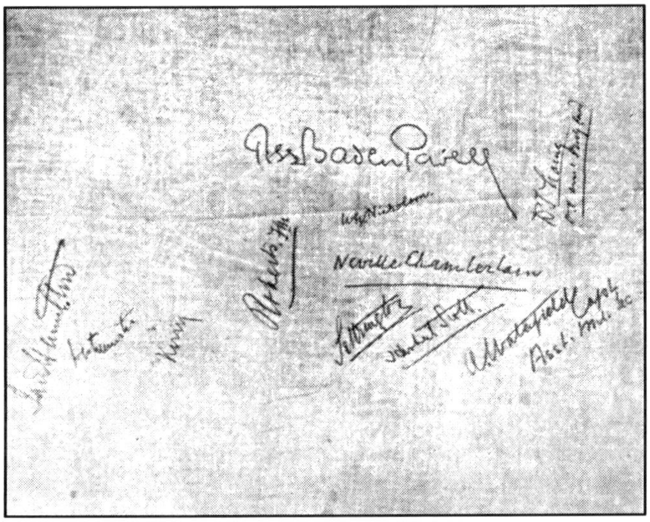
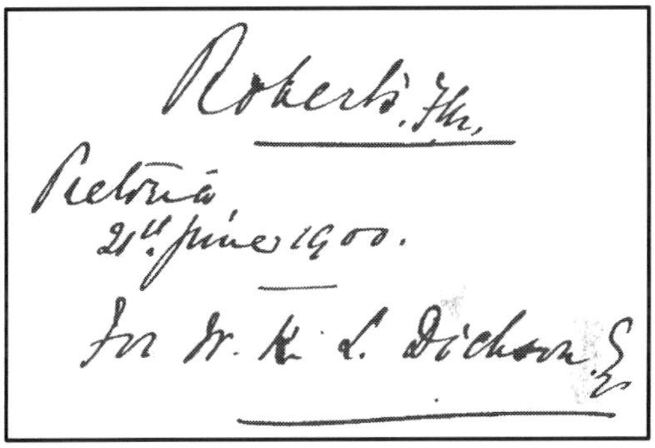

Flag of Truce:[466] What is most instructive is how the flag of truce was authenticated. It is inscribed with the following signatures for authenticity: Roberts (Vertical in Center), Baden Powell, Roberts' Secretary Neville Chamberlain,[460] Kerry, Nicholson, Scott, and others. Roberts' Autograph for comparison.[467]

Notes on interview between Sir Redvers Bailer and Assistant Commandant-General C. Botha, near Laing's Nek, 2nd June, 1900.[461] [462]

Sir R. Buller told [Chris] Botha[463] that he had asked him to meet him with a view to avoid, if possible, the useless sacrifice of brave lives. Lord Roberts was in Johannesburg. There was a large army with many guns in Natal, and he hoped Botha would see that to go on fighting here was only throwing lives away to no purpose.

[460] Sir Neville Francis Fitzgerald Chamberlain (13 January 1856 – 28 May 1944) was an officer in the British Indian Army. He was later Inspector-General of the Royal Irish Constabulary, and resigned in the aftermath of the 1916 Easter Rising in Ireland. He is credited with having invented the game of snooker while serving in Jubbulpore (Jabalpur), India, in 1875. https://en.wikipedia.org/wiki/Neville_Chamberlain_(police_officer)

[461] Proclamations in SA, Interview Notes. https://babel.hathitrust.org/cgi/pt?id=hvd.32044089026009&seq=299&view=1up

[462] The Times' History of the War in South Africa 1899-1902 by Amery, L. S. (Leopold Stennett), Volume 5 pages 604- 606.

[463] Christiaan Botha (also Chris Botha and in English language literature mistakenly Christian Botha, 6 October 1864 – 29 October 1902) was a younger brother of Louis Botha (1862–1919) and Philip Botha (1851-1901) , but an older brother of Theunis Jacobus Botha (1867-1930), and likewise a Boer general in the Second Boer War. Chris Botha - Wikipedia

Commandant Botha asked Sir Redvers to give him his proposals in writing so that he could lay them before his Government. He cannot take upon himself not to fight any more; if he did so he would be a traitor to his country.

Sir R. Buller replied that Botha had received, four days ago, a letter laying the case before him.

Commandant Botha admitted that he did receive Sir Redvers' letter, but the information given in it was not so definite as this conversation was.

Sir R. Buller said that he had stopped all firing since he sent the letter, but in the meantime the Boers had gone on digging trenches. This state of things cannot continue.

Commandant Botha replied that he also had not fired. He then asked Sir Redvers to make a proposal.

Sir R. Buller said he expected the proposal from Botha. He had done the best he could, but he cannot wait long, and he understands that the Boer Government has moved to Lydenburg.

Commandant Botha said he had no proposition to make.

Sir R. Buller said he hates war—it is a miserable thing. A brave man knows when the time comes when to continue fighting means only useless loss of life, and slaughter.

Commandant Botha said it was not for him but for his Government to decide such a question.

Sir R. Buller said he had not asked Lord Roberts about it; but he had many friends in the force opposed to him, and he wished them well. He has so many men that Botha must see it is useless to go on fighting.

Commandant Botha said if he gave in now, Sir Redvers would have a big advantage and he would have great loss.

Sir R. Buller said that on the contrary if war goes on, the Boers' stock would he lost, their homes destroyed, and their property would suffer a great deal of damage, and he wants to avoid that.

Commandant Botha replied that all they have to fight for is their independence. Their stock, &c., is only a secondary consideration to them.

Sir R Buller said "But behind independence is life."

Commandant Botha said he would deserve to be hated by his country if he gave in.

Sir R. Buller said that the Boer Government has been telling the burghers that the English wish to destroy their farms. This is not true.

Commandant Botha said he knows it is not true, and that it is only a rumour and a fiction.

Sir R. Buller said that he was afraid that there was nothing more to be said, and if Botha must go on fighting it could not be helped.

Commandant Botha replied, "If it has to be, it must be." He would heartily welcome a peace if possible, but in present circumstances he would rather not be there to see it made.

The conference was then on the point of breaking up, when—

Commandant Edwards asked Sir Redvers for a definite proposition.

Sir R. Buller replied that the burghers should disperse home, leaving their artillery behind; they might take their small-arms with them. The question of the ultimate surrender of these arms would have to be settled afterwards.

Sir Redvers does not want to take away their arms now, that would be a question for Lord Roberts and Mr. Kruger to settle. He, personally, had no power to make peace. He has met the deputation because he deplores the war. He has had many friends in the Transvaal, and he would like them to live happily and peacefully for the remainder of their lives.

Commandant Botha said he had the same wish, but it was impossible for him to accept or offer terms; he would be a traitor if he did.

Commandant Edwards asked whether Sir Redvers could not make definite proposals for peace.

Sir R. Buller asked whether the commandos here would authorise him to ask Lord Roberts to offer terms.

Commandant Botha agreed, on condition that they retained their independence.

Sir R. Buller replied that he was not in communication with the Government and he could not speak officially, but he had reason to believe that the present Government will not be allowed to continue, that there will be some English form of government, and that the burghers will come in time to participate in the government of the country.

Commandant Edwards said that if the burghers knew what they are to expect, it would facilitate matters, and suggested that Sir Redvers should communicate with Lord Roberts, and ask definitely what the terms of peace are to be.

Sir R. Buller said he would telegraph to Lord Roberts and enquire as to whether he was now in a position to formulate terms of peace.

Commandant Edwards said that his Government knew of this meeting, and they would inform it of Sir Redvers proposals.

Sir R. Buller said he could not wait long, but on being told that a reply could be received in three days, he agreed to an armistice up till Tuesday evening, 5th June.

Commandant Botha said that if he entered into this agreement the English forces must remain where they are.

Sir R. Buller said he could not agree to this as he cannot stop movements now in progress, but he will not fire, nor will he advance his troops in [the] front line beyond their present outposts.

Commandant Botha replied that all he wanted was an assurance that no forces will advance to attack his forces.

It was then agreed that Commandant-General Botha will communicate with his Government as to whether Sir Redvers' proposal should be accepted, and Sir Redvers said he would ask Lord Roberts whether he was yet in a position to formulate his terms of peace.

The meeting then ended.

Buller to Roberts

From General Buller, Newcastle, to Lord Roberts, Front. 2nd June, 1900.[468]

On the 30th May I sent to [the] Commandant on Laing's Nek, and told him that Lord Roberts had crossed the Vaal and that further fighting was useless. This eventually produced a meeting between our outposts today of Assistant Commandant-General Chris Botha and myself.

He asked me what terms of peace you would offer, and I undertook to make no advance for three days while he was communicating to his Government. Can you let me know your terms of peace for individual and separate commandos?

I told him that any terms were that his men should surrender all guns and return to their farms, and if they did that they could take their rifles with them, subject to the understanding that Lord Roberts will later probably order their disarmament.

I thought this would not be wrong, as they can at this moment, if they wish, run away and take both guns and rifles with them. I said that to talk about independence was nonsense, if hereafter they behave themselves they might become an independent colony, that was the only chance they had.

I think they are inclined to give in, and that I have in front of me about half the Transvaal forces now in the field. If you think it worth while, please let me know if may (sic) mention any terms of peace to them. I think that even if assisted from the Orange Free State it will cost me about 500 men killed and wounded to get out of Natal.

Roberts' Reply

From Lord Roberts, Johannesburg, to General Buller, Newcastle. 3rd June, 1900.[469]

Your telegram of yesterday. My terms with the Transvaal Government are unconditional surrender. With regard to troops, those who deliver up their arms and riding animals are allowed to go to their homes on signing pledge that they will not fight again during present war.

The exceptions to this rule are those who have commanded portions of the Republican forces or who have taken an active part in the policy which brought about the war, or who have been guilty of or been parties to wanton destruction of property, or guilty of acts contrary to the usages of civilised warfare.

Principal Officers should remain with you on parole, until you receive instructions regarding their disposal.

Buller's Report

From the General Officer Commanding in Natal to the Secretary of State for War. (Through the Field-Marshal Commanding the Forces in South Africa.) Laing's Nek, Natal, 19th June, 1900.[470] [464]

Sir, The armistice that I reported in my letter of the 4th June[465] ceased on the 5th, as Assistant Commandant-General Chris. Botha wrote on that day declining the terms I had offered. ....

[464] PF: On the 11th, General Buller was able to outflank the Boer position and they retreated. Buller's forces suffered minimal casualties during this operation.

Oath of Neutrality -Proclamation XLI.

Oath of Neutrality.[466]

I, the undersigned, in the District of ________________________

Do hereby solemnly make Oath and declare that I have handed in and given up all the Arms and Ammunition demanded of me by the British Authorities, namely, all Rifles and Rifle Ammunition of whatsoever description they may be.

And I solemnly swear that I have no Rifle or Rifle Ammunition remaining, and that I know of none such being concealed or withheld by anybody whatsoever.

And I further swear that I will not take up Arms against the British Government during the present War, nor will I at any time furnish any member of the Republican Forces with assistance of any kind, or with information as to the numbers, movements, or other details of the British Forces that may come to my knowledge.

I do further promise and swear to remain quietly at my home until the war is over.

I am aware that if have in any way falsely declared in the Premises, or if break my Oath or Promise as above set forth I, shall render myself liable to be summarily and severely punished by the British Authorities.

I make the above declaration solemnly believing it to be true,

So Help Me God.

____

Before me

____

See the Century of Wrong Volume for other proclamations by Roberts relating to the conduct of military operations.

Buller & Botha June Newspaper Coverage June 11

The Commanders' Interview. [Reuter's Special Service.] Mount Prospect, June 2.[467]

General Buller met Christian Botha at O'Neill's Farm, between Umquela and Majuba.

Botha, who met General Buller because his brother Louis could not be found, said that he was not in a position to discuss anything, but as Sir Redvers Buller said "Very well," and turned to go, he pressed him for a proposal.

Sir Redvers said he had no proposal to make, but wished to point out that the Boers were surrounded, and that it was useless to shed more blood.

"What do you want us to do?" asked Botha beseechingly.

"Go back to your farms, leave your big guns, and await Lord Roberts's decision," replied General Buller, at the same time impressing on the Boer officer that lie was not committing Lord Roberts to any definite step.

Botha said he would have to refer the question to the Burghers. General Buller agreed to his appeal for time, and the armistice was arranged.

[465] This letter is not reproduced in the document. Only the letter of the 2nd.

[466] Proclamations in SA, Nov. 18th 1900: https://babel.hathitrust.org/cgi/pt?id=hvd.32044089026009&seq=377&view=1up

[467] Morning Post, Buller & Chris Botha 11 June https://www.britishnewspaperarchive.co.uk/viewer/bl/0000174/19000611/058/0007

Operations in Natal. Proposal to Surrender. From Our War. Newcastle, June 4.[468]

Christian Botha, commandant of the forces at Volksrust and brother of the Boer Commandant-General, met Sir Redvers Buller, at the British commander's request, at **Mount Prospect on Saturday**.

Botha was asked to surrender, and it was agreed that active hostilities should cease for three days, expiring at noon on Tuesday, pending the receipt of Botha's answer.

It is understood that the terms offered were easy.

Before General Hildyard left the vicinity of Utrecht a Proclamation was read to the Landdrost advising the Burghers to return to their farms and promising lenient treatment.

Terms Rejected. Ingogo, June 5.

The terms of surrender have not been accepted by Botha.

Hostilities were renewed on Wednesday, the Creusot on Pogwani Hill firing a few shots at Mount Prospect Camp.

The naval guns replied.

No casualties occurred on our side.

The enemy are still holding their former positions.

The Capture of Van Wyk's. Ingogo, June 7.[469]

The South African Horse were smartly engaged yesterday in taking Van Wyk's Hill, which commands Botha's Pass. They held a front of several miles against a strong force of the enemy until the afternoon, when they were reinforced by the 11th Brigade.

Our casualties were slight.

Through The Drakensberg. Headquarters, June 10.[470]

Yesterday we were engaged in getting our transport through the pass. General advance was begun to-day. We met with opposition eight miles out, but eventually forced the enemy to retire.

Numbers were seen trekking.

A squadron of South African Horse, in trying to cut the enemy off, was engaged all the afternoon at close quarters with a superior force of Boers, probably covering the main retreat.

We are now at the junction of Gans Vlei and the Klip River.

468 Morning Post, Buller & Chris Botha 11 June https://www.britishnewspaperarchive.co.uk/viewer/bl/0000174/19000611/058/0007

469 Morning Post, Buller & Chris Botha 11 June https://www.britishnewspaperarchive.co.uk/viewer/bl/0000174/19000611/058/0007

470 Morning Post, Buller & Chris Botha 11 June https://www.britishnewspaperarchive.co.uk/viewer/bl/0000174/19000611/058/0007

Homeward-Bound

Rhodes Letter

Winston wrote the following letter dated July 12th, 1900 to Cecil Rhodes.[471] [472]

> My dear Mr. Rhodes,
>
> Abe Bailey[473] has spoken to me about a plan to send a small private expedition from Cape Town to Cairo, and has suggested my coming with him. Of course I must think first of all of getting into the House of Commons, but I daresay the general election will be over before the expedition would start and were that the case I daresay I could get away.

He then goes on to say that such an expedition could provide him with another opportunity to write letters to a London Newspaper and ultimately compile them into a book as he has done four times already. His and Abe's efforts would also go towards attracting attention to the project. He actually follows up on this in 1908 when he travels to Africa and writes 'My African Journey.'[474]

He closes by remarking about his visit to Cecil's Brother, Frank[475] at Groote Schuur.[476]

> I lunched and dined with Frankie at Groote Schuur and much admired your beautiful house. I am sorry not to have seen you in South Africa, but the Boers interfered with most peoples' arrangements.
>
> Yours sincerely,
> Winston S. Churchill.
> 12 July 1900

Francis W. Rhodes.[471]

Francis W. Rhodes from the dedication page of 'The River War.'

[471] Lead Source: RMS Dunottar Castle - Wikipedia Primary Source: Cape to Cairo: tracing an imperial imaginary (archive.org) Baderoon, Gabeba; Christopher Roper and Hermann Wittenberg (eds) 1996. Inter Action 4. Proceedings of the Fourth Postgraduate Conference. Bellville: UWC Press, pages 95-97. Archived from the original on 18 December 2007. Retrieved 2 August 2007.

[472] According to the OB (page 515) Winston arrived back from South Africa on July 20th in order to attend his mother's wedding to George Cornwallis-West.

[473] Abe Bailey - Wikipedia [His son John would marry Diana Churchill in 1932] Diana Churchill - Wikipedia

[474] My African Journey; 1908 By Winston Churchill Publisher Toronto, William Briggs. https://archive.org/details/myafricanjourney0000chur

[475] Recall that Winston had dedicated the first volume of "The River War" to Frank Rhodes. See: Colonel Francis William Rhodes CB DSO (9 April 1850 – 21 September 1905). Frank Rhodes (British Army officer) - Wikipedia

[476] Groote Schuur is Dutch for "big shed." Groote Schuur - Wikipedia

Groote Schuur.[472]

Cecil Rhodes.[473]

Entrance.[474]

Library.[475]

Dining-Room at Groot Schuur.[476]

View Seen from the Back of the Hon. Cecil Rhodes' Palatial Home.[477]

Groote Schuur.[478]

Backyard and Gardens.[479]

Barns and stables.[480]

Group Photos

After the occupation of Johannesburg and Pretoria and the scattering of the Boer forces as a result of the battle of Diamond Hill, for journalists and their editors, it was thought that since major combat operations had concluded, the newspapers should recall their best war correspondents for other duties. Churchill, during the upcoming Khaki election[477] would win his seat in Parliament from the constituency of Oldham. He returned to England on the same ship as he had arrived (the Dunottar Castle) along many other journalists and military officers. I have identified two group photos of the passengers, the first with correspondents and the second with servicemen and other dignitaries. Churchill is the only individual to appear in both photographs.

Returning from the Boer War on the R.M.S. Dunottar Castle, July 1900, a Group of Famous War Correspondents.[481 478]

Top row (from left to right):
1. William Dinwiddie, Harper's Weekly; William Dinwiddie - Wikipedia
2. Alister Campbell, Laffan's Agency;
3. J. B. Atkins, Manchester Guardian;
4. Douglas Story, Daily Mail;
5. G. H. Seull, New York Commercial Advertiser;
6. R. C. Booth, Pearson's War News;
7. R. M. B. Paxton, the Sphere;
8. Major Pollock, the Times.

Middle Row:
9. Basil Gotto, Daily Express;
10. Winston Churchill, Morning Post;
11. F. W. Walker, Daily Express;
12. M. H. Donohoe, Daily Chronicle;
13. H. J. Wigham, Daily Mail.

Bottom Row:
14. W. B. Wollern, R.I., the Sphere;
15. J. O. Knight, Times and Herald, Chicago;
16. Ernerst Prater, The Sphere. Ernest Prater - Wikipedia

[477] In Westminster systems of government, a khaki election is any national election which is heavily influenced by wartime or postwar sentiment. See: 1900 United Kingdom general election - Wikipedia. Held between 26 September and 24 October 1900

[478] For more information about individuals in this photograph and others see the Correspondent section of ABW.com. Whether to award QSAs to correspondents was discussed in 1901 with the decision was made to issue medals under certain circumstances. There are 194 records in this article, from both sides. Many with extensive references. https://www.angloboerwar.com/?option=com_content&view=article&id=1861 Also: By entering correspondent into the Name Search field for Unit.

Second Group Photograph

Frederick R. Burnham Photograph.[482]

Back row, standing, left to right:

1. Sir Byron Leighton;
2. Claud Grenfel;
3. Major Frederick Russell Burnham; Frederick Russell Burnham - Wikipedia
4. Captain Gordon Forbes;
6. Abe Bailey Abe Bailey - Wikipedia [his son John would marry Diana Churchill in 1932]; Diana Churchill - Wikipedia
7. & 8. (names of next two gentlemen not recalled [by Mr. Burnham]);
9. [John Weston Brooke] Lord Brooke-Earl of Warwick. John Weston Brooke - Wikipedia

Second row, sitting;

10. Major Bobby White;
11. Lord Downe; Hugh Dawnay, 8th Viscount Downe - Wikipedia
12. General Sir Henry Colville, Henry Edward Colvile - Wikipedia [a year later Churchill as MP would demand an inquiry over his dismissal from South Africa];
13. Major Harry White;
14. Major Joe Laycock; Joseph Laycock - Wikipedia
15. Winston Churchill;
16. Sir Charles Bentinck. Charles Henry Bentinck - Wikipedia

Sitting on deck:

17. (Name of first gentleman not recalled);
18. Colonel Maurice Gifford, who lost his arm in the second Matabele War. Maurice Gifford - Wikipedia

RMS Dunottar Castle[483]: Tonnage: 5,625 GRT Length: 433 ft. (132 m) Beam: 49 ft. 8 in (15.14 m) Draught 25 ft. (7.6 m)

Other Journalists.

Some of the Men Who Have Described the South African War. The Sphere August 4th, 1900.[484]

Mr. E. F. Knight, "The Morning Post" (Lost An Arm), Mr. Winston Churchill The Morning Post" (Wounded And Captured), Mr. Angus Hamilton The Times (In Mafeking), The Late Mr. G. W. Steevens The Daily Mail (Died of Enteric Fever) Mr. Bennet Burleigh The Daily Telegraph, Mr. Julian Ralph The Daily Mail, Mr. Charles Hands The Daily Mail (Wounded Near Mafeking), Mr. M. H. Donohoe The Daily Chronicle (Captured And Released), Mr. Melton Prior The Illustrated London News, Mr. W. T. Maud The Graphic And Daily Graphic, Mr. Frederic Villiers A Literary And Art Agency, Mr. A. G. Hales (an Australian) The Daily News (Captured And Released), Mr. J. B. Atkins The Manchester Guardian, Mr. Rene Bull Black And White, Mr. Ernest Smith The Morning Leader And Star, Mr. Filson Young The Manchester Guardian.

Frederick Burnham's Recollections

Frederick Russell Burnham's[479] recollections:[480]

I was invalided home on the S.S. Dunottar Castle. On the voyage to England were many invalid officers, and some high in the councils of the Empire, who were called home because the political clouds had shifted from South Africa to Fashoda,[481] China,[482] and Constantinople. I recall a graphic review of the world's condition given by young Winston Churchill, who even then had a clear premonition of the coming storm [World War I]. He explained to me why, in his thrilling escape from the Boer prison, he had been compelled to do certain things which I, as a scout, had criticised. His moves were restricted by the handicap of physical weakness which made a 20-mile run at night entirely beyond his power.[483]

Frederick Russell Burnham.[485]

Frederick Russell Burnham.[484]

But the thing that marked him in my memory for life was his solemnly calling a meeting in the cabin and demanding that such men as General Colville,[485] Lord Bentinck,[486] and others should be brought to trial for

[479] Frederick Russell Burnham DSO (May 11, 1861 – September 1, 1947) was an American scout and world-traveling adventurer. He is known for his service to the British South Africa Company and to the British Army in colonial Africa, and for teaching woodcraft to Robert Baden-Powell in Rhodesia. He helped inspire the founding of the international Scouting Movement. Frederick Russell Burnham - Wikipedia See: Burnham, Frederick Russell (1926). Scouting on Two Continents. Garden City, New York: Doubleday, Page & Company. ISBN 978-0-86920-126-8. OCLC 407686. #11 - Scouting on two continents, by Major Frederick Russell Burnham ... - Full View | HathiTrust Digital Library Pages #439-441 of compilation. Pages 349-351 of bound book.

[480] PF: I've been unable to find a second source for this vignette, but as we have seen in his very first despatch Churchill found sea travel tedious during his inbound journey and likely felt the same on the outward one. He did not enjoy the festivities surrounding crossing the Equator or other diversions aboard. This seems like something he would conjure up to liven up an evening and Mr. Burnham is a fair witness and has plenty to talk about without making up stories.

[481] PF: (This incident was actually over by the time Burnham sailed home; but the Boxer Rebellion was still continuing.) The Fashoda Incident was the climax of imperial territorial disputes between Britain and France in East Africa, occurring in 1898. A French expedition to Fashoda on the White Nile river sought to gain control of the Upper Nile river basin and thereby exclude Britain from the Sudan. The French party and a British-Egyptian force (outnumbering the French by 10 to 1) met on friendly terms, but back in Europe, it became a war scare. The British held firm as both empires stood on the verge of war with heated rhetoric on both sides. Under heavy pressure, the French withdrew, ensuring Anglo-Egyptian control over the area. Fashoda Incident - Wikipedia

[482] Because of the Boxer Rebellion taking place. Boxer Rebellion - Wikipedia

[483] A feat which Burnham had accomplished during his first of two escapes while acting as a scout during the Boer War.

[484] Burnham after his investiture with the cross of the Distinguished Service Order by King Edward VII. The black armband was worn in mourning for the recent death of Queen Victoria. London, 1901. File:Major Frederick Russell Burnham DSO 1901.jpg - Wikimedia Commons Also: Richard Harding Davis, Real Soldiers of Fortune. Charles Scribner's Sons, New York 1906 Best Images: #7 - Real soldiers of fortune, by Richard Harding Davis. - Full View | HathiTrust Digital Library Photo by Elliot & Fry.

[485] Colville H (Sir) Major General MID LG: 26 January 1900, page: 503. Source: Lieutenant General Methuen. 16 November 1899. Re: Belmont.

[486] PF: Unable to identify this particular individual amongst others listed in this article: Lord Bentinck - Wikipedia

misappropriation of the sport funds.[487] There was a great buzz throughout the ship, including the crew. Churchill was amply cursed as a bounder, an upstart, a silly ass, a swell-headed "Leftenant," etc., etc. High Dignity appeared offended. Yet the trial had full attendance. Some of the famous legal talent on board was commanded to represent the accused.

This play was all new to me, a Western American. We abuse our gods quite often, but we do not torment them in just this jocular way. I had a creepy, goose-flesh feeling for Churchill, such as one might have for a child innocently gamboling before an onrushing herd of cattle. I was so sure they would turn the tables on him.

It was all wasted sympathy. The cabin was resolved into some sort of parliament and passed a vote of censure on the accused by a safe majority. It was either young Brooke (later Earl of Warwick)[488] or Sir Byron Leighton who remarked to me, "Don't you worry about Churchill. K. of K.[489] sat on him to no purpose, and all hell won't keep him from being Premier some day."

General Herbert Kitchener of Khartoum.[486]

His Brother General Walter Kitchener.[490]

[487] This is explained by Rosslyn in Twice Captured. I spoke elsewhere of the ridiculous and unfair system of raising money for sports and then spending two-thirds of it on the ship's band and other servants of the Company. [Page 88 of the book and 129 of scan.] It appears that this ship brought out the managing director of the Castle Steam Packet Company, Mr. Mirrielees, that he was elected chairman of the Sports Committee, and that he allowed the same system to be carried out, whereby a very large proportion of the £100 collected went for the purpose I am so strongly opposed to. [of tipping everybody and everyone on board for any service they provide from the Bathman to the Smoking Room Steward] Page 31-32 of book and 62-63 of scan.

[488] Leopold Guy Francis Maynard Greville, 6th Earl of Warwick, CMG, MVO (10 September 1882 – 31 January 1928), styled Lord Brooke between 1893 and 1924, was a British officer. Leopold Greville, 6th Earl of Warwick - Wikipedia

[489] General Viscount Kitchener G.C.B. O.M. G.C.M.G. Chief of Staff, South Africa Jan.-Nov., 1900 Commander-in-Chief, South Africa, Nov., 1900 – June, 1902 Lord Kitchener of Khartoum. Herbert Kitchener, 1st Earl Kitchener - Wikipedia

[490] Lieutenant-General Sir Frederick Walter Kitchener KCB (26 May 1858 – 6 March 1912), known as Walter Kitchener, was a British soldier and colonial administrator. He was the younger brother of Herbert Kitchener. Walter Kitchener - Wikipedia Image Source: SA War Vol 7 p. 88 Photo: London Stereoscopic Co.

Jennie's Marriage

Apart from the fact that major military operations in South Africa had concluded and his employer, The Morning Post had other stories of popular interest to cover; notably the Boxer Rebellion, he had family obligations that were calling him home. His mother, Jennie, who had been a widow for over five years;[491] was in love with a dashing young officer, George Cornwallis-West, and she had accepted his proposal of marriage.

Winston, along with his brother Jack and Sunny, were supporting their mother in this new union, and it was scheduled for July 28th, 1900.

He also must have known that the Conservative Party, then riding wave of jingoistic enthusiasm, would call for a Khaki Election[492] to capitalize on the military success and try their hand at the polls for a five-year term.

Engagement announced on July 21st, 1900.[487] Lady Randolph gave a garden party in the grounds of the Royal Normal College on Saturday, Mr. Choate, the United States Ambassador is standing at her side.[488]

Marriage of Lady R. Churchill And Mr. George Cornwallis-West. Published Monday 30 July 1900.[493]

A large congregation assembled in St. Paul's Church, Knightsbridge, on Saturday morning for the wedding of Lady Randolph Churchill with Mr. George Cornwallis-West, Scots Guards, only son of Colonel and Mrs. Cornwallis-West, of Ruthin Castle, Denbighshire. The church was most tastefully decorated — tall palms being arranged outside the handsome rood screen that divides the choir from the nave, with Annunciation lilies and brilliant spikes of crimson gladioli between; while the altar itself was covered with pure white blossoms.

The ceremony was fixed for eleven o'clock, and shortly after that hour the bride entered the church on the arm of the Duke of Marlborough, who afterwards gave her away. She wore a gown of chiffon of a pale sea-blue shade over silk, the bodice made with a bolero of Cluny lace, showing a white chiffon vest, draped across with pale blue silk, fastened with a turquoise brooch set in diamonds. The elbow sleeves were of transparent lace, with full under- sleeves of pale blue chiffon, and the waist belt was of soft blue silk. ....

The bridegroom was supported by a brother officer, Mr. H. C. Elwes, as best man. As the bride and her escort passed along the nave the hymn "Lead us, Heavenly Father, lead us" was sung, after which the service commenced. The Rev. Edgar Sheppard, sub-dean of the Chapels Royal, and the Rev. H. Montagu Villiers, vicar of the parish, were the clergy. After the address "Now thank we all our God" was given, and at the conclusion of the service, while the register was being signed, the anthem "Thou art the King of Glory, O Christ," from the "Dettingen Te Deum" (Handel), was effectively sung by the choir alone under the direction of the Rev. James Baden-Powell, the Precentor.

Those present at the ceremony included the Duchess of Marlborough, the Duchess of Roxburghe and the Ladies Innes-Ker, Mr. Winston Spencer Churchill, the United States Ambassador, the Portuguese Minister, Count Albert Mensdorff, Viscount and Lady Georgiana Curzon, Lady Sarah Wilson, Lady Tweedmouth, the

[491] After 21 years of marriage, her late husband, Randolph Churchill, had died in January of 1895 at the age of 45.

[492] In Westminster systems of government, a khaki election is any national election which is heavily influenced by wartime or postwar sentiment. See: 1900 United Kingdom general election - Wikipedia

[493] Morning Post Jennie Randolph Marriage 30 July p. 5
https://www.britishnewspaperarchive.co.uk/viewer/bl/0000174/19000730/042/0005

Marchioness of Blandford and Lady Norah Spencer Churchill, the Marquis and Marchioness of Londonderry, the Marchioness of Granby, Lord and Lady Churchill, Lord and Lady St. Oswald, Lady Newtown Butler, Lady Gerard, Lady Norreys, Lady Lister Kaye, Lady de Trafford, Lady Sassoon, the Marquesa Jancourt, Mrs. and Miss Moreton Frewen, the Hon. Dudley and Mrs. Leigh, the Hon. Colonel Egerton, the Hon. Evan Charteris, the Hon. Mrs. Stirling, Miss Iznaga, the Hon. Mary Hughes, Sir Vincent Caillard, Sir Edward W. Hamilton, Mrs. W. H. Grenfell, Madame Von Andre, Mrs. Arthur Paget, Miss Muriel Wilson, Mrs. Ronalds, Mr. Smalley, Mrs. Hartmann, Mr. and Mrs. Murray Guthrie, and Mrs. Hwfa Williams.

There was no reception, but near relations and a few intimate friends partook of breakfast at Mrs. Moreton Frewen's residence in Chesham-place. Mr. and Mrs. George Cornwallis-West left shortly before two o'clock for Broughton Castle, near Banbury, Oxon, where they will spend a few days before going on the Continent. ....

Lady Randolph was presented with a superb diamond and pearl tiara from the following friends: The Duke and Duchess of Devonshire, Mr., A. J. Balfour, M. de Soveral, the Marquis and Marchioness of Londonderry, Count Albert Mensdorff, Lily Duchess of Marlborough and Lord William Beresford, the Earl of Crewe, Countess de Grey, Countess Cowper, the Countess of Dudley, the Countess of Essex, the Countess of Gosford, the Countess of Mar and Kellie, Viscount and Lady Georgiana Curzon, Viscountess Chelsea, Loid and Lady Wimborne, Lord and Lady Tweedmouth, Lord and Lady Elcho, Lady Feo Sturt, Lady Gerard, Lord Farquhar, Lady Hindlip, Lady Hartopp, Lady Algernon, Gordon Lennox, Lord Charles Montagu, Lady Alice Stanley, the Hon. Mrs. Stanley, the Hon. Evan Charteris, Mr. Cecil and Lady Lilian Grenfell, Mr. and Mrs. W. H. Grenfell, Sir Edgar and Lady Helen Vincent, Sir Edward and Lady Sassoon, Mr. Henry White (U.S.A.), Miss Emily Iznaga, Mr. Ernest Beckett, Sir E. W. Hamilton, Mrs. George Cavendish Bentinck, Mr. Walter Burns, Mr. and Mrs. Guthrie, Mr. L. V. Harcourt, Mr. H. V. Higgins, Mr. and the Hon. Mrs. K. Maguire, Mrs. Arthur Pazet, Mr. Reuben Sassoon, Mr. and Mrs. Arthur Sassoon, and Mr. and Mrs. Von Andre.

The officers of the Scots Guards presented the bridegroom with a massive beaten silver jug and a pair of silver tankards.

Wedding Portrait of Jennie, appearing August 4th, 1900.[494]

After twelve years of marriage they, regretfully separated in 1912 and divorced in April of 1914. They continued to meet socially upon occasion. After their divorce, she returned to her former name: Lady Randolph Churchill. Shortly thereafter Cornwallis-West married Beatrice Stella Tanner Campbell, an actress known (under her previous husband's name) as "Mrs. Patrick Campbell."[495]

494 The Sphere - Saturday 04 August 1900. P. 29 https://www.britishnewspaperarchive.co.uk/viewer/bl/0001861/19000804/026/0029

495 Major George Frederick Myddleton Cornwallis-West (14 November 1874 – 1 April 1951) was a British Army officer, historian and writer. See: George Cornwallis-West - Wikipedia

Jack Churchill

Major John Strange Spencer-Churchill [496] DSO TD (4 February 1880–23 February 1947), known as Jack Churchill, was the younger son of Lord Randolph Churchill and his wife Jennie, and the brother of Winston Churchill.

He was commissioned into the Queen's Own Oxfordshire Hussars in 1898. He served in the South African Light Horse alongside his war correspondent brother in the Second Boer War in 1899–1900. He was mentioned in dispatches, and was shot through the leg in February 1900, during the Battle of the Tugela Heights.[497], [498] part of the campaign for the relief of Ladysmith.[499] After his recovery, he rejoined his regiment in April of 1900 and was posted at Ladysmith.[500] He served until late October, 1900 and "resisted the blandishments of both his brother and his mother to "chuck" the regiment and come home, even when his mother informed him of her forthcoming marriage to George Cornwallis-West. He felt honour bound to see the thing through."

Before he left South Africa Jack presented his Mauser C96 to a brother officer, W. L. Edwards. This would make sense since it would be way more useful to stay in the theatre rather than to return home with it to the UK.

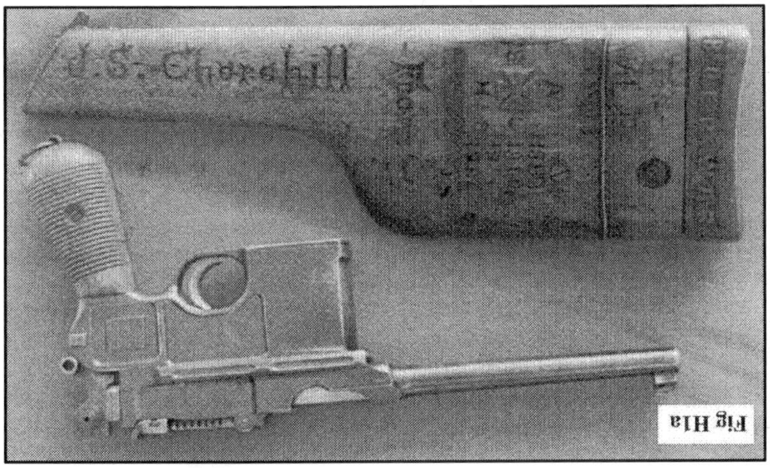

Jack Churchill's Mauser C96. [501]

He fought in World War I, where he was again mentioned in dispatches. He served on the staff of Field Marshal Lord French, General Sir Ian Hamilton (serving as Naval Liaison Officer for the Mediterranean Expeditionary Force) and Field Marshal Lord Birdwood (serving as Camp Commandant, 1st Anzac Corps, and then as Assistant Military Secretary at the headquarters of the Fifth Army).[502]

He reached the rank of major and was awarded the French decorations of the Croix de guerre and the Légion d'honneur and the British Distinguished Service Order in 1918. After the war, he became a businessman in the

[496] This brief biographical sketch is a quote from Jack Churchill's Wikipedia page and was retrieved in January of 2022. It has been slightly augmented with additional source material. Jack Churchill (1880–1947) - Wikipedia

[497] Churchill J S S Lieutenant Wounded at Hussar Hill. 12 Feb 1900. Source: Natal Field Force Casualty Roll, page 230 line 6 And: Churchill J S Lieutenant MID LG: 8 February 1901, page: 973. Source: General Buller. 9 November 1900. Re: Lof C – Anglo Boer War - Name search Also: Churchill J S Lieutenant MID LG: 8 February 1901, page: 973. Source: General Buller. 9 November 1900. Re: Lof C – Natal Anglo Boer War - London Gazette

[498] Battle of the Tugela Heights - Wikipedia

[499] South African Light Horse - details of John S Churchill and Thomas Walden's entitlement to the South African Medal and Clasps, [3 copies]. Personal: South Africa: Correspondence and Papers. https://www.churchillarchive.com/explore/page?id=CHAR%201%2F30%2F34-36#image=0

[500] OB p. 505.

[501] David George has made a record that Winston's brother, John 'Jack' S. Churchill, also carried a Mauser C96 Carvings from the Veldt, Part 2 by David C. George. page 288 Credit Rob D.at: Anglo Boer War - Small Arms and Ammunition - Page 6 - Boer War Forum The 'To' is at the right-hand side of the stock with the initials W L E below them. The 'From' is in the center of the stock. On can also make out the initials SALH and several years, which probably relate to years of service during the war.

[502] Roberts, Andrew (31 October 2007). "Winston Churchill: The 'secret' brother". The Daily Telegraph. Retrieved 30 April 2018.

City of London firstly as a partner at stockbrokers Nelke, Phillips & Bendix from 1906 and then at Vickers, da Costa, making partner in 1921.[503]

During World War II, after the widowed John lost his house during the Blitz, he lived in 10 Downing Street (where he used the bedrooms on the top floor formerly used by Churchill and his wife) or in the No 10 Annex.

He married, in Oxford on 8 August 1908, Lady Gwendoline Theresa Mary Bertie (20 November 1885–7 July 1941), the daughter of Montagu Bertie, 7th Earl of Abingdon, and Gwendoline Mary Dormer. Lady Gwendoline had been raised as a Roman Catholic.[504]

John and his wife Gwendoline had three children:

John George Spencer-Churchill (1909–1992)

Henry Winston (known as Peregrine) Spencer-Churchill (1913–2002), who married Yvonne Henriette Mary Jehannin (1924–2010).

(Anne) Clarissa Churchill, Countess of Avon (b. 1920-2021), the wife of former prime minister Anthony Eden.

Jack died on 23 February 1947, aged 67, of heart disease. He is buried near his parents and brother at St. Martin's Church, Bladon, near Woodstock, Oxfordshire.[505]

The Churchill's in Canada, 1929.[489]

Jack Churchill.[490]

[503] Lee, Celia; Lee, John (2007). Winston and Jack: The Churchill Brothers.

[504] Lee, Celia; Lee, John (2010). The Churchills: A Family Portrait. New York: St. Martin's Press. p. 149. ISBN 9780230112209.

[505] See Also Jack Churchill: The Faithful Brother - Finest Hour 176 Spring 2017 p. 20 (winstonchurchill.org) Jack Churchill: The Faithful Brother - Finest Hour 176 (winstonchurchill.org)

Epilogue

Election and Tour

The war would continue for another two agonising years as a guerrilla campaign but because major combat operations had concluded, the newspapers recalled their first chair war correspondents for other duties.

Winston returned to the England in late July to attend his mother's wedding to George Cornwallis-West,[506] and then stood for a seat in Parliament during the Khaki election[507] that was held between 26 September and 24 October 1900. He won his seat from Oldham, just as Daniel Dewsnap had predicted. He also found time to publish two books, London to Ladysmith and Ian Hamilton's March before the year was out. It would also see him travel to America to engage in an extensive speaking tour as well.

"Churchill in 1900 around the time of his first election to Parliament."[508] "Churchill in 1901."[491]

M.P. for Oldham Photo by Elliot & Fry, Ltd.[492]

Mr. Winston Leonard Spencer Churchill.[493]

[506] He wrote Chapter: VIII 'Ian Hamilton' and dated it "London: August 10, 1900." Two weeks after his mother's wedding on July 28th.

[507] In Westminster systems of government, a khaki election is any national election which is heavily influenced by wartime or postwar sentiment. See: 1900 United Kingdom general election - Wikipedia

[508] PF: Both of these photographs appear to have been taken in the same sitting. File:Winston Churchill 1874 - 1965 Q113382.jpg - Wikimedia Commons

The Two Winstons

Winston Spencer Churchill was introduced to the American public at large in December of 1900 when he arrived from the UK. He had just won his seat in Parliament during the Khaki Election of 1900 as a Conservative MP from Oldham; a seat he failed to gain on his first try prior to his Boer War odyssey. The campaign and election took place during September and October and he sailed for America in November to engage in a speaking tour of the US and Canada.[509]

At that time, the American novelist, Winston Churchill,[510] was a best-selling author, currently known for his 1899 publication of 'Richard Carvel'[511] an American historical novel, that would go on to sell more than two million copies.

Both men literally embraced this coincidence[512] As described below, there are downsides to their mutual fame. Neither of them can get all of their mail because it is being sent to the other; and there is an article to inform everyone about who is who in the duo.[513]

THE REPUBLIC: SUNDAY, DECEMBER 9, 1900.

THE TWO WINSTON CHURCHILLS.

Winston Churchill
The American

Winston Churchill
The British

BY COURTESY OF "THE CRITIC."

[509] See: Churchill and the Great Republic | Exhibitions - Library of Congress (loc.gov) An Age of Youth - Churchill and the Great Republic | Exhibitions - Library of Congress (loc.gov)

[510] Winston Churchill (novelist) - Wikipedia

[511] Richard Carvel is a historical novel by the American novelist Winston Churchill. It was first published in 1899 and was exceptionally successful, selling around two million copies and making the author a rich man. The novel takes the form of the memoirs of an eighteenth-century gentleman, the Richard Carvel of the title, and runs to eight volumes. It is set partly in Maryland and partly in London, England, during the American revolutionary era. Richard Carvel - Wikipedia

[512] 'Books and Authors' section of the New York Daily Tribune Saturday, December 29th, 1900. 0536.pdf (loc.gov)

[513] Winston Arrived on December 8th, per CAC.

> One who has recently returned from Boston had an amusing account to give of the public curiosity about "the two Winstons" who have been together lately in that city. Winston Spencer Churchill and Winston Churchill, it seems, have struck up a friendship, and were wont to go about arm in arm. The public gathered in crowds, lining their path from the Hotel Touraine to the Somerset Club, across the Common, mostly armed with kodaks, and bent on obtaining snapshots of "the two Winstons."

By courtesy of 'The Critic' a pseudonym, not a magazine, the following article was published on December 9th in the St. Louis Republic. As the article states, Winston Spencer Churchill distinguishes his name in order to avoid confusion with the more famous American.[514] [515]

[514] Two Winston's Article Search: Search Results « Chronicling America « Library of Congress (loc.gov) Document: 0199.pdf (loc.gov) The St. Louis Republic. December 09, 1900, Magazine Section, Image 47.

[515] While Winston Spencer Churchill had been receiving periodic mentions in American Newspapers as a result of his Boer War reporting for the UK's Morning Post there was little confusion because his despatches were war related.

Book Advertisements

Advertisements appearing in American newspapers as early as June of 1900 and again during Churchill's American speaking tour.

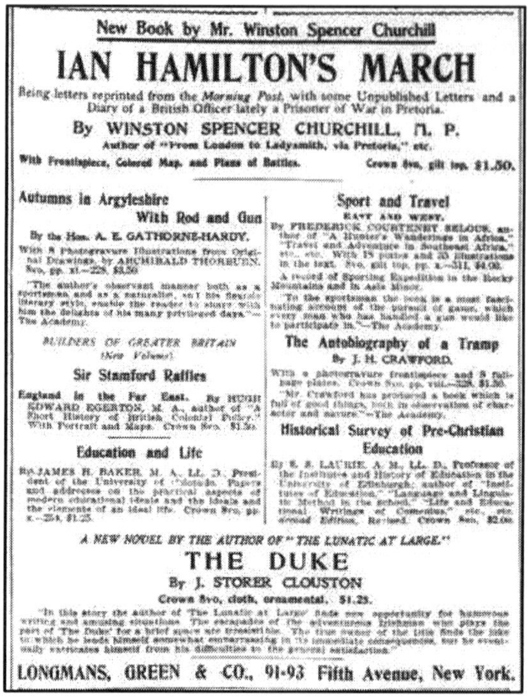

The Sun (NYC) Friday June 8th, 1900.[516] Ian's March Advertisement New York Daily Tribune, Saturday November 10th, 1900.[517]

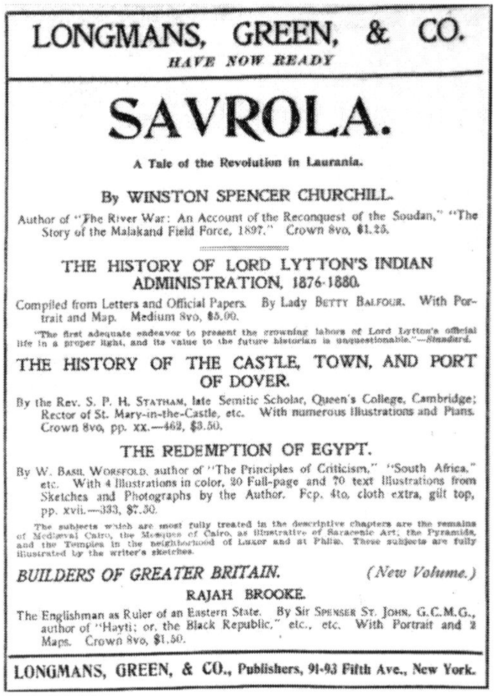

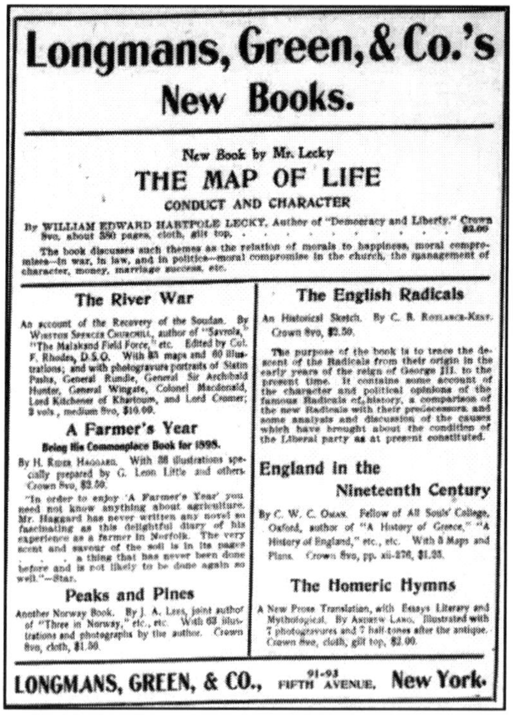

Savrola Advertisement. New Your Daily Tribune. Saturday, January 20th, 1900.[518] River War Advertisement New-York tribune. November 18, 1899.[519]

[516] The Sun (NYC) Friday June 8th, 1900. The Sun NYC 0110.pdf (loc.gov) Search The sun. [volume] (New York [N.Y.]) 1833-1916, June 08, 1900, Page 7, Image 7 « Chronicling America « Library of Congress (loc.gov) See: The Sun (New York City) - Wikipedia
[517] New York Daily Tribune, Saturday November 10th, 1900. 0168.pdf (loc.gov)
[518] New Your Daily Tribune. Saturday, January 20th, 1900 in the Books and Publications section. Savrola: 0329.pdf (loc.gov)

Ian's 1941 Telegram

[This quote from 'Listening for the Drums'[520] demonstrates how enduring his mutual friendship with Winston is.]

It was during the operations in the Bara Valley[521] that I first saw young Winston enjoying himself amongst the bullets: Care and Fear have had no hand in his make-up for which may Providence be praised. ………..

On the 6th of January '41 Bardia had fallen:[522]—red-hot news. Before it began to cool it must be hammered into all sorts of shapes and handed out through many channels leading to Financc, Parliament and the World. Every second was priceless yet he paused to let his mind fly back forty-one years to send a special message to an old comrade of the wars who had long since ceased to interest Press, Parliament or Finance.

Post Office Telegram
London 12.54
General Sir Ian Hamilton—Blair Drummond—Perthshire.

I am thinking of you and Wagon Hill when another January 6th brings news of a fine feat of arms.[523]
Winston

Ian Hamilton, Winston Churchill and Roger Keyes[524] at a Reunion of the Royal Naval Division, Crystal Palace, 1938.[494] Ian Hamilton returning from a Levee in full dress uniform of the Gordon Highlanders.[495]

519 New-York tribune. [volume], November 18, 1899, Page 10, Image 10 0350.pdf (loc.gov) River War Search: Search Results « Chronicling America « Library of Congress (loc.gov)

520 Listening for the Drums by General Sir Ian Hamilton Faber And Faber Ltd 24 Russell Square London 1944 Last Sentence of Chapter XIII and continuing on to Chapter XIV. His entire Chapter XIV is devoted to his friendship with Winston.

521 The valley is located on the bank of Shyok river in the Khaplu, (Ghanche District) of Baltistan. Barah Valley - Wikipedia

522 The Battle of Bardia was fought between 3 and 5 January 1941, as part of Operation Compass. The victory at Bardia enabled the Allied forces to continue the advance into Libya and capture almost all of Cyrenaica, which led to Operation Sonnenblume, German intervention in the fighting in North Africa, changing the nature of the war in the theatre. Battle of Bardia - Wikipedia

523 Battle of Wagon Hill (or Platrand) Siege of Ladysmith - Wikipedia

524 Admiral of the Fleet Roger John Brownlow Keyes, 1st Baron Keyes, October 1872 – 26 December 1945) was a British naval officer. During the First World War Keyes was heavily involved in the organisation of the Dardanelles Campaign. Roger Keyes, 1st Baron Keyes - Wikipedia

Platrand Battle Summary

By early January the siege of Ladysmith had been going on for two months. The Tugela river was in flood and this would keep the British south of the river for some time. Some of the Boer leaders persuaded General Joubert to order an assault on the British defenses and take Ladysmith on the night of January 5th, 1900.[525]

The British line south of Ladysmith ran along a ridge known as the Platrand. The occupying British troops had named its features Wagon Hill to the west and to the east Caesar's Camp. Under the supervision of Ian Hamilton, they had constructed a line of forts, sangars and entrenchments on the reverse slope of the Platrand, of which the Boers were unaware.

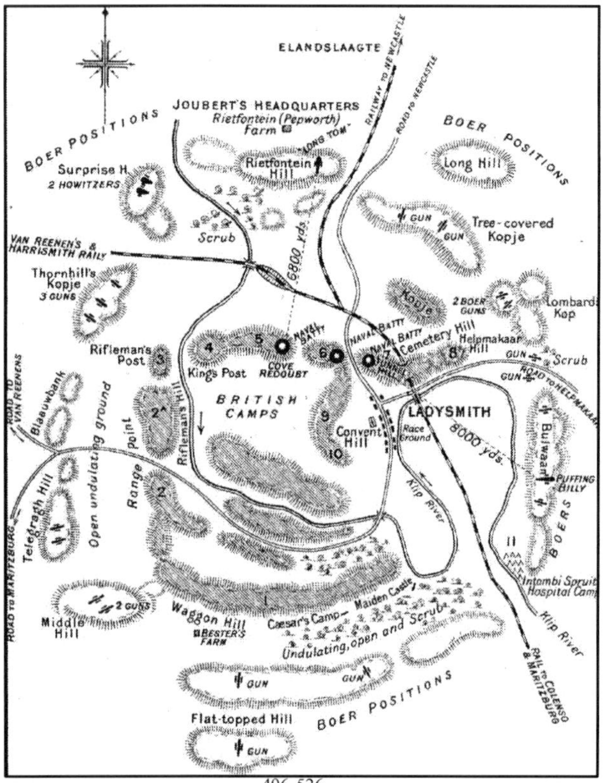

Ladysmith Siege Map.[496 526]

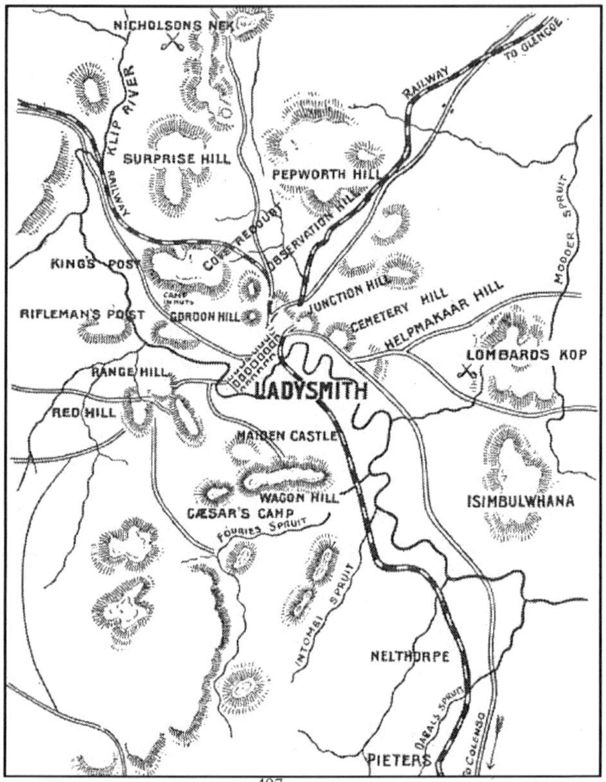

Ladysmith Area Map.[497]

In the early hours of January 6th 1900, Boer storming parties under General C. J. de Villiers began climbing Wagon Hill and Caesar's Camp. They were spotted and engaged by British working parties who were emplacing some guns. The Boers captured the edge of both features, but could not advance further. British counterattacks also failed.

525 See Battle of Wagon Hill (Or Platrand) section of article: Siege of Ladysmith - Wikipedia

526 Explanation to Sketch Map of Positions Round Ladysmith, November 1899.

1. Maiden's Castle, Cæsar's Camp, and Waggon Hill--held by Manchester Battalion, 2nd King's Royal Rifles, and one battery; Gordons in support.
2. Range Post Ridge--held by two companies of Royal Irish Fusiliers.
2A. Rifleman's Ridge--held by King's Royal Rifles.
3. Rifleman's Post detached signal station--held by King's Royal Rifles.
4. King's Post detached signal station--held by King's Royal Rifles.
5. Cove Hill--held by Rifle Brigade; Cove redoubt; at its eastern end is a battery for one 4.7-in. naval gun.
6. Junction Hill--held by Leicesters and naval 12-pounder.
7. Tunnel Hill and Cemetery Hill--held by naval battery in redoubt, 4.7-in. gun, two companies of Gloucesters, and the Liverpool Regiment.
8. Helpmakaar Hill--held by 1st Devons, who have entrenched themselves strongly, and one battery field artillery, protected by epaulements, traverses, etc.
9. Convent Hill.
10. Headquarters.
11. Intombi Spruit, camp for sick and wounded and non-combatants, close to Boer lines.

At noon, de Villiers made another attack on Wagon Hill. Some exhausted defenders panicked and fled, but Hamilton led reserves to the spot and recaptured some empty gun pits. Late in the afternoon, a terrific rainstorm broke, and the Boers withdrew under cover of it.[527]

Imperial Light Horse.[498]

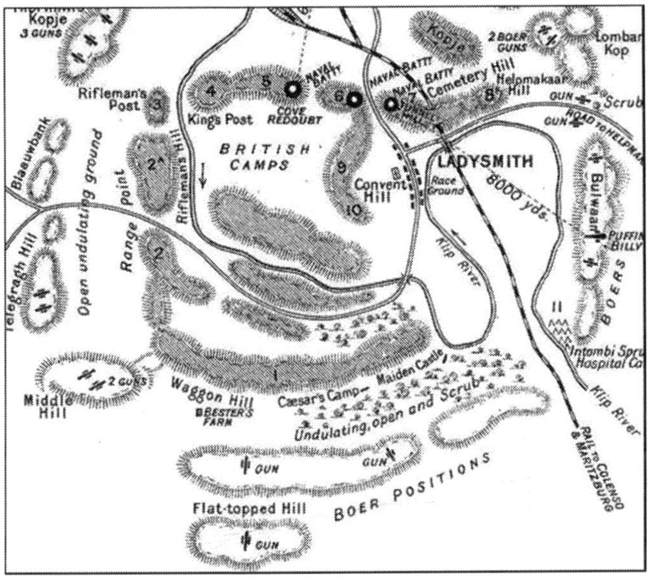

. Caesars Camp Area.[499]

These are WSC's observations in London to Ladysmith.

> At two o'clock, according to Sir George White, the Boers in great numbers, evidently reinforced from Colenso, surprised the pickets and began a general attack on the outpost line round the town, particularly directing their efforts on Caesar's Camp and Waggon Hill. The fighting became very close, and the enemy, who had after all hardened their hearts, pushed the attack with extraordinary daring and vigour. Some of the trenches on Waggon Hill were actually taken three times by the assailants. But every time General Hamilton—the skilful Hamilton as he has been called—flung them out again by counterattacks. At one place, indeed, they succeeded in holding on all day, nor was it until the dusk of the evening, when the rain and thunderstorm which we saw hanging over Colenso broke on Ladysmith, that Colonel Park led forth the Devon Regiment—who, having had half their officers killed or wounded by a shell some days before, were probably spiteful—and drove the Dutchmen helter skelter at the point of the bayonet. So that by night the Boers were repulsed at every point, with necessarily great slaughter, greater at any rate than on our side. Their first experience of assaulting! Encore!

[527] Spiers, Edward, ed. (2010), Letters from Ladysmith: Eyewitness Accounts from the South African War (illustrated ed.), Frontline Books, p. 77–84, ISBN 9781848325944.

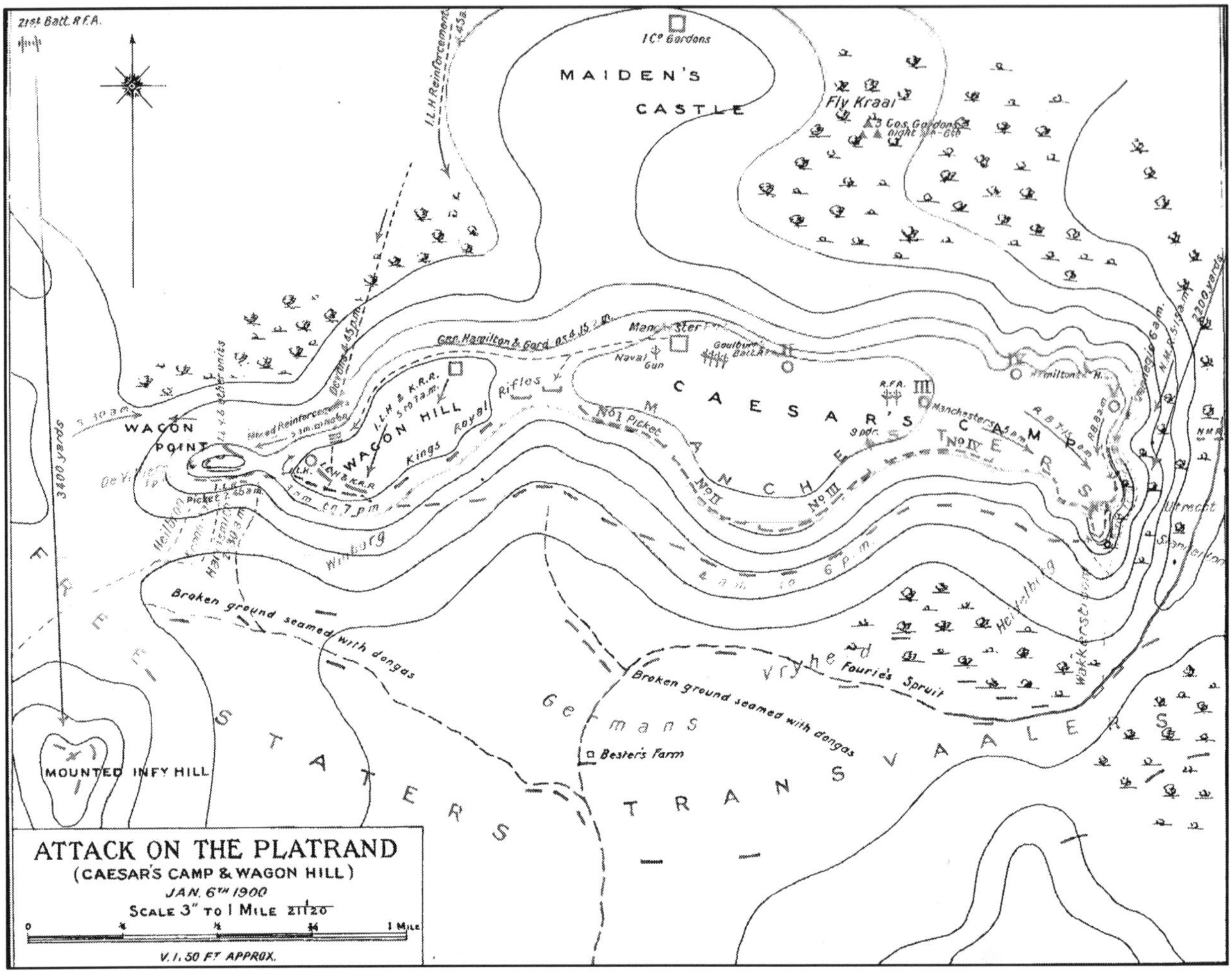

Attack on Platrand.[500] Waggon Hill on the Platrand

The British suffered 175 killed and 249 wounded. 52 dead Boers were left in the British positions, but their total casualties were not recorded.[528]

Casualties In Action	Killed		Wounded		Missing and Prisoners		
Ladysmith Seige & 1/6	Officers	Men	Officers	Men	Officers	Men	Total
Wagon Hill/ Platrand	14	164	33	287	0	2	500

Casualties per Conan Doyle's 'Great Boer War'.

[528] By 4 PM Ian Hamilton surmised that the overall attack had failed and that the Boers would withdraw after dark but White insisted this costly attack be made. Pakenham p. 275.

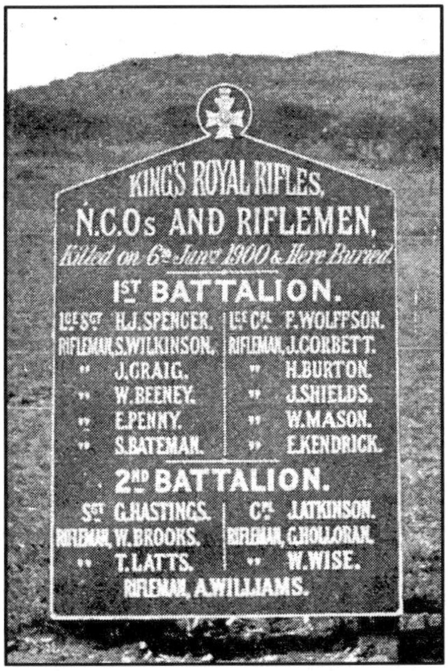

The Men Who Fell at Wagon Hill.[501] The Grave of Lieut. Colonel Dick Cunyngham, V.C.[529] Commanding the 2nd Gordons; killed at Ladysmith January 6th.[502]

Original Imperial Light Horse Monument.[503]

Imperial Light Horse Memorial in 2008.[530]

[529] William Dick-Cunyngham - Wikipedia

[530] Memorial on Platrand outside Ladysmith to those members of the Imperial Light Horse that died during the Siege of Ladysmith. File:Platrand-Imperial Light Horse Memorial-001.jpg - Wikimedia Commons Author: NJR ZA. See Also: Herman Albrecht - Wikipedia and Robert Digby-Jones - Wikipedia, Both posthumously awarded the Victoria Cross for their actions during the engagement.

Morning Post Articles

Annotator's Introduction

The following are the original articles published by the Morning Post from the telegrams they regularly received from Churchill at the front. Telegrams from South Africa could be sent and published in London within with two or three days. The longer 'War Letters', which would later make up the text of London to Ladysmith, and Ian Hamilton's March took a month or so to arrive and be published.[531] Since the purpose of my efforts was to investigate the telegrams, I have only noted a few of the War Letters to show the general trend. There are some newspaper articles not written by Churchill that are included for their probative value.[532] I had originally intended to intersperse these reports in the narrative but it would interrupt the flow of the book.

London To Ladysmith 'War Letters'	Written	Published
The State of the Game	Cape Town: November 1, 1899	Monday 27 November 1899
Along the Southern Frontier	East London: November 5, 1899.	Monday 27 November 1899
The Fate of the Armoured Train	Pretoria: November 20, 1899	Monday 01 January 1900
Prisoners of War	Pretoria: November 24, 1899	Monday 01 January 1900
Back To the British Lines	Frere: December 24, 1899	Saturday 27 January 1900
Christmas and New Year	Frere: January 4, 1900.	Monday 05 February 1900
Trichardt's Drift and Acton Homes	Venter's Spruit: January 22, 1900	Saturday 17 February 1900
A Fresh Effort and an Army Chaplain	Spearman's Hill: February 4, 1900	Monday 05 March 1900
After the Siege	Durban: March 10, 1900.	Monday 16 April 1900

It is surprising that communications achieved the bandwidth necessary to support civilian traffic in addition to the military traffic. There were few undersea or overland cables, and they were simplex, one way, one at a time.

Telegraph operating room at Johannesburg.[504]

The Field Telephone.[505]

[531] The original War letters have previously been published comprehensively by Frederick Woods, in Young Winston's Wars. The original Despatches of Winston S. Churchill War Correspondent 1897-1900. Viking Press. 1972 SBN 670-79515-1.
See also Fredrick Woods, A Bibliography of the works of Sir Winston Churchill. 2nd revised ed. 1975. ISBN 0-906795-001 Pages C 15 et seq. where the titles and dates of publication for despatches and war letters are listed.

[532] The telegrams comprise ~45k words. London to Ladysmith is ~100k, IHM is ~50k excluding Frankland's account of 20k. It amounts to ~20k words a month for 10 months; October. to July. He also wrote extensive correspondence with his friends and family. See OBC Vol I Part 2 1896-1900.

Introduction

Churchill's last War Letter titled, "After the Siege," is dated from Durban: Saturday, March 10, 1900.

Winston sends his last despatch to the Morning Post regarding the Natal operations on March 30th 1900 as well. His 'Riposte' was in reply to his letter regarding the treatment of 'Rebels.'

The War Letters in Ian Hamilton's March begin with "A Roving Commission" which is datelined from Pieters, Natal: March 31st 1900.

His next very terse newspaper despatch is from Bloemfontein on April 13th. The interim is covered by his chapter in Ian's March relating the events occurring during his residence in Cape Town in order to obtain his credentials. It's only two weeks but in his whirlwind life that is a long time.

Some of the circumstances that prevented him from obtaining authorization to proceed are detailed in the Cape Town Background section supra.

Lord Robert's Army: 16 April

In The Orange State. Lord Roberts's Army. From Winston Spencer Churchill, Bloemfontein, April 13, 6:10 P.M.[533]

The general situation here gives no cause for anxiety. Lord Roberts's great Army is widely spread over the country in convenient camps around the town, or is strongly holding the line of communications with Cape Colony. The exertions which the troops have recently made render a period of rest desirable.

Clothing kits and boots are needed for some units. For instance, the Mounted Infantry, which is being formed into one great corps of over 10,000 strong, under General Ian Hamilton, require reorganisation.

Reinforcements are continually arriving. The main necessity, of course, is horses for the Cavalry and Artillery. The waste of horseflesh in the past series of operations was prodigious, but it resulted in an economy of human life.

Now that the relief of Kimberley and Ladysmith has let our heads out of chancery there is no reason why we should hurry.

Time has come back to our side.

Meanwhile Bloemfontein presents a very gay and animated appearance. The streets are crowded with officers of all regiments and all countries. In the evening a band plays in the market square, and everyone gathers to discuss the past or to forecast the future.

Lord Roberts often walks there himself, and many other notable generals attend. The townsfolk watch and listen curiously in the background. Underneath this merry aspect preparations to deal another staggering blow are tirelessly, almost feverishly, proceeding. The common opinion here seems to be that we have reached the half-way house in this war. Much, however, depends on the result of the operations now impending in the right-hand bottom corner of the Free State.

After the occupation of Bloemfontein the colonials, always sanguine, but usually wrong, declared that the war, so far as it concerned the Free State, was practically over.

Encouraged by Lord Roberts's Proclamation[534] many of the Burghers took the oath [of neutrality] and returned to their farms.

[533] Lord Roberts' Army 16 April https://www.britishnewspaperarchive.co.uk/viewer/bl/0000174/19000416/038/0005

[534] Proclamation I; Feb. 19, 1900. The relevant portion being:

To protect these men our lines were much extended. Hence the sudden strong advance of the Boers brought disaster to small detachments of our troops. Many of the pacified farmers have been compelled to take up arms and to rejoin the commandos, and there will be an undoubted prolongation of the war.

I have no wish to be pessimistic, but I believe that 250,000 men will be needed before the end is attained. At any rate, our only logical policy is to pour troops into South Africa in an unceasing stream until the resistance of the enemy is overwhelmed.

The public must have no doubt about the determination of the Boers to fight to the bitter end. Many of the weaklings or lukewarm Burghers will give in, but the implacable nucleus will have to be killed or captured, and this is likely to be an expensive business.

However, it is worth while pushing the matter through now that we have suffered and achieved so much.

Need for Remounts. Bloemfontein, April 14. 4.10 P.M.

The question of the supply of remounts is, and will continue to be, of vital importance. Great numbers of horses are now arriving, but owing to the fact that no time can be given them to recover from the effects of the voyage, and that they must be put to work forthwith, their condition is low and the death-rate among them high.

Cavalry officers are loud in their complaints, declaring that on several recent occasions, had they had strong fresh horses which could develop a fast gallop, they would have produced decisive results with the lance and the sabre.

It is difficult to see how when time presses new horses can have a good chance, but there is an opportunity for the public, always ready to criticise afterwards, to be wise before the event. Probably the war will last a long time, and thousands more horses will be wanted. The public should insist that they are ordered now and shipped to South Africa at once.

We should form great resting depots in which they may get fit to begin work, and provide an ample staff to nurse and exercise them. If that be done, about four or five months hence you will be able to give your Cavalry a new lease of life and strength.

Rundle's Flying Column: 23 April

Rundle's Flying Column. From Winston Spencer Churchill, Oorlogs Poort, April 19, 8 P.M.[535]

General Rundle, with a strong flying column, marched from Reddersburg yesterday to Rosendal, reaching here to-day. There have been heavy rains, and in this roadless country supply is difficult. The rains have now ceased.

The Boers are reported to be in force on both sides of our line of march. A commando 1,000 strong is in laager ten miles to the eastward. Our Cavalry during the afternoon came in contact with small parties of Boers, who retired after exchanging shots.

Boer Positions Captured. Wakkerstroom, April 20, 8:45 P.M.

General Rundle's column advanced to-day on Dewetsdorp. He found the enemy occupying a position of uncertain strength athwart the road.

General Brabazon, with the Cavalry, cleared an extensive plateau, driving back the Boers by a clever turning movement. The Imperial Yeomanry seized an important hill, holding it in spite of a bombardment by five guns, supported by Infantry. The British Artillery silenced the Boer guns.

The force afterwards bivouacked.

The British Government believes that this act of aggression was not committed with the general approval and free will of a people with whom it has lived in complete amity for so many years. ……The British Government, therefore, wishes the people of the Orange Free State to understand that it bears them no ill will, ……and I undertake that any of them, who may so desist and who are found staying in their homes and quietly pursuing their ordinary occupations, will not be made to suffer in their persons or property on account of their having taken up arms in obedience to the order of their Government. Those, however, who oppose the forces under my command, or furnish the enemy with supplies or information, will be dealt with according to the customs of war. Proclamation I: Vol II of Roberts' Despatches from SA: https://babel.hathitrust.org/cgi/pt?id=hvd.32044089026009&seq=356&view=1up

[535] Rundle's Flying Column 23 April https://www.britishnewspaperarchive.co.uk/viewer/BL/0000174/19000423/037/0005

Enemy's Rear Reconnoitred. Wakkerstroom. April 21, 10 P.M.

Pending the arrival of the 16th Brigade, General Rundle entrenched himself to-day on the captured plateau, maintaining an intermittent bombardment of the enemy's position.

Meanwhile General Brabazon moved round the Boer left in widely-extended formation, drove in the enemy's outposts, and thoroughly explored the rear of their position, locating the Boer laagers.

The enemy finally opened fire with four guns. The reconnaissance withdrew skilfully. Our loss was slight.

Combined Operations: 25 April

Combined Operations. From Winston Spencer Churchill, Wakkerstroom, April 23, 9 P.M.[536]

As combined operations on a great scale are now in progress General Rundle's force is still containing the Boer position before Dewetsdorp. Meantime the desultory artillery duel and picket firing continue.

General Brabazon, with the mounted forces, demonstrates actively, scouring the country on the flanks, and keeping the Boers in constant movement.

The Fighting at Thabanchu: 2 May

The Fighting at Thabanchu. From Winston Spencer Churchill, Thabanchu, April 28.[537]

General Rundle's Division and General French's Cavalry Brigades reached here yesterday.

The Boers who had retreated from Wepener were known to be occupying positions to the north of Thabanchu Mountain with a view to covering the withdrawal of their waggon convoys from the south. They were not believed to be in force, however, and the theory was popular that they were discouraged.

During the afternoon of April 27 demonstrations were made on both flanks of the enemy's widely-extended position by General Dickson's Cavalry Brigade, which threatened the Boer left, and part of General Hamilton's Mounted Infantry Division, which occupied a prominent hill on the extreme right.

It was decided, however, to retire at night. The Boers thereon pressed to close quarters, and Kitchener's Horse were unable to evacuate their position until midnight.

Meanwhile some Boers crept to within ten yards, and a sharp fight ensued.

The Boers then retired, leaving one fierce greybeard dead behind them, and probably carrying others away.

Kitchener's Horse then returned to camp safely, where their arrival relieved the growing anxiety.

To-day's operations were intended to compel the Boers to retire from the vicinity of Thabanchu, and if possible to intercept them or part of them.

The enemy occupied a wide horseshoe of mountains with its convex face towards us.

General Gordon's Cavalry Brigade moved round the Boer left, and General Rundle's Division held the centre.

General Ian Hamilton with General Smith-Dorrien's Infantry Brigade and Mounted Infantry turned the right.

The movement was begun at dawn, General French directing it.

General Gordon, finding the enemy in a strong position near the Boer left at the extremity of the horse-shoe, was unable to debouch on to the open plain behind, and remained during the day shelling and knocking at the door.

The centre remained motionless as arranged.

General Hamilton's advance crushed the Boer right away and opened a road for General Dickson's Cavalry Brigade, which then dashed through and hunted the enemy from ridge to ridge, shelling them freely with Horse Artillery.

At last we arrived in the rear of the horseshoe, and the Boers in parties of two hundred could be seen within the enclosed space running about like rats in a trap.

[536] Combined Operations 25 April https://www.britishnewspaperarchive.co.uk/viewer/bl/0000174/19000425/058/0007

[537] The Fighting at Thabanchu 2 May: https://www.britishnewspaperarchive.co.uk/viewer/bl/0000174/19000502/056/0007

General Dickson hoped to make a bag. He signalled to General Hamilton regarding the situation, and General Hamilton came forthwith, bringing up every soldier he could find.

Suddenly, at about half-past four, the Boer Army, nearly four thousand strong, moved out of the horseshoe and began marching north-east. I have never before seen such an array of Boers.

Their order was so regular that at first it was believed that they were General Gordon's Cavalry Brigade.

They quickly opened with artillery on General Dickson, and at the same time the Boers who had escaped earlier in the day from our imagined trap returned in force on General Dickson's left and rear with two guns.

In these circumstances General Dickson resolved to retire, and he only just withdrew in time.

The Recent Movements: 4 May

The Recent Movements. From Winston Spencer Churchill, Bloemfontein, May 2, 4.25 P.M.[538]

Now that the operations in the south-east corner of the Free State have been completed it will be interesting to take a general survey of their course.

Before the main advance was possible it was necessary to clear the Boers from this quarter. Accordingly three columns were ordered to march eastward from the railway.

The first column, under the command of General Ian Hamilton, was to demonstrate near the waterworks.

The second, under Generals French and Pole-Carew, was to move on to Leeuw Kop.

The third, under Generals Rundle and Chermside, was to occupy Dewetsdorp and relieve Wepener.

Three powerful thrusts were designed into the invaded territories, getting deeper and stronger as they went nearer the south, and thus shouldering the Boers northward.

Generals Hart and Brabant co-operated from the southward.

Circumstances altered the plan, but not the result.

General Rundle on advancing found himself confronted by a strong position before Dewetsdorp, and, as the Boers extended and entrenched their flanks when he proposed to turn them, he felt that severe loss must attend his attack. He communicated with the chief, who therefore directed Generals French and Pole-Carew to move past Leeuw Kop to his assistance.

After forcing Leeuw Kop General French arrived behind Dewetsdorp, and joined hands with General Brabazon, General Rundle's Cavalry commander.

The enemy, fearing that they would be surrounded, retreated towards Thabanchu.

General Brabazon was sent to relieve the Wepener garrison, accomplishing the task successfully and bringing the troops back safely.

General Pole-Carew then returned to Bloemfontein.

Meanwhile, General Ian Hamilton, who had been directed to demonstrate near the waterworks, obtained permission to attack. He seized a position, called up General Smith-Dorrien's Brigade to help him, and marched on Thabanchu, hoping to intercept the retreating Boers. The enemy barred his advance at Israel's Poort, but General Hamilton with his usual tactical skill forced a passage with slight loss.

General Hamilton entered Thabanchu.

The Boers, however, were already north of the town.

Generals Rundle and French concentrated there from Dewetsdorp.

After some fighting round Thabanchu Mountain the Boers withdrew northward in good order, driving enormous herds of cattle and flocks of sheep with them. But the south-eastern corner of the Free State has now been practically cleared with little loss of life.

[538] The Recent Movements 4 May https://www.britishnewspaperarchive.co.uk/viewer/bl/0000174/19000504/042/0005

Hamilton's March to Winburg: 10 May

Hamilton's March to Winburg. From Winston Spencer Churchill, Winburg, May 10, 12.55 P.M.[539]

On April 30 General Ian Hamilton, with a strong force of Cavalry and Mounted Infantry, including Broadwood's, Smith-Dorrien's, Bruce Hamilton's, and Ridley's commands, marched from Thabanchu by way of Jacobsrust and Isabellafontein on Winburg.

The enemy, who numbered four thousand and had thirteen guns, opposed him strongly at Houtnek on May 1 and 2, and at Roetofsfontein on the 4th.

The two days' action at Houtnek fought by General Ian Hamilton on May 1 and 2 was well contested, and for a time its issue was doubtful.

The Boers occupied a strong position running along a rolling grassy ridge with a perfect glacis in front, their right flank resting on Thaba Mountain, and their left on ridges and kopjes of indefinite extent.

The attack was pressed by General Hamilton on a total front of eleven miles, and resisted by the Dutch along an equal line.

The troops showed great intelligence in taking advantage of cover, and their handling by the regimental officers as well as the general conception of the action showed that earlier events have taught them a lesson.

By the end of the day, however, scarcely any progress had been made, though a regular advance of the German corps, under Colonel Maxinmoff, was repulsed with great gallantry by a small party of Gordon Highlanders and Kitchener's Horse under Captain Towse.

Night fell, with the enemy still in possession of their ground.

At dawn the action was renewed.

General Hamilton now turned his attention to Thaba Mountain, and began to turn the enemy's right from that direction. At the same time he sent the 8th Hussars and Lancers right round Thaba Mountain to threaten the Boer line of retreat.

The crucial moment could not, however, be avoided, and at one o'clock General Smith-Dorrien advanced across the Thaba plateau with the Gordons and Shropshires in fine dashing style.

The Boers fired heavily for a few minutes, but with both flanks turned and a heavy Infantry assault bearing down on their front their hearts failed them, and on the loud cheer which preceded the actual charge they rushed to their horses and fled.

Forthwith the guns were galloped on to the captured position and the fugitives were heavily shelled.

Moreover the 8th Hussars charged into the tailings and the action was won.

Forty prisoners and a position of great strength were the prize of the victory.

The British losses were under a hundred.

The Capture of Winburg II: 11 May

The Capture of Winburg. From Winston Spencer Churchill, Winburg, May 7.[540]

The action near Welkom Drift, which was fought on the 5th of May, was remarkable for a very brilliant Cavalry incident.

While General Ian Hamilton's main force was advancing to engage the enemy, who occupied a range of high cliffs rising from the smooth plain, a new force of Boers from Brandfort, with guns, suddenly appeared on his left flank.

General Broadwood thereupon, without a moment's indecision, galloped four squadrons to the edge of along ridge at the angle of the junction of the Boer forces, and secured it by scarcely five minutes.

The Boers, finding their forces divided and their plan foiled, began a general retreat, which did not cease until they were beyond Winburg.

[539] Hamilton's March to Winburg 10 May https://www.britishnewspaperarchive.co.uk/viewer/bl/0000174/19000510/050/0007

[540] The Capture of Winburg II 11 May https://www.britishnewspaperarchive.co.uk/viewer/bl/0000174/19000511/031/0005

On May 6 General Ian Hamilton marched on fifteen miles towards the town, which he summoned to surrender to avoid a bombardment.

Captain Balfour, the bearer of the flag of truce, reached the market square, and was about to receive the submission of the mayor, when Philip Botha, with five hundred Hollanders and Germans, arrived fresh and truculent. An angry and excited scene now ensued.

General Louis Botha's brother, Philip Rudolph Botha.[506 541]

Botha ordered the arrest of Captain Balfour, whereon that officer pointed indignantly to the white flag under which he had come. The mayor and townsfolk took his part, and in the end the Boer commando left the town at a furious gallop.

The large silver keys of the town were tendered to General Ian Hamilton by the mayor and the landdrost, and the surrender was accomplished.

It is interesting to note that the four generals with the force—Ian Hamilton, Bruce Hamilton, Broadwood, and Smith-Dorrien—average only 43 years of age.

This war brings young men to the front.

Enemy's Stores Seized. Winburg, May 7, Later.

General Ian Hamilton captured here stores of food and forage, together with some prisoners and waggons.

The Boers retreated towards Senekal and Kroonstad, but they have since been strongly reinforced.

General Ian Hamilton has issued orders praising the troops, which include Canadians and City Imperial Volunteers, who marched a hundred miles and fought on nine days out of thirteen.

Our losses have been moderate.

Winburg is an important town, comprising several streets and a large market square.

It has a thousand inhabitants, and is connected with Smaldeel by a railway which has been temporarily cut by the British.

Advance to Bloemplaats. Bloemplaats, May 9.

General Ian Hamilton's force marched here today in accordance with the general scheme of advance.

A long Boer convoy was observed trekking towards Ventersburg, but it is reported that the enemy intends to descend the line of the Zand River.

About 3,000 Boers with several guns are holding ridges on the northern bank.

Desultory cannonading and musketry occurred during the afternoon.

[541] Botha Philip Rudolf, was born in 1851. While in command of the Harrismith burghers, he made a breakthrough at Sprinkaan's Nek. He was killed in action while attacking the British position near Ventersburg on March 6th, 1901. See also: Philip Botha - Wikipedia

Occupation of Lindley: 22 May

Occupation of Lindley. From Winston Spencer Churchill. Lindley, May 13, 8.15 a.m.[542]

General Ian Hamilton's Cavalry brigade occupied Lindley yesterday.

The leading Infantry brigade arrived with the general to-day.

Boers, in varying strength, are reported in all directions, and there has been slight firing along the picket line.

Rumours are prevalent that Piet De Wet's commando wishes to surrender, and indirect approaches have been made to General Broadwood asking that the men may be allowed to return to their farms.

Lord Roberts, however, requires unconditional surrender.

Therefore their surrender is improbable.

Mr. Steyn's Flight. H. F. Prevost Battersby, Our War Correspondent. Kroonstad, May 20, 8.50 a.m.

The enemy are retreating from Harrismith northward.

Van Reenen's Pass, is reported to be open and Botha's to be lightly held by the Vrede commando.

Mr. Steyn has fled from Heilbron to Pretoria.

The congestion of traffic at Smaldeel is delaying the advance.

The railhead is expected to be at Kroonstad on Tuesday, and the deviation railway has been almost completed.

Prisoners state that Johannesburg is policed by neutrals under Dupont.

The Irish Brigade, which assisted in the removal of "Long Tom" from Modder's Spruit, lost twenty-three killed at Vaal Kranz.

The Capture of Heilbron: 24 May

The Capture of Heilbron. From Winston Spencer Churchill, Heilbron, May 22.[543]

The town of Heilbron fell to-day to General Ian Hamilton's column, which marched here from Lindley by rapid stages in spite of continual opposition in front and heavy attacks on the rearguard.

General Broadwood pursued the retreating enemy with great vigour, breaking through their rear and driving back their artillery.

He captured fifteen waggons and nine prisoners.

The pursuit is proceeding.

Press Censorship 4 June

Press Censorship An exchange of letters between Churchill and **Lord Stanley, The Press Censor**, concerning the fate of three of Churchill's despatches and their transmission to London.: 4 June.

Press Censorship Strange Adventure of Despatches. [544]

Our letters from South Africa include one addressed as [Urgent. .ate. Urgent.]

To: The Press Censor, Bloemfontein. On Behalf of the "Morning Post," 346, Strand, London. With Mr. Winston S. Churchill's compliments.

It bears the stamp of the Field P. O., British Army, South Africa, April 26, and a Bloemfontein stamp whose date is illegible; and it is marked "found open and officially sealed." Two notes and two envelopes containing

[542] Occupation of Lindley 22 May https://www.britishnewspaperarchive.co.uk/viewer/bl/0000174/19000522/054/0007

[543] The Capture of Heilbron 24 May https://www.britishnewspaperarchive.co.uk/viewer/bl/0000174/19000524/058/0007

[544] Press Censorship An exchange of letters between Churchill and Lord Stanley, The Press Censor, concerning the fate of three of Churchill's despatches and their transmission to London. 4 June https://www.britishnewspaperarchive.co.uk/viewer/bl/0000174/19000604/054/0006

despatches are enclosed. The notes are to Lord Stanley,[545] the Press Censor, and to Mr. H. F. Prevost Battersby,[546] one of our War Correspondents, and run thus:

H. F. Prevost Battersby.[547]

Military Postmaster, Cape Town (Major Treble).[507]

To Press Censor.

My Dear Stanley,—Please see enclosures, which I beg may be sent to England at Press and full rates as Battersby will arrange, if you pass the messages. We suffer greatly from the limit, and these wires having been carried by special despatch rider have cost the *Morning Post* money. This is my excuse for troubling you. I have asked Battersby to call on you at once.

Yours sincerely,

Winston S. Churchill.

To H. F. P. Battersby.

Dear Battersby,—Please go and see Press Censor, and have the wires I have sent pushed through as quickly as possible. I have had another horrible adventure here, and my hair is quite grey. The Boers have captured my mackintosh![548] Will you try and get me another for my return?

Yours sincerely,

Winston S. Churchill.

One envelope is marked:

First Wire. Note to Battersby. 280 words, of which first 150 full rates, rest Press rates, 130. [549]

Add words "continuing," and re-address where split occurs.—W. S. C.

[545] Edward George Villiers Stanley, 17th Earl of Derby (4 April 1865 – 4 February 1948), styled The Hon. Edward Stanley from 1886–93 and Lord Stanley from 1893 to 1908, was a British peer, soldier, Conservative politician, diplomat and racehorse owner. He was twice Secretary of State for War and also served as British Ambassador to France.
On 11 January 1899, he was commissioned a lieutenant in the reserve of officers, and on 17 May, was made honorary colonel of the 2nd Volunteer Battalion, Loyal North Lancashire Regiment. Lord Stanley served on the staff in the Second Boer War, and **was appointed Chief Press Censor at Cape Town**, graded as assistant adjutant-general, on 18 January 1900. He accompanied Lord Roberts's headquarters as Press Censor when he left Cape Town, and was mentioned in despatches of 31 March 1900 by Roberts for his "tact and discretion" in that role.[9] He was appointed Roberts' private secretary on 25 July 1900. and was again mentioned in despatches of 2 April 1901 for his "thorough knowledge of men and affairs". https://en.wikipedia.org/wiki/Edward_Stanley,_17th_Earl_of_Derby
His father had established the 'Stanley Cup'. https://en.wikipedia.org/wiki/Stanley_Cup

[546] Harry Francis Prevost Battersby (1862 – 1949) most well known as H. F. P. Battersby was a poet, novelist journalist and psychical researcher, who published under the name Francis Prevost. Battersby was born in 1862, the son of a major-general. He graduated with distinction from the Royal Military College at Sandhurst and initially joined the Royal Irish Rifles before moving into journalism. He served as the Boer War correspondent for the Morning Post. https://en.wikipedia.org/wiki/Francis_Prevost

[547] The journalist and psychical researcher H. F. Prevost Battersby. From: In the Web of a War. Methuen & Co, 1901. https://commons.wikimedia.org/wiki/File:H._F._Prevost_Battersby.jpg

[548] A waterproof raincoat. Mackintosh - Wikipedia

[549] Press Censorship An exchange of letters https://www.britishnewspaperarchive.co.uk/viewer/bl/0000174/19000604/054/0006

The message it contains is as follows:

Rundle's column resumed advance towards Dewetsdorp to-day through pleasant grassy country with many undulations and occasional hills. Weather was fine and transport made better progress.

At eight o'clock McNeil's scouts forming advance point Brabazon's Cavalry screen drew fire of Boer outposts on high and extensive plateau about four miles from Dewetsdorp.

Brabazon continued to advance and by clever turning movement executed by Staffordshire and Hampshire Yeomanry obtained possession of entire feature and moreover seized prominent hill to eastward. This hill enemy appeared anxious to hold and as Cavalry advancing on both flanks compelled them disclose their main position they opened sharp fire on hill with five field guns and two Vickers-Maxims but Brabazon managed to find good cover for men and horses and maintained himself without much loss until Infantry advanced guard arrived when Rundle delighted secure this important feature so little loss promptly reinforced him with Worcester Regiment which now entrenching.

Three o'clock British Artillery came into action and after brief duel Boer guns became silent. Column bivouacks on and around captured plateau enemy's position is of uncertain strength but they are strong in artillery and evidently prepared to dispute occupation of Dewetsdorp.

General engagement will probably be fought to-morrow. Very pleasing features in to-day's affair were graceful handling of Cavalry by Brabazon and neat way in which Imperial Yeomanry manoeuvre round enemy's left. Their high order intelligence and initiative compensates for much military training. Indeed they may turn out the surprise packet of whole war high-class man behind rifle very powerful factor.

Churchill

[Censored - Leslie Rundle, Lt.-General.]

Dewetsdorp, 21 April, 1900.[550]

The other envelope is marked: Second Wire. Note to Battersby. 330 words at Press rates. 170 go with first message. Rest next day.

Please add word "separate" between wires. W.S.C.

The terms of the second message are as undernoted:

[April] 21st. Rundle's column having made good its position on plateau captured yesterday by Brabazon, maintained desultory bombardment of Boer position pending arrival reinforcements under Campbell which reached here midday.

Meanwhile Brabazon with whole mounted force and guns moved round Boer left and making wide detour turned their position driving in Boer flank outposts and locating their laager. Enemy fall back without much opposition for some distance being unable to assail Cavalry on account of their scattered formation which covered great extent of ground.

About eleven reconnaissance obtained possession of long ridge almost in rear of Boer position. Enemy who had rallied around very prominent kopje behind their main line of entrenchments made bold effort to prevent further intrusion of Cavalry and despatched 200 men who galloped across open to seize rocky feature and thus outflank outflanking Cavalry.

McNeil with fifty scouts raised by Montmorency raced to anticipate movement but was obstructed by wire fences and Boers reached coign of vantage first shooting two horses and pouring in so heavy fire that escape of scouts without loss life extraordinary.

Due to Boers being breathless and excited. Scouts fell back to second position and thence checked further advance enemy who now brought four guns into action against Brabazon's dismounted men making really sharp shellfire wounding general's orderly causing few other casualties and giving many of Yeomanry their baptism of fire.

Rundle having ordered that no serious engagement should be fought by Cavalry this morning Brabazon began rearguard action noon and by one o'clock using his guns to check enemy who pressed somewhat withdrew his

550 Press Censorship An exchange of letters https://www.britishnewspaperarchive.co.uk/viewer/bl/0000174/19000604/054/0006

force and returned safely to camp without much loss and having explored large part of enemy's position and line of retreat. Rundle's force being now complete general action will be fought to-morrow.

Churchill. [Censored to Bloemfontein. Geo. S. Milne. D.A.A.G. In., A Headquarters. 21st April.]

Sent from Ladysmith.[551]

(We are indebted to the courtesy of Captain S. H. Pollen, the Press Censor, Natal, for the following letter:

Headquarters, Ladysmith, Natal.

May 8, 1900.

The Editor "*Morning Post.*"

Dear Sir,—It may interest you to have this cable (intended for your paper).

It was found sewn up in a packet, the bearer of which (a native runner) was sent out from Ladysmith about January 19, and was either shot or captured by the enemy.

The bundle was found about four miles from here on May 3, 1900.

I am, &c,

S. H. Pollen, Capt., Press Censor. Natal.

The message is enclosed in an envelope marked:

Cable for transmission.

To the Telegraph Officials, Colenso, Estcourt, or Frere.

Kindly give bearer receipt.

E. A. Altham, Major.

The message itself reads: Cable message. Receiver to pay full rates. Kindly give bearer a receipt.

From Stuart. To *Morning Post*, London. Ladysmith [May] 18th.

Singularly devoid interest from last message till yesterday when heavy firing heard from direction Potgieter's Drift from small hours till late evening also intervals last night.

During dark possible see from Waggon Hill flash our shells bursting in Boer laager three miles this side drift. During day saw from Cove Redoubt shrapnel bursting over two Boer camps. Certainly artillery much heavier any yet heard here. To-day firing considerably lighter. Possible from Waggon Hill see shell bursting among groups Boers.

Fancy Boers take Colenso situation more seriously large reinforcements travelling. Small bodies moved south-west last three days also quantity cattle removed behind berg. Few waggons also moved south. Regard important fact Boers lately spent considerable time large number men strengthening breastwork other fortifications [round Ladysmith….. Hill Rifleman's….] to west on Buller's ………………….. march thoroughly breast worked new gun emplacement whence high velocity throwing shrapnel busily last two days into old camp also cattle grazing Gun Hill. Strengthened also new emplacement whence similar gun firing greater part day. Working parties visible all round. Our artillery more usually active this week. Consider final desperate effort take Ladysmith highly possible. General spirits wonderfully good. Later Boers placed howitzer donga foot Surprise Hill shrapnelling Observation Hill.

[Approved—E. A. Altham, Major.]

[551] Press Censorship An exchange of letters https://www.britishnewspaperarchive.co.uk/viewer/bl/0000174/19000604/054/0006

Fighting for Johannesburg: 6 June

Fighting for Johannesburg. From Winston Spencer Churchill, Doornkop. May 31.[552]

On May 29 General French, advancing northward on the western side of Johannesburg, found himself confronted by about seven thousand Boers with fifteen guns, occupying strong positions in advance of the main ridge or reef of Witwaters Rand.

After an artillery duel which lasted all day, and was at times very loud, but was fortunately not accompanied by heavy losses, he withdrew further to the west.

On the 30th General Ian Hamilton advanced, according to orders, and found General French still obstructed by a part of the enemy's advanced position.

General Hamilton, being unable to move further owing to the scarcity of supplies, resolved to attack forthwith.

General French, with the Cavalry Division and Broadwood's command, lent by General Bruce Hamilton, co-operated on the left.

During the morning the artillery on both sides fired heavily, but at midday, while the Infantry were getting into positions for attack, there was a strange lull.

At two o'clock General Hamilton launched his whole force on the enemy.

General Smith-Dorrien, commanding the Infantry Division, directed the actual attack, the 21st Brigade being on the left and the 19th on the right.

The ground favoured the enemy, and the bare grass slopes terminating in rocky kopjes seemed to threaten heavy loss; but the attack was well conceived, and the advance of General Bruce Hamilton on the left and of General French beyond him materially weakened the Boers opposed to the British right.

The City Imperial Volunteers, who were the leading battalion of the left attack, were sharply engaged, but moving forward with great dash and vigour they pressed the enemy hard, and drove him from all his positions in advance of the main ridge.

On the right the fighting was more severe.

The Gordon Highlanders strongly attacked a kopje, and in spite of the terrible fire, which caused heavy losses, carried it at the point of the bayonet with their customary devotion.

The Cornwalls supported them.

The Boers fought stoutly, and managed to save their gun, but the whole of their position fell into the hands of the British, and the enemy were driven back towards the main ridge of the range. Night closed in while the musketry and cannonade continued, and the stubborn combatants fought on for nearly an hour by the glare of the burning veldt, which shells had set on fire.

The Boers suffered heavily in their retreat, and our troops bivouacked on the captured ground.

General Hamilton spoke a few words to the Gordon Highlanders, "the regiment my father commanded and I was born in;" and told them that all Scotland would ring with the tale of their deeds. There is no doubt that they are the finest regiment in the world.

Their unfaltering advance across the naked plain on a concealed enemy in spite of frontal and enfilading fire, which spotted the ground on all sides with bullets, their machine-like change of direction as the attack closed, and their fine charge with the bayonet constitute their latest feat of arms the equal of Elandslaagte or Dargai.

[552] Fighting for Johannesburg 6 June https://www.britishnewspaperarchive.co.uk/viewer/bl/0000174/19000606/042/0005

Awaiting the Surrender. Johannesburg, June 1.[553]

The town of Johannesburg is now at Lord Roberts's mercy. Consequent on General Ian Hamilton's victory at Doornkop the Boers withdrew from all the positions they held near the town and retreated to the north in good order towards Pretoria.

General Hamilton entered Florida, [A suburb of Johannesburg] and he is drawing what supplies can be obtained there and in Neavasburg.

Lord Roberts, advancing on the east of the railway, has sent Cavalry brigades and the Seventh Division northwards to join hands with General French, who should be at Drietfontein.

The capture of rolling stock by Lord Roberts in the surprise of Germiston has greatly modified the problem of supply, which has caused much anxiety to the commanders and imposed some privations on the troops.

The railway now works from here to Vereeniging, and from Rhenoster River northward the intervening gap is being rapidly bridged.

Last night I cycled through the town from the headquarters.

Johannesburg is quiet and orderly.

A few armed Boers still ride about the streets, but they were too much occupied in their own affairs to question a stray passer-by.

Yesterday Dr. Krause visited Lord Roberts, and promised to give a definite reply about the surrender of the town at ten o'clock this morning.

Meanwhile our heavy guns have moved into position against the town. It is now practically certain that the occupation will be unaccompanied by street fighting.

The Boers have not destroyed nor damaged the mines, and the crumbling Government seems resolved to fall with dignity.

Among the prisoners captured by General Ian Hamilton is Commandant Botha, not the General, but the leader of the Zoutpansberg commando, and also a burgher who was present at my capture in November.

I had a long talk with both men, who are most intelligent fellows, and have fought bravely throughout the war.

Both of them agreed that the end was near, but Commandant Botha believes that Pretoria will stand some siege.

In any case, further fighting is certain. These men, who are a good type of Boer, spoke without any bitterness, and seemed prepared to acquiesce in British rule.

If the war be prosecuted rigorously to its end, and if afterwards the only policy sensible and humane men can support—the policy of shaking hands and making friends—be pursued, there are good hopes of lasting peace and returning prosperity.

Occupation Ceremony. [Reuter's Special Service.] Johannesburg, June 1.[554]

The entry of the British troops into Johannesburg took place yesterday. Dr. Krause prepared the city for our entry, not by clearing out the undesirable population, but by preventing, with the British Commander-in-Chief's permission, the premature entrance of any of the members of our force.

Soon after ten o'clock Lord Roberts with his staff moved from camp, followed by the Guards and Stevenson's Brigade. Preparations on a large scale had been made to welcome the Commander-in-Chief. The space round the Court House was crowded, and the balconies of the building were filled with ladies. From the flagstaff in front of the Court House the Vierkleur was still flying.

About ten o'clock distant cheers announced the approach of the Commander-in-Chief. Lord Roberts then appeared and acknowledged the cheers of the crowd. When he and his staff arrived in front of the Court House the people became almost unmanageable, trying to break through the cordon of troops in order to obtain a closer view. Amid the cheers and shouts cries of welcome such as "Well done, Bobs," could be distinguished. The

[553] Fighting for Johannesburg 6 June https://www.britishnewspaperarchive.co.uk/viewer/bl/0000174/19000606/042/0005

[554] Fighting for Johannesburg 6 June https://www.britishnewspaperarchive.co.uk/viewer/bl/0000174/19000606/042/0005

women were even more demonstrative than the men. They waved their handkerchiefs and pressed forward against the horses' flanks to see the Boer chiefs salute the victorious British commander.

The officials then came forward, and Lord Roberts, having dismounted, led the way to a room inside the Court House. There he made a short speech, formally accepting the surrender of the town and requesting all the officials to continue the exercise of their functions until further arrangements had been concluded. In this the officials acquiesced. These formalities over, Lord Roberts left the building and remounted his charger.

The next ceremony was the hauling down of the Transvaal flag. More than one spectator among the Burghers standing round appeared touched to the heart when the Vierkleur was lowered. But the nondescript population cheered repeatedly at the downfall of the flag which many among them formerly professed to cherish.

General Pole-Carew then arrived at the head of the Guards. At the same moment the flag worked by Lady Roberts was hoisted, and the fifes played the National Anthem. When the music ceased a great roar of cheering and shouting went up.

While the National Anthem was being played a tall Free Stater, an artillerist, refused to remove his hat, and a fellow-spectator, a small, stunted man, attempted to pull it off, whereon a British soldier standing near pushed the aggressor away, saying: "Leave him alone. He fought for his flag. You fight for none."

Then followed one of the finest displays ever made by a British Army, the whole of General Pole-Carew's and General Tucker's divisions marching past. The force camped to the north of the town.

Lord Roberts took up his quarters in a small inn with the sign "Orange Grove."

Early in the evening, soon after Lord Roberts reached his quarters, one of the officers of his staff approached him in order to discuss a matter of importance. He found the field-marshal with one of the innkeeper's little children on his knee trying to teach the mite to trace the letters of the alphabet with a pencil. When the officer entered the room Lord Roberts looked up with a smile and said: "Don't come now; can't you see I'm busy?"

Boer Report. [Reuter's Special Service.] Johannesburg, May 30 (via Pretoria, June 3).[555]

About half-past nine this morning three British officers with orderlies carrying a white flag rode into Government-square. Major Karri Davis, who was one of them, acted as spokesman. He said they brought a message from Lord Roberts demanding the unconditional surrender of the town. Dr. Krause, the specially-appointed commandant, rode out with the officers to Lord, Robert's headquarters. During Dr. Krause's absence Commandant Vanaswegen insisted on hoisting the Transvaal flag in the Government-square.

Though an armistice was said to have been arranged, firing was heard all the afternoon in Bezuidenhout Valley, east of the town. About noon four British prisoners were brought in, together with two waggons. They had all been captured close to Prospect.

The township of Bertram's Town was visited by a large number of scouts belonging to Nesbitt's Horse, and later in the afternoon the troops of the Fourteenth Brigade began to arrive at Bertram's Town. They will probably camp there to-night, and the other forces on the different hills round the town.

The Fight at Elandsfontein. [Laffan's Telegram.] Johannesburg (via Kroonstad, June 3, 4:20 p.m.)[556]

The seizure of Elandsfontein resulted in an amazing scene of excitement and confusion. Elandsfontein is a little tin-roofed mining suburb of Johannesburg, lying along a valley below Elsburg. Beyond it are great heaps of cyanide whitened tailings thrown up from the mines. These heaps formed a fine military position commanding the railway station, which become the centre of the struggle for the town when a section of forty Victorian Mounted Rifles rushed the building.

Lieutenant Walker had previously shifted the points of the line, thus preventing the progress of the trains which were to convey the retreating Boers from the town. A commando of 600 Boers, thus arrested, opened a sharp fire from a train which was ready to depart. They then left the carriages, and scaled the heaps of tailings. A thrilling half hour followed. There was rapid rifle fire in all directions. The fight spread to the town. The flying bullets threw up little puffs of dust in the streets. Women and children ran hither and thither, some dumb with

[555] Fighting for Johannesburg 6 June https://www.britishnewspaperarchive.co.uk/viewer/bl/0000174/19000606/042/0005

[556] Fighting for Johannesburg 6 June https://www.britishnewspaperarchive.co.uk/viewer/bl/0000174/19000606/042/0005

excitement, others screaming as they tried to dodge the bullets. Some of the inhabitants watched the battle from their doors.

Some of the Burghers who had left the train succeeded in escaping to the north-west. Many others descended the shafts of a mine. These were subsequently captured by the Guards. Thirteen of the Boers who took refuge in the mine were killed, and many were wounded. The Russian ambulance remained and attended the injured. The Kaffirs, seeing that the Boers were repulsed, rushed in and tried to loot the shops of the Jews. The British sharply checked this movement, and thrashed several of the natives before order was restored.

Large crowds from Johannesburg watched the fighting from the north-west corner of the town. They say to-day that they had expected there would be weeks of fighting, especially on the south-west outskirts, before Johannesburg could be captured.

The fall of Elandsfontein was a complete surprise to them. Its first result was considerable disorder in Johannesburg during the evening. It was caused chiefly by the foreign contingents, who invaded houses and stores, and threw from the doors and windows goods held there for needy Burgher families. Dr. Krause met the rising with firm measures, in which he was assisted by a self-constituted committee of Burghers. He issued a Proclamation ordering all men under arms to join their commandos before daybreak under penalty of the severest punishment. The result was that 2,000 joined Commandant Grobler's commando and retreated rapidly to the north-west.

I have visited several of the mines to-day. Their local managers report that the property under their charge has, on the whole, been better protected during the war than it was before hostilities began.

When General Pole-Carew's signal officer and interpreter reached Elandsfontein Station before the attack the interpreter was captured and a guard was placed over him during the night. The guard in the course of conversation said he was sick of the war. "Then," replied the interpreter, "let's change places. You come with me and give up your arms." "All right," replied the Boer, and both appeared in our camp in the morning, our man carrying his captor's gun.

The telegraph operator at Elandsfontein when taken prisoner said that Commandant-General Botha's last telegram to Kruger was: "Burghers won't fight."

A Proclamation, signed by Kruger, is posted in Johannesburg, which forbids any damage being done to the mines.

Rand Opinion. [Laffan's Telegram.] Johannesburg (via Kroonstad), June 3, 2 P.M.[557]

The consensus of opinion here is that so far as the military are concerned their work is practically finished, and that further operations may safely be entrusted to the Mounted Police.

British detectives arrested O'Reilly in the Market square to-day. O'Reilly was unarmed; he will be charged with treason.

The *Diggers' News* has ceased publication.

All business at Johannesburg is still at a standstill. Banks remain closed. Everything has been found in good order, and rights of property have been respected. Even unoccupied houses have suffered neither damage nor loss. Everyone is agreed that great praise is due to Dr. Krause for his firmness in preserving order.

Hamilton's Advance North: 11 June

Hamilton's Advance North. From Winston Spencer Churchill. Heilbron, May 22.[558]

General Ian Hamilton's column left Lindley at daylight on [the 20th.]

No encouragement had been given to Piet de Wet to surrender, and some attack on our rearguard was expected.

No sooner had our troops begun their march northward than a brisk musketry fire developed along our outpost line.

[557] Fighting for Johannesburg 6 June https://www.britishnewspaperarchive.co.uk/viewer/bl/0000174/19000606/042/0005

[558] Hamilton's Advance North 11 June https://www.britishnewspaperarchive.co.uk/viewer/bl/0000174/19000611/042/0005

Our rearguard consisted of Major Bainbridge's Corps of Mounted Infantry, the 84th Battery, and the Derby Regiment under Colonel Cunningham.

General Ian Hamilton remained himself with the rearguard.

The enemy pressed on boldly in spite of heavy rifle tire, with the vigour of Afridis, but the 84th Battery, posted by General Hamilton on a commanding hill, checked them sharply, making excellent practice and enabling our troops to withdraw quietly after the march of the convoy had been secured.

The musketry, however, spread round our right flank, when Legge's Corps of Mounted Infantry and Smyth's Regiment became severely engaged, their being turned and their ponies stampeded.

Reinforcements from the rearguard soon arrived, and the Boers were driven off with some loss.

Over forty casualties had, however, occurred.

While this rearguard action was in progress the Cavalry, advancing from the other end of the column, were checked by a force of 1,500 Boers, who were in position with two guns, which fired shrapnel effectively.

Several casualties occurred.

The 10th Hussars and General Broadwood's Brigade halted pending the arrival of Infantry and field guns.

Those General Bruce Hamilton soon brought up, and after a short affair the Boers retreated in disorder, neither molesting nor obstructing us further during the day.

Our force, marching steadily on, crossed the Rhenoster River before night, though the transport was late in getting through the drift.

The next morning the advance was resumed and another good march was made.

The enemy threatened but did not successfully attack the rearguard, which, under Colonel de Lisle, assumed a wide front and inflicted some punishment on the Dutch.

Three Australians were wounded.

To-day the force moved to Heilbron.

At nine o'clock the Cavalry came in sight.

The town, which lies in a deep fold of the rolling downs, is a picturesque patch of grey houses and dark green trees, all grouped around the tall spire of the church.

Lieutenant Clay, of the 2nd Life Guards, was sent in to demand the town's surrender.

He was treated with much rudeness by a tall, fine-looking foreign captain in a smart uniform, who was apparently in command.

The Landdrost pleaded to be allowed to surrender, and, as black swarms of British Cavalry poured over the hills, the inhabitants prevailed on their unwelcome defenders to depart.

General Broadwood forthwith pushed on past the town and discovered the Boer convoy retreating northward.

He pursued with the greatest vigour, thrusting his horse batteries and "pom-poms" forward and shelling the enemy heavily.

The Boers at first resisted bravely, and the waggons were preserved in spite of our shells; but at eleven o'clock the Boer rearguard was scattered and driven off, when the 10th Hussars and Household Cavalry, galloping forward, captured fifteen waggons and seventeen prisoners.

The remainder of the Boers made good their disorderly retreat.

Heilbron is the fifth town taken by Hamilton's force, which is now called "The Army of the Right Flank."

The force has marched in the last month nearly three hundred miles, and has been engaged with the enemy on fifteen different occasions.

It moves on again to-morrow.

Baden-Powell Interviewed: 27 June

Baden-Powell On His Siege. [In Mafeking] Special Interview. From Winston Spencer Churchill, Pretoria, June 19.[559]

Yesterday afternoon I had a long conversation with Major-General Baden-Powell, who had just arrived from Mafeking. The general, who was in excellent health and spirits, was kind enough to give me some details of the defence.

The position of their small isolated force, many weeks distant from the earliest succour or refuge, had filled them all with anxiety even before the outbreak of war. Yet all precautions had to be taken secretly.

"I was not allowed by the Cape Government to fortify myself. When at last we were compelled to begin building forts we pretended that the first one was a rifle range, and set up a target on the embankment." By great exertions, however, the town was put in a defensible condition on the outbreak of war.

Outwitting the Boers.

Little flags were dotted about the outskirts, and the general issued numerous warnings to the townsfolk not to let their cattle stray on to the areas sown with mines. The news of these precautions was carried to the enemy, and his complete belief that the mines were real saved both digging and dynamite.

Not until after the siege had begun did he discover that there was an enormous truck-load of dynamite sufficient to blow down the whole town in the railway station.

Since the stationmaster was believed to be a Fenian, the general determined to have this truck sent out of the town without delay.

When the Boers fired on the truck, and a terrific explosion resulted, they were naturally filled with respect for the reported mines.

"Perhaps if it had not been for this," said General Baden-Powell, "they would have stormed the place in the first few days."

"Could they?" I inquired.

"Certainly," he replied. "At that time they were very, very strong—nearly 9,000, and full of confidence. They only needed to make up their minds."

Husbanding All Resources.

So soon as the investment closed arrangements were made to husband and organise all resources.

Four trainloads of provisions for Bulawayo were luckily intercepted at Mafeking.

Stores far in excess of Government provision had already been laid in.

A regular system of vegetable gardens was started.

Every drop of liquor or ounce of food was requisitioned, and the whole town, black and white, civil and military, went on strict rations.

"We had to look far ahead," said the general gravely.

Incidents of the Siege.

He alluded to various incidents of the siege.

After the severe losses which attended the repulse of his attack on Game Tree Fort he dared not risk more assaults, not having enough men. It was necessary, however, to take the place, since its rifle fire commanded the town.

Prolonged sapping operations were then began, and after two months the fort was almost reached. The Boers now replied with counter sap, and for a fortnight the two forces remained scarcely seventy yards apart, throwing hand grenades, bottles, and abuse.

The general brought all the machine guns he could spare into the foremost trench, and "laced into them" at such a short range that the bullets drove through the Boer sandbags, killing many men.

Owing, however, to the British being provided with iron loopholes the Boer bullets could not reach them.

[559] Morning Post, Baden-Powell Interview 27 June https://www.britishnewspaperarchive.co.uk/viewer/bl/0000174/19000627/035/0007

At last the Boers evacuated the fort.

"We knew they were going, and I gave strict orders that no one was to go near it until our scouts had been all round it. It was just as well we thought of this, because we found a tremendous mine of nitro-glycerine with wires laid ready for us. We used the nitro-glycerine to make two mines of our own. They were both blown up by lightning, but they served to increase the enemy's belief that the whole of the ground was mined."

Baden-Powell Interviewed 28 June

The Transvaal. Baden Powell Interviewed. Pretoria. June 19th.[560]

The following is the continuation of Mr. Winston Spencer Churchill's telegram describing his interview with Major-General Baden-Powell:

Major-General Baden-Powell told me in the course of the interview of all the things they had contrived to manufacture during the siege of Mafeking: ball cartridges out of blank, gunpowder, cannon, shells, and fuzes ("the fuzes were a great success"), postage stamps, bank-notes, and a strategic railway, and much besides.

"They kept us busy and made life less monotonous," he said.

"The bombardment was wearying.

"The enemy used to fire three or four shells every night, no more.

"When the shells had come one could go to sleep in peace, but it was a nuisance to have to wait for them.

"Towards the end of the siege the difficulty was to make people take cover—all were callous, and as the whole town was exposed to musketry, accidents frequently occurred."

The Darkest Hour.

I asked what was their darkest hour.

The general replied, "It was very painful to get all the bad news—Spion Kop, for instance.

"I used to hear these things, and I had to keep them to myself, and look cheerful.

"It was worth something to have been an actor.

"He laughed in a way that showed that months of anxiety had, after all, made some impression even on his unyielding spirit. "Things got rather bad towards the end—the last three months that is to say.

"You see we kept on extending our lines, pushing out trenches, and capturing their forts.

"We had to do this to secure grazing and breathing space.

"In consequence our lines became greatly extended.

"At last I found myself holding ten miles of perimeter with 700 men.

"Of course, we had an inner line, but the outer ring of forts was that which we relied on, nor was it a sure defence, and, besides, the men were so weak from want of food. "The Boers could have rushed us at any time if only they had come in force.

"That was why I could not help being a little anxious sometimes, in the nights especially."

Boers Came at Last.

"Well, at last they did come."

"What happened at first? "I asked.

He replied: "I did not know how many had entered my lines. They reached the orderly room, which stood just outside the Kaffir stad, and the clerk telephoned up to me, 'Boers are all in among us.' We thought it was an exaggeration, but I put my ears to the telephone and heard the Dutch voices, so that there was no doubt it had come at last.

"It would take too many words to describe the fight that followed in any detail.

[560] Morning Post, Baden-Powell Interview (continued) 28 June https://www.britishnewspaperarchive.co.uk/viewer/bl/0000174/19000628/050/0007

"Luckily," said Major-General Baden-Powell, "only the storming party came in, three hundred all told, mostly foreigners. When the five hundred Burghers in support heard the firing, and knew they would have to fight, they sat still to see what happened, and so we were able to reoccupy our old line, and round them up."

Baden-Powell Interview 29 June

Baden-Powell Interview Conclusion.[561]

The following is the conclusion of Mr. Winston Spencer Churchill's telegram describing his interview with Major-General Baden-Powell: Pretoria. June 19.

Morning broke on this situation: the Boers who had got inside Mafeking were unable to get out, but were defending themselves stubbornly; the Boers outside were doubting whether to fall on the town or not.

It had been the general's practice for the previous fortnight to send out a flag of truce every day to inquire after the health of Major M'Laren, the wounded prisoner in Snyman's camp.

Though fighting within the town was at its height, he punctually observed his usual practice, and in his letter to M'Laren, which the Boers, of course, read, he expressed the hope that the firing of the night had not alarmed him, adding that all danger was now past.

(South African War—Continued from Page 5.) [top of column 3] Page 3 conclusion.[562]

This produced a profound impression on the waverers. But though the original stormers were not supported the issue hung in the balance all day.

The Convicts Assist in the Defence.

He had to release the European convicts from the prison, including the murderer who was under a life sentence, and to invite them to take part in the defence.

Armed with rifles, these men had fired from the roof of their prison on the Boers.

When a bullet killed the gaoler the murderer took the command of the prisoners, and their influence in the struggle was effective.

The general remarked that he could make representations on their behalf.

Not till the evening, when the water tank on which the Boers depended was riddled, and so emptied, would they surrender.

Some tried to escape, "but," said the general, "that was very dangerous business."

The rest became prisoners of war.

They complained bitterly of the food they were given, though it was precisely the same as that of the garrison.

He therefore undertook not to touch any food that Commandant Snyman might send in to them.

Snyman Declines to Feed the Boer Prisoners.

Snyman declined to send any, however, being delighted that Eloff and the foreigners would be hungry.

The general said he tried to organise a field force to help the relieving column to come in, but the men could scarcely walk; still they went out and fired off their guns in the enemy's rear, which, though, not within range, perhaps disconcerted them.

Finally he said, "It is very strange, after being asleep, like Rip Wan Winkle, for so many months, to come back into the world again and to find everyone anxious to hear what had happened in Mafeking. We are all astonished to find so much interest taken in the defence of Mafeking."

Then we parted.

Later on in the afternoon one of his officers, speaking of him, said that many of the Boers who had surrendered on the march to Pretoria from Mafeking had asked to shake hands with Baden-Powell, and departed quite pleased.

Apparently Boers and British have something in common.

[561] Morning Post, Continuation of Baden-Powell Interview 29 June https://www.britishnewspaperarchive.co.uk/viewer/bl/0000174/19000629/022/0005

[562] Baden Powell Interview Conclusion 29 June https://www.britishnewspaperarchive.co.uk/viewer/bl/0000174/19000629/009/0003

The Capture of Pretoria: 29 June

The Capture of Pretoria. From Winston Spencer Churchill, Pretoria, June 6 (By Post to Cape town).[563]

Two days' marching from Johannesburg carried Lord Roberts's Army within striking distance of Pretoria.

On June 4 information was received that the Boers did not intend to stand a siege, and no opposition was expected, but about midday Colonel Henry's corps of Mounted Infantry became sharply engaged with the enemy in strong positions south of the capital.

The Commander-in-Chief ordered General Ian Hamilton to support him with all his mounted troops.

This secured the arrival and the deployment of the three Infantry divisions, and by two o'clock a line of 50 guns was developed against the Boer position and forts, which latter remained significantly silent. Though General Ian Hamilton's batteries on the left could not come into action because of the precipitous ground, the cannonade grew loud and sustained. The heavy artillery on the right bombarded the forts, and, throwing their shells over the hills, flung them into the town itself.

As soon as Hamilton's Infantry had deployed he sent all his Cavalry and Mounted Infantry in a circuit round the Boer right.

This operation was executed with great vigour, and, combined with a cannonade, compelled the Boers to evacuate their position, which the Infantry then occupied with little loss.

Meanwhile Colonel de Lisle, who was in command of Hamilton's leading corps of Mounted Infantry, pushed on with the Australians and seized a position within 2,000 yards of the town, which he forthwith summoned to surrender.

Scene in the Town.

The Boers were now flying in confusion through the streets, and the civil authorities, with the sanction of Botha, made an informal submission.

At daybreak on the 5th the whole Army was ordered to march on to Pretoria.

General Hamilton swept round to the west of the town, and Pole-Carew with a guard seized the railway station.

Here, in violation of the surrender, the Boers succeeded in sending away a train under the eyes of the troops. The staff officers and the orderlies pursued the train vainly with revolvers.

Three other trains, however, were captured after ten volleys and scattered shots had been fired.

The Grenadiers then began cautiously to enter the town through the main streets to the great square.

Release of the Prisoners.

The Duke of Marlborough, procuring a Dutch guide, galloped ahead to the cage where the British officers were confined, and required the Boer commandant to release them forthwith.

Though the demand was not supported by any force the Dutch officer complied, and after a little demur the gate was opened, and the officers, rushing out, seized the rifles from their gaolers, and in five minutes 129 British officers were free, and 48 Burghers, lately their gaolers, were confined in their stead.

The Dutch flag was hauled down, and the prisoners raised wild cheers as they hoisted the Union Jack, made secretly during their confinement out of the Vierkleur, in its place.

These officers had had a terrible experience.

Weakened by long imprisonment, tortured by false hopes and bitter delays, they were finally ordered at two o'clock that very morning to start at once under escort to Middelburg in a special train waiting at the station, but they decided unanimously to defy the Boers.

They seized the commandant and his lieutenant, and refused to let them go until they were promised that they should not be moved.

They also refused absolutely to leave Pretoria, saying they would rather die if necessary.

The commandant then withdrew, and the prisoners did not know what would follow or when the British troops would enter.

[563] The Capture of Pretoria 29 June https://www.britishnewspaperarchive.co.uk/viewer/bl/0000174/19000629/022/0005

When they saw us galloping towards them they were at first uncertain whether we were not leaders of a Boer commando coming to compel them to go.

When they found that we were friends their relief was very apparent.

Their faces were white with excitement and trouble.

President Kruger and his Officials.

All that is over now, however.

The seat of the Boer Government has been moved to Machadodorp.

Kruger left Pretoria on Friday last, taking with him upwards of a million in cash, all the guns in the forts, and some hundreds of British prisoners, leaving behind Mrs. Kruger, whose health is good, Mrs. Botha, and all the understrappers.

These latter were all paid their salaries to date by cheques on the Standard Bank, and as the President, had taken the precaution of previously removing all money therefrom they are considerably annoyed.

The British Entry.

At two o'clock the Field-Marshal, with the great staff of foreign attaches and Generals Kitchener and Brabazon, entered Pretoria. They proceeded to the main square, where the flag was hoisted over the Town Hall.

The victorious Army then paraded before him, Hamilton's force coming from the west, Pole-Carew's and Tucker's divisions from the south.

The Guards were the first to enter.

For three hours a broad river of khaki and bayonets flowed majestically through the great square, while the sunlight refracted from the red gravel of the streets and the red stone facade of the Town Hall threw a golden glitter on the scene.

The townspeople gazed awe-stricken and astonished on those stately soldiers whose order and discipline neither perils nor hardships had disturbed, whose march no obstacles could prevent.

Thus at a heavy cost in blood and treasure the experiment of the Presidents has been brought to an end.

Republics have vanished.

The war, however, continues.

Operations against Botha: 30 June

South African War. Boer Guerilla Tactics. Skirmish Near the Rand. Pursuit of Raiders.

Operations Against Botha. From Winston Spencer Churchill, Pretoria (delayed) (via Bloemfontein, June 28).[564]

Though the Boers had yielded Pretoria almost without resistance, they were encouraged by the news that communication with Cape town was cut, and they held strong positions, threatening the capital, and displayed considerable activity.

Consequently, to relieve Pretoria from a state of semi-siege, and to get more elbow room, Lord Roberts decided to move against them, intending to clear the neighbourhood and hoping… [The rest of the page is whited out.]

Protecting the Railway

Orange River Colony. Protecting the Railway. Cape town, June 28th, 1.5 p.m.[565]

The following is the conclusion of Mr. Winston Spencer Churchill's telegram describing the incidents of his journey from Pretoria to Cape town. The first portion concluded with the announcement of his arrival at the scene of the disaster to the Derbyshire Militia:

[564] Operations against Botha (Only a portion legible) 30 June https://www.britishnewspaperarchive.co.uk/viewer/bl/0000174/19000630/044/0008

[565] Protecting the Railway 30 June https://www.britishnewspaperarchive.co.uk/viewer/bl/0000174/19000630/044/0008

Beneath the kopjes were the graves of officers and men of the Derbyshire Militia, and also of several Burghers.

One glance shows the utter hopelessness of their position, when morning discovered them camped on a smooth open plain, with Boer riflemen and artillery crowning the height, scarcely 500 yards away.

The affair is still shrouded in uncertainty, but it is scarcely possible to believe that any man responsible for the safety of the troops could have halted within range of those forbidding rocks without being absolutely sure of holding them.

Roodeval is the scene of grim devastation, whole acres of ground being covered with charred fragments of letters, papers, and clothing.

Among these lie huge shells of 9 inch. guns, some exploded, others strangely battered but still alive.

The station buildings, signals, water tanks, trees, and adjoining houses have vanished, scarcely a single sign of them remaining.

To complete the dismal picture of waste and ruin—a dark stain on the bare and lonely veldt—was a long row of graves.[566]

Boers Attack the Train.

After passing Roodeval we were breakfasting when a servant entered with news that the bridge in front of us was burning, and almost immediately the train stopped.

Everyone got out of the carriages, two hundred soldiers who were in the train going south for remounts and many Kaffirs running stupidly along the line to look at the bridge without arms of any kind.

The next thing to arrive was a very well-aimed shell, which burst in a ditch five yards from our carriage, and had the effect of sending all the foolish sightseers scurrying back to their trucks.

Two Boers were cantering off about 400 yards away, and fifty or sixty were galloping towards the train from the south. In these circumstances the engine was reversed, and the train steamed slowly back to Roodeval.

The second shell from the gun was ill-directed, and the range rapidly increased. As soon as the nearest Boers saw the train retreating they dismounted and loosed off their magazines into it, to which we replied with carbines and Mauser pistols, no harm being done on either side.

It was now necessary to wait until an escort could be obtained to carry the mails through to the other end of the line.

All day long we waited, with continual rumours of attack and the distant sound of guns. The Boers meanwhile were attacking the Honing Spruit post. The garrison of this post ordinarily consisted of only two companies, who should have had two guns. These guns had been removed on the order of the Chief of the Staff, however, so that another battery might replace them.

Colonel Bullock's Determined Defence.

By great good fortune, however, General Smith Dorrien had ordered Colonel Bullock's provisional battalion, composed of released prisoners of war, to reinforce this post, thus unexpectedly rendered insecure.

These men, though armed only with Martini rifles, firing black powder, and one in three jamming, fought with great stubbornness, declaring that they would never be taken again, and maintained themselves under the fire of the well-posted guns until the arrival of Yeomanry and a battery from Kroonstad.

Colonel Bullock's determined defence was much praised by all who saw it.

The next day, General Smith-Dorrien having arrived with two reconstruction trains, communication was restored, and, escorts being provided, we reached Honing Spruit Station safely, and there entrained.

[566] The Battle of Roodewal Station took place on 7 June 1900 during the Second War of Independence. Roodewal Station is three kilometres south of the Rhino River in the Free State. A force of Gen. Christiaan de Wet, together with Commander Piet Fourie, attacked the British garrison, who were guarding the fully loaded trains and a supply depot. After De Wet had first cut off the telegraph's wires, he sent a message to the British force asking them to surrender. They refused, whereupon De Wet began his attack. The battle lasted for five hours and swung in De Wet's favor when General Froneman arrived with two guns. The British surrendered immediately as the bombs began to fall between them. The Boers looted more than 500 crates full of ammunition. Later that night, the Boers blew up the remaining ammunition, thousands of lignite bombs, and two bags of mail. The explosion leaves behind a crater measuring 30 x 18 x 6 meters.
The British losses were 45 killed, 123 wounded, and 645 captured. The Boers lost 8 civilians and 24 were wounded.
https://af.wikipedia.org/wiki/Slag_van_Roodewalstasie

I have described these incidents at length, not with the view of exaggerating their importance, but to explain the great difficulties that are encountered in keeping the span of a tremendous line of communications, exposed everywhere to the flank attack of an enterprising enemy.

The establishment of the Natal line will greatly diminish the strain, but it is probable that a period of railway war lies before us.

Operations Against Botha - Diamond Hill July 2

The Transvaal Operations Against Botha.[567]

The following is the continuation of Mr. Winston Spencer Churchill's telegram describing the preliminary stages of the operations of Lord Roberts against the Boers who had retreated from Pretoria:

Bloemfontein, June 29.

At dawn on the 11th the whole Army was in motion.

General French immediately encountered opposition and was unable to make much progress, though he was persevering when night fell.

The Eleventh Division with Lord Roberts moved slowly forward and engaged in an artillery combat between our heavy guns and those of the Boers, including one mounted on a railway truck.

But in the belief that the enemy were retreating, and not wishing to weary the troops if the strategic object was attained, Lord Roberts did not press the main attack.

Hamilton Strongly Opposed.

Meanwhile, on the right and south of the rail-way, General Ian Hamilton soon came into contact with the Boers, who not only opposed his advance, but also swept round his right and rear, obliging him to use the Gordons, a Cavalry brigade, one battalion, and eight guns to contain them, while he thrust forward with the rest.

When General Ridley's Brigade of Mounted Infantry was holding the enemy off on the left and the Gordons on the right, General Broadwood determined to make a bold dash to pierce through the Boer flank and local centre, and to reach the railway. Though the country was suitable for Cavalry, the Boers offered severe resistance. In spite of a converging artillery fire from seven guns General Broadwood continued to force his way forward, and at about eleven o'clock, finding his further advance in the open barred by the Boers, who made a fierce attack on his guns, repulsed only by case shot, he ordered the 12th Lancers and the Household Cavalry to charge.

Gallant Charge of The Cavalry.

Thereupon these regiments in fine style galloped forward; and though their horses were so exhausted that the Household Cavalry had to flog them along with their sabres, and each regiment was scarcely as strong as an ordinary squadron, they reached the Boers with sword and lance, and hurled them out of their position, driving them across the plain.

But now, while pursuing, they came under a heavy flanking fire, and were compelled to retire to the ground they had won, suffering severe loss in officers and men and particularly in horses.

In this gallant charge the Earl of Airlie and other officers lost their lives.

This was the boldest bid for victory that has yet been made by the British Cavalry in this war.

Its results showed the great difficulties which hamper the horsemen in offensive action against Boers.

General Broadwood's conduct of the operation is greatly admired.

Battle of Diamond Hill.

Seeing the Cavalry thus severely engaged General Hamilton felt bound to attack, and about mid-day began the action of Diamond Hill. Covered by two battalions and two 5-in. guns, General Bruce Hamilton's brigade advanced against the Boers, who occupied strong rocky positions.

Our troops clambered up the steep and lofty line of bluffs called Diamond Hill, and by skilful movements drove the Boers back by nightfall to their main position.

[567] Operations Against Botha 30 June https://www.britishnewspaperarchive.co.uk/viewer bl/000174/19000702/044/0005

In retiring the Dutchmen were compelled to cross some open ground at a range of 1,200 yards from our firing line, and paid the penalty accordingly, though they returned bravely for their wounded.

So determined was their attitude, that it being now imperative to carry the matter through, Lord Roberts instructed General Pole-Carew to support General Hamilton with the Brigade of Guards and artillery.

Boer Shrapnel.

The morning of the 12th was chiefly occupied with intermittent artillery practice, pending the approach of reinforcements, but the Boers shelled the 72nd Battery severely with shrapnel, several shells bursting over the heads of the members of the staff.

One of the bullets struck General Ian Hamilton in the shoulder, causing a contusion. Fortunately it did not penetrate.

At noon General Bruce Hamilton moved his brigade forward, the Derbys on the right, the City Imperial Volunteers in the centre, and the Sussex on the left, and effected lodgments all along the line of the Diamond Hill plateau.

Fierce Fighting.

Here the fighting became fierce, and the musketry and artillery fire was very loud.

The City Imperial Volunteers, admirably handled, pressed forward eagerly, but could make no headway.

The 72nd Battery, dragged up a steep hill, came into action against about a thousand Boers at a range of 1,800 yards, losing considerably from rifle fire in men and horses, but by their noble behaviour greatly moderating the severity of the enemy's musketry.

The Coldstream Guards now began to come into line in prolongation of our left, and the Scots Guards, between the Derbys and the City Imperial Volunteers, with two more batteries, also came up.

Night fell on a heavy hammer-and-tongs fight on the plateau, and preparations were made to renew the action at daybreak.

Retreat of the Enemy.

During the darkness the Boers, who saw two-thirds of their position captured, and forward movements of artillery indicating a determination of attack, being besides threatened on their line of retreat by Colonel de Lisle, who had also effected lodgments in decisive positions to the eastward on the ridge, retreated, and are now marching eastward ten miles off.

Pursuit is being arranged forthwith.

Our losses were greatly lessened owing to the plateau being rocky and affording good cover.

Publication of Despatches: 4 July

Cape Colony. Publication of Despatches. From Winston Spencer Churchill, Bloemfontein, July 3[568] (Delayed by the Censor).

The following statement regarding the recent publication of the Spion Kop despatches may be accepted with confidence:

After their respective reverses Lord Methuen and General Gatacre both transmitted reports to the Commander-in-Chief in South Africa. Lord Methuen's despatch reached General Buller while he was still in supreme command. General Buller forwarded it to the Secretary of State without comment.

General Gatacre's despatch, on the contrary, miscarried in the post, and was delayed more than a week. It therefore reached General Buller when he was no longer Commander-in-Chief, and was passed on to Lord Roberts, who commented on it as published.

Meanwhile Lord Methuen's despatch had proceeded to England, where it was found that he had made a compilation of the various reports of brigadiers and colonels, instead of writing a general narrative embracing the reports of his subordinates, which course is prescribed and is the usual form.

[568] Publication of Dispatches 3 July https://www.britishnewspaperarchive.co.uk/viewer/bl/0000174/19000704/060/0007

No exception whatever was taken to the spirit of Lord Methuen's despatch, but, in order that the regular custom should be observed, he was requested to rewrite it in the correct and formal style.

The despatch therefore came back to South Africa, where, since it was a matter already finally disposed of by General Buller, Lord Roberts could not in fairness intervene. He therefore made no observations on it, and none such exists or has ever been put on record.

After the War Churchill Interview 23 July.

Mr. W. S. Churchill Interviewed.[569] [570]

Bronzed and healthy, as full of energy as ever, Mr. Winston Spencer Churchill returned from South Africa on Friday, looking none the worse for the exciting personal experiences which have fallen to his lot. He received on Saturday a representative of the *Morning Post*, with whom he had a long conversation, continually broken by the arrival of telegrams congratulating him on his return from the war.

"As you have returned, I suppose, Mr. Churchill, that we may consider that the fighting is practically over?"—"Not at all. It would be quite a mistake to think that the war has come to an end. Of course, the great operations are over. So far as I can gather there will not be any large operations, Armies and divisions on the march, and so on. The scale of the war is altered, and we have entered on a period of flying columns, brigades being practically the units, and they will act in combination. The operations, if not on so large a scale, are as extensive as ever, but in South Africa we cannot get a good general view of them. Recently you have been in a better position to do so at home."

To the Bitter End.

"Then the Boers will not give in?"—"I never disguised the fact that I believed that a certain proportion of the Boer Army is absolutely implacable, and that these men will fight until they are either killed or placed in some position in which further resistance is hopeless. The vast size of the country makes it very difficult to grapple with these men, but now that we can move troops up and down the railway, the question of communications is no longer of such vital importance, particularly as we hold the Natal line as well as that in the Free State. The next move we are likely soon to see will, I think, be one on the Delagoa Railway."

Lessons of the War.

"Yes; and I suppose energetic men like General Baden-Powell, with his Canadians and Bushmen, can do a good deal. But what lessons do you draw from the war, so far as your experience goes?"—"Well, it has certainly shown us a good many points in which we can improve our Armies. The Cavalry have found it a terrible hardship to have to fight men with rifles, when they themselves are armed with miserable carbines."

"But what of the lance and sword?"—"In fighting on the flat or on the rolling grass of the veldt they have been only an encumbrance. I would arm the Cavalry soldier with a good firearm and teach the men who can ride to shoot. Albrecht said: 'Don't teach the Infantry to ride, teach the Cavalry to shoot.' It is all nonsense to say that the Boer is a mounted infantryman. He is nothing of the kind. He is essentially a cavalryman: he is a beautiful horseman, with a graceful seat, his body sways with every movement, showing his complete sympathy with his mount. Then, too, he is a scout, trained by nature in matters of reconnaissance. Mounted Infantry is simply a mobile rifle corps, and we used it for duties to which it was not trained. To that were due some of the unpleasant incidents you have heard of. If a corps of Mounted Infantry had to retire, the instinct of the officers was to send the horses out of the way and make an Infantry retirement to the horses."

"So you think the day of the Cavalry is gone forever?"—"Not at all, don't mistake me, but sharp sticks and long knives have no more part in modern warfare, though it is probable that just as the Chinese Army is exercised in archery, though armed with Mausers, so we shall still arm our Cavalry with the weapons of the savage."

The Artillery.

"The work done by the Artillery was good, I think you have said?"—"It was more satisfactory in the Free State than in Natal, so far as my personal observation went. There was a false idea of the use of Artillery, and our generals seemed afraid to push it to the front after poor Long's misfortune at Colenso. But I have seen Ian

[569] After the War Churchill Interviewed 23 July https://www.britishnewspaperarchive.co.uk/viewer/BL/0000174/19000723/056/0005

[570] After the War Churchill Interviewed 23 July https://www.britishnewspaperarchive.co.uk/viewer/BL/0000174/19000723/056/0006

Hamilton push his guns to the firing line with great success since then. The outranging of our guns is due to the fact that the Boers had more modern weapons than we had. The Creusot guns are good, and yet I have been told by the French military attaché that the French Army discarded this pattern in 1896. Yet they are better than ours."

The Rifle.

"So we are behind Continental nations in this respect?"—"Undoubtedly, both in rapidity of fire and in type of ordnance. But in nothing are we so far behind our friends on the Continent as in respect of rifles. The Lee-Metford is a beautiful shooting weapon for a trained marksman, nothing can surpass it for accuracy. But all men are not good marksmen. The Mauser is better for the average shot. Then, again, I can't help thinking that the magazine action of our rifle is wrong. You can charge your magazine and carry it about till you want it, but when the soldier has fired his ten cartridges it is quite improbable that he will in action find time to replenish the magazine, and practically the Lee Metford is used as a single loader. The clip action of the Mauser is greatly in its favour."

"You think, then, that the Mauser is the better rifle?"—"Yes, for the ordinary shot. But please remember that it is inferior to the rifle of several European Powers. The German Army is completely rearmed with a weapon which experiment and experience show to be superior to the Mauser. But to return to the comparison between that rifle and ours. The manner in which the ammunition of the Mauser is put up, in its clips of five and nice cardboard boxes, shows it to be the product of a superior manufacture; ours is simply done up in brown paper parcels like so much sugar. Really, taking it all round, 1 should be inclined to say that our Army is seven years behindhand in its armament."

The Bayonet.

"And has the bayonet played any great part in the campaign?"—"There is no doubt it has. Not that it has killed many, but it has robbed the Boer of many of the advantages of his courage and cunning. It has been the feeling that they had something definite and tangible before them that has enabled our Infantry to drive the enemy from position after position. No doubt our men were better men than the Boers when in close personal contact. The bayonet gave them a superiority which all the cleverness of the Boer could not make up for. By-the-bye, I have had many letters about my remarks, telegraphed to the *Morning Post*, comparing the Briton and the Boer as fighting men. What I actually said was that in his own country the veldt Boer on his horse was equal to anything between three and five regulars—regulars, mind that—of any nationality. In the light of further experience I am inclined to consider that an understatement of the case. It is really the town mouse against the country mouse in his own territory, the wild cat against the tame cat in the wood."

The Hospitals.

"You have, of course, heard of the sensation caused by the attacks made on the hospitals at the front?"—"Yes, I have, but thank goodness I had no opportunity of going to the hospitals. I was fighting with the troops throughout. Of course, there has been a deal of misery and suffering, and some of it might have been avoided, but at the same time I do not suppose that in any other campaign such a large number of sick and wounded recovered. I should imagine that great as the misery was, it was enormously below the percentage of other wars. The general opinion, so far as I could gather it from officers cognisant of the state of affairs, was that the hospital arrangements on the whole were better than up the Nile, though it would have been easier to arrange them up the Nile than in South Africa."

The Staff as a Machine.

"You have alluded to unpleasant incidents on the line of communications. What do you attribute them to?"—"To bad staff work in many cases. The machine went wrong. Take one instance. Colvile was told to march via Lindley and Heilbron on fixed dates, and that both places were garrisoned by our troops, whereas Hamilton, who was seven days in front of him, was told to take on all his troops and leave no men behind for garrisons. Then Spragge with his men was sent by mistake to Kroonstad, instead of Winburg, with other troops going up in the general advance. Colvile was told to wait for him, consequently Spragge missed his connection. Colvile had difficulty in forcing his way through without Spragge's Cavalry, and Spragge was cut off separately before he could join Colvile. These mistakes do not arise from force of circumstances, such as the difficulties of the difficulties of the country or the action of the enemy. Orders are given to a man to meet another, and that man get orders to march on which prevent him being at the appointed place. There is no more excuses for that than for trains missing connection or being on the same line. The cause, I will not say the fault, is due to this. The control of the Army has been personal. Kitchener instead of being used strictly as a Chief of the staff whose business it is to present everything in a finished condition to the Commander-in-Chief whose work the staff as a machine, has been used as 'Bobs's Handy Man.' He has been sent tearing off to different points on the line of communications, hustled about wherever the local situation required to be grappled with, and meanwhile the chief of the staff's department has been handed over to

Page 6

General Kelly, and, not through any fault of either officer, the machine did not work smoothly. Imagine a man of great energy but no special knowledge going down to your office and saying to your colleagues: 'Come along, we will get the paper out.' It might come out, but there would be a muddle. I will not say that the Army has not gained as well as lost by the fact that the chief of the staff's department has practically been in abeyance."

Lord Roberts's Energy.

"The fact would have been far more serious had it not been for the amazing personal energy of Lord Roberts. Probably no Commander-in-Chief has ever done so much labour with his own hands as Lord Roberts. Why, the official day begins at quarter to five in the morning! His extraordinary physical and mental activity almost compensated for the defects of the staff machine, while the personal force which he and in a lesser degree Lord Kitchener have infused into the operations has practically enabled us to gain an advantage which could never have been obtained by generals whose character led them to trust in a smooth working staff machine, a no-little-point-forgetting staff machine. I don't say this, mind you, with a view to making any hostile criticism, but it is worth while to note these facts."

The Future.

"And as to the future, Mr. Churchill?"—"Oh, as to the future, there are a few points on which one sees one's way so clearly as in connection with the conduct of affairs in South Africa. This country must keep Lord Roberts's Army at full working strength by continued drafts of men and horses, and give him whatever he may demand in the matter of troops and supplies. He must be supported in any measures, whatever their severity, which he may consider necessary to stamp out what remains of the war. Anything is better than its continuance, and in the interest of humanity it is now necessary to be severe. With the Dutch Lord Roberts's name is not without meaning, Before I left South Africa I went to see Mr. Schreiner, that honest man. Naturally he is very gloomy and despondent about the affairs of his country, but be told me that he was quite sure from what he had seen of Lord Roberts that any measures which he might propose or think necessary were not needlessly severe. 'Only,' he added, 'I am not sure that he always knows about everything.' This, mind you, is a typical Dutch opinion."

Held by The Enemy

For the integrity of 'Ian Hamilton's March' Lieutenant Frankland's narrative is included. It was previously included in the first volume where it integrated so well with Haldane and Hofmeyr's memoirs.

Extracts from the Journal of Lieutenant Thomas Hugh Colville Frankland, Royal Dublin Fusiliers, lately prisoner of war at Pretoria.

"Lieutenant Frankland was captured by the Boers when the armoured train was ambushed at Chieveley, in Natal, on the 15th of November, 1899. He was carried as a prisoner to Pretoria, where he arrived on the 19th of November, and where he remained until the 5th of June, 1900, when Pretoria fell and the greater part of the prisoners were set free by their victorious comrades." [508]

Thomas Hugh Colville Frankland.[571]

Frankland's sketch of the armoured train disaster, with his signature.[572]

November

November 19th—To wake up and find oneself enclosed in the space of a few acres for an indefinite period is scarcely pleasant; however, one cannot always be miserable. The monotony will, I have no doubt, become very trying, but for the first few days I have a good deal to do. The State Model School, which has been turned into a prison for the officers, is a building of rectangular shape. A long corridor runs through the centre, and on both sides of this are the rooms, where the officers sleep. They are supplied with a spring bed and two blankets apiece, while the whole place is lighted by electricity. At one end is the dining-room and gymnasium.

In front is the road, from which the building is separated by iron railings. Behind there is a sort of back garden where the police and soldier servants live in tents, and where the kitchen and the bath-room are situated. This piece of ground is surrounded on three sides by a six-foot fence of corrugated iron, and the whole place is watched by a cordon of armed police, about fifteen being on duty always.

[571] Photo Credit: Thomas Hugh Colville Frankland (unknown-1915) - Find A Grave Memorial Memorial Id: 56009464 Photo added by Nancy Wright: Member Profile: Nancy Wright - Find A Grave

[572] A sketch by Mr. T. H. C. Frankland, an officer in the train, forwarded by Mr. Churchill from the States School Prison at Pretoria to the Editor of The Graphic with the comment: "The enclosed sketch seems to me to be a very accurate presentation of what actually occurred." By permission of the "Daily Graphic," From: Winston Spencer Churchill by A. MacCallum Scott, Methuen, London 1905.

The Government here generously supplies the officers with bread and water, half a pound of bully beef a day, and groceries. We have a small piece of ground and a gymnasium for exercise. As there are, alas! about fifty officers here, we have formed a sort of mess, and for the sum of three shillings a day we improve our scanty allowance of food.

They have supplied us with a suit of clothes each, but mine was much too big for me. I began to write my diary this evening, and had a long talk with Garvice[573] in my regiment, who told me how he had been captured. Dinner 7:30; bed, and sleep.

November 20th—It looks as if the rest of my diary for several months would contain each day the words, "the same as usual." I have only been here forty-eight hours, but the monotony has already begun to show itself. Not the monotony only, but the want of freedom, the want of news, the knowledge that the rest of the war will be carried out without my share in its victories, when, had it not been for some unhappy fate, I might yet have seen many an action—all these combine to oppress and irritate my mind. I tried to make a sketch of the armoured train, but it was not a success, and I must begin again to-morrow. The very length of empty time in front of me makes me quite patient.

November 21st—It is getting extremely hot. The lack of open space to walk in makes me feel lazy, and one gets quite tired after going a few times around the building. What one most looks forward to are the meals, and these are not very satisfying. But of course I am still suffering from the appetite of freedom, and I have no doubt that a month or so of this sort of life will make me feel less ravenous. I wrote some of my diary, and commenced another sketch of the armoured train, which I hope to be able to send to the "Graphic." Churchill has written asking to be released, but he does not expect any result. The mosquitoes here are very troublesome, and I have been constantly bitten.

November 23rd—The mail was supposed to go to-day, so I found occupation in a few letters. It is still very sultry. I succeeded in getting through a good deal of my diary, and, after writing nearly all day, played a game of rounders in the evening. This last occupation appears to cause much annoyance to the police, who frequently get hit by the ball. Another game here is fives, which we play with a tennis ball in the gymnasium. There seems to be some news about, but we can get nothing out of these people. By these people I mean Malan[574]—a

[573] PF: I am confident enough in the identity of this individual to include his sketch biography here, despite the fact of his captivity not being mentioned in the summary.
Chudleigh Garvice, Lieutenant, Royal Dublin Fusiliers, 2nd Battalion. He was born 12 January 1875, at Cookham, Bucks, son of Charles and Elizabeth Garvice. He was educated at Blundell's School, Tiverton, and was gazetted to the Royal Dublin Fusiliers 9 December 1896, and was promoted to Lieutenant, 30 June 1898. He served in the South African War from the commencement of hostilities until peace was signed, employed with Mounted Infantry, and was present at operations in Natal in 1899, including the action at Talana; operations in the Transvaal, east of Pretoria, July to 22 November 1900; operations in Orange River Colony, May to July 1900; operations in the Transvaal 30 November 1900 to January 1902. He was mentioned in Despatches [London Gazette, 10 September 1901]; received the Queen's Medal with three clasps, the King's Medal with two clasps, and was created a Companion of the Distinguished Service Order [London Gazette, 27 September 1901]: "Chudleigh Garvice, Lieutenant, Royal Dublin Fusiliers. In recognition of services during the operations in South Africa." He was promoted to Captain 28 May 1902. In 1903 he served in the Aden Hinterland operations, and in the Sudan (Egyptian Army) from 1904 to 1912, in command of the Arab Battalion in various patrols, etc. (Sudan Medal, 1910); retired from the Royal Dublin Fusiliers 9 April 1913, and became Major, Reserve of Officers, 31 August 1914. In the Great War he served on the Western Front of Egypt in 1915-16, and later on the Staff in Egypt. Major Garvice was mentioned in the Despatches of General Maxwell, dated 1 March, 1916, and of General Allenby in the London Gazette, dated 15 January 1918. He was awarded the OBE in 1919, and had the 1914-15 Star, the General Service Medal, the Victory Medal, the 4th Class Osmanieh, 3rd Class Order of the Nile, 1st Class Order of the Redeemer (Greece). He married, 31 January 1920, at Alexandria, Egypt, Isabel, daughter of Andrew Ormiston. He died 3 March, 1921. His father was the famous novelist, of whom his friend, Mr. Douglas Sladen, writes as follows: "The late Charles Garvice, father of this distinguished soldier, was the most popular novelist of his time, and a great figure in the literary world. He was Chairman of the Authors' Club, one of the most prominent members of the Garrick, and Chairman of the Committee of the new After Dinner Club, of which he was co-founder with Mrs. Baillie-Reynolds and Mr. Douglas Sladen. He was the life and soul of the clubs to which he belonged, for, in addition to his sociability and wit in the smoking-room, he was an admirable, after-dinner speaker, and when it was necessary for the club to assert itself on any occasion, he was always willing to take the responsibility, although it might make enemies. He was fearless in the performance of a duty." QSA (3) Tal OFS Trans (Lt DSO R Dub Fus). Holditch 1985 £285. Source: DSO recipients (VC and DSO Book) Also: Mounted Infantry. MID LG: 10 September 1901, page: 5953. Source: Field Marshal Roberts. 4 September 1901.

[574] Field Cornet Malan was a grandchild of President Kruger. (Sandys p. 82) This was not exceptional because Kruger had 17 children and over 100 grandchildren. He was captured in August of 1901 and exiled to Ceylon. (Haldane) Over 5,000 Boers were interned in Ceylon (Sri Lanka) during the Boer War. The Ceylon camps were the second set of overseas POW camps after those in St. Helena had filled. http://libcom.org/library/pow-camps-ceylon-during-boer-war

spiteful, objectionable animal—who ought to be at the front, were he not a coward; [R. L. W.] Opperman,[575] a slightly more agreeable person, of large dimensions, and Dr. Gunning,[576] a much more amiable fellow. It seems absurd that they do not allow us to buy papers. What harm could we do with them?

Some of the restrictions are so childish, and tend to make life here so sickening, that I am sure if curses could harm the Transvaal Government it would not be long-lived.

This morning Churchill was visited by De Souza, the Secretary of War, by the Under-Secretary for Foreign Affairs, and others, and there followed a very animated discussion about the causes and the justice of the war. It was a drawn game, and they all talked at once at the end, especially Churchill. I am afraid for his sake he is not likely to be exchanged or released. The Boers have got to hear of the part he played in the armoured train episode.

Louis de Souza.[509]

Louis de Sousa and his spouse, Marie.[510]

November 24th—There is some news abroad to-day. The Free Staters have been attacked at Belmont by the British, probably under Buller, but the result is uncertain. Of course the Boers report a victory on their side, but one gets quite accustomed to their "victories." Dundee was a [British] victory, likewise Elandslaagte. I am getting on slowly with my diary, and manage to make it occupy a great deal of time.

November 25th—Evidently we have won a victory at Belmont; its results are immediately apparent here. They have suddenly become much more lenient and complacent. We are actually allowed newspapers, and the President is considering the question of beer. The papers admit that the British drove the Free Staters from their position at Belmont, but with great loss, while that of the Boers is practically nil. Rumours say that General Joubert is cut off between Estcourt and Mooi River; how I hope it is true!

November 26th—The Rev. Mr. Hofmeyr is a prisoner here, and held service this morning, when he delivered a most eloquent address. There is a harmonium[577] in one of the rooms, and Mr. Hofmeyr, who sings very well, gives us some very good music. He knows a lot of old English songs, which are pleasant to hear, although they rather suggest the Psalm beginning "By the waters of Babylon." Hofmeyr, though a Dutchman, is an ardent supporter of the Imperial cause, and he has in consequence been very cruelly treated by the Boers before he came here.

[575] "Commandant Opperman, who was directly responsible for our safe custody, was in times of peace a Landdrost or Justice. He was too fat to go and fight, but he was an honest and patriotic Boer, who would have gladly taken an active part in the war. He firmly believed that the Republics would win, and when, as sometimes happened, bad news reached Pretoria, Opperman looked a picture of misery, and would come to us and speak of his resolve to shoot his wife and children and perish in the defence of the capital." (WSC London to Ladysmith.)

[576] Dr. J. W. B. Gunning is one of the administrators of the State Model School Dr. J. W. B. Gunning, per Brian Davis, Churchills in Africa p. 201 See also: The Strand Magazine, No. 112, April 1900; University of Michigan and Google Books for an article on his Staats Museum of which he is the director.

[577] A pump organ.

Adrian Hofmeyr.[511]

Mason & Hamlin Organ.[512]

It is quite touching to see how the Boers try to hide their defeat. All the accounts are cooked, but even De Souza acknowledges that if things go on as at present the war will soon be over. There have been several days' fighting south of Kimberley, and Buller is advancing steadily. On the Natal side Joubert passed Estcourt, and reached Mooi River, where he was attacked by the new division and defeated. In retiring he was attacked by part of the Estcourt garrison, result unknown. He will probably retire on Colenso.

November 27th—Not much news to-day. According to the "Volksstem" British lost fifteen hundred at Belmont, and the Boers nine killed and forty wounded. However, they can't deny that the Free Staters were licked, and De Souza admits that Kimberley will probably be relieved shortly. Moreover, Khama[578] is said to have risen. This has disturbed them all exceedingly, and Opperman is highly indignant.

November 30th—I find nothing to record here except the scraps of news one gets in the newspapers, all else is monotonous—appalling monotony. In the evening one feels it most, and sometimes I don't think I can endure it for another month. All sorts of absurd rumours are spread about here by that intelligent paper the "Volksstem." The latest is that four British regiments have refused to fight, being in sympathy with the Republican cause. I wonder whether Buller will desert to the Boer side? The fact remains that the papers give no news whilst there must be plenty, and this looks as if the untold news must be bad for them. We hear that General Forestier-Walker[579] has been killed, and that Lord Methuen is seriously wounded. This morning the rumour runs that our troops have occupied Colenso. The regiment is sure to be there. How I wish I were with it!

December

December 4th—No real news, but various and contradictory rumours. The Boers have begun to acknowledge their losses, and the paper have long lists of killed and wounded. Major ——, of the West Yorks, arrived to-day, having been captured near Estcourt. From him I learned that all was well there. A few days ago three battalions—West Yorks, Borderers and Second Queen's—went out and attacked the Boers. Apparently the engagement was indecisive, and the losses on either side not very great. The rumour goes that Buller is in Natal, and not in the Free State after all. Of course he is advancing to the relief of Ladysmith. We all think that his plan will be to hold the Boers in front of Colenso while he takes a large force around by the flank. The Boers have retired beyond the river, and have blown up the Tugela railway bridge. On the other side, Lord Methuen's

[578] Khama III (1837?–1923), referred to by missionaries as Khama the Good also called Khama the Great, was the Kgosi (meaning *king*) of the Bangwato people. Khama III - Wikipedia

[579] PF: This does not appear to be correct. See: General Sir Frederick William Edward Forestier-Walker, GCMG, KCB (17 April 1844 – 30 August 1910) was a British senior military officer and Governor of Gibraltar. In 1899 he again returned to Africa, becoming GOC Cape Colony and acting as lieutenant general in command of Lines of Communication, South Africa Field Force, 1899–1901. He was thus responsible for disembarkation of troops and military stores and sending them to the front. In a despatch dated 31 March 1900, the Commander-in-Chief in South Africa, Lord Roberts, wrote how Forestier-Walker carried out his duties "with credit to himself and with advantage to the public service." He was promoted to Knight Grand Cross of the Order of St Michael and St George (GCMG) in November 1900 for his services in South Africa. Following the end of the war, Forestier-Walker was promoted to the rank of general on 6 July 1902. Frederick Forestier-Walker - Wikipedia See also: Anglo Boer War - Name search

Division is having severe fighting; he has defeated the Boers at Modder River, and the relief of Kimberley is imminent.

The papers do not publish much news themselves, but occasionally publish some of the English cuttings with sarcastic editorial comments. In the Dutch version of the "Volksstem" they slate[580] the Free Staters unmercifully for having run away at Modder River.

Oh, that we might be exchanged. Joubert has wired via Buller to England advocating such a step.

December 15th—"Tempus fugit,"[581] and it has not been quite so dull as usual. First, and most important of all, Churchill has escaped. Whether he has made it good or not is still uncertain; but he has now been gone two days, and I have great hopes. Besides the excitement there has been a very amusing side to the affair. Of course Churchill was the very last person who ought to have gone. He was always talking and arguing with the officials, and was therefore well known, and, indeed, scarcely a day passed without Dr. Gunning or Mr. de Souza inquiring for him. His plans for escape were primitive; but, being still in prison, I must not write anything about this part of the affair. Let it suffice that Churchill got away without any trace left behind. Next morning, as it chanced, it was the day for the barber to come and shave him, and having only just woke up I put the barber off rather feebly by saying that Churchill had gone to the bath-room, and would not need shaving. What should the detective who accompanied the barber do but wait outside the bath-room, and, finding no Churchill, began to suspect. Gunning then came upon the scene, closely followed by Opperman, both asking and seeking anxiously for their captive. Their distress at finding him gone was really pathetic. They immediately put on all kinds of restrictions. No papers, calling rolls, not allowing anyone into the yard outside the building after 8 P.M., and stopping all beer. I am reminded of the fable "Le Corbeau et le Renard,"[582] which ends, "Le Corbeau ... jura, mais un peu tard, qu'on ne l'y prendroit plus."[583] Curiously enough, the day after Churchill had escaped an order is said to have come from General Joubert for his release. However, I have no doubt but that this was all made up to excuse themselves for not being able to catch him. I do hope he gets away.

Our spirits are constantly on the rise and fall. At one time we are about to be exchanged, at another nothing has been heard of it; at one time there is a brilliant British success, greatly modified, of course, by the enlightened "Volksstem" editor, at another a crushing British defeat, with all the Generals and thousands of soldiers killed and wounded.

Yesterday we heard of the splendid achievement of the British troops in Ladysmith in smashing up the 84-pounder at Lombard's Kop, several Howitzers and a Maxim.[584]

Then came the defeat of General Gatacre at Stormburg, and the capture of 600 prisoners, and on the top of this the victory which the Boers claim at Magersfontein.

All this is very terrible. I think I feel almost as miserable as I did the night I was captured. Are the British troops ever going to drive the Boers back? Will they ever come and take Pretoria? or will they, on the other hand, be driven back, and the people at home get sick of the war, like in 81, and—no, impossible—and yet who will dare predict? It is too awful to hear all these shocking reports, and to be able to do nothing oneself. One always imagines on these occasions one's presence at the scene of fighting absolutely indispensable if there is to be a victory. However, these miserable days cannot last for ever. Perhaps they are even now at an end. De Souza, with a faltering voice, has confessed that Buller is advancing at last in great force. He must win.

December 19th—Worse than ever. Buller has attacked in full strength at Colenso and has been defeated with a loss of ten guns and many hundred men. This is too awful—I could have cried. The hand of fate seems to be raised against us. The only thing to do is to wait patiently till the next disaster. The Stormburg prisoners have arrived, the Colenso prisoners are expected to-morrow. Everybody is cursing the Generals; but they always think they could do better themselves. I hear that Hart's Brigade, with our regiment in it, were caught in quarter

[580] From Middle English, "to split, break, or cut up."

[581] "meanwhile, the irreplaceable time escapes," literally "but it flees meanwhile: irretrievable time flees."

[582] The Fox and the Crow (Aesop) - Wikipedia In the fable a crow has found a piece of cheese and retired to a branch to eat it. A fox, wanting it for himself, flatters the crow, calling it beautiful and wondering whether its voice is as sweet to match. When it lets out a caw, the cheese falls and is devoured by the fox.

[583] The Raven ... swore, but a little late, that he would no longer be taken there.

[584] December 8th, 1899. British sortie from Ladysmith, Lombard's Kop being carried. Creswicke Chronology Source.

column at close range. They must have suffered terribly. Never mind; Methuen has relieved Kimberley. The officials all deny it, but it must be true.[585]

December 23rd—No more news. The authorities are getting more and more silly and disagreeable; all kinds of babyish restrictions are invented to annoy us. Churchill has got to Delagoa Bay, and has wired his safe arrival to De Souza. Hurrah![586]

I have not dared until now, when all is a failure, to set down in this book any account of the one occupation that has prevented us from going mad with disappointment in these sad times. About the middle of the month Haldane devised a plan of making a tunnel from under our room across the road. The five fellows in our dormitory and Le Mesurier, who shifted his abode for the purpose, began about ten days ago. First, we thought of cutting a hole in the floor, but, on looking round, we suddenly found a trap-door already made. Beneath the floor there is a curious place. The rafters are supported by stone walls, so that underneath there is a series of compartments about twenty-four feet by four, with access from one to another by means of man-holes in each wall. We commenced digging in the compartment next to the one under the trap-door. The ground at first was very hard, but with chisels and implements taken from the gymnasium, we managed to get down four feet of the shaft in about four days. It was a queer sight to see two half-naked figures digging away by candle light, for we used to work in reliefs of two—one to dig and the other to cast away the earth in boxes or jugs. Suddenly, one day, we broke through the hard crust and came to some soft clay soil. We were delighted at this, and expected to get through it in no time; but, alas! with the soft earth came water, and without pumps, bale as we would, we could not get rid of it. Every morning the shaft was completely bilged; so, having dug down six feet, our plan was brought to an end, and we had to screw up our trap-door again in bitter disappointment. The officers of the Gloucester Regiment are digging too, but they are sure to find the same difficulties.

Christmas Day, 1899—I can scarcely realise that it is Christmas, the day I have hitherto spent at home with family and friends. I can see the rooms decorated with holly and "Merry Christmas" cut in white paper and pasted on red Turkish twill hanging over the doorway. A Merry Christmas! What irony! The time, of course, was bound to come when the circle at home would be broken; but little did I dream where or under what unhappy circumstances. A Merry Christmas! to a prisoner—not when his countrymen, victorious and full of enthusiasm, are marching rapidly to his release, but when the armies of his country, beaten back, lie far away; when, helpless himself, despair seizes his heart; when reverses grow into disasters and the might of the dear old land in which he trusted seems to have weakened and died. A Happy Christmas! with the New Year black, uncertain, and unknown. Of course we drank the health of the Queen at dinner—in lime-juice. 'Twas all we had; but we meant it none the less.

December 30th—They say there were only 1,200 casualties at Colenso; but we have just heard that —— and —— of our regiment have been killed. O, God! it seems too awful. To hear of all one's friends crippled or dead; all the best are picked off, and here are we tied up quite safely with our beastly skins unhurt, and not likely to run into the slightest danger while our comrades are losing their lives. We must win this war.

January

January 1st—I have had many arguments as to whether this is the commencement of a new century or not, and after much reasoning I have decided that as it is the year 1900, or the nineteen hundredth year, it is the last of the nineteenth century and not the beginning of the twentieth. Whatever it may be, this is a hateful place to spend the beginning of anything in. The "Volksstem" printed a list of casualties to-day, and I see that our regiment lost forty-two killed at Colenso. What must the numbers of the wounded have been? {Here follows a list of wounded officers.}[587] Sergeant Gage[588] was killed, and they say he was one of the first to cross the waggon bridge. This looks as if the regiment had stormed the bridge, which is much better than being mown

[585] The siege of Kimberley would not be relieved until February 15th, 1900. Siege of Kimberley - Wikipedia

[586] In the wire Churchill made sure to mention that the guards at the prison were not involved with his departure since they had come under fierce suspicion of having been bribed or colluded in some manner.

[587] The list of names was not included in the original publication.

[588] Gage J C V 3467 Cr Sergeant 2nd Btn Royal Dublin Fusiliers, Died of wounds at Colenso. 20 Dec 1899. Source: Natal Field Force Casualty Roll, page 186 line 4 Also: Source: In Memoriam by S Watt Anglo Boer War - Name search

down in quarter column. All these losses are terrible, but I believe that Colenso is only a reconnaissance in force. What must a battle be like?

Passage of Infantry over a bridge of Wagons.[513]

The last week has been, if possible, more dreary than usual. One of the fellows in our room has made himself very obnoxious lately, and has had to be sat upon severely. I have never met such an ungentlemanlike creature. It is all the more unpleasant in a place like this, where we are so closely packed. There are rumours of fighting near Colesburg, probably by General French. The Boers say the action is indecisive, which means a victory for us.

January 7th—Nothing of importance has occurred lately. There has been a bit of a fight with Opperman, who tried to take away from Boscher,[589] the local grocer, his contract for the supply of our mess, on the ground that Boscher had helped Churchill to escape: Result a complete victory for us and the reinstatement of Boscher. More Zarps, as the policemen who guard us are called, and poor little Gunning has been commandeered. He prepares himself to go. His reason is peculiar. Should his children, in after years, ask him if he fought for the freedom of the State, he would like to be able to say "Yes." However, if he goes I hope he will find a large rock to get behind and so come back safely.

This afternoon a most alarming rumour was started by somebody, namely: that Ladysmith had fallen. Though I did not actually believe it, we are always having such frightful disasters that I felt very uncomfortable. Later, however, we learned that all was well.

January 10th—Ladysmith has not fallen. The news of the defeat of the Boers on the Platrand has been confirmed, and, in spite of their lies, we know their losses were heavy.[590] At Colesburg there was a night attack, and a half battalion of the Suffolks got much knocked about. [see infra] Two of their officers came in as prisoners yesterday; they say the Boers have received large reinforcements at Colesburg. There is a rumour that Dr. Leyds[591] has been arrested in Germany for trying to enlist German Reservists. A British force is said to be at Douglas, west of Kimberley. They made a night attack and captured some stores and ammunition. The

[589] This is Mr. Bosscher.

[590] The Battle of Waggon Hill; January 5th and 6th, 1900. The Boers around Ladysmith were also growing weak from lack of forage. With little action, many fighters took unauthorised leave or brought their families into the siege encampments. Eventually, with the Tugela in flood, preventing Buller from giving any support, some younger leaders persuaded Joubert to order a storming attempt on the night of 5 January 1900, before another relief attempt could be made. The British line south of Ladysmith ran along a ridge known as the Platrand. Under Ian Hamilton, they had constructed a line of forts, sangars and entrenchments on the reverse slope of the Platrand, of which the Boers were unaware. In the early hours of 6 January 1900, Boer storming parties under General C.J. de Villiers began climbing Wagon Hill and Caesar's Camp. They were spotted and engaged by British working parties who were emplacing some guns. The Boers captured the edge of both features, but could not advance further. British counter-attacks also failed. At noon, de Villiers made another attack on Wagon Hill. Some exhausted defenders panicked and fled, but Hamilton led reserves to the spot and recaptured some empty gun pits. Late in the afternoon, a terrific rainstorm broke, and the Boers withdrew under cover of it. The British suffered 175 killed and 249 wounded. 52 dead Boers were left in the British positions, but their total casualties were not recorded. Siege of Ladysmith - Wikipedia See also: Platrand Monument. Johannesburg Light Horse Regiment - Wikipedia

[591] Willem Johannes Leyds (1 May 1859 – 14 May 1940) was a Dutch lawyer and statesman who served as state attorney and state secretary of the South African Republic. From 1898 to 1902, during the crucial period of the Second Boer War, he was the Republic's special envoy and minister plenipotentiary in Brussels, accredited to several European states. Willem Johannes Leyds - Wikipedia

Transvaalers in their excitement succeeded in firing into the Free Staters, shooting, among others, Opperman's nephew. We offered our sympathies, but after all it is one the less. This evening we received a most excellent rumour that the Boers had lost 900 men near Colenso. I hope it is true, and that the Tugela has, therefore, been crossed. This will be a step towards the relief of Ladysmith. At Colesburg the Boers are in a critical position. Things seem to be looking up a bit. I wish that we could get just a little truth. These rumours torture and deceive.

January 14th—All kinds of startling rumours have been about to-day: The British fighting in overwhelming numbers around Ladysmith; Buller surprised and taken prisoner at Pieters Station. Boers in a tight corner at Colesburg. What can one believe? All men are liars—in Africa! Life is getting very unbearable. I am sure we shall be a lot of lunatics when we are set free.

January 29th—How we clamour for news, and how our spirits rise and fall as the rumours are favourable or bad. The other day the prisoners arrived from the Spion Kop fight. The result of the attack on Spion Kop is not known. We took the hill, but, for some reason, the rumour goes that we have left it again and re-crossed the river. Can this be another lie? We hear that the regiment did not cross the waggon bridge, but tried to swim the river at Colenso last month. Very few got over. Hensley was killed the other day at Spion Kop. One can scarcely realise these losses, and I don't think we shall until we join the mess and see the sad gaps among familiar faces.

Capt. C. A. Hensley.[592]

Also thought to be Captain Hensley pictured on the train at Estcourt.

February

February 5th—We have been getting a fair share of good news lately, or, at least, good rumours. The relief of Kimberley is an established fact.[593]

Colesburg is on its last legs, though news of its surrender to French needs confirmation. There is fighting at the Tugela, concerning which the latest bulletin is "British have taken a position—Vaal Krantz." Nor is this all, other factors are at work besides the British Army. There is considerable dissension between the Transvaalers and the Free Staters. The former complain that they are always put in the fore front of the battle, while the latter rejoin that not only are they invariably sent to the more exposed kopjes, but that while they are aiding the Transvaalers to fight in Natal they are receiving no help in the defence of the Free State.

[592] Captain Charles Albert Hensley. He died January 20th 1900, of wounds received in the operations on the Upper Tugela, near Venter's Spruit. He was the elder son of Albert W Hensley of Halifax, Nova Scotia, was born in Prince Edward Island September 1865, and educated at the Collegiate School, Windsor, S, and the Royal Military College, Kingston, Canada. Being an enthusiastic and successful sportsman, he was known in his regiment as "the young Shikari." He entered the army September 1885, being promoted Captain July 1895. Having served in Egypt and India, he proceeded to Natal with his battalion September 1899, and was present at the battle of Talana and in the retirement on Ladysmith. Just before the siege began, he was sent with part of his battalion to garrison Colenso, and was present at the battle of that name and the fighting on the Tugela, being twice wounded. Captain Hensley, whose family was well known in Canada, married eighteen months before his death the daughter of H Wylde Brown of Pietermaritzburg. He was mentioned in despatches, LG February 8th 1901, for his services. Source: Donner. See also: Vol. 6 No. 6 of SAMHS South African Military History Society - Journal - LETTERS OF CAPT C.A. HENSLEY TO HIS FATHER (samilitaryhistory.org)

[593] The siege will not be lifted until February 15th. See cite supra. On this date per Creswicke: General French reached and relieved Kimberley, captured Boer laager and supplies, and forced the enemy to withdraw. The Boers evacuated Majersfontein and Spyfontein, retreating to Koodoosrand Drift.

February 12th—It would take too long, even when time is nothing but a curse, to record all the items of news we have lately received. So many startling rumours have been confirmed and denied that I long to know what is the real truth, but in the Capital of this doomed country—in the very metropolis of lies and liars—we shall never learn the truth until our friends come to bring it with them.

I have just finished reading "Esmonde,"[594] which I enjoyed very much. One advantage of my forced sojourn in this country is that I may improve my education. Indeed, reading occupies the greater part of our time, though I myself cannot fix my attention on a book for very long under these miserable circumstances. The State Library has a fair selection of books, and by paying a small subscription the prisoners are allowed to take out books therefrom. The only forbidden fruits are the books of South Africa; for these volumes, recording the evil wrought by the British race on this chosen people, are carefully stowed away for fear of the English trying to destroy the histories of their crimes.

This morning an officer of the South African Light Horse was buried.[595] To all intents and purposes he was murdered by the Transvaal Government. Although he had typhoid fever he was thrown into prison, and not until the authorities were pretty certain he would die was he sent to the hospital. Ten officers on parole went as pall-bearers and we all subscribed for a very pretty wreath.

Patience[596] is played as a game here largely by ancient Colonels and Majors, and practised by us all with indifferent success as a cruel necessity.

February 17th—Good news at last! Kimberley has been relieved! Boers are retiring in all directions. Lord Roberts, with the British Army, has entered the Free State. Warrenton has been occupied, there is great consternation in Pretoria. Opperman is furious. Perhaps the tide has begun to turn.

To explain how we get news: Brockie, a Sergeant-Major in the Imperial Light Horse, knows a Zarp here and gets a certain amount of news from him, which is not, however, very trustworthy. When we first came here an Englishman named Patterson, employed in the Government telegraph office, used to pass by the railings and whisper the news. He only used to come when there was good news to tell, and generally ended with the words, Hurrah, hurrah! Since he was always accompanied on these occasions by a large St. Bernard, we called him the Dogman. Lately he has elaborated and improved his system of giving us news and has begun to signal with a flag from the passage of Mr. Cullingworth's house opposite. Either he or one of the Misses Cullingworth stands some way back in the passage so as not to be visible to the Zarps and sends messages, which are read by Captain Burrows[597] from the gymnasium window. As he is in the telegraph office and sees all that passes, the Dogman sends very truthful information.

February 18th—More good news this morning. Cronje is lost, strayed or stolen. The Boers have been driven back at Dordrecht.[598]

The British Army is within forty miles of Bloemfontein. Buller has taken the Tugela position.[599] All this needs no comment. "*Quo plus—eo plus—*." I meant to quote a Latin phrase—the only one I ever knew—but I cannot risk the tenses and moods of the verbs. It means, however, the more we have the more we want.[600] We live, as it were, from news to news. Two officers arrived from Colesburg this morning. They say Colesburg has never been quite surrounded, only hemmed on three sides. General French began to withdraw his Cavalry about three

594 Unable to identify this work or author. Rapporto di due cause mosse dal Rndo padre Bartolomeo Esmonde della Compagn... - Google Books inauthor:Esmonde - Google Search

595 This is Lt. Charles A Percy Tarbutt, Lieutenant: He died of typhoid fever at Pretoria, February 13th, 1900. He was the eldest son of Percy Tarbutt of the Consolidated Gold Fields of South Africa. Lieutenant Tarbutt was twenty six years of age, and was educated at University College School (where he was in the football XV), and at the Bedford Grammar School. When Lieutenant Tarbutt landed in South Africa he was given a commission as Lieutenant in the South African Light Horse. After the battle of Colenso he was reported missing, but it was subsequently discovered that he had been wounded and captured. Source: Donner. Anglo Boer War - Name search

596 Solitaire. Patience (game)-Wikipedia

597 Possibly: Arnold Robinson Burrowes. Captain, (Princess Victoria's) 1st Battalion Royal Irish Fusiliers QSA (4) QSA known to exist Source: List of QSAs with the clasp Talana from the Talana book. Anglo Boer War - Name search

598 2/17/1900 Cronje's force overtaken and surrounded at Paardeberg. General Brabant engaged the enemy near Dordrecht.

599 2/16/1900 Capture of Cingolo Hill by Sir Redvers Buller's force. Flight of Cronje's force and occupation of Majersfontein by the Guards. General Kelly-Kenny, in pursuit of Cronje retiring east with 10,000 men on Bloemfontein, captured 78 waggons with stores, 2 waggons with Mauser rifles, and 8 waggons with shell belonging to Cronje's column.

600 *Quo plus habent eo plus cupiunt.* (The more they have, the more they want.)

weeks ago, sending away detachments every night until only an Infantry Brigade was left to sit in front of Colesburg, occupying exactly the same extent of front as before. The Boers never spotted this, so that French and his Cavalry succeeded in joining the Free State column, and the Infantry Brigade, by making a great show of their forces, was able to keep up the ruse until the other day, when it was decided to retire. Everything went well with the retirement except for two companies of the Wiltshire who were cut off and captured after a gallant fight. I suppose all Governments lie to a certain extent about their defeats, but this Boer one takes the cake.

February 19th—I have caught the patience disease. I spent most of the day at this interesting game, but found by 7 P.M. I was rather sick of it. Le Mesurier told me to-day that Haldane, Brockie, Grimshaw and he had thought of a plan of escape. The idea was to put out the electric light in the house and in the yard by cutting the wire as it entered the building in the roof above the entrance. The sudden extinguishing of the lights on a dark night would enable them to creep to the back wall and climb over unobserved by the Zarps, whose eyes would not have become accustomed to the sudden darkness. They had made small ladders, by means of which they could climb over the corrugated iron more easily and with less noise. Once outside, they were going to trek for Mafeking, which is only about one hundred and eighty miles off. They had meant to go to-night, but, though it was wet, there was too much lightning.

February 21st—More good news both from Stormburg [area] and the Tugela. Our friend Opperman is getting excessively polite. I think one can best describe him as a greasy, unwashed bully, oily physically and morally, cruel to anyone in his power, cringing to those he fears.

February 22nd—We hear that Cronje is completely surrounded. De Wet tried to break the encircling cordon, but was defeated with great loss. Buller has taken the Boschkop and all the British troops have crossed the Tugela.

A very amusing article appeared in one of the papers the other day, in which Napoleon was termed "the Botha of the early '10's." Botha the Napoleon of these days is presumption, but Napoleon, the Botha of the early '10's! I cannot help pitying the editor of the "Volksstem," as he is only allowed to publish good news, and must really be at his wit's-end to know what to put in now.

Haldane and the others had arranged to go to-night, but unfortunately the sentry was walking about the place which had been chosen for getting over, so that the escape was prevented.

February 24th—Haldane and Co. have tried again. This time they were determined to go. Clough,[601] the servant, was sent up via the gymnasium on to the roof to cut the wire. I gave the signal by going into the room under the main switch and asking for a map. The light went down temporarily but came up again almost immediately. We were much alarmed lest Clough should have got a shock, but he came down all right, surprised that the lights had not gone out. Of course the escape was off.

February 25th—We were all sure that Clough had not cut the wires at all last night. He had received a slight shock and then left it, so it was arranged that Cullen[602] should try. However, the position of the sentry again prevented any attempt.

February 26th—Best,[603] of the Inniskilling Fusiliers, arrived to-day from the Tugela. He said that all were well down there, though the fighting had been very severe, and that the troops were beyond Pieters. Cronje had no food and must surrender shortly.

This evening the lights went out without any mistake. Opperman was greatly alarmed, and the electrician could not find out what was up. They all believed a football must have hit the wire outside and put the light out. Probably Clough had partially severed the wires, and the football had completed the damage. Now, however, the wire being broken before it was quite dark, the advantage of surprise would be lost. It was, moreover, a bright night, and we noticed that the light in the streets shone on the wall where we had meant to climb over it.

The sentries were doubled, so we finally gave up the plan and tried to think of another. We are told that they will remove us to a new place on the 1st of March, and, perhaps, this will give us a better chance.

[601] Unable to identify this individual. Too many possibilities without rank or a first name. Anglo Boer War - Name search

[602] Unable to identify this individual. Too many possibilities without rank or a first name. Anglo Boer War - Name search

[603] Possibly T.A.D. Best, but the date of release does not match. 2nd Lieutenant 1st Btn. Missing - released at Natal. 23 Feb 1900. Source: Natal Field Force Casualty Roll, page 98 line 16. Royal Inniskilling Fusiliers.

When I went into my room at about 9.30 I found that Le Mesurier, Haldane, and Brockie were having a discussion. As we were to move in two days to the new prison they argued "why not go to earth now." The authorities would think they had escaped under cover of the light going out and would, if anything, hasten the removal of the prisoners, leaving these three under the floor to depart in peace when opportunity offered.

February 27th—This morning Opperman came into our room as usual to count the number of prisoners in bed, and on seeing three beds empty he fairly staggered with astonishment. I was looking at him with one eye and chuckled to myself at his dismay. He went and asked Brett[604] if he knew anything about it. Brett asked innocently, "About what?" Then I pretended to wake up and ask Opperman what the hell he meant by disturbing us at this hour. He left the room in a fury, but presently returned with Gunning and later with Du Toit, the Chief of the Police, who examined everything *à la* Sherlock Holmes, and expressed, with a smile, his confidence in the recapture of the flown birds. After breakfast the whole house was cleared and searched. The rooms, the cupboards, the roof—everywhere except under the floor. Then they brought in a dark lantern, and I really thought they had discovered the fugitives at last, but Sherlock Holmes never thought of the floor; his reasoning did not carry him there.

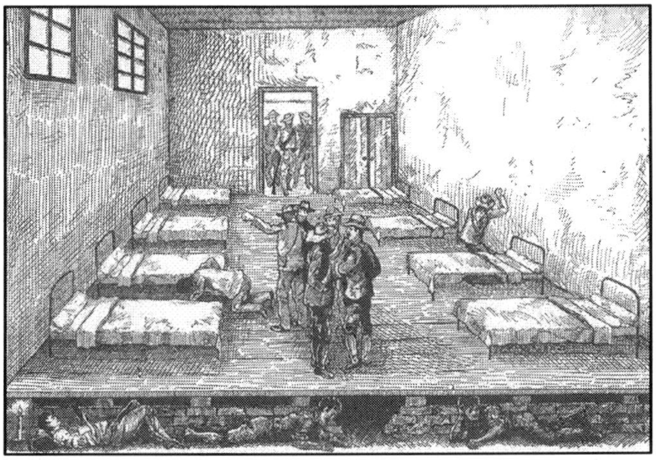

Hiding under the floor. "We hold our breath as they approach the roof of our humble dwelling."[514]

He found Haldane's saw made out of a table knife, and connecting this with the hole in the roof of the gymnasium, and the wires cut, he was sure they had gone away in the darkness. The rest, such is their mutual trust of one another in this country, were quite sure somebody had been bribed. The theories of the other officers in the prison are diverting. The discussions as to how the escaped had got out and where they had gone were full of imagination, but quite off the mark. In the afternoon Opperman and Sherlock Holmes came in with a hat and said the prisoners had been seen going over the hills towards Mafeking and had dropped the hat in question. By nightfall they had been tracked to Koodoosburg, about thirty miles out; and, indeed, the remains of their midday meal had been found. O wise detectives! This evening the Dogman[Patterson] went into Cullingworth's house in a great state of excitement and lit a candle at the verandah—a sign of good news, and on Majuba day too!

February 28th—We received the good news which the Dogman's excitement last night portended. Cronje has surrendered. This was received through the British Consul at Delagoa Bay. Buller has also driven back the Boers, and Botha wired: "No use; Burghers here won't face British." In the afternoon we received the following

[604] Charles Arthur Hugh Brett, Captain Suffolk Regiment, 1st Battalion Wounded and prisoner. Colesberg, 6 January 1900. Source: South African Field Force Casualty Roll. He was born at Muttra, India, 28 March, 1865, son of Lieutenant Colonel Arthur Brett, APD (formerly 2nd Dragoon Guards, Queen's Bays), and Georgina, his wife, daughter of Hugh Hannay, Paymaster, RN. He entered the Army as Lieutenant in the Suffolk Regiment 9 May 1885. He served in the Hazara Expedition in 1888, and received the Medal with clasp; was promoted Captain 15 June, 1894. From January 1895 to January 1899, he was Adjutant of the 1st Battalion of the Suffolk Regiment, and Adjutant, 4th Battalion, 15 November 1900 to 14 November 1905. He again saw active service in the South African War, taking part in the operations south of the Orange River, 1899-1900, including the action at Colesberg when he was severely wounded. Later he saw service in the Transvaal, east of Pretoria, and in the Orange River Colony from July to October 1900. He was mentioned in Despatches 10 September 1901, and received the Queen's Medal with three clasps. For his gallantry at Colesberg, in 1900, when, though shot through the lungs, he kept up command of his company actually charging the Boer trenches in this state until he fell exhausted from loss of blood. He was promoted Major, 24 February 1906. When European hostilities commenced he left the Curragh, August 1914, in command of the 2nd Battalion, to go to France, and fell in action at Le Gateau 26 August 1914, whilst leading his men. Major Brett married, in 1909, Enid Geraldine, daughter of Lieutenant Colonel H H St George, Senior Ordnance Officer, Scottish District, and they had one daughter, Ione Moncrieff St George, born 10 October 1909. Source: DSO recipients (VC and DSO Book).

wire: "Cronje's surrender unconditional.[605] Boers retreating on the Biggarsburg," and in the evening we heard that the British were entering Ladysmith.

Three more officers replaced the three escaped in my room. We did not let them know about those underground, but I managed to send food, news, and water down as usual, also some hot cocoa at night.

March

March 1st—Ladysmith is relieved. Joubert wires: "On Lancers coming out of Ladysmith my mounted men retired leaving waggons and stores behind them." This afternoon the Cullingworths signaled over: "No more news, furthest telegraph station Elandslaagte." Kruger has gone to the front to exhort his burghers with texts. He was preceded by a telegram which was sent to all laagers. It is too long and too profane for me to copy out. Nothing but texts and psalms, showing that they are bound to win "though the enemy compass them about," as the Almighty is their own exclusive and peculiar property. The "Volksstem" says: "There seems to be some foundation for the rumour that Cronje has surrendered, but the report that Ladysmith has been relieved is quite untrue, our burghers are still fighting bravely south of that town. Should, however, Ladysmith be relieved, the war will only enter upon a new phase. We will then have to defend our borders against the greedy grasp of an unholy race. Now will the British see what fighting with the Boers really is. Now will the war begin in earnest."[606]

(Sherlock Holmes & Co. are completely off the track and all is well below.)

March 2nd—There are no signs of our moving into our new prison. This is very disconcerting as our friends cannot stay below much longer without getting ill. The Zarps' tents have been moved into the road. Opperman says because the yard was damp, but I fancy they are afraid of an attack on the Zarps. With the dumbbells in the gymnasium it might be possible to overpower them. The day was wet and dreary; I wrote letters, Mr. Hofmeyr prayed for the escaped. I have had to divulge the secret to No. 12 room, owing to one of them unfortunately seeing the trap-door open. They were very nice about it, and will do nothing to compromise the chances of success.

March 6th—Our signals this morning informed us that the President [Kruger] had wired to Lord Salisbury,[607] "Is it not time bloodshed ceased? Will send peace proposals." These people have got some nerve. First they declare war against an Empire, and then they expect that when they have had enough they can demand a cessation of hostilities.[608] There are no signs of moving.

March 7th—The Ides of March [sic[609]], but I don't expect Kruger will be murdered in the forum of Pretoria. Those below are still all right, though their condition is not enviable.

March 8th—The following telegrams were received to-day by our signaler-in-chief Burrows: (1) Fighting with De Wet; (2) Occupation of Bloemfontein on the 6th.[610] I busied myself in drawing a picture of Kruger going to the front to exhort his burghers, on the wall my room. There seems no chance of moving. Opperman says they have not even put down the floor in our new abode. Haldane wants to try to make them move. He thought that if Grimshaw vanished too it might alarm the authorities, and make them anxious to move us to a more secure place, but I feel sure—and Grimshaw agrees with me—this would only lead to the discovery of everything.

[605] Battle of Paardeberg - Wikipedia

[606] 2/23/1900 The Boers' position at Grobler's Kloof attacked.

2/26/1900 Finding the passage of the river near Colenso commanded by strong entrenchments, Sir Redvers Buller sent his guns and baggage back to the south side of the Tugela, and found a new crossing.

2/26/1900 Colesberg and Rensberg, having been evacuated by the Boers, were occupied by General Clements, while Jamestown was occupied by General Brabant.

2/27/1900 (on anniversary of Majuba, 1881) Cronje, with 44 commandants and other officers of all grades, and over 3500 men, surrendered unconditionally to Roberts at Paardeberg. Sir Redvers Buller's force captured the Boer position at Pieters. This action opened the road to Ladysmith. Boers retired north to Ladysmith.

2/28/1900 Relief of Ladysmith after 120 day's investment.

[607] He is the Prime Minister of the UK at this time. Robert Gascoyne-Cecil, 3rd Marquess of Salisbury - Wikipedia

[608] 3/11/1900 Presidents Kruger and Steyn received reply from the Prime Minister refusing to entertain their overtures for peace.

[609] The Ides of March is the 74th day in the Roman calendar, corresponding to 15 March. Ides of March - Wikipedia

[610] 3/13/1900 Roberts occupied Bloemfontein. His despatch ran:—"The British flag now flies over the Presidency vacated last evening by Mr. Steyn, late President of the Orange Free State. The inhabitants gave the troops a cordial welcome."

March 11th—I drew another large picture on my wall, a sequel to the first. It represents Kruger just escaping from Lord Roberts, who with drawn sword appears to be running after him at a good pace. My picture No. 1 is entitled "President Kruger goes to front to exhort his burghers;" No. 2 "But returns on urgent business."

Similar Cartoon

No record of Thomas Frankland's cartoon exists, but Clive M. Dixon sketched a similar cartoon about an event that was observed outside of the siege of Ladysmith.

Cartoons showing a Boer Commandant inspecting a position and then having his horse bolt when the gun is fired.[611] [515]

March 12th—The man who came for grocery orders reported this morning that Bloemfontein had fallen, but our signal was that the British were within seven miles of the Free State capital. Opperman saw my portraits of Kruger this morning; I am afraid he did not appreciate them as he should have done. However, I told him that with a pail of whitewash and a brush he might obliterate them if he chose. (N.B.—Such is the procrastinating nature of these Boer-Hollander people that Opperman never had the pictures removed, and this with other things had, I believe, a good deal to do with his own eventual removal.)

No. 12 decided to have nothing more to do with the digging plan. We have therefore arranged that Grimshaw, Garvice, and I shall take part in the operation. Garvice has not been informed of Le Mesurier's whereabouts, but has decided to dig. The Colonials in No. 20 room are also digging, but theirs is to be a deep tunnel and I doubt if they can master the water question.

March 13th—Tragedy. The Dogman [Patterson] and Cullingworth have been commandeered as undesirables, but intend, I fancy, to escape to the British lines. We signaled to him, "Good-bye, eternal gratitude, God bless you!" The Dogman replied, "British twenty miles north of Bloemfontein; Good-bye; speedy release; will return with Bobs." [General Roberts][612]

We started our shaft under the big room No. 16. Apparently we made a good deal of noise, for the old Colonels were very much alarmed and threatened to stop all digging, though they did not know who the culprits were. Opperman came into the room when mining was in full swing below, and it was all the occupants could do to hustle him outside, drowning the noise of the pick by stamping. We were rather distressed and decided to wait a few days. Garvice was very much startled when he saw Le Mesurier.

[611] Major Clive MacDonnell Dixon (10 February 1870 in Middlesbrough – 5 November 1914 in Ypres) was an English illustrator and soldier, best known for the charming images in his book The Leaguer of Ladysmith, created during the four-month Siege of Ladysmith in South Africa. This material also appeared in the Ladysmith Lyre at the time of the siege. Clive MacDonnell Dixon - Wikipedia And: The leaguer of Ladysmith, by Captain C.M. Dixon. London, Eyre and Spottiswoode [1900]. Catalog Record: The leaguer of Ladysmith | HathiTrust Digital Library

[612] He is referring to Lord Roberts who is commanding those forces in the Orange Free State.

Commando mustering (Commandeering) at Bloemfontein.[516]

He describes his feelings vividly. On going down by the trap-door he remarked what an awful hole it was. Suddenly, in the flickering candle-light he saw a gaunt, bearded, unwashed face, and a half-naked body. At first he could not make out what it was, but when he at last realised it was a brother officer he said you could have knocked him down with a feather had it not been that he was already crawling on his stomach.

The new shaft is a long way off; when I went down I had to crawl on hands and knees along passages and through man-holes, backwards and forwards in a regular maze of compartments, and, indeed, had the candle gone out one could easily have been lost. Haldane looked very ill, but the others, except for being covered with dirt, seemed well enough.

March 14th—Grimshaw went down this evening to hold a confab. They have managed to dig without making a noise by wetting the earth. Grimshaw and I made the trap-door into one piece by securing the planks together and also made it so as to batten down from underneath. I sent them down jugs of water during the day to wash in.

March 15th—All went as usual this morning. Grimshaw descended and did a little digging. In the afternoon Opperman brought the news that we were to be moved to-morrow! Most of the officers were very annoyed, but Grimshaw and I sent the information below with gladness. Well, there was no time to be lost. Food enough to last them a week, all the bottles filled with water, and everything that could possibly be of any use to the cave-men was sent down. We heard, however, and not to our surprise, that others were thinking of going into their respective holes so as to escape after we had moved. As this could have had no other effect than to cause the discovery of all, we were determined if possible to stop it. We told Colonel Hunt,[613] and he managed to persuade all concerned to abandon their schemes.

This settled, we set to work, after final good-byes and handshakings, to putty up the cracks between the boards of the trap-door, which had already been fastened down from underneath. This we succeeded in doing to perfection, and after covering the place well with dust, the trap-door could scarcely have been located by anyone; certainly not by those who did not know of its existence.

March 16th—The Staats Model School at an early hour was more than usually busy. We were all packing up such belongings as we had. I rolled everything in my mattress and rugs, and secured with rope. Then the gates were opened and all baggage was moved out on the road ready to be packed on the trolleys provided for the occasion. To be outside those gates was to breathe fresh air; to pass those barriers which had so long defied our efforts and our wits was like going out into another world. I went back into my room, and by prearranged taps on the floor Grimshaw signalled that all was well. I then sang "For Auld Lang Syne" as a parting farewell.

The Government had generously provided cabs for the convenience of the officers (who afterwards found they had to pay), and at about 10 A.M. the first cabs rolled off amid the friendly farewells of many neighbours. The

[613] Mr. H. Hunt Lt Col. Royal Regiment of Artillery. Severely wounded at Colenso. 15 Dec 1899. Source: Natal Field Force Casualty Roll, page 27 line 34. Anglo Boer War - Name search

long column of vehicles was escorted by a motley guard, consisting of very old men and tiny boys armed with Sniders[614] and sporting guns of ancient pattern.

Officers leaving the Model School.[517]

We soon passed out of the town and, crossing a small river, began to crawl up a steep hill. The roads outside of Pretoria appear very much neglected, but, of course, the money that should have been devoted to general improvements was all spent in secret service or in preparations for the war. We soon arrived at our destination. The building stands halfway up the side of a hill, and is probably a much healthier place than the Model School. Besides, the view is really pretty. To the north, indeed, it is limited by the tops of two hills. Southward lies Pretoria, a collection of large Government buildings and of small villas amid masses of trees, nestling beneath a high range of hills, along the crest of which rise the famous forts. The view on the west is merely a vast plain which reaches to the horizon, and a large hill obliterates any view to the east.

The place itself consists merely of a long white shanty with a fairly large compound enclosed by formidable barbed-wire entanglements. Outside are Opperman's house and the Zarps' tents. There are electric lights all round the enclosure, making escape a matter of considerable difficulty. Inside, the place looked more like a cattle-shed than anything else.

Pretoria from the East.[518]

A long galvanised-iron building, divided into a servants' compartment and kitchen, eating rooms, sleeping room, and four small bath-rooms. The sleeping-hall is eighty-five by thirty yards long and accommodates 120 officers, our beds being, roughly, a yard apart. There is no flooring. The drains consist of open ditches, while the sanitary arrangements are enough to disgust any civilised being. A strong protest was at once sent in to the authorities, but I doubt that it will have any effect.

[614] The British .577 Snider–Enfield was a breech-loading rifle. The American Jacob Snider invented this firearm action, and the Snider–Enfield was one of the most widely used of the Snider varieties. The British Army adopted it in 1866 and used it until 1874 when the Martini–Henry rifle began to supersede it. Snider–Enfield - Wikipedia

The Birdcage, British Officer's Prison, Pretoria.[519]

Inside Camp.[520]

Inside the Waterval prison.[521] Note the extensive electric light poles.

How the British prisoners were guarded at Waterval.[522] Note the high fence in the background and light pole.

The Bathing Tank for the Privates at Waterval.[523]

British Prisoners at Waterval.[524]

March 18th—The greatest disadvantage of this place over the Staats Model School is that we can get no news.

March 22nd—Gunning gave us a small baboon the other day, which was very fierce at first, but has tamed wonderfully. There are many different kinds of curious insects here, not curious for this country, of course, but which I have never seen before. The "Praying Mantis" or "Kaffir God" is one of the queerest. The whole place seems to be a large ants' nest, and we have often witnessed great fights between the different kinds. Snakes also abound. A night-adder was killed the other day. It was about thirteen or fourteen inches long and very poisonous, so Gunning says.

We hear Gunning and Opperman are going to the front to-morrow. I am very sorry for the former, though the departure of the latter is a great advantage.

March 23rd—The Zarps and Opperman departed for the front this morning. Their place was taken by a new guard selected from the Hollander Corps. The Commandant is a pleasant fellow and a great improvement on Opperman. [Westernant] [615]

March 25th—We had service as usual this morning. This evening an attempt to escape was going to be made by Ansell[616] and Co., but it never came off. There has been no news of Haldane and the others, so I suppose they are well away by now. This evening the new Commandant had roll-call. We call him "Pyjamas," because he wears a suit of clothes for all the world like a pair of pyjamas. His real name is Westernant.

March 30th—There has not been anything very important to record for some days. On Tuesday an attempt to escape was made by Best. While one sentry was gossiping with another he crept under the barbed wire. As luck would have it, when Best had got half way through, the sentry finished his *tête-à-tête* and returned to his post. At first he thought Best was a dog and called out *footsack*,[617] but seeing he was a human being, merely told him to go back. He might have shot him with some excuse, so Best was lucky in striking a kind-hearted man.

On Wednesday Joubert died.[618] In respect to him we sent a wreath. I don't think this will have any effect on the war, as (and the papers say as much) his moderate attitude in the recent crisis had taken away much of his popularity.

There was great excitement this evening caused by an attempted escape. The electric wires had been tampered with, and at about 10.30, by some device, Horne, a colonial, who is also an electrician, made the current travel on a shorter circuit, thus blowing out the main fuse and extinguishing all the lights round the building. Hardly had this happened than two shots were fired in quick succession, and then another. The escape failed, but all got back into the building unwounded. Apparently the lights had gone down, then up for a second, then finally out.

During the momentary flash Hockley,[619] [one] of the escapees, had been seen and fired at. However, "All's well that ends well," though some say that two bullets went through the dining-room. Sentries were doubled for the night and patrols sent out.

April

April 3rd—Hurrah! the papers this evening report the safe arrival of Haldane, Le Mesurier and Brockie at Lourenço Marques, having travelled through Swaziland. We were so glad to hear this news. Alas! We also hear that sixteen officers arrive tomorrow, and that seven guns were captured with them.

The Cullingworth girls came up this evening and signalled with a handkerchief that Mafeking had been relieved.[620] I hope it is true. We all admire the pluck of those girls. We have already collected a large subscription to get them and the Dogman handsome presents.

[615] Unable to identify this individual.

[616] Possibly: George Kirkpatrick Ansell, Captain & Brevet Major QSA (3). Prisoner, Arundel, 20 Feb 00. Source: QSA medal rolls 6th (Inniskilling) Dragoons. Also: MID LG: 17 January 1902, page: 374. Source: General Kitchener. 8 November 1901. Re: General mentions.

[617] Note in Frankland's Original, it would be Dutch or Afrikaans for "Be off."

[618] 3/27/1900 Death of General Joubert.

[619] Probably: W. Hockley, Lieutenant, Prisoner. Schoeman's Kop, 23 February 1900 Attached FID Source: South African Field Force Casualty Roll, Montmorency's Scouts.

[620] Not until 5/17/1900 Colonel Mahon, at the head of the relief force, entered Mafeking.

There was a large swarm of locusts yesterday. So thick was the cloud that it quite obliterated the view of the distant hills. They continued passing over nearly all day.

April 5th—The prisoners arrived this morning. They mostly belong to U Battalion, R.H.A.; [Royal Horse Artillery] some to the M.I. [Mounted Infantry] and Cavalry. I have not quite gathered the circumstances of their capture, but they seem to have been caught in a trap, owing to the want of the ordinary precautions. The convoy and one battery were practically held up without firing a shot, but the other battery got away. When marched off they heard that another British force was pursuing so that the guns may be recaptured.[621]

They bring very little news; apparently they have heard nothing about the relief of Mafeking, though Warren[622] was on his way thereto. Roberts has been delayed in his advance for the want of horses, but as this has been remedied the forward movement should begin shortly.

Had the horses not been so done after Abrams Kraal, they say De Wet would have been caught and the war over. Such is the fashion of war. If so-and-so had happened—always "if"!

April 6th—How the fortunes of war vary! We seem to be going through a series of small disasters. To-day the papers have the report of a "Brilliant Boer Victory, thirty-six miles south-east of Bloemfontein;[623] 450 prisoners!!!" The only hope is that the account is not "official." But we must be ready for the worst. The leading article says: "Within a few days Roberts will be forced to evacuate the Free State. *His retreat from Bloemfontein will be like Napoleon's retreat from Moscow*."

April 11th—The prisoners reported captured some time ago have not arrived yet. They always seem to be "expected to arrive somewhere," but apparently have not yet been actually seen by anybody. On Saturday their capture was reported officially. On Thursday English wires said that 300 Royal Irish were surrounded. To-day they say the prisoners are expected at Pretoria to-morrow! Well, we shall see.

The last few days we have had many good rumours about the capture of Boers and British victories. To-day the papers say that Lord Methuen is advancing on Boshof (he must be there by now), and that Colonel de Villebois[624] has been killed. He apparently and his men (100, so they say—probably 500) were all killed, wounded, or taken prisoners. A distinguished ex-French officer and his foreign legion is a good bag.

The next piece of information is, quoting from Boer paragraphs or head lines, "Fifteen hundred English in a corner;" "Brabant's Horse in a trap." Then, again, "There is every hope of their surrender." So much for this. But on the Dutch side we read that all telegraphic communication with Ladybrand and the south has been cut, so I rather fancy the Boers have over-reached themselves for once.

The Boers have attacked our camps at Elandslaagte, and because, when they shelled, our camp tents were struck, they report that the British fled. I wonder if Le Mesurier was in this show.

In all these fights, as usual, the Boers "By the grace of God had (about) one man killed and four wounded." This is heavy; generally it is one horse and three mules. "The enemy," of course, "must have lost heavily." So the paragraphs run on. Many are the funny expressions. "One brave burgher succumbed to the explosion of a bomb." "One of our guns *in firing* damaged its sight and one of its wheels!" They always end up with "Our burghers are full of courage, and determined to withstand the enemy to the last."

[621] 4/4/1900: Capture of British troops by the Boers near Reddersburg.

[622] General Sir Charles Warren, GCMG, KCB, FRS (7 February 1840 – 21 January 1927) was an officer in the British Royal Engineers. He was one of the earliest European archaeologists of the Biblical Holy Land, and particularly of the Temple Mount. Much of his military service was spent in British South Africa. Previously he was police chief, the head of the London Metropolitan Police, from 1886 to 1888 during the Jack the Ripper murders. His command in combat during the Second Boer War was criticised, but he achieved considerable success during his long life in his military and civil posts. Charles Warren - Wikipedia

[623] Sanna's Post (a.k.a. Korn Spruit) was an engagement fought during the Second Boer War on March 30th And 31st, 1900.
In all, the British suffered 155 men killed or wounded. 428 men, seven field artillery pieces and 117 wagons were captured. The Boer force suffered three killed and five wounded. But even more serious than the losses in the action was the loss of Bloemfontein's water supplies. This greatly aggravated an epidemic of enteric fever dysentery and cholera among the occupying British army, which eventually caused 2,000 deaths. Sanna's Post - Wikipedia

[624] George Henri Anne-Marie Victor count de Villebois-Mareuil or by his shortened name George de Villebois-Mareuil (22 March 1847, in Montaigu, Brittany, France – 6 April 1900, in Boshof, OFS. George de Villebois-Mareuil - Wikipedia

Various officials came up the day before yesterday to inquire into the causes of the protest we had sent in, signed by all the officers here. They promised that everything would be seen to; but they are all—well they are Boer officials, and I doubt if our lot is to be in any way improved.

The weather is getting much colder now, though the sun is still hot by day. A few stray shots whistled over the building to-day, probably "accidentally on purpose." I hope they do not begin sniping regularly.

Rosslyn Twice Captured

James Erskine, 5th Earl Rosslyn[625] April 18th just before primrose day.[626]

"Mr. De Souza left me after telling me that Winston Churchill's warrant of release was in their hands the night he escaped, and that they were glad to be rid of him. "He talked so much," he added, "that we all avoided him" I could not restrain a smile; but his escape and Mr. Hellawell's[627] unsuccessful attempt have certainly made matters "difficult" for war correspondents and officers too, since these have been removed to more guarded quarters ever since three of their number escaped about a month ago."

"Talking of escapes, the story of Winston Churchill's was already known to me through his own description, and the officers were most interested to hear about it, though, I am afraid, he was not a persona grata with his fellow prisoners, and, as far as I could ascertain, did not play quite fairly with the others who concocted the plot, but, according to them, followed the principle of sauve qui peut[628] rather than "shoulder to shoulder."[629] [630]

May

May 8th—We have had an immense amount of news lately. Roberts has begun his big advance. Brandfort is in our hands, also Winburg. The force advancing via Boshof has reached Hoopstad, while the British have crossed the Vaal at Fourteen Streams. De Wet has not been heard of for a considerable time. So much is acknowledged in the papers. Rumours say that we are behind Kroonstadt!! That De Wet, Steyn, and 8,000 Boers have been taken!![631] The English in the town think we shall be released by the 24th of May. A panic seems to have seized the Boers, and excited meetings have been held. Kruger summoned the Volksraad on Sunday, and addressed them in stirring words, which, while acknowledging the serious nature of the situation, exhorted the burghers to

[625] Earl of Rosslyn, James St Clair-Erskine, 5th Earl of Rosslyn Twice Captured: A Record of Adventure During the Boer War. United Kingdom, William Blackwood and Sons, 1900. Page 295-296. See: Twice Captured - Google Books And: James St Clair-Erskine, 5th Earl of Rosslyn - Wikipedia File:James Francis Harry St Clair-Erskine, 5th Earl Rosslyn,.jpg - Wikimedia Commons

[626] Page 335 Twice Captured.

[627] Possibly, A. Hellawell Source: QSA and KSA medal rolls. Western Province Mounted Rifles.

[628] Every man for himself - more literally "save yourself if you can"; most literally "may he save himself, whoever can."

[629] Page 335 Twice Captured.

[630] Rosslyn will pay dearly for publishing this observation. Churchill and others will force a retraction.

[631] Saturday, April 28, 1900 Fighting near Thabanchu Mountain.
Tuesday, May 1, 1900 General Hamilton captured Houtnek.
Saturday, May 5, 1900 British occupation of Brandfort. Roberts's advances towards the Vet River.
Sunday, May 6, 1900 The Vet River (Tributary of the Vaal River) passed and Smaldeel occupied.
Monday, May 7, 1900 General Hunter occupied Fourteen Streams.

continue the struggle trusting in the Lord. General Schalk Burger,[632] while addressing the townspeople, said that a stand might yet be made, if not, the independence of the Republic was at an end. The Church of Pretoria has addressed petitions for peace to the Churches of Great Britain and of Europe and America. They pray that this unholy bloodshed may cease. Kruger says "Continue the struggle to the end." Is it for England or for Kruger to give in?

We have started a newspaper; it is progressing. We call it the "Gram," because at the Staats Model School all our news came in under the popular names of signal-gram (when news was signalled), Kaffir-gram (when brought through the Kaffir),[633] Brockiegram (when [Adam] Brockie succeeded in getting information from the Zarps), and so forth. Rosslyn is editor; Major Sturges[634] sub-editor. White, R.A., Wake, 5th Fusiliers, and I, are the artists. The paper has been all written out by Rosslyn, and is now being hectographed.[635] We hope to bring out seventy good copies of the first number.

Rosslyn at his desk. Note initials THCF for Thomas Hugh Colville Frankland.[636]

[632] Schalk Willem Burger (6 September 1852 – 5 December 1918) was a South African military leader, lawyer, politician, and statesman who was acting president of the South African Republic from 1900 to 1902, whilst Paul Kruger was in exile. Schalk Willem Burger - Wikipedia

[633] During the Dutch and English Colonial Period the word Kaffir was used to describe monotheistic peoples (Nguni ethnic groups in particular) of South Africa, who were not of a Christian or of an Islamic religious background, without derogatory connotations, until the early twentieth century. It appears in many historical accounts by anthropologists, missionaries and other observers, as well as in academic writings. See Also: Kaffraria and British Kaffraria (Queen Adelaide's Province) were historical regions along the coast north or Port Elizabeth. See: File:Eastern Frontier, Cape of Good Hope, ca 1835.png - Wikimedia Commons Subsequently it became a racial slur. Kaffir (racial term) - Wikipedia See also the British military unit Kaffrarian rifles which were white and black units which were still organized as late as 1947. East London, 1 March 1947. King George VI inspecting 'Kaffrarian Rifles' guard of honour. - Atom site for DRISA And: Kaffrarian Rifles AKA: Buffalo Volunteer Rifles Buffalo Volunteer Rifles - Wikipedia Also: a historical region in Afghanistan and the setting of most of Rudyard Kipling's famous 1888 novella "The Man Who Would Be King." Kafiristan - Wikipedia And: an Arabic and Islamic term which, in the Islamic tradition, refers to a person who disbelieves in God as per Islam. Kafir - Wikipedia PF Note: I have no authority to change the wording used by original authors. There are instances where a pejorative or admiring modifier is used. While this is racist in and of itself, to identify someone by their complexion, in this narrative, identifiers such as Dutch, Boer, English, Uitlander, Indian, and many others are used as means of identification sometimes with pejorative or admiring modifiers.

[634] W. E. Sturges Major Prisoner. Stormberg, 10 December 1899 Northumberland Fusiliers 2nd Battalion. Source: South African Field Force Casualty Roll.

[635] The hectograph, gelatin duplicator or jellygraph is a printing process that involves transfer of an original, prepared with special inks, to a pan of gelatin or a gelatin pad pulled tight on a metal frame the special aniline dyes for making the master image came in the form of ink or in pens, pencils, carbon paper and even typewriter ribbon. Hectograph pencils and pens are sometimes still available. Hectograph - Wikipedia

[636] "The Gram" A social Magazine Founded by British Prisoners of War in Pretoria, Facsimile Edition Eyre & Spottiswoode, Her Majesty's Printers, London #1 - The Gram ; a social magazine founded by British prisoners ... (no.1-3). - Full View | HathiTrust Digital Library Scanned from the University of Minnesota Copy. Image: #60 - The Gram ; a social magazine founded by British prisoners ... (no.1-3). - Full View | HathiTrust Digital Library Cover Image: #61 - The Gram ; a social magazine founded by British prisoners ... (no.1-3). - Full View | HathiTrust Digital Library

No. 0 "Cheapside."[637]

May 13th—Though two or three prisoners have arrived lately, we can get no particular details of the news. There is no doubt that a general advance has been begun, but what point our troops have reached is uncertain. Also, it is still a question whether De Wet is captured or not. This morning the most serious rumour came in, to the effect that Mafeking had fallen, but I can scarcely believe it.[638]

Yesterday Mr. Hofmeyr received the welcome order to pack up his things and go. He seemed very affected at saying good-bye and nearly broke down. We all liked him very much, and bade him a hearty farewell, cheering him as he left the enclosure, and singing "He's a jolly good fellow." We shall miss him as well as his services.

Our paper came out yesterday and was a great success. We hope to bring out a new one on the Queen's Birthday, though it is an awful labour.

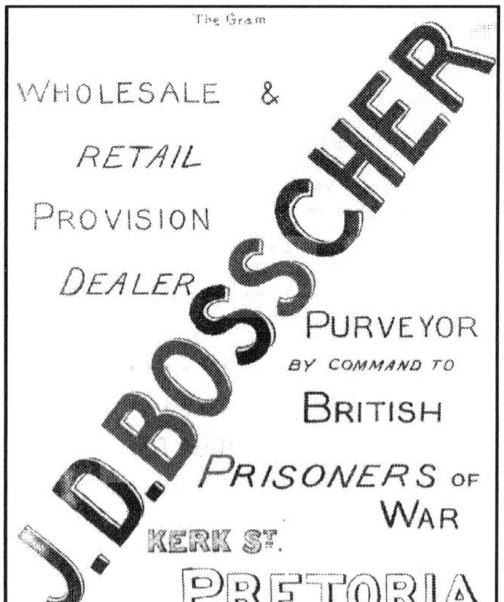

Cover page and an advertisement Mr. Bosscher took out in the prisoners' newsletter 'The Gram.'[639] [525]

[637] Believed to be Rosslyn's own sketch of his quarters. p. 330 of Twice Captured and p. 407 of the scan.

[638] Tuesday, May 8, 1900 Ladybrand deserted by the Boers.
Wednesday, May 9, 1900 Capture of Welgelegen. Mafeking Relief Force reached Vryburg.
Thursday, May 10, 1900 Battle of Zand (Sand) River. Occupation of Ventersburg.

[639] PF: Mr. Bosscher was the only advertiser, even though he need not have done so, since he still had the concession. See Hofmeyr, for particulars, and Rosslyn. The magazine did not begin publishing until May 1900 well after the prisoners had been transferred to a purpose-built prison outside of Pretoria.

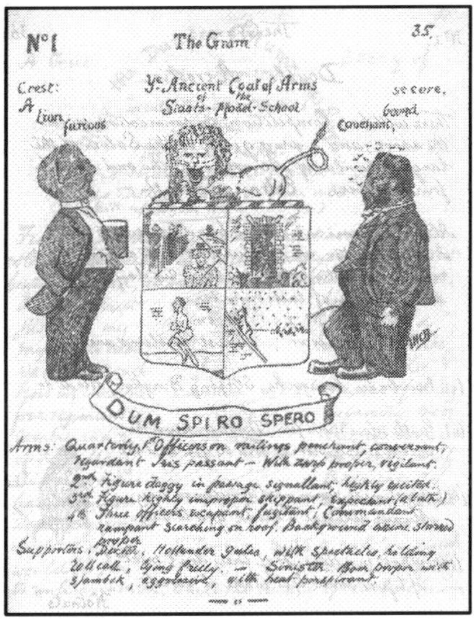

Image of "Captain Dunham" by Thomas Frankland.[640] Page from the "Gram" Note THCF Author.[526]

Life has not been so bad lately. Buoyed up with hope of a speedy release, and occupied with the "Gram," time has passed, in my case, more quickly. We had a selling lottery the other day for the day of our release. The dates ranged from the 15th of May to the 15th of August. The Queen's Birthday was much in request, while "the field" (any day after August 15th) went for six pounds.

The "Volksstem," of course, progresses as usual. Having exhausted all other insults on England, they commenced lately on the Queen! During the present British advance the mendacious powers of the editor are once more brought to trial, and once more he has not been found wanting. The burghers are full of courage (running everywhere); even the women wish to fight! There was, indeed, a rumour that our present guard was to be commandeered and the women put here to look after us. Poor time for us! I fancy we should be all shot! The Volksraad sat the other day, and after Kruger and others quoting a few scriptures the session of 1900 was closed after sitting two days!

May 14th—So much news has arrived today, that I think I had better inscribe it, while I remember. This morning came the rumour that a good many Boers actually did get into Mafeking, but, being unsupported, still remain there. This evening's "Volksstem" is truly a wonder. It gives more news than it ever has given before. An attack was made on Mafeking. The Boers took a "fort," but were attacked by night, and lost seven killed and "some" wounded and prisoners.[641] At present Carrington[642] and Plumer[643] are proceeding to Mafeking by train, so that it must have been relieved. Everywhere the Boers fly, and the British troops entered Kroonstadt on the 11th inst. Hunter,[644] with his 25,000 men, drove the enemy back at Warrenton,[645] and "the Boers are unable to resist the advance of the forces at Vryburg."[646]

[640] Image and text per Neville Chamberlain "Frankland drew the illustrations on pages 1, 9, 22a, 35, 36, 39 & 41 of 'The Gram' (7 of the 18 drawings)." Anglo Boer War - Armored Train Ambush 11-15-99 Chuchill & Haldane - Page 17 - Boer War Forum

[641] Saturday, May 12, 1900 Commandant Eloff attacked Mafeking, and was captured by Col. Baden-Powell. Roberts occupied Kroonstad without resistance.
Sunday, May 13, 1900 General Buller advanced towards the Biggarsberg.

[642] Frederick Carrington. In February of 1900 he was appointed to command of the Rhodesian Field Force. This was a force established to prevent attacks on Rhodesia from the south, and the soldiers were drawn from the various colonial contingents in the war, including the South Australian Citizen Bushmen. Carrington left for South Africa in March 1900 on the SS Dunottar Castle, and arrived in Cape Town the following month. Frederick Carrington - Wikipedia

[643] Herbert Plumer. In 1899 Plumer went to Southern Rhodesia where he raised a force of mounted infantry and led them during the Relief of Mafeking during the Second Boer War. Herbert Plumer, 1st Viscount Plumer - Wikipedia

[644] This is: Archibald Hunter - Wikipedia

[645] Warrenton is situated in the north-eastern part of the Northern Cape province, on the south bank of the Vaal River, approximately 70 kilometers (40 mi) north of Kimberley. Warrenton, South Africa - Wikipedia

[646] Vryburg is situated halfway between Kimberley and Mahikeng. Vryburg - Wikipedia

"But," says the "Volksstem," "the fact that Kroonstadt is in the hands of the enemy need create no alarm.[647] As we retire our line of defence becomes less and our commandos can be concentrated to resist more effectually the advance of the British forces. Besides, many things may happen which will put an entirely new face on the war. Our delegation has reached America, &c., &c. Lord Roberts's hastened advance is said to be caused by his desire to reach Pretoria on the Queen's Birthday, but might not the real reason be the fear of foreign intervention? Lord Roberts wishes to strike a decisive blow before his forces are needed elsewhere. Every day's delay is, therefore, an advantage to our cause. Courage is all that is needed, &c., &c."

The above is a *precis* of the "Volksstem" leading article. Still they harp on foreign intervention, but from what I gather from recent Continental criticisms on the war, I fancy their chances in this line are less than at the beginning of the war. As to the burghers' courage, I doubt if the majority of them have much left. For many months the Transvaal Government have whipped their subjects to the fight; but even the worm will turn, and to the simplest, or the most ignorant, the Government promises and hopes must seem vain.

The day of our release is, perhaps, approaching; but it does not do to be too sanguine; one never knows where a check may occur. Still I "plump" on the end of the present month.

May 20th—The month is drawing to a close, and the day of our release is still a matter of speculation. News is pretty plentiful; even the "Volksstem" tries to hide nothing. Roberts has made a great advance, but whether he has halted at Kroonstadt or not is uncertain. We all hoped he would not stop until he had reached Pretoria.[648]

We have been very much alarmed lately at the rumoured intention of the Government to move us to Lydenburg,[649] but at present it is only a rumour. If we are moved we shall have every prospect of being shunted about the country with guerilla bands of Boers who would keep us merely as hostages, if, however, we are kept here we shall have every chance of being released during the siege of Johannesburg. The Boers, it is said, have decided to hold that place and are not going to blow up the mines. The defence of Pretoria would be impossible with the troops at their disposal.

Life goes on as usual. The only diversion that has lately occurred was the athletic sports, which were got up by some energetic people. The event took place yesterday, and, on the whole, was a decided success. The chief feature, however, of the day was the betting. Several enterprising officers kept books, but Haig,[650] of the Inniskilling Dragoons, cut the best figure in that line, and it was chiefly owing to his amusing performance that the day was a success. White has made an excellent sketch of "Our Bookie" for the next "Gram" number.

The sermon this morning is worth recording. The Rev. Mr. Bateman delivered a most extraordinary speech as part of his service. Whether it was meant for our spiritual edification, or merely intended to convey news to us under the disguise of a text, was not quite certain; but, by preaching on the text that begins "as cold water is to the thirsty soul, so is good news, &c.," he led us to believe that we were to be released in a very short time. Roulette has been going very strong. Large sums have been lost and won.

May 25th—Yesterday we, prisoners of war, joined with the British Empire all over the world in the celebration of the Queen's Birthday. In our little enclosure we have quite a representative British Empire—English, Scotch, and Irish soldiers, Colonials, South Africans, Australians, and civilians, and, indeed, we only require a Canadian to complete the list.

Yesterday evening we drank the Queen's health in light port (rather nasty). The first drops of wine or spirit I had tasted since the 18th of November. This was followed by "God Save the Queen," sung by all with a heartiness and feeling that I never heard before. It must have sounded very well outside. To us it was as it were "giving

[647] Saturday, May 12, 1900 Roberts occupied Kroonstad without resistance. Commandant Eloff attacked Mafeking, and was captured by Col. Baden-Powell.

[648] Tuesday, May 15, 1900 Christiana occupied. Christiana, South Africa - Wikipedia

Thursday, May 17, 1900 General Ian Hamilton occupied Lindley. Colonel Mahon, at the head of the relief force, entered Mafeking. Methuen entered Hoopstad.

Friday, May 18, 1900 Occupation of Newcastle, in Natal by Buller's forces.

[649] Lydenburg, officially known as Mashishing, is a town in Thaba Chweu Local Municipality, on the Mpumalanga highveld, South Africa. It is situated on the Sterkspruit/Dorps River tributary of the Lepelle River at the summit of the Long Tom Pass. Lydenburg - Wikipedia

[650] Neil Wolseley Haig Lieutenant. Prisoner. Vredes Verdrag, 10 May 1900 Released Source: South African Field Force Casualty Roll 6th (Inniskilling) Dragoons MID LG: 10 September 1901, page: 5929. Source: Field Marshal Roberts. 4 September 1901. Re: General mentions Captain QSA (3). Source: QSA medal rolls.

vent" to our imprisoned feelings, while we also found in it a link with our country, from which we have for so many months been severed.

It is now pretty certain that Roberts is resting his troops, and rumours have it that the Boers have asked for an armistice. Whether Lord Roberts celebrated the Queen's Birthday by a victory or a peaceful armistice remains to be seen.

The "Volksstem" considers that it would be a graceful act on the part of the State President if he were to wire the Queen and offer her as a birthday present the unconditional release of all the British prisoners of war. As the "Volksstem" is the official organ, this may quite possibly be merely a feeler to the public (if public there be in this country). At any rate it would be an act worthy of the wily Boer. He finds it a source of trouble and expense feeding and guarding 5,000 prisoners, so he gives them away with a pound of tea—I mean as a graceful act. Whether the offer would be accepted is uncertain. But we at any rate will be very happy if the Transvaal Government puts us over the border.

The weather (by day) is simply perfect. Every morning the lovely air makes one long for a walk or ride, and causes one to chafe at the inability to roam beyond the one hundred yards' enclosure. We are henceforth to be allowed to have wine, but personally I shall wait for freedom before I indulge in that luxury again. The second number of the "Gram" came out yesterday, and, I believe, was much appreciated.

May 26th—Two prisoners of war arrived this morning. They were caught at Lindley, which the Boers have apparently reoccupied. They were taken across country to the Natal railway, and then conveyed straight to Pretoria. They say they have heard firing at the Vaal, so I suppose Lord Roberts is there. The Boers hold a strong position south of Johannesburg, and they also intend defending that town. One of the De Wets is still on the right rear of our army, but will be dealt with by Rundle's division which is coming up that way. It is said that De Wet at one time offered to surrender on condition that he himself should not be made a prisoner. But Roberts would receive none but an unconditional surrender. Buller has been ordered to force Laing's Nek at all costs. The "Volksstem" says that Lord Roberts's headquarters are at Honningspruit, some way north of Kroonstadt, but this is probably news of some days' standing. There is also a rumour that our troops have occupied Potchefstroom.[651]

May 29th—At last our release seems near at hand. Yesterday and to-day big guns were heard plainly in the direction of Johannesburg, which is now in our hands.[652] Boscher, the grocer, has just arrived, having come up by the last train. He says that the Dragoons were actually in the streets when he left. [Johannesburg] I fancy to-morrow or next day will see us out. Everybody is in the best of spirits and full of excitement.

[651] Tuesday, May 22, 1900 General Ian Hamilton occupied Heilbron after a series of engagements. The main army, under Roberts, pitched its tents at Honing Spruit, and General French crossed the Rhenoster to the north-west of the latter place.
Wednesday, May 23, 1900 Rhenoster position turned.
Thursday, May 24, 1900 British Army entered the Transvaal, crossing the Vaal near Parys, unopposed.

[652] Tuesday, May 29, 1900 Battle of Doornkop: Boers defeated. Battle of Doornkop - Wikipedia

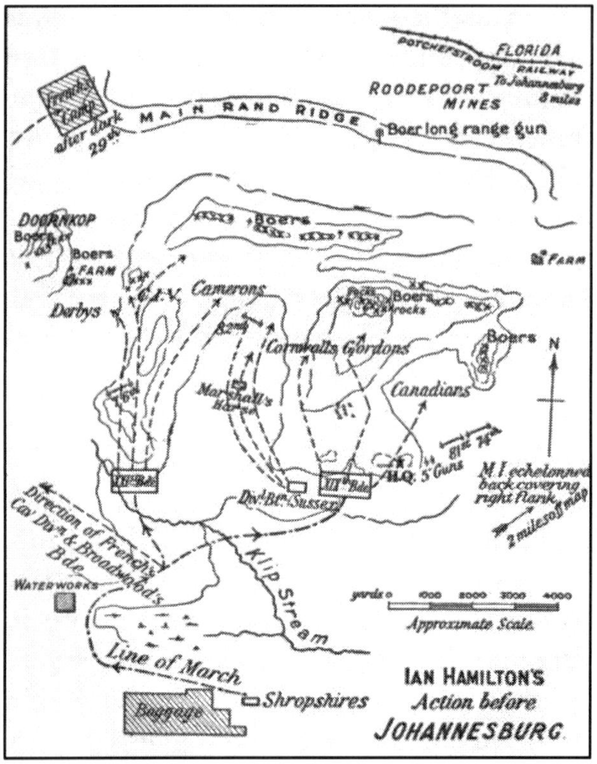

Ian Hamilton's Action Before Johannesburg.[527] Dorset Yeomanry entering Johannesburg, 1900[653] Also published as The City [London] Volunteers Entering Johannesburg.[654]

Greatest excitement during dinner. Mr. Hay and Mr. Wood came in and asked Colonel Hunt to send twenty-four officers to Waterval to look after the men.

Adelbert S. Hay U.S. Consul at Pretoria.[528 655] The Consul for The British Prisoners at Pretoria.[529 656]

[653] File:Dorset Yeomanry Boer War.png - Wikimedia Commons Credit: Axel Ravera. Public domain.

[654] Navy & Army Illustrated Vol 11 November 17th, 1900.

[655] Adelbert Stone Hay (November 1, 1876 – June 23, 1901) was an American consul, politician and son of U.S. Secretary of State John Hay. Adelbert S. Hay - Wikipedia

Richard Harding Davis,[657] Adelbert S. Hay, [658] Consul and J.G. Coolidge Vice-Consul. Referred to as Gardner F. Coolidge of Boston. American consulate, Pretoria. [530 659]

Kruger has gone to Holland.[660] The British are expected here to-morrow, and we shall be free! We sang "God Save the Queen" and cheered Hay and the Commandant, who made a very nice speech, saying he hoped to shake hands with us outside. Oh! how I longed to see the old regiment once more! The Commandant says that there is still fighting at Klip drift, but a force of 4,000 men has broken through and come here. I believe there is a lot of looting going on in the town now. Roulette is at an end. I can scarcely write coherently, so excited am I. Fancy being free; I can scarcely believe it! Six and a half months' imprisonment, and about to be freed! Thank God!

May 31st—Too premature were our hopes. Yesterday and to-day have been spent in awful suspense. Distant guns have been heard, Boers have been seen riding about, and rumours of all kinds and descriptions are rife. It is too awful this final suspense. We do nothing in hope of a speedy release, and we pass the day anxiously scanning the horizon for the approach of troops.

All day commandos have gone through the town, and one was seen on the plain coming in from Mafeking. One commando came up our way, and we were rather surprised that they made no attempt to shoot us. Indeed there was nothing to prevent them. Three prisoners came in. They were caught in or near Johannesburg. That town was officially surrendered at 10 A.M. this morning. The Boers intend making a sort of stand (one of their usual ten-minute affairs I suppose) at Irene, a place six miles south of Pretoria, and a fight is expected there tomorrow. Their line of flight is past our abode and Waterval, and I should not be surprised if, unable to face and shoot armed men, some of these foreign ruffians shoot a few prisoners.

The town is evidently to be handed over quietly. The "Volksstem" is still covering a sheet of paper with print, but seems to take not the slightest interest in the war. They speak of giving up Pretoria as one of our papers might of a concert. Well, I suppose it will come at last, but I shall heave a sigh of relief when it does!

[656] "Hon. Adelbert S. Hay, American Consul-General at Pretoria, who attended to the wants of the British prisoners and acted for British subjects during the war. He stands by the railway carriage, behind him is Mr. John G. Coolidge, Vice-Consul; they are bidding goodbye to attaché's leaving Pretoria, and Mr. Hay is talking to Captain Slocum, United States Army, American attaché.' Other travelers seen in the picture are the Japanese and Russian attaché's" Hillegas, Boers in War, 1900.

[657] Richard Harding Davis With Both Armies in South Africa. Charles Scribner's Sons, New York 1900 Richard Harding Davis (April 18, 1864 – April 11, 1916) was an American journalist and writer of fiction and drama, known foremost as the first American war correspondent to cover the Spanish–American War, the Second Boer War, and the First World War. His writing greatly assisted the political career of Theodore Roosevelt. He also played a major role in the evolution of the American magazine. Richard Harding Davis - Wikipedia

[658] Webster Davis. "[upon] my return to Pretoria from the visit to the battlefields and camps of the fighting Boers, I had the pleasure of meeting United States Consul Hay, who had just arrived. His predecessor, Consul Macrum, after having made his complaint to his government that his mail had been opened by British officials through whose hands it had passed, was relieved of his office and Mr. Hay appointed in his place." Quote from Davis' book, John Bull's Crimes. Hay was the eldest son of John Milton Hay, who served as United States Secretary of State and the United States Ambassador to Great Britain at the time. John Hay - Wikipedia

[659] During the time intervening between Mr. Macrum's dismissal and Mr. Hay's arrival, Mr. W. J. Hollis, United States Consul at Lorenzo Marquez, was stationed in the consulate at Pretoria. Webster Davis' John Bull's crime; or, Assaults on Republics 1901 p. 23.

[660] Not Until Tuesday, September 11, 1900 "Kruger, having fled from the Transvaal, arrived in Portuguese territory, and proceeded to Lorenzo Marques."

June

June 1st—No sign of the British! But we expect to hear guns tomorrow. There are plenty of rumours about—Roberts captured, French killed, &c. There was a good deal of looting in the town yesterday, and five men were shot. Our hopes of a few days ago have been somewhat damped, and most of us put our release down at a week hence.

Boer Retreat from Pretoria.[531]

The "Volksstem" is remarkable. The editor is evidently wishful to avoid his tarring and feathering, and scarcely speaks of the war at all.

June 3rd—I have almost given up looking forward to our release, and have fallen back into the ordinary monotonous life. No guns have been heard, and therefore no serious fighting can have taken place anywhere near Pretoria. Rundle has been reported as having received a check in the Free State, and Lord Roberts is said to be still in Johannesburg; otherwise there is no news at all. Botha has taken matters into his own hands, has kicked out the officials appointed by Kruger, chosen a committee of his own, and has arranged the defence of the positions outside the town. He has therefore made himself practically President of what remains of the Transvaal. Kruger went off with a million of hard gold, paying the Government officials with dishonoured cheques on the National Bank, from which he has removed all the money. Every one of his ministers thirsts for the old man's blood, and perhaps it were best for him to go further than Middelburg.

President Kruger's private car used as the capital after evacuation of Pretoria.[532] Mr. Kruger…at Lorenzo Marques[533] (October 1900).

June 4th—At about 8.30 this morning firing was heard at no great distance, in the south-west direction—field-guns, "pom-poms," Maxims, and even musketry. At about nine o'clock a shell was seen to burst on an earthwork on a ridge of hills south of the town. Field-glasses and telescopes were immediately brought out, and we were well entertained for the rest of the day. Shrapnel burst all along the ridges, and presently lyddite[661] shells were planted on the hills. The firing seemed very unmethodical, and the Boers made little or no reply. On the western kopjes shrapnel was seen bursting all over the place, and we expected the Infantry to attack them. But the lyddite shells were certainly the most interesting. They burst with a tremendous noise, throwing up clouds of brownish earth. For some time the forts seemed the mark our gunners were aiming at, and these costly erections certainly received their share—four shells pitching well inside the west fort; but, later, the shells were directed on the eastern outskirts of the town. Whether these were intended for the railway station, we could not make out; but, otherwise, they seemed to have no object.

At about 4.30 the Boers were seen leaving the western ridges and trekking at a remarkable pace across the plain, disappearing along the northern road. The day's action was ended by a kind of *feu de joie*[662] of lyddite shells, which struck the two forts and the surrounding hills. Then peace ensued. The last few shots seemed to have been fired by guns which were much closer than at the commencement of the bombardment, and the flight of the projectiles, which we could distinctly hear, passed from west to east, so that we hope our troops have occupied the hills on the west.

The hills are burning to-night, and the scene is strangely illuminated in honour of our approaching rescue.

June 5th—A day of strangely mingled hopes and fears. This morning at about 1.30 the Commandant awoke us and ordered us to pack up at once and prepare to march to the railway, whence we were to be transported by train down the Delagoa Bay line to some station beyond Middelburg. All were filled with consternation. To be hurried away when release was so near at hand seemed too awful. Words cannot express my feelings. At last we decided to refuse to go. Let them massacre us if they dared. We reminded the Commandant of the promise made to the officers the week before that if they restrained the men in Waterval neither they nor the men should be transported. The Commandant replied that he had his orders and must execute them, and he rose to leave the building, but we refused to let him or his lieutenant go, and held them both prisoners. The Commandant said that the guards would soon come in to rescue him, but he eventually promised to do his best to save us from being deported, if we set him free.

Then, by Colonel Hunt's advice, for we did not know when a commando might appear, we returned to bed—you cannot shoot men in their beds. And so passed the anxious hours away till dawn. With the first streaks of daylight we scanned the hills anxiously for the British troops. We could see lines of men moving on the race-course, but it was impossible to make out what they were. Presently, at about half-past eight, two figures in khaki came round the corner, crossed the little brook and galloped towards us. Were they Boers come to order our removal?—The advance scouts, perhaps, of a commando to enforce the order! or were they our friends at last? Yes, thank God! One of the horsemen raised his hat and cheered. There was a wild rush across the enclosure, hoarse discordant yells, and the prisoners tore like madmen to welcome the first of their deliverers.

Who should I see on reaching the gate but Churchill, who, with his cousin, the Duke of Marlborough,[663] had galloped on in front of the army to bring us the good tidings. It is impossible to describe our feelings on being freed. I can scarcely believe it, after seven months' imprisonment; the joy nearly made up for all our former troubles, and, besides, the war is not yet over.

[661] A type of high explosive. Picric acid - Wikipedia

[662] Literally "fire of joy."

[663] Charles Richard John Spencer-Churchill, 9th Duke of Marlborough, (13 November 1871 – 30 June 1934), was a British soldier and Conservative politician, and a close friend of his first cousin Winston Churchill. He was often known as "Sunny" Marlborough after his courtesy title of Earl of Sunderland. Charles Spencer-Churchill, 9th Duke of Marlborough - Wikipedia

Waterval after the liberation of the prisoners.[534]

"Boer Military Prison at Waterval.[535]

"Boer Prisoners in the Prison Prepared for the British Officers Just Outside Pretoria."[536]

To close the scene we hoisted the Union Jack which Burrows (one of the prisoners) had made by cutting up a Vierkleur,[664] on the staff whence the Transvaal colours had so long reminded us of our condition.

I will not write about the triumphal entry of Lord Roberts and the army into Pretoria, because that has been already told by so many others.

The Dogman [Patterson] and Cullingworth shared our good fortune, both being speedily released from the gaol where they had languished since their attempt to get through to the British lines, and with this happy fact let me end my record of so many weary days passed in uncertainty, disappointment, and monotony, but borne, I hope, with patience, and ending at last in joy.

[664] PF: The flag of the South African Republic was a vertical green stripe with horizontal red white and blue stripes. Flag of Transvaal - South African Republic - Wikipedia

The Model School, Pretoria, Now an Hospital for British Soldiers, Recently the Prison of Our Officers.[537]
Three Thousand British Soldiers Released from Prison at Pretoria Being Re-Armed in the Public Square.[538]

Biographical Summary

Thomas Hugh Colville Frankland[665] was born on 17 October 1879 in Cork, County Cork Ireland.[666] He was the son of Colonel Colville Frankland[667] and Mary Jay Dawson.[668] He attended Cheltenham College from September 1889 to July 1890[669] and continued his education at the Charterhouse School, Godalming from September 1893 to March 1895.[670]

He fought as a Lieutenant in the Royal Dublin Fusiliers during the Boer War between 1899 and 1902. He earned the Queen's South African medal with three bars and the King's medal as well. [671] His commanding officer, Captain Aylmer Haldane, described Frankland's actions during the ambush of the armoured train in his after action report as: "a gallant young officer who carried out my orders with coolness."

He served as adjutant to the Headquarters staff of the British Army in 1911 and was posted to Ahmidnagar, Deccan India that same year. He gained the rank of Captain and Brevet Major in the Royal Dublin Fusiliers.

He gained the rank of Flying Officer in the Royal Flying Corps.[672] He used a Vickers Biplane at the Vickers School, Brooklands to earn an aviators certificate on November, 5th, 1913. [673]

He most likely trained in a Royal Aircraft Factory F.E.2 of which 1,939 examples were constructed according to an evolving design.[674]

The first F.E.2 Pusher Biplane in 1911.[675]

F.E.2a with original undercarriage.[676]

He fought in the First World War, and was mentioned in despatches.[677]

He died on April 25th 1915 at age 35, killed in action at the Dardanelles, Gallipoli, Çanakkale, Turkey.[678]

665 Person Page #53145 citing [S37] BP2003 volume 3, page 4292. Person Page (thepeerage.com)

666 Source:8476612 Source:8476606. Life story: Thomas Hugh Colville Frankland | Lives of the First World War (iwm.org.uk) And: Brevet Maj T H C Frankland And Capt R C C Frankland | War Imperial War Museums (iwm.org.uk)

667 Colonel Colville Frankland was born on 6 November 1839. He is the son of Sir Frederick William Frankland, 8th Bt. and Katherine Margaret Scarth. He married Mary Jay Dawson, daughter of William Dawson, on 22 September 1870.2 He died on 22 December 1913 at age 74. He fought in the Indian Mutiny. He gained the rank of Colonel in the 2nd Battalion, Royal Dublin Fusiliers. Person Page #36657 @ Person Page (thepeerage.com)

668 Mary Jay Dawson was the daughter of William Dawson. 1 She married Colonel Colville Frankland, son of Sir Frederick William Frankland, 8th Bt. and Katherine Margaret Scarth, on 22 September 1870.1 She died on 25 January 1914. Person Page #53414 @ Person Page (thepeerage.com)

669 Source:8510474.

670 Source:8510475.

671 Frankland Thomas Hugh Colville Lieutenant QSA (3) bars Source: List of QSAs with the clasp Talana from the Talana book. Also: QSA and KSA 1st Battalion and 2nd Battalion medal rolls Royal Dublin Fusiliers https://www.angloboerwar.com/name-search

672 Person Page #53145 @ Person Page (thepeerage.com)

673 Record Card #679 List of pilots awarded an Aviator's Certificate by the Royal Aero Club in 1913 - Wikipedia

674 Between 1911 and 1914, the Royal Aircraft Factory used the F.E.2 ("Farman Experimental 2") designation for three quite different aircraft that shared only a common "Farman" pusher biplane layout. Royal Aircraft Factory F.E.2 - Wikipedia

675 An FE 2 pusher biplane, on the ground with pilot and passenger. File:Aviation in Britain Before the First World War RAE-O763.jpg - Wikimedia Commons

676 F.E.2a with original undercarriage. File:Aviation in Britain Before the First World War RAE-O597.jpg - Wikimedia Commons

677 Source:8476604 of Commonwealth War Graves Commission.

The following photograph was taken by the Royal Aero Club when he gained his flying license. Comparing the signature to the initials on his sketches, particularly the 'F' in Frankland it is his autograph.

Thomas Hugh Colville Frankland.[679] Thomas Hugh Colville Frankland Captain (Bvt. Maj) Royal Dublin Fusiliers Died 25 April 1915 35 years old, Remembered with Honour - Helles Memorial Panel 192 to 197. Turkey (including Gallipoli).[680]

His older brother Robert Cecil Colville Frankland[681] was born on 7 July 1877. Robert fought in the Boer War between 1899 and 1901.[682] He continued in the military and fought in the First World War, attaining the rank of Captain in the 3rd Battalion, North Staffordshire Regiment. He died on August 7th 1915 at age 38, killed in action at the Dardanelles,[683] Gallipoli, Çanakkale, Turkey.[684]

Immediate Family: The children of Mary Jay Dawson and Colonel Colville Frankland:

William Jay Colville Frankland b. 14 Apr 1876, d. Nov 1896
Robert Cecil Colville Frankland b. 7 Jul 1877, d. 7 Aug 1915
Thomas Hugh Colville Frankland b. 17 Oct 1879, d. 25 Apr 1915
Eleanor Colville Frankland d. 10 Mar 1946
Katharine Marian Colville Frankland d. 17 Sep 1950
Margaret Lee Colville Frankland d. 26 Nov 1874
Beatrice Colville Frankland d. 11 Oct 1959
Mary Olive Elsie Colville Frankland d. 26 Mar 1960

678 Captain Thomas Hugh Colville Frankland | War Casualty Details | CWGC The Commonwealth War Graves Commission | CWGC | CWGC And: Thomas Hugh Colville Frankland (unknown-1915) - Find A Grave Memorial Memorial Id 56009464

679 Photo credit: Lives of the First World War Life story: Thomas Hugh Colville Frankland | Lives of the First World War (iwm.org.uk) Photo credit to Royal Aero Club. Contributed by JOHN14663.
And Captain Thomas Hugh Colville Frankland Royal Dublin Fusiliers. Died Sunday 25 April 1915 - A Street Near You

680 The Commonwealth War Graves Commission | CWGC | CWGC

681 History Life story: Robert Cecil Colville Frankland | Lives of the First World War (iwm.org.uk) Also: Person Page #53415Person Page (thepeerage.com)

682 R C C Frankland 1st Battalion Source: QSA and KSA medal rolls South Staffordshire Regiment.

683 Imperial War Memorial #56284 Located on the south pillar of the chancel arch, St. John the Baptist Church Palmeira Square, Hove, East Sussex. Brevet Maj T H C Frankland And Capt R C C Frankland | War Imperial War Museums (iwm.org.uk)

684 Helles Memorial Panel 171 to 173. Turkey (including Gallipoli). Captain Robert Cecil Colville Frankland | War Casualty Details | CWGC

Robert Frankland

Thomas' older brother was Captain Robert Cecil Colville Frankland b. 7 Jul 1877, d. 7 Aug 1915 who also served in two of the same campaigns as his younger brother.

London Gazette: 2nd Lieutenant, 3rd & 4th Bn the South Staffordshire Regiment 14 July 1896, Lieutenant, 2 June 1897, Lieutenant, 3rd Battalion, the South Staffordshire Regt, Reverted to be 2nd Lieutenant, South Staffordshire Regiment 4 May 1898, 2nd Lieutenant South Staffordshire Regiment, 30 December 1899.

Resigned Commission South Staffordshire Regt, 25 February 1903.

Assigned to be Captain 10th (Service) Battalion, North Staffordshire Regiment, 24 October 1914,

Killed In Action: August 7th, 1915, No Known-Grave.[685]

Robert Cecil Colville Frankland Medals.[539]

1. QSA, Transvaal, Wittebergen, Cape Colony
2. KSA South Africa 1902 and South Africa 1901
3. 1914 - 1918 War Medal - Captain R. C. C. Frankland

[685] Data and image credit: QSAMIKE, Past-President Calgary Military Historical Society O.M.R.S. 1591: Anglo Boer War - Armored Train Ambush 11-15-99 Chuchill & Haldane - Page 18 - Boer War Forum

Doing for Queen & Country

What the Churchills are doing for Queen and Country. [686]

The Duke of Marlborough Captain Imperial Yeomanry (Elliott Fry) The Duke of Roxburghe (his cousin) Lieutenant in the Royal Horse Guards (Mayall) Hon. D. C. Marjoribanks (his cousin) Lieutenant in the Royal Horse Guards (Mayall) Winston Churchill (his cousin) Who escaped from Pretoria (Ellis & Walery)

No fewer than thirteen members of the historic House of Churchill are involved in the war, twelve of them (including one woman, and the head of the family) being at the front. The fact seems ironical when you remember that the first Duke, born in an impoverished Royalist's house, rose to greatness at the hands of our Dutch King, William III., and fought many great battles for the Dutch from 1689 to 1713.

But he really began his career in 1673, by helping France (under Turenne) to reduce the Dutch fortresses. And as he began. his descendants of to-day are acting, by fighting against the Dutch in South Africa.

The accompanying table puts the position of the Churchills in a nutshell, and links America, willy nilly, with us in the struggle, for the Duke's wife (nee Miss Vanderbilt) and his aunt (Lady Randolph Churchill) are New Yorkers, born and bred. It is appropriate that these women should be representing the Churchills at the front, for the first Duchess was adventurous at a time when women were un-emancipated.

Lady Sarah Wilson, her name sake, long immured in Mafeking with her husband, tried to carry out despatches, and was taken prisoner by the Boers in December, to be exchanged for a notorious horse thief. She was slightly wounded on Jan. 25, when the Boers shelled, the convent which has been turned into a convalescent hospital. Her sister, Viscountess Curzon, has been the heart and soul of the committee to equip the Yeomanry Hospital, and her sister-in-law, Lady Randolph Churchill, is at Durban looking after our wounded in the hospital ship Maine, the gift of her American country women besides which she has given her only children to the war—the plucky Winston and John, whom she is now nursing on the Maine.

Hon. Ivor Guest Imperial Yeomanry and the new M.P. for Plymouth (Lafayette) Hon. F. E. Guest 1st Life Guards (Lafayette) Hon. C. H. C. Guest 1st Royal Dragoons. The Duke of Marlborough's Cousins: Sons of Lady Wimborne.

686 Doing for Queen & Country, The Sphere - Saturday 03 March 1900 p. 20 https://www.britishnewspaperarchive.co.uk/viewer/bl/0001861/19000303/023/0020 The next page (21) and subsequent pages are missing from the print copy used for digitization in the BNA Archive, as of 12-17-24.

The Duke of Marlborough himself, whose mother was a Hamilton, and who is thus linked with the fifteen members of that family at the front, has six cousins fighting. They include the Duke of Roxburghe Lord Tweedmouth (the Liberal's) son and Lady Wimborne's three sons.

His second cousin, Mr. E. G. Spencer-Churchill (who, though belonging to an older generation, is his junior by four and a half years), is also out. It is not merely a question of being in it. The Churchills have distinctly distinguished themselves. Mr. Winston Churchill has not merely bravery he has brains too. Though he is just six and twenty he has seen a great deal of exciting service. He began his career in the 4th Hussars and entered the Egyptian army in 1895.

Then he served with the Spanish forces in Cuba, went through the Malakand expedition (attached to the 31st Punjab Infantry), served as orderly officer to Sir William Lockhart up in Tirah, and took part with the 21st Lancers in their heroic charge at the Battle of Omdurman. He has written three books, including a novel. When the war broke out in South Africa he went off to represent The Morning Post, and was captured, slightly wounded in the armoured train which was sent out from Estcourt to reconnoitre in November. He was taken to Pretoria, and escaped by scaling the wall of his prison. After many adventures he reached Durban, taking five days on the journey. He has recently taken a commission in the Imperial Light Horse. His young brother John, who entered the South African Light Horse, was shot through the leg during a cavalry skirmish at Hussar Hill, Chieveley, on Feb. 13.[687]

Captain Gordon Wilson (of the Blues) (Russell) And His Wife. Lady Sarah Wilson Now in Mafeking (Ellis Walery) The Duke's aunt. Viscountess Curzon: has raised Yeomanry Hospital (Starling) E. G. Spencer-Churchill The Duke's second cousin: Lieut., 3rd Grenadiers (Stuart)

In 1909 Sarah Wilson[688] dedicated her book, 'South African Memories,' To:

> The Memory of My Beloved Sister, Georgiana, Countess Howe, To Whose Efforts And Unceasing Labours In Connection With The Yeomanry Hospitals, During The War In South Africa, The Early Breakdown of Her Health, And Subsequent Death, Were Undoubtedly Due, This Book, Containing Recollections of That Great And Mysterious Land, The Grave of So Many Brave Englishmen, Is Affectionately Dedicated.

[687] The next page (21) and subsequent 4 pages are missing from the print copy used for digitization in the BNA Archive, as of 12-17-24.

[688] Lady Sarah Wilson, DStJ, RRC (born Lady Sarah Isabella Augusta Spencer-Churchill; 4 July 1865 – 22 October 1929) became one of the first woman war correspondents in 1899. Sarah Wilson (war correspondent) - Wikipedia
Her work: South African Memories Social, Warlike & Sporting from Diaries Written at the Time by Lady Sarah Wilson with Illustrations London Edward Arnold 1909. Cornell U. Scan: https://archive.org/details/cu31924028617367/mode/2up
PF: Georgiana is the youngest of 13 children. See: John Spencer-Churchill, 7th Duke of Marlborough - Wikipedia And George Spencer-Churchill, 8th Duke of Marlborough - Wikipedia Accordingly Winston had a total of 17 uncles and aunts and many 1st cousins.

Sarah Wilson in self-designed Mafeking outfit.[540] Sarah Wilson at doorway[541] Sarah Wilson's bomb shelter.[542]

Siege of Mafeking

Other than Winston's very good interview with Robert Baden-Powell it is very interesting that there is no other mention of events relating to the Siege of Mafeking in his book or his reporting.

The investment of the town was one of the earliest military actions of the war; commencing two days after hostilities began and starting on October 13th 1899; concluding 7 months and 4 days later on May 17th, 1900.

As in the case of the nearby siege of Kimberley, the Boers decided that the town was too heavily defended to take. On 19 November 1899, 4,000 Boers out of the ~9,000 who initially surrounded the town were redeployed elsewhere, although the siege remained and shelling continued.

The siege was finally lifted when a flying column of some 2,000 British soldiers, including many South African volunteers from Kimberley, commanded by Colonel B. T. Mahon with Prince Alexander of Teck as his Aide-de-camp, relieved the town after fighting their way in.[543]

The Relief of Mafeking: The procession through the market square.[544]

The relief of Mafeking was considered such a signal event in the prosecution of the war that its liberation became a celebratory occasion. In fact the celebrations were so exuberant that they engendered a their own word, 'mafficking' as well as the verb to maffick[545] meaning to celebrate both extravagantly and publicly.

Mafeking Day in Natal. The mass meeting in the Albert Park, Durban, "Hurrah for Baden-Powell" – God Save the Queen."[546] The principal streets adorned for Mafeking Day.[547]

With the children's procession; about four thousand children marched through the streets.[548] Baden-Powell being honored in Durban celebration of Maefking's relief.[549 689]

For Baden-Powell, and in the British media, the siege was thought of as a victory, but at a high cost. The townspeople and garrison suffered 212 killed and over 600 wounded. Baden-Powell is sometimes criticized for not remaining as a mobile force and becoming besieged; as well as for not breaking out when the Boer forces were reduced, but with the supplies, civilians, and infrastructure that would be either exposed to danger or exploited by his adversaries it is difficult to see what other course he had other than to hold out until help arrived.

Mafeking Defenders.[550]

Mafeking Artillery.[551]

Lord Nelson Cannon Dated 1770.[552]

Casting Shells for Lord Nelson.[553]

This was an old gun presented to Montsloa, father of Wessels, the chief. It had lain buried for twenty years, but at the commencement of the siege was unearthed and handed over to the military by Wessels and by them used throughout their later operations. [In defense of Mafeking] It was a smooth bore muzzle loader, firing round balls.

[689] Robert Baden-Powell, 1st Baron Baden-Powell - Wikipedia

Colonel Baden-Powell and his military and civil staffs.[554]

Top Row: Left to Right: Major Panzera (artillery, Captain Ryan (Commissariat) Captain Greener, Lord Edward Cecil[690] (Chief Staff Officer), Captain Wilson, A.D.C., Lieut A. Hanbury Tracy (press censor, Military Intelligence Officer). Captian Cowan (Bechuanaland Rifles).

Second Row: Major Godley (Commanding Western Outposts), Colonel Vyvyan, Mr. C. Bell (Civil Commissioner), Col Baden-Powell, Mr. Whiteley (Mayor), Colonel Hore (in command of the Western Fort), Dr. Hayes (Principal Medical Officer. In front, Lieut. Moncrieff.

One Pound Siege Note.[555] The figures, supposed to represent a Boer and a Briton, are rough reproductions of drawings by Colonel Baden-Powell.[556]

See: Solomon Tshekisho Plaatje's Mafeking diary referenced below.

Solomon Tshekisho Plaatje.—'Mafeking diary: A Black man's view of a white man's war.'[557 691 692]

[690] Lord Edward Cecil - Wikipedia

The R.C. Convent [in Mafeking] as hospital showing shell holes.[558] Graves at Mafeking.[559]

Substantial efforts were made to feed all residents of the town. Even the historian Thomas Pakenham came to agree with this assessment.[693]

Boer Fort at Mafeking.[560] Horses being conveyed into town to be made into soup.[561]

"Natives waiting for their rations of horse-meat soup."[562]

[691] Solomon Tshekisho Plaatje (9 October 1876 – 19 June 1932) was a South African intellectual, journalist, linguist, politician, translator and writer. Plaatje was a founding member and first General Secretary of the South African Native National Congress (SANNC), which became the African National Congress (ANC). Sol Plaatje - Wikipedia

[692] Mafeking diary: A Black man's view of a white man's war. https://archive.org/details/mafekingdiarybla0000plaa/page/n5/mode/2up And: https://archive.org/details/boerwardiaryofso0000plaa And: The Boer War Diary of Sol T. Plaatje: An African at Mafeking Solomon Tshekisho Plaatje, Mafeking diary: A Black man's view of a white man's war. Meridor Books & Ohio University Press, Athens. 1973. https://archive.org/details/mafekingdiarybla0000plaa/page/n5/mode/2up AKA: The Boer War diary of Sol T. Plaatje; an African at Mafeking. Ed. John L. Comaroff. MacMillan 1973 https://archive.org/details/boerwardiaryofso0000plaa

[693] Pakenham, Thomas (2001). The Siege of Mafeking. And Jeal, Tim (1989). Baden-Powell. Hutchinson. ISBN 0-09-170670-X.

Trial

Trial.[563] "The photograph represents the Court of Summary Jurisdiction in Mafeking at which a native was tried before Lord Edward Cecil (with black band on arm, and whose face is a study of utter absorption in the question of the moment) and Mr. G. H. Bell (resident magistrate) for stealing a goat. The reason why extreme measures had to be taken in such cases has already been mentioned (p. 615). In this case the thief was sentenced to death, and the warrant for his execution, duly certified as carried out, appears on the next page. The man standing on the right of the photograph is Heald, the jailor, who was killed in the fight with Eloff. On the left, in front of the bare -headed prisoner, stands the attorney, J. W. de Kock; and Melemo, the Baralong Chief, in the centre of the picture, is watching the case with anxious interest."

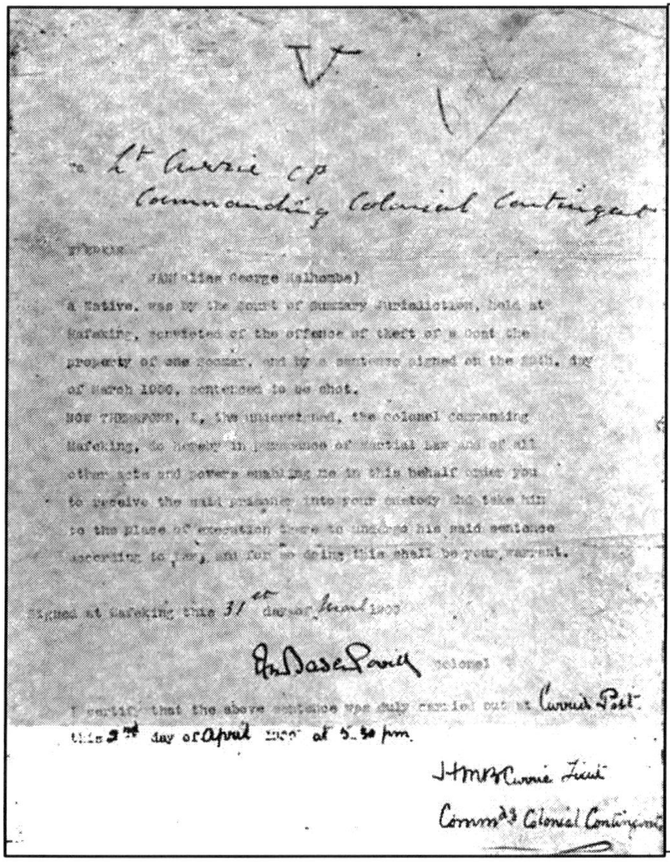

To: Lieutenant Currie (Sp?) Commanding Colonial Contingent.

Whereas: Janfulina Geroge Malhombe)

A Native, was by the Court of Summary Jurisdiction, held at Mafeking, convicted of the offence of theft of a goat the property of one Bookan (Sp?), and by a sentence signed on the 20th, day of March 1900, sentenced to be shot.

NOW THEREFORE, I, the undersigned, the Colonel Commanding Mafeking, do hereby in pursuance of Martial Law and of all other acts and powers enabling me in this behalf order you to receive the said prisoner into your custody and take him to the place of execution there to undergo his said sentence according to Law and for so this shall be your warrant.

Signed at Mafeking this the 31st day of March 1900

(Robert (Sp?) Baden-Powell) Colonel.

I certify that the above sentence was duly carried out at Carried (sp?) post this the 2nd day of April, 1900 at 5:30 PM. JTMB Currie, Lieut.

Death Warrant.[564] Dated March 31st, 1900. The siege will not lift for another month and a half, on May 17th, 1900. Signed by Baden Powell.

Oaths of Neutrality & Allegiance

Proclamation XLI.

Oath of Neutrality.[694]

I, the undersigned, in the District of ________________________

Do hereby solemnly make Oath and declare that I have handed in and given up all the Arms and Ammunition demanded of me by the British Authorities, namely, all Rifles and Rifle Ammunition of whatsoever description they may be.

And I solemnly swear that I have no Rifle or Rifle Ammunition remaining, and that I know of none such being concealed or withheld by anybody whatsoever.

And I further swear that I will not take up Arms against the British Government during the present War, nor will I at any time furnish any member of the Republican Forces with assistance of any kind, or with information as to the numbers, movements, or other details of the British Forces that may come to my knowledge.

I do further promise and swear to remain quietly at my home until the war is over.

I am aware that if have in any way falsely declared in the Premises, or if break my Oath or Promise as above set forth I, shall render myself liable to be summarily and severely punished by the British Authorities.

I make the above declaration solemnly believing it to be true,

So Help Me God.

____

Before me ___________________________

Oath of Neutrality Informal

In the absence of any official forms, hand-written "oaths" were penned and signed by surrendering Boer combatants. This example was, signed at Bloemfontein on 29 March 1900, two weeks after the capital fell to the British forces.

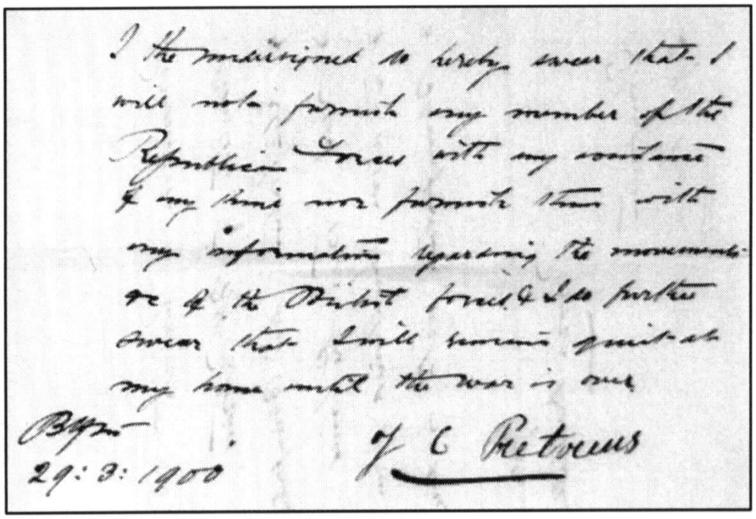

I the undersigned do hereby swear that I
will not furnish any member of the
Republican Forces with my assistance
of my kind nor furnish them with
any information regarding the movements
etc of the British forces & I do further
swear that I will remain quiet at
my home until the war is over

Bfn
29: 3: 1900

J C Pretorius

Handwritten Oath of Neutrality.[695]

It reads, "I the undersigned do hereby swear that I will not furnish any member of the Republican Forces with my assistance of any kind nor furnish them with any information regarding the movement etc. of the British Forces and I do further swear that I will remain quiet at my home until the war is over." J. C. Pretorius.

[694] Robert, Frederick Sleigh Roberts, 1st earl. Proclamations Issued in South Africa. London, 1900. Vol I & II 1901 820 pages.
https://babel.hathitrust.org/cgi/pt?id=hvd.32044089026009&seq=15
Neutrality Oath November 18th 1900. : https://babel.hathitrust.org/cgi/pt?id=hvd.32044089026009&seq=377&view=1up

[695] Picture: https://www.angloboerwar.com/forum/11-research/31054-a-hand-written-oath-of-neutrality#gallery-2
Topic: https://www.angloboerwar.com/forum/11-research/31054-a-hand-written-oath-of-neutrality Credit: Rory
Location: de Wet\Three Years' War.

Revocation of Oath

Form of Oath:

I, ________________, the undersigned burgher of the South African Republic, declare under oath that the oath of neutrality taken by me, in the hands of the enemy, was taken by me without the sanction of my military officers, and I consider the same null and void."[696]

Oath of Allegiance to Boer Forces

Before a foreign volunteer was allowed to join a commando, and before he received his equipment, he was obliged to take an oath of allegiance to the Republic. Only a few men who declined to take the oath were allowed to join the army. The oath of allegiance was an adaptation of the one which caused so much difficulty between Great Britain and the Transvaal before the war. A translation of it reads—

"I hereby make an oath of solemn allegiance to the people of the South African Republic, and I declare my willingness to assist, with all my power, the burghers of this Republic in the war in which they are engaged. I further promise to obey the orders of those placed in authority according to law, and that I will work for nothing but the prosperity, the welfare, and the independence of the land and people of this Republic, so truly help me, God Almighty." [697]

[696] Michael Davitt The Boer Fight For Freedom, 3rd ed., Funk & Wagnalls Company 1902. P.580–581.
[697] Howard Hillegas, Boers in War, 1900. Chapter 9.

Roberts' Proclamations

Proclamation I

To the Burghers of the Orange Free State.[698] [699] February, 1900

The British troops under my command having entered the Orange Free State, I feel it my duty to make known to all Burghers the cause of our coming, as well as to do all in my power to put an end to the devastation caused by this war, so that, should they continue the war, the inhabitants of the Orange Free State may not do so ignorantly, but with full knowledge of their responsibility before God for the lives lost in the campaign. Before the war began the British Government, which had always desired and cultivated peace and friendship with the people of the Orange Free State, gave a solemn assurance to President Steyn that, if the Orange Free State. remained neutral, its territory would not be invaded, and its independence would be at all times fully respected by Her Majesty's Government.

In spite of that declaration, the Government of the Orange Free State was guilty of a wanton and unjustifiable invasion of British territory.

The British Government believes that this act of aggression was not committed with the general approval and free will of a people with whom it has lived in complete amity for so many years It believes that the responsibility rests wholly with the Government of the Orange Free State, acting, not in the interests of the country, but under mischievous influences from without.

The British Government, therefore, wishes the people of the Orange Free State to understand that it bears them no ill will, and, as far as is compatible with the successful conduct of the war, and the reestablishment of peace in South Africa, it is anxious to preserve them from the evils brought upon them by the wrongful action of their Government.

I therefore warn all Burghers to desist from any further hostility towards Her Majesty's Government and the troops under my command, and I undertake that any of them, who may so desist and who are found staying in their homes and quietly pursuing their ordinary occupations, will not be made to suffer in their persons or property on account of their having taken up arms in obedience to the order of their Government. Those, however, who oppose the forces under my command, or furnish the enemy with supplies or information, will be dealt with according to the customs of war.

Requisitions for food, forage, fuel, or shelter, made on the authority of the Officers in command of Her Majesty's troops, must be at once complied with; but everything will be paid for on the spot, prices being regulated by the local market rates. If the inhabitants of any district refuse to comply with the demands made on them, the supplies will be taken by force, a full receipt being given.

Should any inhabitant of the country consider that he or any member of his household has been unjustly treated by any Officer, soldier, or civilian attached to the British Army, he should submit his complaint, either personally or in writing, to my Headquarters or to the Headquarters of the nearest General Officer. Should the complaint on enquiry, be substantiated, redress will be given.

Orders have been issued by me, prohibiting soldiers from entering private houses, or molesting the civil population on any pretext whatever, and every precaution has been taken against injury to property on the part of any person belonging to or connected with, the Army.

Roberts, Field-Marshal, Commanding in Chief, South Africa. February 1900

[698] All citations from: Robert, Frederick Sleigh Roberts, 1st earl. Proclamations Issued in South Africa. London, 1900. Vol I & II 1901 820 pages. https://babel.hathitrust.org/cgi/pt?id=hvd.32044089026009&seq=15

[699] Proclamation 19 on page 356 https://babel.hathitrust.org/cgi/pt?id=hvd.32044089026009&seq=356&view=1up

Proclamation IXX

Army Head-quarters, South Africa, Pretoria, 16th June 1900.[700] [701]

Whereas small parties of raiders have recently been doing wanton damage to public property in the Orange River Colony and South African Republic by destroying railway bridges and culverts and cutting the telegraph wires, and whereas such damage cannot be done without the knowledge and connivance of the neighbouring inhabitants and the principal civil residents in the districts concerned;

Now, therefore, I, Frederick Sleigh, Baron Roberts, of Kandahar and Waterford, K.P., G.C.B., G.C.S.I., G.C.I.E., V.C., Field-Marshal, Commander-in-Chief of Her Majesty's troops in South Africa, warn the said inhabitants and principal civil residents that, whenever public property is destroyed or injured in the manner specified above, they will he held responsible for aiding and abetting the offenders. The houses in the vicinity of the place where the damage is done will be burnt and the principal civil residents will be made prisoners of war.

Roberts, Field-Marshall, Commander-in-Chief, South Africa.

Proclamation XX

Army Head-quarters, South Africa, Pretoria, 19th June 1900.[702]

Referring to my Proclamation, dated Pretoria, 16th June 1900, I, Frederick Sleigh, Baron Roberts, of Kandahar and Waterford, K.P., G.C.B., G.C.S.I., G.C.I.E., V.C., Field-Marshal, Commander-in-Chief of Her Majesty's troops in South Africa, do hereby declare, proclaim, and make known, that, should any damage be done to any of the lines of railway or to any of the railway bridges, culverts, or buildings, or to any telegraph lines or other railway or public property in the Orange River Colony, or in that portion of the South African Republic for the time being within the sphere of my military operations, the following punishment will be inflicted;

1. The principal residents of the town and districts will be held, jointly and severally, responsible for the amount of damage done in their district.

2. In addition to the payment of the damage above mentioned, a penalty depending upon the circumstances of each case, but which will in no event be less than a sum of 2s. 6d. per morgen[703] on the area of each farm, will be levied and recovered from each burgher of the district in which the damage is done in respect of the land owned or occupied by him in such district. Furthermore, all receipts for goods requisitioned in such district on behalf of the military authorities will be cancelled, and no payment whatsoever will be made in respect of the same.

3. As a further precautionary measure, the Director of Military Railways has been authorized to order that one or more of the residents, who will be selected by him from each district, shall from time to time personally accompany the trains while travelling through their district.[704]

4. The houses and farms in the vicinity of the place where the damage is done will be destroyed, and the residents in the neighbourhood dealt with under Martial Law.

5. The military authorities will render every facility to the principal residents to enable them to communicate the purport of this Proclamation to the other residents in their district, so that all persons may become fully cognizant of the responsibility resting upon them.

Roberts, Field-Marshal, Commanding in Chief in South Africa.

[700] All citations from: Robert, Frederick Sleigh Roberts, 1st earl. Proclamations Issued in South Africa. London, 1900. Vol I & II 1901 820 pages. https://babel.hathitrust.org/cgi/pt?id=hvd.32044089026009&seq=15

[701] Proclamation 19 on page 364. https://babel.hathitrust.org/cgi/pt?id=hvd.32044089026009&seq=364&view=1up

[702] Proclamation 20 on Page 365. https://babel.hathitrust.org/cgi/pt?id=hvd.32044089026009&seq=365&view=1up

[703] The South African Law Society published a conversion factor of 1 morgen = 0.856532 hectares to be used for the conversion. Morgen - Wikipedia

[704] Just over a month later, on July 27th, this portion of the directive was repealed. See Proclamation XXIII. https://babel.hathitrust.org/cgi/pt?id=hvd.32044089026009&seq=366&view=1up&q1=accompany

Proclamation XXVI

Proclamation[705] [706] 14th day of August, 1900.

To the Inhabitants of the South African Republic

Whereas by Proclamation No. 1 of 1900, burghers who had not taken a prominent part in the hostilities were allowed, upon taking an oath, to return to their homes, and were not dealt with as prisoners of war, and

Whereas by Proclamation No. 2 of 1900, burghers to whom passes and permits had been granted might retain their stock, or take them to the winter veld, and

Whereas many burghers have taken the said oath, but have, notwithstanding this oath, taken up arms against the forces of Her Majesty the Queen, and

Whereas many burghers who have taken the said oath have aided and abetted the enemy in raiding trains and destroying property belonging to the Forces of Her Majesty the Queen, or have acted as spies for the enemy, and

Whereas the Government of the South African Republic considers such oath immoral, and has issued a notice warning all burghers against taking the said oath, and

Whereas it is manifest that the leniency which has been extended to the burghers of the South African Republic is not appreciated by them, but, on the contrary, is being used as a cloak to continue the resistance against the Forces of Her Majesty the Queen, and Whereas there are no means of distinguishing the combatant from the non-combatant portion of the population:

Now, therefore, I, Frederick Sleigh, Baron Roberts, of Kandahar and Waterford, K.P., G.C.B., G.C.S.I., G.C.I.E., V.C., Field-Marshal, Commander-in-Chief of Her Majesty's troops in South Africa, do hereby proclaim and make known as follows:

1. That from and after this date Sections 1 and 2 of Proclamation No. 1 of 1900 are repealed, except in respect of those burghers who have already taken the said oath.

2. That Proclamation No. 2 of 1900 is repealed.

3. That all such persons who have taken the said oath, and who have in any way broken such oath, will be punished with either death, imprisonment, or fine.

4. That all burghers in the districts occupied by Her Majesty's Forces, except such as have already taken the said oath, shall be regarded as prisoners of war, and shall be transported or otherwise dealt with as I may determine.

5. That all buildings and structures on farms on which the scouts or other forces of the enemy are harboured will be liable to be razed to the ground.

6. That the fine mentioned in Proclamation No. 6 of 1900, Section 2, shall be rigorously exacted where any damage is done to the railway, and persons are hereby warned to acquaint Her Majesty's Forces with the presence of the enemy upon their farms, and, failing to do so they will be regarded as aiding and abetting the enemy.

God Save the Queen.

Given under my hand and seal, at Pretoria, this 14th day of August, 1900.

Roberts, Field-Marshal, Commander in Chief in South Africa

With reference to paragraph 6, the fine therein mentioned shall in no event be less than a sum of .2s .6d per morgen on the area of each farm.

[705] Proclamation 26 on page 368: https://babel.hathitrust.org/cgi/pt?id=hvd.32044089026009&seq=368&view=1up&q1=Proclamation

[706] See also Proclamation 28 on page 369 regarding the status of forces in the field in the Orange River Colony: https://babel.hathitrust.org/cgi/pt?id=hvd.32044089026009&seq=369&view=1up&q1=Proclamation

Proclamation XXVIIII

Proclamation.

To the Inhabitants of the Orange River Colony.[707] [708]

Whereas by Proclamation a warning was issued on the 1st June 1900, that persons in arms against Her Majesty within the Orange River Colony would be liable to be dealt with as rebels. And whereas it was directed that all arms and ammunition should be surrendered. Now, therefore, I, Frederick Sleigh, Baron Roberts, of Kandahar and Waterford, K.P., G.C.B., G.C.S.I., G.C.I.E., V.C., Field- Marshal, Commander in-Chief of Her Majesty's Forces in South Africa, do hereby proclaim and make known as follows:-

1. That all persons residing in the Orange River Colony, heretofore burghers of the Orange Free State, have, in consequence of the annexation of the territories of that State to the dominions of Her Majesty the Queen, become the subjects of Her Majesty, except those burghers who were attached to some commando prior to the annexation, and who have continuously since then been in arms against Her Majesty.

2. If any burghers of the Orange Free State be found in arms against Her Majesty, the burden of proving that they were attached to some commando prior to the annexation, and that they have continuously since then been in arms against Her Majesty, will rest upon such burghers.

3. Those burghers of the Orange Free State who were attached to some commando prior to the annexation, and who have been continuously in arms against Her Majesty, shall be regarded as prisoners of war, and be transported or otherwise dealt with as I may determine.

4. Those burghers of the Orange Free State who have taken the so-called Oath of Neutrality or submission, and who have in any way broken such oath, shall be punished with either death, imprisonment, or fine.

5. All buildings and structures on farms on which the scouts or other forces of the enemy are harboured will be liable to be razed to the ground.

6. Where any damage is done to the railway, a fine of two shillings and sixpence sterling (2s. 6d.) per morgen will be levied on the area of the farm or farms on which the damage was done, and the farms in the immediate neighbourhood of the damaged place may also, if I deem fit be subjected to a similar fine.

7. All persons are hereby warned to acquaint Her Majesty's Forces of the presence of the enemy upon, or in the neighbourhood of their farms, and if they fail so to do they will be regarded as aiding and abetting the enemy, and will be liable to be treated as rebels. God Save the Queen.

Given under my hand and seal, at the Head- quarters of the Army in South Africa at Belfast, this first day of September, nineteen hundred.

Roberts, Field- Marshal, Commander in Chief in South Africa

September 1st, 1900

[707] All citations from: Robert, Frederick Sleigh Roberts, 1st earl. Proclamations Issued In South Africa. London, 1900. Vol I & II 1901 820 pages. https://babel.hathitrust.org/cgi/pt?id=hvd.32044089026009&seq=15

[708] Proclamation 28 on page 369 https://babel.hathitrust.org/cgi/pt?id=hvd.32044089026009&seq=369&view=1up

Roberts to Botha

On September 2 Lord Roberts communicated his intentions to General Botha: [709]

Sir,—I have the honour to address Your Honour regarding the operations of those comparatively small bands of armed Boers who conceal themselves on farms in the neighbourhood of our lines of communication and thence endeavour to damage the railway, thus endangering the lives of passengers travelling by train who may or may not be combatants.

2. My reason for again referring to this subject is that, except in the districts occupied by the Army under the personal command of Your Honour, there is now no formed body of Boer troops in the Transvaal or Orange River Colony, and that the war is degenerating into operations carried on by irregular and irresponsible guerillas. This would be so ruinous to the country and so deplorable from every point of view, that I feel bound to do everything in my power to prevent it.

3. The orders I have at present issued, to give effect to these views, are that the farm nearest the scene of any attempt to injure the line or wreck a train is to be burnt, and that all farms within a radius of 10 miles are to be completely cleared of all their stock, supplies, &c. Granting that the penalty is legal at all, it must be allowed that it is put in a minimum form, since only one farm in each case is to be destroyed; and the further clearing of stock is undoubtedly justified, since it would tend to cripple the mobility of Boer raiders approaching the line. Yet one farm for each attack becomes a formidable total when the attacks are on an average of one per day.

We have treated two causes for which farms were burned: (1) For being used as cover for snipers; (2) as a punishment for the cutting of railways. A third cause now comes to the front. A large number of burghers had taken the oath of neutrality and had been allowed to return to their farms by the British. These men were persuaded or terrorised by the fighting commandos into breaking their parole and abandoning those farms on which they had sworn to remain. The farmhouses were their bail, and Lord Roberts decreed that it was forfeited.

On August 23 he announced his decision to General Botha:

Your Honour represents that well-disposed families living on their farms have been driven from their houses, and that their property has been taken away or destroyed. This no doubt is true, but not in the sense which your letter would imply. Burghers who are well-disposed towards the British Government, and anxious to submit to my authority, have had their property seized by the Boer commandos, and have been threatened with death if they refused to take up arms against the British forces. Your Honour's contention that a solemn oath of neutrality which the burghers have voluntarily taken in order to remain in unmolested occupation of their farms is null and void, because you have not consented to it, is hardly open to discussion. I shall punish those who violate their oath and confiscate their property, no burgher having been forced to take the oath against his will.

It is quite certain that the Boer Government committed a very clear breach of the Conventions of the Hague in compelling, or even in permitting, these men to rejoin the ranks. In such, cases, says Article X., 'their own government shall not require of, nor accept from, them any service incompatible with the parole given.' This is clear as regards the government.

But in the case of the men it is different. Their promise was in a sense conditional upon effective protection from our troops. We had no right to place a man in so terrible a position that he had to choose between breaking his parole and death at the hands of his own countrymen. If we were not sure that we could protect them, we could have retained them in guarded camps, as we eventually did.

If we chose to turn them loose upon the wide veldt, then it was our fault more than theirs that they were forced into the ranks of the enemy. To their credit be it said that even under such pressure many of them were true to their oath.

But if their guilt is indeed no greater than our own, then how are we justified in burning down their houses? It seems to me that these cases are very different from those in the two other categories, and that the question of compensation to these men should be at least considered. I take it that the numerous cases where 'on commando' is marked against a burned farm on the official list, means that he had returned to commando after giving his parole.

[709] Doyle, Arthur Conan, 1859-1930 The war in South Africa, its cause & conduct; 1902 McClure, Phillips & Co. U. Cal. Copy: https://archive.org/details/warinsouthafrica00doylrich

The destruction of his house under those circumstances is, in the peculiar conditions of the case, a harsh measure, but if 'on commando' means simply that the man was away doing his duty to his country, without any question of parole, then our conscience can never permit that man to go without compensation.

We can trace in this account of the communications between the leaders the growth of those harsher measures which have been so generally deplored in this country. So long as the war was regular it is certain that nothing could be more regular than the British conduct. When, however, the war became irregular upon the part of the Boers, and their army dissolved into small bands which harried the lines of communications, the small posts, and the convoys, there was a corresponding change upon the part of the troops.

Towards the end of the year 1900 that change was pushed to considerable lengths. Certain districts which had been Boer centres, where they habitually collected time after time, were devastated and destroyed. Such districts were those of Kroonstad, Heilbron, Ventersburg, and Winburg. In these four districts about one hundred and seventy houses were destroyed.

The village of Bothaville, which was a depot of the enemy, was also destroyed. It consisted of forty-three houses.

In the Transvaal the number of houses actually destroyed for Strategic purposes seems to have been very much smaller. In the official returns only about twelve houses are so mentioned. Altogether the houses which have been burned for reasons which are open to dispute, including those of the men upon commando, do not appear to exceed two hundred and fifty.

It must be confessed that the case of these houses is entirely different from the others which have been destroyed, because they were used for active warlike operations. Of the 630 buildings which we know to have been destroyed, more than half have been used by snipers, or in some other direct fashion have brought themselves within the laws of warfare. But it cannot be said that these others have done so.

The cost of the average farmhouse is a mere trifle. A hundred pounds would build a small one, and 300*l.* a large. If we take the intermediate figure, then the expenditure of 50,000*l.* would compensate for those cases where military policy and international law may have been at variance with each other.

The burning of houses ceased in the year 1900 and, save in very special instances, where there was an overwhelming military necessity, it has not been resorted to since.

In the sweeping of the country carried out by French in the Eastern Transvaal and by Blood to the north of the Delagoa Railway, no buildings appear to have been destroyed, although it was a military necessity to clear the farms of every sort of supply in order to hamper the movements of the commandos.

The destruction of the crops and herds of the Boers, distasteful as such work must be, is exactly analogous to the destruction by them of our supply trains on which the army depended for their food. Guerilla warfare cannot enjoy all its own advantages and feel none of its own defects. It is a two-edged weapon, and the responsibility for the consequences rests upon the combatant who first employs it.

Botha's Reply to Roberts' Letter

On September 2, [1900]Lord Roberts addressed a communication to General Botha, in which he denied the existence of organized resistance except by the army commanded by Botha, declared his intention to destroy farms and confiscate stock in districts where any act of aggression was made, and threatened to transport from Pretoria and Johannesburg all women, including Mrs. Kruger, and Mrs. Botha, to the Republican lines, because as he alleged, the women were transmitting information to their people. Four days later. General Botha sent the following reply, which is reprinted *verbatim*, and the dignity of which speaks for itself..[710]

Botha's Letter

Commandant-General's Camp,

September 6, 1900.

To Lord Roberts,

Field Marshal, Commander-in-Chief, of the British troops in South Africa:

Your Excellency—In reply to your Excellency's letter of the 2nd inst., I have the honor to state the following:

Inasmuch as our entire armed force is only a small one in comparison with that of your Excellency, it cannot, of course be expected that strong commandos should be in the field everywhere, and it naturally follows that now, as during the war, what is incumbent upon us must be done by small forces. Moreover we have been compelled to still further scatter our commandos in order to be able to check the looting patrols under your Excellency's chief command, who scour the country to carry off cattle and provisions from the different farms.

2. As regards your contention that, with exception of the burgher forces under my command, no other Boer forces should be in existence, I most strongly deny this, since our armed forces are still disposed and directed in the same manner as in the beginning of the war, and in accordance with the country's laws.

3. In paragraph 3 of your letter, with which I am now dealing, it is already known to me that barbarous actions of this kind are committed by your troops under your command, not only alongside or near the railway, but also in places far removed from railways. Wherever your troops move, not only are houses burned down or blown up with dynamite, but defenseless women and children are ejected, robbed of all food and cover, and all this without any just cause existing for such proceedings.

4. With regard to paragraph 4 of your Excellency's letter, I extremely regret to learn that my burghers' and my own determination to persevere in the struggle for our independence, is to be visited on our wives and children, and this is the first instance of this kind known to me, in the history of civilized warfare. I can only protest against your proposed measures as being in opposition to all principles of civilized warfare, and excessively cruel toward the women and children, cruel especially toward elderly women, and above all toward the wife of his Honor, the President of the state, who as you must be well aware, is not able to travel without risk to her life, so that it would be simply murder to compel her to take such a journey. The pretext alleged by you, viz.: that by so doing your Excellency desires to protect yourself, against transmission of information to us, lacks all substance, since such proceedings were not considered necessary at a time when our troops were encamped in the immediate neighborhood of Pretoria. It is needless to state that we have never, by means of women and children, received information regarding operations of war.

5. If your Excellency still intends to persevere in carrying out your Excellency's plans, which I hope will not be the case, I request your Excellency to give me timely notice of the period and particulars of the expulsion, as I wish to arrange for the direct transport of the families to Europe. With regard to your Excellency's remark about proper accommodation, I am prepared to send proper carriages to a place to be indicated, and also, if required, a Cog-wheel engine, for the track between Waterval Boven and Waterval Onder, provided that your Excellency guarantees the safe return of such carriages and engine.

6. In conclusion, I desire to give you the assurance that nothing you may do to our women and children, will deter us in continuing the struggle for our independence.

I have the honor to be, etc. Louis Botha, *Commandant-General.*

[710] Alan R. I. Hiley and John Arthur. Hassell The Mobile Boer Being the Record of the Observations of Two Burgher Officers. By Alan R. I. Hiley and John Arthur. Hassell, Captains of Scouts, In the Boer Army. Grafton Press, New York. (c 1902)
Cite as: Hiley and Hassell, The Mobile Boer, Grafton Press, New York, (c. 1902) Cornell Copy: pages 224-226
https://archive.org/details/cu31924028616682

Proclamation XXXI

September 14, 1900.

Resignation of Mr. S. J. P. Kruger.[711]

The late President, Mr. Kruger, and Mr. Reitz, with the archives of the South African Republic, have crossed the Portuguese frontier, and arrived at Lourenco Marques with a view to sailing for Europe at an early date.

Mr. Kruger has formally resigned the position he held as President of the· South African Republic, thus severing his official connection with the Transvaal.

Mr. Kruger's action shows how hopeless in his opinion is the war which has now been carried on for nearly a year, and his desertion of the Boer cause should make it clear to his fellow burghers that it is useless for them to continue the struggle any longer.

It is probably unknown to the inhabitants of the Transvaal and Orange River Colony that nearly 15,000 of their fellow-subjects are now prisoners of war, not one of whom will be released until those now in arms against us surrender unconditionally.

The burghers: must by this time be cognisant of the fact that no intervention on their behalf can come from any of the Great Powers, and, further, that the British Empire is determined to complete the work which has already cost her so many valuable lives, and to carry to its conclusion the war declared against her by the late Governments of the South African Republic and Orange Free State, a war to which there can be but one ending.

If any further doubts remain in the minds of the burghers as to Her Britannic Majesty's intentions, they should be dispelled by the permanent manner in which the country is gradually being occupied by Her Majesty's Forces, and by the issue of the Proclamations signed by me on the 24th May and 1st September 1900, annexing the Orange Free State and the South African Republic respectively, in the name of Her Majesty.

I take this opportunity of pointing out that, except in the small area occupied-by the Boer army under the personal command of Commandant General Botha, the war is degenerating, and has degenerated, into operations carried on in an irregular and irresponsible manner by small, and in very many cases by insignificant bodies of men. I should be failing in my duty to Her Majesty's Government and to Her Majesty's Army in South Africa if I neglected to use every means in my power to bring such irregular warfare to an early conclusion.

The means which I am compelled to adopt are those which the customs of war prescribe as being applicable to such cases. They are ruinous to the country, and entail endless suffering on the burghers and their families, and the longer this guerilla warfare continues the more vigorously must they be enforced.

Roberts, Field-Marshal, Commanding in Chief in South Africa. Pretoria, September 14, 1900.

[711] Proclamation 31 on Page 371 https://babel.hathitrust.org/cgi/pt?id=hvd.32044089026009&seq=371&view=1up&q1=Proclamation

Proclamation XXXV

Government Notice No. 115 OF 1900.[712] [28th September 1900]

Uniformity of Treatment.

As it seems to me advisable to insure uniformity in the treatment of the inhabitants of the various districts of the Transvaal and Orange River Colony, the following instructions are hereby published to assist General Officers Commanding in carrying out the conditions of recent Proclamations issued by me:

1. By Military Secretary's telegram, No. C, 4825, dated 20th September, General Officers Commanding have been empowered to promise that Burghers who surrender voluntarily will not be sent out of South Africa, provided they have been guilty of no acts, other than fighting against us which should debar them from this privilege. This concession does not apply to: those who have taken the prominent military or political part in the war, nor to those who have broken their oath of neutrality, nor to foreigners. In the event of a military or political leader enquiring as to terms of surrender, the question is to be referred direct to Army Head–quarters.

2. All stock, supplies, &c., of those on commando, or of those who have broken their oath, are to be taken and no receipt given.

3. In cases where some members of a family, who all live on one farm, have broken their oath and gone on commando, those remaining are to be warned that, unless the former surrender within a reasonable period, all stock, supplies, &c., will be taken and no receipt given.

4. In cases of sniping, leaders of bands are to be informed that, unless it ceases, their (the leaders') houses will be burned, A few days should be allowed for this to become known, notices being sent to the resorts of the leaders and to their wives. In the event of its being necessary to burn the farms, further notice should be issued stating why this has been done, and giving a list of other houses that will next be burned.

5. All stock, supplies, &c., of snipers are to be taken and no receipt given.

6. Protection is to be given to all inhabitants who have kept the oath of neutrality, remaining quietly at home and taking no further part in the war. Any stock or supplies taken from them are to be paid for or receipts given in full for them.

7. Wives and families of men who are prisoners of war are to be protected, and receipts in full given for anything taken When a man surrenders a protection pass is to be given to his wife.

8. Widows of men killed in the war and all lone women are to receive protection passes, and to be paid or given receipts for every thing taken.

9. Burghers are to be informed that as soon as their leaders submit, and when every cannon has been surrendered, peace will be declared, and all prisoners of war will then be sent back to their homes. Exceptions only will be made in the cases of members of the late Governments of the South African Republic and Orange Free State, who are responsible for the war and its present disastrous prolongation, and of those who may be proved to have been guilty of acts contrary to the customs of war.

Roberts, Field-Marshal, Commanding in Chief in South Africa.

Army Head-quarters, South Africa, 28th September 1900

[712] Proclamations in SA, Page 373–4: https://babel.hathitrust.org/cgi/pt?id=hvd.32044089026009&seq=373&view=1up&q1=customs

Proclamation XL

Order issued by: Field-Marshal Lord Roberts, dated 18th: November. [1900][713]

As there appears to be some misunderstanding with reference to burning of farms and breaking of dams, Commander-in-Chief wishes following to be lines on which General Officers Commanding are to act:- No farm is to be burnt except for act of treachery, or when troops have been fired on from premises, or as punishment for breaking of telegraph or railway line, or when they have been used as bases of operation for raids, and then only with direct of General Officer Commanding, which is to be given in writing, the mere fact of a burgher being absent on commando is on no account to be used as a reason for burning the house. All cattle, wagons, and foodstuffs are to be removed from all farms; if that is found to be impossible, they are to be destroyed, whether owner be present or not.

[713] Proclamation 40 on Page 377. https://babel.hathitrust.org/cgi/pt?id=hvd.32044089026009&seq=377&view=1up

Queen Victoria's Proclamation

It is difficult to underestimate the moral sway which Queen Victoria held over her subjects and the devotion which they brought to their admittedly Imperial goals in the latter part of the 1800s

It was more than just the national pastime to go too far flung places in order to spread the glad tidings of British Rule. Not only with missionary work but with military force as well. As we have seen, when practically the whole Churchill family goes to South Africa to assist in carrying forward Milner's 'Forward Policy', it takes on the character of a national duty to participate in what the Queen and her government is asking of its citizens. The moral basis for this was established with the Queen's Proclamation of 1858 which is worth addressing here.

After the suppression of the Indian Mutiny of 1857-58 the punishment for any soldier in British service who had previously sworn allegiance to the Queen included "Blowing from the Gun" This was not random terror. In the UK, subjects who revolted were subjected to humiliating punishments throughout its military history. That said, this particular form of punishment was not repeated. Contrasting the more realistic with the romanticized.

Blowing from the Gun.[565]

Suppression of the Indian Revolt by the English.[714]

After this undoubtedly barbaric exhibition of retribution came Queen Victoria's proclamation of 1858 which recognized all people in British Dominions as right holders and stakeholders in the Empire with civil rights. This proclamation and amnesty applied to those who had laid down their arms and not committed murder during the rebellion.[715]

At the conclusion of most of the hostilities, in November of 1858, the British government revoked the British East India Company charter and made India a Crown Colony of the empire under the direct control of the sovereign. The document that effectuated part of this change granted an amnesty and also signalled a major shift in British policy towards its peoples.

There are many foundational documents that have shaped the British constitutional monarchy, Magna Carta, being the most notable. There are also significant documents and policies[716] that extended aspects of this system across the British Empire, such as the Balfour declaration regarding a Jewish homeland; with respect to India it

[714] Suppression of the Indian Revolt by the English, which depicts the execution of mutineers by blowing from a gun by the British, a painting by Vasily Vereshchagin c. 1884. Note: This painting was allegedly bought by the British crown and possibly destroyed (current whereabouts unknown). It anachronistically depicts the events of 1857 with soldiers wearing (then current) uniforms of the late 19th century. https://en.wikipedia.org/wiki/Suppression_of_the_Indian_Revolt_by_the_English

[715] Se Ramanlal Kakalbhai Dharaiya (1970). Gujarat in 1857. Gujarat University. p. 120. And: (https://en.wikipedia.org/wiki/Indian_Rebellion_of_1857)

[716] It is important to note that it was the British navy who established and maintained the blockade off the coast of Africa to prevent slave trafficking for 62 years from 1808 to 1870. The United States banned the importation of slaves beginning in 1807 but the U.S. Navy did not materially participate in the blockade until 1819 and the British formed the overwhelming majority of the ships employed previously and thereafter. It was slave transportation to America that provided the motivation for vessels to attempt running the blockade. The blockade represents a three-generation effort by a country, at great expense, to eliminate a social evil. https://en.wikipedia.org/wiki/Blockade_of_Africa. See Also: The Suppression of the Atlantic Slave Trade : British Policies, Practices and Representations of Naval Coercion ISBN: 071908511X ISBN13: 9780719085116 Robert Burroughs and Richard Huzzey Royal Navy Versus the Slave Traders: Enforcing Abolition at Sea 1808-1898 ISBN-10: 1844156338 ISBN-13: 978-1844156337 by Bernard Edwards See: Drachinifel Channel: Anti-Slavery Patrols - The West Africa Squadron - YouTube

is Queen Victoria's proclamation by which the British Crown imposed direct sovereignty and in effect nationalized the Indian possessions of the British East India Company and reads in part:

Proclamation by the Queen in Council to the princes, Chiefs, and People of India (published by the Governor-General at Allahabad, November 1st, 1858).[717]

> Victoria, by the Grace; of God of the United Kingdom of Great Britain and Ireland, and of the. Colonies and Dependencies thereof in Europe, Asia, Africa, America, and Australasia, Queen, Defender of the Faith.
>
> Whereas, for divers weighty reasons, We have resolved, ... to take upon Ourselves the Government of the Territories in India heretofore administered in trust for Us by the Honorable East India Company:
>
> We desire no extension of Our present territorial Possessions; .... We shall respect the Rights, Dignity, and Honour of Native Princes as Our own; and We desire, that they, as well as Our own Subjects, should enjoy that Prosperity and that social Advancement which can only be secured by internal Peace and good Government.
>
> We hold Ourselves bound to the Natives of Our Indian Territories by the same obligations of Duty which bind Us to all Our other Subjects; and those Obligations, by the Blessing of Almighty God, We shall faithfully and conscientiously fulfil.
>
> Firmly relying Ourselves on the truth of Christianity, and acknowledging with gratitude the solace of Religion, We disclaim alike the Right and the Desire to impose our Convictions on any of Our Subjects. We declare it to be Our Royal Will and Pleasure that none be in any wise favored, none molested or disquieted by reason of their Religious Faith or Observances; but that all shall alike enjoy the equal and impartial protection of the Law: and We do strictly charge and enjoin all those who may be in authority under Us, that they abstain from all interference with the Religious Belief or Worship of any of Our Subjects, on pain of Our highest Displeasure.
>
> And it is Our further Will that, so far an may be, Our Subjects, of whatever Race or Creed, be freely and impartially admitted to Offices in Our Service, the Duties of which they may be qualified, by their education, ability, and integrity, duly to discharge............

This Proclamation....laid the foundation for Indian secularism and established the semi-legal framework that would govern the politics of religion in colonial India for the next century. ... It promised civil equality for Indians regardless of their religious affiliation, and state non-interference in Indians' religious affairs. Although the Proclamation lacked the legal authority of a constitution, generations of Indians cited the Queen's proclamation in order to claim, and to defend, their right to religious freedom," [718]

"When the governance of India was transferred from the East India Company to the Crown in 1858, she (Queen Victoria) and Prince Albert intervened in an unprecedented fashion to turn the proclamation of the transfer of power into a document of tolerance and clemency. ... they ... insisted on the clause that stated that the people of India would enjoy the same protection as all subjects of Britain. Over time, this led to the-Proclamation of 1858 becoming known in the Indian subcontinent as 'the Magna Carta of Indian liberties', a phrase which Indian nationalists such as Mohandas Gandhi later took up as they sought to test equality under imperial law" [719]

These inherent natural rights of equality, social advancement, peace, good government, religious freedom, equal protection of the law and admission to civil service were actualized most effectively in India itself over the course of the next 89 years, before India gained independence on August 15th, 1947; an event that Mohandas Gandhi saw to fruition, if only for five months, before his assassination on January 30th 1948.

[717] 1858 Proclamation obtained from: www.csas.ed.ac.uk/mutiny/confpapers/Queen%27sProclamation.pdf (Retrieved 9/19).

[718] Adcock, C.S. (2013), The Limits of Tolerance: Indian Secularism and the Politics of Religious Freedom, Oxford University Press, pp. 23–25, ISBN 978-0-19-999543-1.

[719] (pages 38–39) Taylor, Miles (2016), "The British royal family and the colonial empire from the Georgians to Prince George", in Aldrish, Robert; McCreery, Cindy (eds.), Crowns and Colonies: European Monarchies and Overseas Empires, Manchester University Press, pp. 38–39, ISBN 978-1-5261-0088-7.

Marching on Pretoria

British troops on the march.[566]

Infantry Marching.[567]

Royal Lancasters January 19th, 1900.[568]

Sherman's March to the Sea.[720]

Marching on Pretoria[721] is the earliest recording in the flatinternational archive and probably dates from around 1902. Recorded in London, the song comes from the Second Anglo-Boer War and is sung by Ian Calquhoun, obviously from a British, patriotic perspective. This Anglo version of the song is actually an update of an 1865 American Civil War marching song Marching Through Georgia by Henry Clay Work.[722]

[720] An 1868 engraving by Alexander Hay Ritchie depicting Sherman's March to the Sea. The engraving shows Union soldiers destroying telegraph poles and railroads, and freed slaves assisting Union soldiers and making their way to safety. Marching Through Georgia - Wikipedia
[721] flatinternational - south African audio archive - Ian Colquhoun - Marching on Pretoria. flatinternational - south african audio archive
[722] United States of America (1776-) Civil War song "Marching Through Georgia" (1864) - YouTube

Marching on Pretoria.[723] [724]

When the call to arms went forth our hearts were blithe and glad.
Castle Gate and Cottage School, the bravest blood they had,
Every English meadow yielded up its yeoman lad;
to March, March March! on Pretoria.
Hurrah, Hurrah from Landsdale they come.
Hurrah, Hurrah they've made the business hum.
Round the world you heard the beat of Britain's ceaseless drum, when we went marching on Pretoria.

Far from overseas our brothers flocked to join the flag,
showing thus that British heart was neither bluff nor brag.
Foot to foot they stood with us beneath that dear old flag,
When we went marching on Pretoria.

Hooraah! Hooraah! Oh' play us bonnie songs.
Hooraah! Hooraah! Let's mark an [aid in tongue]
They have [shown it to their face the bond [blood] of Britain run]
When we went marching on Pretoria.

Cheer up boys and sing good luck to all our gallant men.
Who fought for the Empire out in Africa and when,
they have crossed the seas once more we welcome home again.
Conquerors of proud Pretoria.

Hooraah! Hooraah! Foot-soldiers of the King.
Hooraah! Hooraah! Their praises let us sing.

Victory they have brought us back, wherever they have been,
And now they are coming home from Pretoria.

Marching Through Georgia.[725]

Bring the good old bugle, boys, we'll sing another song
Sing it with a spirit that will start the world along
Sing it as we used to sing it, 50,000 strong
While we were marching through Georgia.

Hurrah! Hurrah! we bring the jubilee!
Hurrah! Hurrah! the flag that makes you free!
So we sang the chorus from Atlanta to the sea
While we were marching through Georgia.

How the darkeys shouted when they heard the joyful sound
How the turkeys gobbled which our commissary found
How the sweet potatoes even started from the ground
While we were marching through Georgia.

Yes and there were Union men who wept with joyful tears,
When they saw the honored flag they had not seen for years;
Hardly could they be restrained from breaking forth in cheers,
While we were marching through Georgia.

"Sherman's dashing Yankee boys will never make the coast!"
So the saucy rebels said and 'twas a handsome boast
Had they not forgot, alas! to reckon with the Host
While we were marching through Georgia.

So we made a thoroughfare for freedom and her train, sixty miles of latitude, three hundred to the main;
Treason fled before us, for resistance was in vain
While we were marching through Georgia.

[723] Rare British South African war song | marching onto Pretoria. And: https://www.youtube.com/watch?app=desktop&v=geKCr-SYNM8. PF Transcription of the recording, with assistance of Ms. Katharine Thomson, archivist of the Churchill Archive Centre. The Churchill Archives Centre – Churchill College (cam.ac.uk)
[724] flatinternational - south african audio archive - Ian Colquhoun - Marching on Pretoria With liner notes and alternative transcription.
[725] Audio: "Marching Through Georgia" - Union Civil War Song - YouTube And Alternative lyrics: United States of America (1776-) Civil War song "Marching Through Georgia" (1864) - YouTubeLyrics: American Folk - Marching Through Georgia lyrics (lyricstranslate.com) Article: Marching Through Georgia - Wikipedia See Leviticus 25 for description of Jubilee.

Modern Version[726]

I'm with you and you're with me and
So we are all together, so we are all together, so we are all together.
Sing with me, I'll sing with you and so we will sing together
As we march a long.

Chorus:
We are marching to Pretoria, Pretoria, Pretoria,
We are marching to Pretoria, Pretoria, Hurrah.

We have food, the food is good, and
So we eat together, (3 times)
When we eat it will be a treat, and
So we will sing together,
As we march along.

Sing with me, I'll sing with you, and
So we will sing together (x3)
Sing with me, I'll sing with you and so we will sing together
As we march along
We are marching to Pretoria, Pretoria, Pretoria
We are marching to Pretoria, Pretoria, Hooorah!

Walk with me, I'll walk with you, and
So we will walk together (x3)
Walk with me, I'll walk with you, and so we will walk together
As we march along
We are marching to Pretoria, Pretoria, Pretoria
We are marching to Pretoria, Pretoria, Hooorah!

Dance with me, I'll dance with you, and
So we will dance together (x3)
Dance with me, I'll dance with you, and so we will dance together
As we march along
We are marching to Pretoria, Pretoria, Pretoria
We are marching to Pretoria, Pretoria, Hooorah!

Drink with me, I'll drink with you, and
So we will drink together (x3)
Drink with me, I'll drink with you, and so we will drink together
As we march along
We are marching to Pretoria, Pretoria, Pretoria
We are marching to Pretoria, Pretoria, Hooorah!

[726] PF compilation from several versions including, The Highwaymen: Marching To Pretoria - YouTube, Merv Griffin: Marching To Pretoria - YouTube and The Weavers' Marching To Pretoria - YouTube

Casablanca Conference

Since we have delved this far into Winston's attitude towards warfighting and peace making we will take half a page to look forward forty years to the conclusion of the Casablanca Conference[727] in February of 1943. There are two sides to the story that Winston was 'startled' by Franklin Roosevelt's public announcement, at a press conference, that the Allies would only countenance 'unconditional surrender' for the cessation of hostilities against the German, Italian and Japanese forces.

While there were differences of opinion between the Allies as to the terms to be offered, Franklin Roosevelt borrowed the term from US Army General Ulysses S. Grant (known as "Unconditional Surrender" Grant) and Churchill had cabled the War Cabinet four days earlier and they had not objected to this formulation, which was the same formula the British had maintained vis a vie the Boer forces. In this case there was the additional consideration of Stalin's suspicions of the Western Allies making a separate peace. It was felt that Allied morale would be "strengthened by the uncompromising demand, and Stalin's suspicions allayed."[728]

However, it may also be true that the public announcement of this formula at a press conference may have come as something of a surprise to Churchill. Drew Middleton, the reporter for the New York Times, who was in Casablanca at the conference, later revealed in his book Retreat From Victory, that Churchill had been "startled by the [public] announcement [of unconditional surrender]. "I tried to hide my surprise. But I was his [Roosevelt's] ardent lieutenant."[729]

President Franklin D. Roosevelt, Prime Minister Winston S. Churchill and their Combined Chiefs of Staff at the Casablanca Conference. Standing, (left to right): General Brehon B. Somervell; General H. H. Arnold; Admiral Ernest J. King; Major General Sir Hastings Ismay; General George C. Marshall, Admiral Sir Dudley Pound; General Sir Alan Brooke; Sir Charles Portal; and Vice Admiral Louis Mountbatten.[569]

In a similar vein of preferring moderation over vengeance; in 1944 both Churchill and Anthony Eden would object to the adoption of the Morganthau Plan for the deindustrialization of Germany after WWII.[730] Churchill only agreed after it became clear that continuing American economic aid, as opposed to military aid, would only be granted on condition of subscribing to the plan. During the signing of the plan, which coincided with the signing of the economic aid agreement, President Roosevelt proposed that they sign the plan first. This prompted Churchill to exclaim: "What do you want me to do? Get on my hind legs and beg like Fala?"[731]

[727] https://en.wikipedia.org/wiki/Casablanca_Conference Retrieved Dec, 17, 2024.

[728] Roberts, Andrew (2009) [2008]. Masters and Commanders: The Military Geniuses who Led the West to Victory in World War II. Penguin. ISBN 978-0-141-02926-9.

[729] Middleton, Drew, On This Day, "Roosevelt, Churchill Map 1943 War Strategy," January 24, 1943, retrieved August 27, 2012.

[730] See: Morgenthau Plan - Wikipedia

[731] Fala was Franklin Roosevelt Scottish Terrier who knew how to perform tricks. Fala (dog) - Wikipedia

The Plot to Kill Roberts.

Friday, August 10, 1900 Discovery of the plot at Pretoria to kidnap Roberts and the British officers.

War in S. Africa.[732]
Plot against Lord Roberts and British Officers.
A Dastardly Conspiracy.

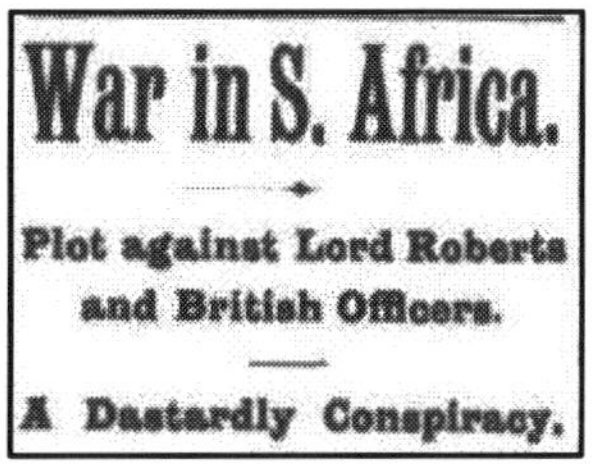

War in S. Africa.

Plot against Lord Roberts and British Officers.

A Dastardly Conspiracy.

It is reported that a plot to shoot all the British officers and capture Lord Roberts has been discovered at Pretoria. Ten of the leaders of the supposed plot have been arrested.

Later. Further details have now been received of the plot to shoot the British officers and capture Lord Roberts, which was reported to have been discovered at Pretoria. It appears that the conspirators intended to set fire to the western extremity of Pretoria, hoping that while the British troops were engaged in extinguishing the fire they would be able to enter the residences of the officers and assassinate the occupants, and also to seize Lord Roberts and convey him to the nearest Boer commando, for which purpose horses were held in readiness in one of the suburbs of Pretoria.

It is asserted that many of the townspeople were aware of, and connived in, the plot, and intense indignation has been aroused in the British army. Every one of the conspirators had heretofore enjoyed entire personal freedom after having taken the oath of neutrality.

London, August 11. Arrangements have been made at Pretoria for the trial by court-martial of the ten leading conspirators, who have been arrested. The prisoners, it is stated, have made recriminating statements.

The discovery of the plot was made on Tuesday last by the new Pretoria Police Force under Major Maxse,[733] who arrested the leaders just prior to the attempt being made to carry out the conspiracy. The other prisoners were arrested subsequently, and all were lodged in separate cells.

A number of Pretoria burghers and Hollanders are implicated in the plot, as it has been found that they maintained communications between the conspirators and the outside Boers.

The Pretoria Plot.
British Uniforms Worn.[734]

Further details have been received of the plot to kidnap Lord Roberts, and to assassinate the British officers in Pretoria. All of the ringleaders were arrested, and are now awaiting trial by court-martial. Two of whom had dressed themselves in British uniforms to facilitate their entry into the officers' quarters.

Lord Roberts and his chief of staff, Colonel Sir Henry Rawlinson.[570]

Lord Roberts' Bodyguard.[571]

[732] Warwick Examiner and Times (Qld.: 1867 - 1919), Wednesday 15 August 1900, page 3 https://trove.nla.gov.au/newspaper/article/82159549

[733] General Sir Frederick Ivor Maxse, (22 December 1862 – 28 January 1958) was a senior British Army officer who fought during the First World War, best known for his innovative and effective training methods. Ivor Maxse - Wikipedia And an extensive military biography for Frederick Ivor Maxse, at: https://www.angloboerwar.com/name-search

[734] Riverine Herald (Echuca, Vic. : Moama, NSW : 1869 - 1954; 1998 - 2002), Monday 20 August 1900, page 2. https://trove.nla.gov.au/newspaper/article/114954476#

The Plot Against Lord Roberts.[735 736]
Proceedings at The Court Martial.
Kruger and Botha Implicated.

The court-martial on the Boers arrested in connection with the plot formed in Pretoria to assassinate the British officers in occupation and kidnap Lord Roberts is proceeding. Major General Marshall, R.A., is president of the court.

The conspirators, about fifteen in number, ten of whom were immediately arrested, had taken the oath of neutrality, which especially stated that any contravention was liable to punishment under martial law.

Lieutenant Hans Cardua is at present on trial charged with breaking his parole and plotting to kill the British officers and kidnap Lord Roberts'. Lieutenant Cardua is a young, well-dressed man. He is. very cool in his demeanor in court. He has the assistance -of a lawyer to defend him. When indicted by the court the prisoner pleaded guilty, but the court advised him to withdraw the plea.

Sensational evidence has been forth-coming.

The trial of the ringleader, Hans Cordua, a youth of twenty-three, formerly lieutenant of the Staats Artillery, was opened on the 16th. Lieutenant Cardua is a young, well-dressed man. He is. very cool in his demeanor in court. He has the assistance of a lawyer to defend him. The prisoner was charged with having broken his parole, and attempted to abduct British officers. When indicted by the court the prisoner pleaded guilty, but the court advised him to withdraw the plea.

Evidence was then brought forward to prove the extent and ramifications of the plot, the complicity of Botha and others therein. The prisoner was ably defended by Mr. Berrange, who endeavoured to show the complicity of the witness,

Detective Gano, a clever Mexican, who was accused by Cordua of having been instigator of the plot. This imputation was denied by Gano, who declared that the plot was already hatched when he was sounded on the subject. The accused, on the other hand, declared his own unwillingness to join in the plot, and pretended that he had been over persuaded by the detective, who represented that he was a pro-Boer in British employ, who, tired of his position, desired to help the Boers.

A Transvaal burgher named Da Toit deposed that early in July Cardua asked him to join in an enterprise that would stagger all Europe. He (Cardua) then described a plot that had been formed to seize all the British officers in Pretoria. Cardua said that he would have the help of 250 residents of the city. While the officers were being seized, Cardua was to rush to Lord Roberts' house, abduct the Commander-in Chief, and hand him over to General Louis Botha, who was to be in waiting behind what was known as Beckett's Kopje.

Cardua showed [the] witness a cypher letter in proof of the genuineness of the plot, and of General. Botha's connection with it. When witness asked him about his (Cardua's) oath of neutrality, Cardua replied that that was nothing- the discovery of the plot and his connection with it would only mean transportation to Ceylon. Witness declined to have anything to do with the plot, and subsequently communicated what he knew to the British military police.

Captain Duban deposed to finding a key to General Botha's letter on the prisoner. Lieutenant Kruger, of the Imperial Transvaal police, translated a cypher message found on one of the conspirators which read: "Proceed with the work. Be courageous. I will not leave you in the lurch. Kruger."

An ex-field cornet in the Boer army, named Marais, gave evidence that he had been approached and had refused to join in the plot.

The prisoner was found guilty on all charges, and was condemned to death.

On the 24th he met his fate like a brave man within the precincts of the gaol, General Maxwell, Colonel Maxse, the chaplain, the doctor, and the firing squad being the only persons present at the execution. So ended one of the most remarkable episodes of the campaign, and the career of a young enthusiast, whose curious ambition renders more convincing the old aphorism that every blackguard is a hero spoilt.

[735] Barrier Miner (Broken Hill, NSW: 1888 - 1954), Monday 20 August 1900, page 2.
https://trove.nla.gov.au/newspaper/article/44262649#
[736] Creswicke, South Africa and the Transvaal War. Vol 6. Pages 83-85.

Kruger In Exile

After the battle of Diamond Hill on June 12th, 1900. Kruger moved East to Waterval Onder, where his small house became the "Krugerhof".[737] [738]

Subsequent to the battle of Bergendal on August 26th 1900 and the Boer retreat to the border of the Transvaal. Roberts announced the annexation of the South African Republic to the British Empire on 1 September 1900. Kruger proclaimed on 3 September that this was "not recognised" and "declared null and void".

It was decided in the following days that to prevent his capture Kruger would leave for Lourenco Marques and there board ship for Europe. Officially he was to tour the continent, and perhaps America too, to raise support for the Boer cause.

Kruger left the Transvaal by rail on September 11th 1900—he wept as the train crossed into Mozambique.

The Komatie Poort Railway Bridge at the border between the Transvaal and Portuguese East Africa.[572]

He had planned to board the first outgoing steamer, the Herzog of the German East Africa Line, but was prevented from doing so when, at the behest of the local British Consul, the Portuguese Governor insisted that Kruger stay in port under house arrest.

About a month later Queen Wilhelmina of the Netherlands concluded a deal with Britain to extricate Kruger on a Dutch warship, HNLMS Gelderland,[739] and convey him through non-British waters to Marseille. Kruger was delighted to hear of this but dismayed that [his wife] Gezina, [also referred to as Mary was] still in Pretoria, was not well enough to accompany him.

President and Mrs. Kruger in their cottage in Pretoria, April, 1900.[573] Four generations of the Kruger family.[574] Mrs. Paul Kruger, Mrs. F.C. Eloff, Mrs. Louis Jacobz, and Baby Jacobz.

[737] Meintjes, Johannes (1974). President Paul Kruger: A Biography (First ed.). London: Cassell. ISBN 978-0-304-29423-7. Meredith, Martin (2007). Diamonds, Gold, and War: The British, the Boers, and the Making of South Africa (First ed.). New York: PublicAffairs. ISBN 978-1586484736.

[738] See Generally: https://en.wikipedia.org/wiki/Paul_Kruger And: https://en.wikipedia.org/wiki/Paul_Kruger#Exile_and_death

[739] HNLMS Gelderland (Dutch: Hr.Ms. Gelderland) was a Holland-class protected cruiser of the Royal Netherlands Navy. During its career in the Dutch Navy it was most notable for being the ship Queen Wilhelmina sent to Portuguese East Africa to transport Paul Kruger to Europe during the Second Boer War. The ship was taken over by the Germans during World War II, rebuilt as an anti-aircraft cruiser and renamed Niobe. Commissioned into the German navy on 1 March 1944, she was sunk in Kotka harbour in Finland on 16 July 1944. https://en.wikipedia.org/wiki/HNLMS_Gelderland_(1898)

The Gelderland, with Kruger aboard, departed on 20 October 1900.

Dutch protected cruiser Gelderland.[575] Kruger leaving for Europe in 1900.[740] At right is his secretary Madie Bredell.

He received a rapturous welcome in Marseille on 22 November. 60,000 people turned out to see him disembark. Accompanied by Leyds, [741] he went on to an exuberant reception in Paris, then continued to Cologne on 1 December. Here the public greeted him with similar excitement, but Kaiser Wilhelm II refused to receive him in Berlin. Having apparently still harboured hopes of German assistance in the war, Kruger was deeply shocked. "The Kaiser has betrayed us", he told Leyds.[742]

They went on to the Netherlands, which was strictly neutral and could not assist militarily, but would feel more like home. After another buoyant reception from the general public, Kruger was cordially received by Wilhelmina and her family in The Hague, but it soon became clear to Leyds that it embarrassed the Dutch authorities to have them stay in the seat of government. The Kruger party moved to Hilversum in April 1901.

Shortly afterwards, Gezina, with whom Kruger had had 16 children—nine sons, seven daughters (of whom some died young) had eight sickly grandchildren transferred to her from the concentration camp at Krugersdorp, where their mother had died, in July 1901. Five of the eight children died within nine days, and two weeks later Gezina also died.[743]

By now partially blind and almost totally deaf, he dictated his memoirs to his secretary Hermanus Christiaan "Madie" Bredell and Pieter Grobler during the latter part of 1901, and the following year they were published.[744]

Kruger and his entourage relocated in December 1901 to Utrecht, and was joined by his daughter Elsje Eloff and her family.

Kruger would not countenance the idea of returning home, partly because of personal reluctance to become a British subject again, and partly because he thought he could better serve his people by remaining in exile.

President Steyn similarly refused to accept the Boer defeat and joined Kruger in Europe, though he did later return to Southern Africa.

[740] Scanned from Meintjes, Johannes (1974). President Paul Kruger: A Biography (First ed.). London: Cassell. ISBN 9780304294237. Between pages 192 and 193. Source given as "Nat. Cult. Hist. Museum", presumably the National Cultural History Museum in Pretoria, South Africa. https://commons.wikimedia.org/wiki/File:Kruger%27s_Last_View_of_Africa.jpg

[741] Willem Johannes Leyds (1 May 1859 – 14 May 1940) was a Dutch lawyer and statesman who served as state attorney and state secretary of the South African Republic. From 1898 to 1902, during the crucial period of the Second Boer War, he was the Republic's special envoy and minister plenipotentiary in Brussels, accredited to several European states. Willem Johannes Leyds - Wikipedia

[742] PF: The Kaiser had no navy with which to challenge the British for control of the southern ocean. This was a deficiency he would make strenuous efforts to correct prior to the outbreak of the First World War. See: https://en.wikipedia.org/wiki/High_Seas_Fleet

[743] PF: Measles was rampant in the camps and the medical authorities were mostly unable to enforce quarantine for children since it would involve separation from their mothers which they naturally declined to permit. This was also the middle of Winter in the Southern Hemisphere and Pretoria while at latitude 25° South is also at an elevation of 1,339 m (4,393 ft) does have cold winters. For comparision Miami Florida is located at 25° N. Latitude and occasionally experiences frost in winter despite being at sea level.

[744] Kruger, Paul (1902). The Memoirs of Paul Kruger, Four Times President of the South African Republic. Trans. Alexander Teixeira de Mattos. Toronto: George N Morang & Co. OCLC 1491193

Louis Botha, Christiaan De Wet and Jacobus De la Rey visited Kruger in August 1902 and, according to hearsay, were berated by Kruger for "signing away independence". The rumours of such a scene were widespread enough that the generals issued a statement denying them.

Two years later, on July 14th, Paul Kruger died in Clarens small village in the canton of Vaud in western Switzerland at the age of 78. His Bible lay open on a table beside him.

Kruger's body was initially buried in The Hague but was soon repatriated with British permission. After ceremonial lying in state, he was accorded a state funeral in Pretoria on 16 December 1904, the vierkleur of the South African Republic draped over his coffin, and buried in what is now called the Heroes' Acre in the Church Street Cemetery.

Funeral Train of the late President, Mr. Kruger seem crossing the Vaal; River bridge at Vereeniging.[576]

Paul Kruger as a young man.[745] “This portrait is from a daguerrotype and is consequently reversed. The hand which has lost the thumb is really the left hand. Mr. Kruger had the member shattered by a gun accident, and amputated it himself with a knife.” PF: As described above regarding his hand, the picture has been reversed. the buttons are on the left rather than the right..

[745] With the Flag to Pretoria, H. W. Wilson Vol 2 June 1900.

Original Appendix to Ian's March

Composition of Lieut.-General Ian Hamilton's Force

Divisional Staff

Lieutenant-General Ian Hamilton. C.B., D.S.O.
A.D.C.s—Captain de Heriez Smith.
Captain Balfour, late 11th Hussars.
Captain Maddocks, R.A.
Captain Duke of Marlborough, I.Y.
A.A.G.—Lieut.-Colonel Le Gallais, 8th Hussars.
D.A.A.G.s—Captain Vallentin, Somerset L.I.
Captain Gamble, Lincoln Regiment.
Captain Atcherley, A.S.C.
Captain Kirkpatrick, R.E.
Provost Marshal—Captain Sloman, East Surrey Regiment.
Div. Signalling Officer—Captain Ross, Norfolk Regiment.
P.M.O.—Colonel Williams, N. S. Wales A.M.C.
Divisional Troops—Rimington's Guides under Major Rimington, Inniskilling Dragoons.
2nd Mounted Infantry Brigade
Brigadier-General R. Ridley
A.D.C.—Captain Hood, Coldstream Guards.
Brigade Major—Lieut-Colonel Mitford, East Surrey Regiment.
Staff Officers—Captain Sir T. MacMahon, Royal Welsh Fusiliers.
Captain Eustace Crawley, 12th Lancers.

2nd Mounted Infantry Corps

Lieut.-Colonel de Lisle, Commanding Durham Light Infantry.
Staff Officer—Captain Fanshawe, Oxford L.I.
6th M.I. Battalion—Captain Pennefather, Welsh Regiment.
New South Wales Mounted Rifles.
West Australians.
1 Pom-pom.

5th Mounted Infantry Corps

Lieut.-Colonel Dawson, I.S.C.
Staff Officer—Captain Ballard, Norfolk Regiment.
5th M.I. Battalion—Major Lean, Warwick Regiment.
Roberts' Horse—Captain Baumgartner, East Lancashire Regiment.
Marshall's Horse—Captain Corbett.
Ceylon M.I.—Major Rutherford,
1 Pom-pom.

6th Mounted Infantry Corps

Lieut.-Colonel Legge, 20th Hussars.
Staff Officer—Captain Hart, East Surrey Regiment.
2nd M.I. Battalion—Major Dobell.
Kitchener's Horse—Major Cookson, I.S.C.
Lovat's Scouts—Major A. Murray.
1 Pom-pom.

7th Mounted Infantry Corps

Lieut.-Colonel Bainbridge, Buffs.
Staff Officer—Captain Hamilton, Oxford L.I.
7th M.I. Battalion—Major Welch.
Burmah M.I.—Captain Copeman.
1 Pom-pom.

P Battery

Ammunition Column—Major Mercer, R.H.A.

Bearer Company and Field Hospital—New South Wales Army Medical Corps.

2nd Cavalry Brigade

Brigadier-General Broadwood
A.D.C.—Captain Aldridge, R.H.A.
Brigade Major—Captain Hon. T. Brand, 10th Hussars.
Signalling Officer—Captain Sloane Stanley, 12th Lancers.
Household Cavalry—Lieut.-Colonel Galley.
10th Lancers—Lieut-Colonel Fisher.
12th Lancers—Lieut.-Colonel Earl of Airlie.
Q Battery, R.A.
Ammunition Column—Captain Kincaid, R.A.
Bearer Company.
Field Hospital.

19th Brigade

Major-General Smith-Dorrien
A.D.C.s—Captain Hood, R.M.L.I.
Lieut. Dorrien Smith, Shropshire L.I.
Brigade Major—Major Inglefield, East Yorkshire Regiment.
74th Battery—Major MacLeod.
2nd Duke of Cornwall L.I.—Lieut.-Colonel Ashby.
Shropshire L.I.—Lieut.-Colonel Spens.
Gordon Highlanders—Lieut.-Colonel MacBean.
Royal Canadians—Lieut.-Colonel Otter.
Bearer Company and Field Hospital.

21st Brigade

Major-General Bruce-Hamilton
A.D.C.—Lieut. Frazer, Cameron Highlanders.
Brigade Major—Major Shaw, Derbyshire Regiment.
76th Battery—Major Campbell.
1st Royal Sussex—Lieut.-Colonel Donne.
1st Derby—Major Gossett.
1st Cameron—Lieut.-Colonel Kennedy.
City Imperial Volunteers—Brigadier-Colonel MacKinnon; Colonel The Earl of Albemarle.
Bearer Company.
Field Hospital.

Divisional Artillery

Lieut.-Colonel Waldron, R.F.A.
81st Battery.
82nd Battery—Major Conolly.
1 Section of Five-inch guns—Captain Massey.
Ammunition Column—Captain Hardman.

Effective Fighting Strength

11,000 Men.
4,600 Horses.
8,000 Mules.
36 Field guns.
2 Five-inch guns
23 Machine guns.
6 Pom-poms.
The force left Bloemfontein, April 22.
Arrived at Pretoria on June 5.
Distance traversed, 401 miles in a straight line.
Time on the march, 45 days. Halts, 10 days.

General actions on nine days:

Israel's Poorte, April 25.
Houtnek, April 30 and May 1.
Welkom, May 4.
Sand River, May 10.
Affair of Lindley, May 20.
Doornkop (Florida), May 29.
Six Mile Spruit (Pretoria), June 4.
Diamond Hill, June 11 and 12.
Eighteen days' skirmishes.

Towns captured:

Thabanchu.
Winburg.
Ventersburg.
Kroonstadt.
Lindley.
Heilbron.
Johannesburg.
Pretoria.

Bibliography

Many works have been footnoted in the text. Restating those references again here would extend this work unnecessarily.

Anyone who has read this work will easily come across the major historical works which chronicle this era and conflict as they follow the leads I have cited. I would rather cite material directly in the location where it is most relevant than bury it at the end of the work or in the endnotes where I primarily include citations to images.

The full and equally idiomatic bibliography for this series is over sixty pages long and will be included in an Appendix volume to be published at the end of the series.

For this volume an abbreviated bibliography is as follows:

Winston Spencer Churchill, Ian Hamilton's March, 1900 The Copp, Clark Company, Ltd. Toronto

Hathi Trust: #1 - Ian Hamilton's March by Winston Spencer Churchill ; together ... - Full View | HathiTrust Digital Library

Internet Archive: Ian Hamilton's March: Churchill, Winston, 1874-1965 : Free Download, Borrow, and Streaming : Internet Archive

Faded Page: Ian Hamilton's March (fadedpage.com)

Project Gutenberg: Ian Hamilton's March by Winston Churchill - Free Ebook (gutenberg.org)

Lieutenant Thomas Hugh Colville Frankland: Held by the Enemy: Extracts from the Journal of Lieutenant Thomas Hugh Colville Frankland, Royal Dublin Fusiliers, lately prisoner of war at Pretoria. Originally Included in Winston Spencer Churchill's book; Ian Hamilton's March, 1900 The Copp, Clark Company, Ltd. Toronto. Ian Hamilton's March : Churchill, Winston, 1874-1965 : Free Download, Borrow, and Streaming : Internet Archive

Randolph Churchill. Official Biography Winston S. Churchill Volume I by Randolph S. Churchill (Houghton Mifflin Co.) 1966. The abbreviation OB is used to refer to this volume.

Official Biography Companion Volume(s) I and II by Randolph S. Churchill, Published by William Heinemann Ltd.[746], 1967. Copyright by C & T Publications Limited. The abbreviation OBC is used to refer to these volumes.

License Links

[746] William Heinemann - Wikipedia And: Heinemann (publisher) - Wikipedia

[747] Creative Commons License: Creative Commons license - Wikipedia And About CC Licenses - Creative Commons

Some documents are reproduced pursuant to the UK Open Parliament License and this applies not only material in which either House owns the copyright but also material published before 1 August 1989 which is the operative Commencement date of the Copyright and Designs act of 1988; and 50 years having elapsed subsequent to such publication. The same terms applying to previously published material that at one time subsisted under Crown Copyright. [748]

Also, this work contains public sector information made available under the South African Promotion of Access to Information Act, 2000 ("PAIA"). https://www.gov.za/documents/promotion-access-information-act

Google Earth License Links

Google Earth License: https://www.google.com/intl/en-GB_ALL/permissions/geoguidelines/

Sales of these annotated, illustrated and supplemented historical narratives are anticipated to be less than the 5,000 copies permitted. If that is not the case the author will contact the Google Cloud Customer Team for additional permission(s). https://cloud.google.com/contact/

Copyright Discussion

This work is a comprehensive republication of the included works which have been compiled, annotated, illustrated and expanded in order to provide historical context and clarity.

Images and text used by permission or by law are clearly identified by citation and should not be reproduced; as to the remainder there is no copyright claimed by myself on any public domain images, text or any common property information contained in this work.

Future creators should continue to follow the permissions, guidelines and requirements from the original source and author/uploader who is cited. Links to Creative Commons, Google Earth and other licenses are noted in the license links section.

So far as can be determined, as of 2025, the text and images used to illustrate the first person accounts included in this volume reside in the public domain in the United States having all been first published prior to January 1st, 1930 as well as in all other jurisdictions which employ any of the following copyright regimens:

1. A copyright regimen that grants an author's works subsisting copyright for seventy (70) years pma[749] or less.
2. A copyright term subsisting for 120 years from the date of creation or first publication.
3. And such jurisdictions that apply the Rule of the Shorter term to an author's pma date or publication date.

This work will not be marketed or deliberately sold in any jurisdiction with a longer copyright term.

Copyright for the assembled compilation of this material, or substantial portions thereof that would comprise separate articles or works unto themselves is retained. This work and the series in sum comprise a derivative work. Literary, dramatic, and adaptation rights of the saga demonstrated and explicated by this series is retained.

Otherwise, with respect to the underlying texts and images recast in a more permanently accessible form I choose to pay forward my creative energy to the people who lived through these experiences and it has been a pleasure to find and assemble their collective words and images in a new synoptic and synergistic arrangement.

Except where noted otherwise, every effort has been made to ensure that no material has been used without an attributed source. However, if any non-attributed source, incorrect source, or copyrighted material used without permission or justification, has been included then I would be grateful for information that would enable any errors or inappropriate inclusions to be corrected or omitted in subsequent impressions.

Starting from creation, to publication, through to purchase, archiving, and digitizing, this annotated, illustrated and expanded edition is the sixth step in the journey of these works. My thanks go to all who have helped preserve and perpetuate these important historical records.

'For those who have come thus far: Thank you again for your unseen but ever felt companionship.'[750]

[748] If the work was commercially published within 75 years of the end of the calendar year in which it was made, Crown copyright will last for 50 years from the end of the calendar year in which it was published.
https://copyrightservice.co.uk/copyright/p01_uk_copyright_law Flow Chart: Crown copyright flowchart 2015

[749] post mortem auctoris or "pma" [post mortem author]

[750] Per, Will Durant, in his eleven-volume work 'The Story of Civilization,' Simon & Schuster, 21st Edition. 1972 ISBN: 0-671-41800-9

Endnotes

[1] Black & White Weekly Vol 1-2 1899-1900 Photo by D. Barnett. Image Credit: Rob Davidson, RobD at Anglo Boer War.com

[2] Anglo-Boer War Album, Cape Town, D. Edwards & co. [1902?]

[3] Without Hat Photo Credits. Cape Argus/Letterhead portrait. Image source: Cape Argus (Roberts)

[4] See: An Age of Youth - Churchill and the Great Republic | Exhibitions - Library of Congress (loc.gov)

[5] Image credit: Rob Davidson at AngloBoerWar.com Anglo Boer War - Armored Train Ambush 11-15-99 Chuchill & Haldane - Page 11 - Boer War Forum

[6] Image credit: Rob Davidson at AngloBoerWar.com Anglo Boer War - Armored Train Ambush 11-15-99 Chuchill & Haldane - Page 11 - Boer War Forum

[7] My Escape from the Boers, Part 1 Strand Magazine Strand Magazine Vol 66 p. 546. And Radio Times, Hulton Picture Library per Frederick Woods, "Young Winston's Wars."

[8] C. E. Bechhofer Roberts, Winston Churchill; being an account of the life of the Right Hon. Winston Leonard Spencer Churchill, P.C., C.H., T.D., M.P. by Ephesian [pseud.] With 11 illustrations Published 1927

[9] Library of Congress image courtesy of Charles Scribner used as the cover of a book showing penciled in moustache. Winston Churchill, ca. 1900. Photograph Prints and Photographs Division.

[10] My Escape from the Boers, Part 1 Strand Magazine Strand Magazine Vol 66 p. 546. And Radio Times, Hulton Picture Library per Frederick Woods, "Young Winston's Wars."

[11] C. E. Bechhofer Roberts, Winston Churchill; being an account of the life of the Right Hon. Winston Leonard Spencer Churchill, P.C., C.H., T.D., M.P. by Ephesian [pseud.] With 11 illustrations Published 1927

[12] Library of Congress image courtesy of Charles Scribner used as the cover of a book showing penciled in moustache. Winston Churchill, ca. 1900. Photograph Prints and Photographs Division.

[13] Photo Credit: Thomas Hugh Colville Frankland (unknown-1915) - Find A Grave Memorial Memorial Id: 56009464 Photo added by Nancy Wright: Member Profile: Nancy Wright - Find A Grave

[14] Black & White Weekly Vol 1-2 1899-1900 Photo by D. Barnett. Image Credit: Rob Davidson, RobD at Anglo Boer War.com

[15] Anglo-Boer War Album, Cape Town, D. Edwards & co. [1902?]

[16] Anglo Boer War - Pictures from Under the Union Jack - Page 2 - Boer War Forum Credit Dr. David Biggins

[17] Anglo Boer War - Pictures from Under the Union Jack - Page 2 - Boer War Forum Credit Dr. David Biggins

[18] File:Colonel Ian Hamilton.jpg - Wikimedia Commons By: John Singer Sargent R.A. PNG Version

[19] File:Sir Ian Hamilton (cropped).jpg - Wikimedia Commons This work has been released into the public domain by its author, Bain News Service. This applies worldwide. Photo by Elliot & Fry

[20] WSC, My Early Life. Photo by Elliot & Fry

[21] New York Daily Tribune, Saturday November 10th, 1900. 0168.pdf (loc.gov)

[22] Advertisement carried in Navy & Army Illustrated Vol 10 Nov, 24th, 1900. p. 271 of PDF

[23] File:SouthAfrica1885.svg - Wikimedia Commons

[24] South African Railway System circa 1900: File:SouthernAfricaRailwayMap 1901.png - Wikipedia

[25] Navy & Army Illustrated Vol 9, October 7th 1899 and Creswicke, Louis South Africa and the Transvaal War, Vol. 1

[26] WSC, My Early Life

[27] Creswicke SA War Vol 5 Map page VIII

[28] Doyle, Arthur Conan. The Great Boer War 17th Edition, 1902

[29] Davitt, Boer Fight for Freedom, 3rd ed. 1902. (Duke Scan)

[30] Doyle, Arthur Conan. The Great Boer War 17th Edition, 1902

[31] Doyle, Arthur Conan. The Great Boer War 17th Edition, 1902

[32] WSC Ian Hamilton's March.by G. Philip & Son, London And Longmans, Green & Co. Since this map is original to WSC's book it is included.

[33] Ian Hamilton's March 1900 The Copp, Clark Co. Toronto. PF: With apologies for the quality but not for lack of trying.

[34] File:10 of 'History of the Boers in South Africa ... with three maps' (11189449154), crop.jpg - Wikimedia Commons Author: The British Library

[35] File:11 of 'History of the Boers in South Africa ... with three maps' (11191958004), crop2.jpg - Wikimedia Commons

[36] Ian Hamilton's March 1900 The Copp, Clark Co. Toronto. And Frontiers & Wars.

[37] File:Boer War Photograph Taken by B W Caney Q115355.jpg - Wikimedia Commons Author: B.W. Caney

[38] File:Boer War Photograph Taken by B W Caney Q115351.jpg - Wikimedia Commons Author: B.W. Caney

[39] File:Boer War Photograph Taken by B W Caney Q115351.jpg - Wikimedia Commons Author: B.W. Caney
[40] Anglo-Boer War Album, Cape Town, D. Edwards & co. [1902?]
[41] With Roberts to the Transvaal. C.N. Robinson [1900]
[42] After Pretoria: The Guerilla War Vol 3 Jan. 1901
[43] Anglo-Boer War Album, Cape Town, D. Edwards & co. [1902?]
[44] Times History War in S.A Vol 3 Photo by Elliott & Fry.
[45] War with the Boers, H. Brown, London [1902?] Vol 2
[46] File:South Africa - Drakensberg (16261357780).jpg - Wikimedia Commons Author: Diriye Amey from Locarno, Switzerland
[47] Anonymous (1910) A Handbook of the Boer War - With General Map of South Africa and 18 Sketch Maps and Plans, London & Aldershot: Gale & Polden Ltd File:NorthNatalWarTheatre, colour, crop.png - Wikimedia Commons
Black and White Original: NorthNatalWarTheatre - File:NorthNatalWarTheatre, colour, crop.png - Wikimedia Commons
[48] SA Boer-British War Vol 1. Hopkins, 1900
[49] Bridge across the Valsch River at Kroonstadt, blown up by the Boers. File:Gezicht op een opgeblazen brug over de Valsrivier bij Kroonstad, Zuid-Afrika Railway Bridge across the Valsch River at Kroonstadt, blown up by the Boers, S.A. (titel op object), RP-F-F09001.jpg - Wikimedia Commons
[50] Atom site for DRISA Page 28. Temporary bridge over Sundays River. - Atom site for DRISA
[51] These men are holding Van Reenen's Pass with General Prinsloo. SA National Archives: http://www.nationalarchives.gov.za/node/1236
[52] Item N03503 N03503.jpg (1400×964) (drisa.co.za)
[53] With the Flag to Pretoria, H. W. Wilson Vol 1 1900 and britishbattles.com Photo by Maj. Brazier-Creagh. R.A.M.C.
[54] Navy & Army Illustrated Vol 11 January 19th, 1901
[55] Navy & Army Illustrated Vol 10
[56] SA War Vol 3 P1900 Photo by Hosking, Cape Town.
[57] With the Flag to Pretoria, H. W. Wilson Vol 1
[58] Image source: britishbattles.com
[59] Richard Harding Davis With Both Armies in South Africa. Charles Scribner's Sons, New York 1900
[60] Navy & Army Illustrated Vol 10
[61]. Heliograph - Wikipedia File:The Second Boer War, 1899-1902 Q72417.jpg - Wikimedia Commons
[62] Image source: britishbattles.com
[63] Davitt, Boer Fight for Freedom, 3rd ed. 1902. (Duke Scan)
[64] Cunliffe, History of the Boer War Vol 1 (NY Library Scan)
[65] Rosslyn, Twice Captured. 1900.
[66] Colenso, 19 March 1900. 'Princess Christian' ambulance train the first to cross the trestle bridge. - Atom site for DRISA P1808.jpg (1400×1016) (drisa.co.za)
[67] Davitt, Boer Fight for Freedom, 3rd ed. 1902. (Duke Scan)
[68] Navy & Army Illustrated Vol 10
[69] Collection of photographs on bridge damage during Boer War. Colenso bridge. - Atom site for DRISA PB1144_001.jpg (1400×974) (drisa.co.za)
[70] Cunliffe, History of the Boer War Vol 1 (NY Library Scan)
[71] Cunliffe, History of the Boer War Vol 1 (NY Library Scan)
[72] Credit: Neville Constantine, Neville C. at Angloboerwar.com
[73] Credit: Neville Constantine, Neville C. at Angloboerwar.com
[74] The Biograph in battle: its story in the South African War related with personal experiences. by William Kennedy Laurie Dickson,, London: T. Fisher Unwin, 1901. Catalog Search: https://archive.org/search?query=title%3A%28The+Biograph+in+Battle%29
[75] South Africa; its History, Mackenzie 1899
[76] SA Boer-British War Vol 1. Hopkins, 1900
[77] Davitt, Boer Fight for Freedom, 3rd ed. 1902. (Duquesene Scan)
[78] After Pretoria: The Guerilla War Vol 3 July 1900
[79] File:The Second Boer War, 1899-1902 Q71966.jpg - Wikimedia Commons Author: William Skeoch Cumming
[80] Anglo-Boer War Album, Cape Town, D. Edwards & co. [1902?]
[81] The reminiscences of Lady Randolph Churchill, by Mrs. George Cornwallis-West. New York: Century 1908

[82] The reminiscences of Lady Randolph Churchill, by Mrs. George Cornwallis-West. New York: Century 1908
[83] File:Henry Fitzalan-Howard (1890).jpg - Wikimedia Commons
[84] File:2nd Duke of Westminster.jpg - Wikimedia Commons Author: George Watmough Webster.
[85] File:Charles Spencer-Churchill, 9th Duke of Marlborough.jpg - Wikimedia Commons Author: William van der Weyde.
[86] With Roberts to the Transvaal. C.N. Robinson [1900]
[87] War with the Boers, H. Brown, London [1902?] Vol 3
[88] Library of Congress: https://www.loc.gov/item/2003681182/ Winston Spencer Churchill, the famous war correspondent, Bloemfontein, South Africa Summary Stereo copyrighted by B.L. Singley (Keystone View Co.).
[89] The Biograph in battle: its story in the South African War related with personal experiences. by William Kennedy Laurie Dickson, London: T. Fisher Unwin, 1901. Catalog Search: https://archive.org/search?query=title%3A%28The+Biograph+in+Battle%29
[90] After Pretoria: The Guerilla War Vol 3 p. 493 1901
[91] After Pretoria: The Guerilla War Vol 3 p. 493 1901
[92] Davitt, Boer Fight for Freedom, 1902 (U. Cal. Scan)
[93] Anglo-Boer War Album, Cape Town, D. Edwards & co. [1902?]
[94] Navy & Army Illustrated Vol 9 Photo by G. W. Wilson, Aberden.
[95] File:The National Archives UK - CO 1069-214-204.jpg - Wikimedia Commons Author: The National Archives UK - Flickr account
[96] File:The National Archives UK - CO 1069-214-205.jpg - Wikimedia Commons Author The National Archives UK - Flickr account
[97] File:AHI Treasures of Southern Africa C 059 (556308917).jpg - Wikimedia Commons Author: Joe Ross from Lansing, Michigan
[98] Prime Minister of Cape Colony, 1898-1900 William Schreiner - Wikipedia
[99] Times History War in S.A Vol 4 (1906) Photo by Elliott & Fry and William Schreiner - Wikipedia
[100] With Roberts to the Transvaal. C.N. Robinson [1900]
[101] War with the Boers, H. Brown, London [1902?] Vol 1
[102] Anglo-Boer War Album, Cape Town, D. Edwards & co. [1902?]
[103] Cunliffe, History of the Boer War Vol 1 (NY Library Scan)
[104] SA War Vol 7 p. 148 Photo by Elliot & Fry.
[105] Anglo-Boer War Album, Cape Town, D. Edwards & co. [1902] Bloemfontein, the Post Office is recognizable in the background.
[106] With the Flag to Pretoria, H. W. Wilson Vol 2 Dec., 1899.
[107] Anglo-Boer War Album, Cape Town, D. Edwards & co. [1902?]
[108] Anglo-Boer War Album, Cape Town, D. Edwards & co. [1902?]
[109] Anglo-Boer War Album, Cape Town, D. Edwards & co. [1902?]
[110] Anglo-Boer War Album, Cape Town, D. Edwards & co. [1902?]
[111] File:Troop Train Crossing Modder River, Boer War (1899).jpg - Wikimedia Commons
[112] Anglo-Boer War Album, Cape Town, D. Edwards & co. [1902?]
[113] Anglo-Boer War Album, Cape Town, D. Edwards & co. [1902?]
[114] Anglo-Boer War Album, Cape Town, D. Edwards & co. [1902?]
[115] Anglo-Boer War Album, Cape Town, D. Edwards & co. [1902?]
[116] Anglo-Boer War Album, Cape Town, D. Edwards & co. [1902?]
[117] Cunliffe, History of the Boer War Vol 2 (Composite of Mich. St. & U. Cal. Scans)
[118] File:JH Olivier.jpg - Wikipedia Afrikaans Cultural Almanac of the FAK
[119] File:Thomas Pilcher (1858-1928).jpg - Wikimedia Commons
[120] File:Black & White Budget00.jpg - Wikimedia Commons And: File:Robert George Broadwood.png - Wikimedia Commons
[121] File:Leslie Rundle in 1916.jpg - Wikimedia Commons Author: Bain News Service
[122] Navy and Army Illustrated Cover of two generals. May 4th, 1901. Brabant and Dartnell. File name: File:Major General Dartnell 1901.jpg - Wikimedia Commons
[123] Churchill Wanted Dead or Alive by Celia Sandys, Skyhorse Publishing, 2019
[124] File:Sir Archibald Hunter.jpg - Wikimedia Commons Source: The River War, Volume I, facing page 322, by Winston Churchill. Published 1899 by Longmans, Green and Co., 39 Paternoster Row, London New York and Bombay.

[125] Doyle, Arthur Conan. The Great Boer War 17th Edition, 1902
[126] WSC, My Early Life
[127] Hillegas, With the Boer forces, 1901 (Drawn by the Author under supervision of General)
[128] Rosslyn, Twice Captured. Blackwoods, 1900.
[129] History of War SA, Birch, 1899
[130] With Roberts to the Transvaal. C.N. Robinson [1900]
[131] File:Charles Spencer-Churchill, 9th Duke of Marlborough.jpg - Wikimedia Commons Author: William van der Weyde.
[132] File:The Second Boer War, 1899-1902 Q72002.jpg - Wikimedia Commons Author: William Skeoch Cumming
[133] Anglo-Boer War Album, Cape Town, D. Edwards & co. [1902?]
[134] War with the Boers, H. Brown, London [1902?] Vol 2
[135] Anglo-Boer War Album, Cape Town, D. Edwards & co. [1902?]
[136] File:The Second Anglo - Boer War, South Africa 1899 - 1902 Q72318.jpg - Wikimedia Commons
[137] File:Boer War Q71976.jpg - Wikimedia Commons
[138] Anglo-Boer War Album, Cape Town, D. Edwards & co. [1902?]
[139] SA War Vol 3 1900 Photo by Vickers Sons & Maxim, Limited.
[140] Identified is 210 Private John McCracken (left) and 320 Private Lucian Franklin Hayward. The centre soldier is unidentified, but believed to be Private William James Neild. All three were involved in the capture of a Maxim gun; for this action Hayward was awarded the Distinguished Conduct Medal (DCM) and McCracken and Neild were both mentioned in despatches. Hayward's name is wrongly entered in the records as Lessian." File:AustralianTroopsWithQF1pounderBoerWar.jpg - Wikimedia Commons
[141] File:The Second Boer War, 1899-1902 Q72267.jpg - Wikimedia Commons Author: William Skeoch Cumming
[142] File:The Second Boer War, 1899-1902 Q72181.jpg - Wikimedia Commons Author: William Skeoch Cumming
[143] The Biograph in battle: its story in the South African War related with personal experiences. by William Kennedy Laurie Dickson, London: T. Fisher Unwin, 1901. Catalog Search: https://archive.org/search?query=title%3A%28The+Biograph+in+Battle%29
[144] The Biograph in battle: its story in the South African War related with personal experiences. by William Kennedy Laurie Dickson, London: T. Fisher Unwin, 1901. Catalog Search: https://archive.org/search?query=title%3A%28The+Biograph+in+Battle%29
[145] Carvings from the Veldt, Part 2 by David C. George. page 288 Credit Rob D.at: Anglo Boer War - Small Arms and Ammunition - Page 6 - Boer War Forum The 'To' is at the right-hand side of the stock with the initials W L E below them. The 'From' is in the center of the stock. On can also make out the initials SALH and several years, which probably relate to years of service during the war.
[146] File:17th Lancers - cap badge, original, antique.jpg - Wikimedia Commons Author: GMJ
[147] Courtesy of Rob Davidson: Anglo Boer War - Artillery and Ammunition - Page 7 - Boer War Forum
[148] Cunliffe, History of the Boer War Vol 2 (Composite of Mich. St. & U. Cal. Scans)
[149] Original lent by 'Navy and Army Illustrated' File:QF 4.7 inch gun Colenso.jpg - Wikipedia
[150] Doyle, Arthur Conan. The Great Boer War 17th Edition, 1902
[151] Anglo-Boer War Album, Cape Town, D. Edwards & co. [1902?]
[152] File:The Second Boer War, 1899-1902 Q72298.jpg - Wikimedia Commons Author: William Skeoch Cumming
[153] File:The Second Boer War, 1899-1902 Q72250.jpg - Wikimedia Commons Author: William Skeoch Cumming
[154] File:The Second Boer War, 1899-1902 Q71977.jpg - Wikimedia Commons Author: William Skeoch Cumming
[155] Per Neville Constantine: Anglo Boer War - Artillery and Ammunition - Page 10 - Boer War Forum
[156] Anglo Boer War - Small Arms and Ammunition - Page 4 - Boer War Forum
[157] Anglo-Boer War Album, Cape Town, D. Edwards & co. [1902?]
[158] Courtesy of Neville Constantine: Anglo Boer War - Artillery and Ammunition - Page 13 - Boer War Forum
[159] Credit Neville Constantine: https://www.angloboerwar.com/forum/21-uniforms-badges-and-equipment/31231-artillery-and-ammunition?start=252
[160] File:The Second Boer War, 1899-1902 Q72266.jpg - Wikimedia Commons Public Domain
[161] With Roberts to the Transvaal. C.N. Robinson [1900]
[162] Anglo-Boer War Album, Cape Town, D. Edwards & co. [1902?]
[163] War pictures. v. 1 (no. 1-16) 1900 #1 - War pictures. v. 1 (no. 1-16) 1900. - Full View | HathiTrust Digital Library And: Catalog Record: War pictures | HathiTrust Digital Library
[164] Navy and Army Illustrated Cover of two generals. May 4th, 1901. File name: File:Major General Dartnell 1901.jpg - Wikimedia Commons

[165] Anglo-Boer War Album, Cape Town, D. Edwards & co. [1902?]
[166] Webster Davis, John Bull's crime; or, Assaults on Republics. 1901 New York The Abbey Press
[167] Webster Davis, John Bull's crime; or, Assaults on Republics. 1901 New York The Abbey Press
[168] File:Horace Smith-Dorrien.jpg - Wikimedia Commons
[169] File:Horace Smith-Dorrien Vanity Fair 1901-12-05.jpg - Wikimedia Commons
[170] War with the Boers, H. Brown, London [1902?] Vol 5
[171] File:The Second Boer War, 1899-1902 Q72193.jpg - Wikimedia Commons Author: William Skeoch Cumming
[172] View from Sehlabathebe House File:Early Morning view from Sehlabathebe House - panoramio.jpg - Wikimedia Commons
[173] Hillegas, With the Boer forces, 1901 (Photograph by the Author)
[174] File:Graham Maclachlan - Thaba N'chu Mountain in the distance.jpg - Wikimedia Commons Author Rana Thakar Source: https://www.panoramio.com/user/33623
[175] File:The Second Boer War, 1899-1902 Q72305.jpg - Wikimedia Commons Author: William Skeoch Cumming
[176] Ian Hamilton's March 1900 The Copp, Clark Co. Toronto. And Frontiers & Wars.
[177] Navy & Army Illustrated Vol 10
[178] Navy & Army Illustrated Vol 9
[179] Davitt, Boer Fight for Freedom, 3rd ed. 1902. (Duke Scan)
[180] Anglo-Boer War Album, Cape Town, D. Edwards & co. [1902?]
[181] File:The Second Boer War, 1899-1902 Q72300.jpg - Wikimedia Commons Author: William Skeoch Cumming
[182] File:The Second Boer War, 1899-1902 Q71993.jpg - Wikimedia Commons Author: William Skeoch Cumming
[183] Ian Hamilton's March 1900 The Copp, Clark Co. Toronto. And Frontiers & Wars.
[184] WSC, My Early Life
[185] War with the Boers, H. Brown, London [1902?] Vol 2
[186] Images - Page 2 (angloboerwar.com)
[187] File:The Second Boer War, 1899-1902 Q72153.jpg - Wikimedia Commons Author: William Skeoch Cumming
[188] File:The Second Boer War, 1899-1902 Q72299.jpg - Wikimedia Commons Author: William Skeoch Cumming
[189] File:The Second Boer War, 1899-1902 Q72291.jpg - Wikimedia Commons Author: William Skeoch Cumming\
[190] File:The Second Boer War, 1899-1902 Q72271.jpg - Wikimedia Commons Author: William Skeoch Cumming
[191] Group of Boer leaders in Kimberley area. General Koos de la Rey is the man with black beard in right foreground with his rifle pointed towards the vertical. File:The Second Boer War, 1899-1902 Q68490.jpg - Wikimedia Commons Author Unknown author
[192] Davitt, Boer Fight for Freedom, 1902 (U. Cal. Scan)
[193] Creswicke Chronology: 4/28/1900 Fighting near Thabanchu Mountain.
[194] File:The Second Boer War, 1899-1902 Q72051.jpg - Wikimedia Commons
[195] Second Anglo-Boer War - 1899 - 1902 | South African History Online (sahistory.org.za)
[196] File:The Second Boer War, 1899-1902 Q72301.jpg - Wikimedia Commons Author: William Skeoch Cumming
[197] Listening for the Drums by General Sir Ian Hamilton Faber And Faber Ltd 24 Russell Square London 1944 p. 126 of PDF.
[198] Listening for the Drums by General Sir Ian Hamilton Faber And Faber Ltd 24 Russell Square London 1944 p. 126 of PDF.
[199] File:The Nile Expedition for the Relief of General Gordon, from The Graphic, 29 November 1894.png - Wikimedia Commons
[200] Listening for the Drums by General Sir Ian Hamilton Faber And Faber Ltd 24 Russell Square London 1944.
[201] War pictures. v. 1 (no. 1-16) 1900 #1 - War pictures. v. 1 (no. 1-16) 1900. - Full View | HathiTrust Digital Library And: Catalog Record: War pictures | HathiTrust Digital Library
[202] The Record of a Regiment of the Line Being a Regimental History of the 1st Battalion Devonshire Regiment during the Boer War 1899-1902, by Mainwaring George Jacson. Hutchinson and Co., London, 1908. Harvard Copy: #13 - The record of a regiment of the line : being a regimental history ... - Full View | HathiTrust Digital Library
[203] Photo by Duffus Bros., Cape Town. Times History Vol 4. 1906.
[204] File:Brucehamilton.jpg - Wikimedia Commons Author: Dormskirk at en.wikipedia
[205] Ian Hamilton's March 1900 The Copp, Clark Co. And Frontiers & Wars.
[206] Frederick Woods, Young Winston's Wars. Viking Press. 1972.
[207] Heroes of the Boer War. Review of Reviews, 1903.

[208] Hiley and Hassell, The Mobile Boer, Grafton Press, New York, (c. 1902)
[209] Image source: Navy–Army Illustrated Vol 10 And britishbattles.com
[210] Hillegas, Boers in War, 1900.
[211] File:Captain Towse shoots at Lt Col Maximov at point-blank range.jpg - Wikimedia Commons Source: The Russians and the Anglo-Boer War 1899-1902, page 48+1.
[212] File:Beachcroft Towse VC.jpg - Wikipedia
[213] They are probably servicemen of the Lothians and Berwickshire Yeomanry. File:The British Army in Pre-1914 Period Q72122.jpg - Wikimedia Commons Author: William Skeoch Cumming.
[214] File:The Second Boer War, 1899-1902 Q72206.jpg - Wikimedia Commons Author: William Skeoch Cumming
[215] Anglo-Boer War Album, Cape Town, D. Edwards & co. [1902?]
[216] Reports on military operations in South Africa and China. U.S. Adjutant-General's Office. 1901 Captain S. L H Slocum 4 reports to page 93 Captain Carl Reichman pages 93-201
[217] Hillegas, With the Boer forces, 1901 (Photograph by Leo Weinthal.)
[218] Davitt, Boer Fight for Freedom, 3rd ed. 1902. (Duke Scan)
[219] Heroes of the Boer War. Review of Reviews, 1903.
[220] War with the Boers, H. Brown, London [1902?] Vol 2
[221] Hillegas, Boers in War, 1900.
[222] Reports on military operations in South Africa and China. U.S. Adjutant-General's Office. 1901
[223] With Roberts to the Transvaal. C.N. Robinson [1900] Underwood and Underwood
[224] Courtesy of Neville Constantine: Anglo Boer War - Artillery and Ammunition - Page 11 - Boer War Forum
[225] Map by G. Philip & Son, London. Original published with book: Kaalfontein is identified on the G. Phillips and Sons Map as being dated on May 17th; Just before Lindley on May 18th.
[226] WSC Ian Hamilton's March.by G. Philip & Son, London And Longmans, Green & Co.
[227] Doyle, Arthur Conan. The Great Boer War 17th Edition, 1902.
[228] File:The Second Boer War, 1899-1902 Q71945.jpg - Wikimedia Commons Author: William Skeoch Cumming
[229] File:The Second Boer War, 1899-1902 Q71965.jpg - Wikimedia Commons Author: William Skeoch Cumming
[230] Vet River, circa 1900. Bridge damaged by dynamite. (Van Hoepen, Pretoria) (Donated by Mr. GJ Krister, Pretoria) - Atom site for DRISA
[231] War with the Boers, H. Brown, London [1902?] Vol 3
[232] War with the Boers, H. Brown, London [1902?] Vol 1
[233] File:FWReitz VA0947.jpg - Wikimedia Commons
[234] File:Randolph Caldecott collection-page 0066 crop-balance-cenhance.jpg - Wikimedia Commons Author: Randolph Caldecott.
[235] Heroes of the Boer War. Review of Reviews, 1903.
[236] With Roberts to the Transvaal. C.N. Robinson [1900]
[237] Ian Hamilton's March 1900 The Copp, Clark Co. Toronto. And Frontiers & Wars.
[238] Ian Hamilton's March 1900 The Copp, Clark Co. Toronto. And Frontiers & Wars
[239] File:The Second Boer War, 1899-1902 Q72267.jpg - Wikimedia Commons Author: William Skeoch Cumming
[240] The leaguer of Ladysmith, by Captain C.M. Dixon. London, Eyre and Spottiswoode [1900] Catalog Record: The leaguer of Ladysmith | HathiTrust Digital Library See Also: Clive MacDonnell Dixon - Wikipedia
[241] File:The Second Boer War, 1899-1902 Q72278.jpg - Wikimedia Commons Author: William Skeoch Cumming
[242] The Biograph in battle: its story in the South African War related with personal experiences. by William Kennedy Laurie Dickson, London: T. Fisher Unwin, 1901. Catalog Search: https://archive.org/search?query=title%3A%28The+Biograph+in+Battle%29
[243] The Biograph in battle: its story in the South African War related with personal experiences. by William Kennedy Laurie Dickson, London: T. Fisher Unwin, 1901. Catalog Search: https://archive.org/search?query=title%3A%28The+Biograph+in+Battle%29
[244] Heroes of the Boer War. Review of Reviews, 1903.
[245] Basuto-Pony - Basuto pony - Wikipedia
[246] File:Howard C. Hillegas.jpg - Wikimedia Commons Source: With the Boer Forces Howard C. Hillegas, Methuen & Co. 36 Essex Street W.C., London, 1900. Author: Thomas Franklin Fairfax Millard for Howard C. Hillegas.
[247] Anglo-Boer War Album, Cape Town, D. Edwards & co. [1902?]
[248] File:The Boer War 1899-1902 Q71943.jpg - Wikimedia Commons Author: William Skeoch Cumming

[249] File:Gezicht op een rivieroever met paardenkarren bij Kroonstad, Zuid-Afrika History Recorders and History Makers, Correspondents' Carts waiting Lord Roberts' entry to Kroonstadt, S.A., crop.jpg - Wikimedia Commons
[250] Anglo-Boer War Album, Cape Town, D. Edwards & co. [1902?]
[251] Anglo-Boer War Album, Cape Town, D. Edwards & co. [1902?]
[252] Anglo Boer War - Images from 'On the Transvaal border' published by W H Gill, 1900 - Page 3 - Boer War Forum Credit Dr. David Biggins
[253] Bridge across the Valsch River at Kroonstadt, blown up by the Boers. File:Gezicht op een opgeblazen brug over de Valsrivier bij Kroonstad, Zuid-Afrika Railway Bridge across the Valsch River at Kroonstadt, blown up by the Boers, S.A. (titel op object), RP-F-F09001.jpg - Wikimedia Commons
[254] Convoy of the 11th Division "outspanned" on the march to Kroonstadt. File:Gezicht op transportkarren onderweg naar Kroonstad, Zuid-Afrika Convoy of the 11th Division outspanned on the march to Kroonstadt, Boer War, South Africa (titel op object), RP-F-F08997.jpg - Wikimedia Commons
[255] File:The Second Boer War, 1899-1902 Q68477.jpg - Wikimedia Commons
[256] File:The Second Boer War, 1899-1902 Q68484.jpg - Wikimedia Commons
[257] Kroonstad. Station building. - Atom site for DRISA P2327_01.jpg (1400×1000) (drisa.co.za)
[258] After Pretoria: The Guerilla War Vol 3 July, 1900
[259] Anglo-Boer War Album, Cape Town, D. Edwards & co. [1902?]
[260] War pictures. v. 1 (no. 1-16) 1900 #1 - War pictures. v. 1 (no. 1-16) 1900. - Full View | HathiTrust Digital Library And: Catalog Record: War pictures | HathiTrust Digital Library
[261] File:Gezicht op brugreparaties aan de spoorlijn tussen Kroonstad en Pretoria, Zuid-Afrika Handiwork of the retreating Boers - destroyed Railway Bridge between Kroonstadt and Pretoria, South Africa (titel op object), RP-F-F09003.jpg - Wikimedia Commons
[262] Engineers repairing a Railway destroyed by the Boers near Kroonstadt. File:Reparaties aan de spoorlijn naar Kroonstad, Zuid-Afrika Royal Engineers repairing a Railway destroyed by the Boers - Kroonstadt, South africa (titel op object), RP-F-F09002.jpg - Wikimedia Commons
[263] Anglo-Boer War Album, Cape Town, D. Edwards & co. [1902?]
[264] Anglo Boer War - Images from 'On the Transvaal border' published by W H Gill, 1900 - Page 3 - Boer War Forum Credit Dr. David Biggins
[265] British troops guarding the Rhenoster River bridge whilst repair work is being done after Boers dynamited it ABW British troops guarding the Rhenoster River bridge whilst repair work is being done after Boers dynamited it ....ABW | Troops, War, Armed conflict (pinterest.com)
[266] File:The Second Boer War, 1899-1902 Q71973.jpg - Wikimedia Commons
[267] File:The Second Boer War, 1899-1902 Q72220.jpg - Wikimedia Commons Author: William Skeoch Cumming
[268] Image source: With the Flag to Pretoria, H. W. Wilson Vol 1 1900 and britishbattles.com Photo by Middlebrook
[269] Anglo-Boer War Album, Cape Town, D. Edwards & co. [1902?]
[270] File:The Second Boer War, 1899-1902 Q72180.jpg - Wikimedia Commons Author: William Skeoch Cumming
[271] File:The Second Boer War, 1899-1902 Q72002.jpg - Wikimedia Commons Author: William Skeoch Cumming
[272] Reports on military operations in South Africa and China. U.S. Adjutant-General's Office. 1901 Captain S. L H Slocum.
[273] File:The Second Boer War, 1899-1902 Q72256.jpg - Wikimedia Commons Author: William Skeoch Cumming
[274] Anglo-Boer War Album, Cape Town, D. Edwards & co. [1902?]
[275] Richard Harding Davis With Both Armies in South Africa. Charles Scribner's Sons, New York 1900
[276] The Biograph in battle: its story in the South African War related with personal experiences. by William Kennedy Laurie Dickson, London: T. Fisher Unwin, 1901. Catalog Search: https://archive.org/search?query=title%3A%28The+Biograph+in+Battle%29
[277] After Pretoria: The Guerilla War Vol 3 Jan-April 1901
[278] SA National Archives: http://www.nationalarchives.gov.za/node/1236
[279] Navy & Army Vol 13 p. 59 Oct. 5th, 1901
[280] Navy & Army Vol 13 p. 59 Oct. 5th, 1901
[281] File:Grass fire on the side of Bryn Glas - geograph.org.uk - 1746681.jpg - Wikimedia Commons Author: Rudi Winter
[282] File:Grass fire near Potlatch, ID. (10489967546).jpg - Wikimedia Commons Author: Robert Ashworth from Bellingham, WA., USA
[283] Anglo-Boer War Album, Cape Town, D. Edwards & co. [1902?]
[284] After Pretoria: The Guerilla War Vol 3 Oct..1900 Photo by Spalding
[285] File:The Second Boer War, 1899-1902 Q72101.jpg - Wikimedia Commons Author: William Skeoch Cumming

[286] http://libcom.org/library/pow-camps-ceylon-during-boer-war
[287] Arthur Mann The Boer in Peace and War, 1900
[288] File:The Second Boer War, 1899-1902 Q72449.jpg - Wikimedia Commons Author: William Skeoch Cumming
[289] Black & White Weekly Vol 1-2 1899-1900 Photo by D. Barnett.
[290] Anglo-Boer War Album, Cape Town, D. Edwards & co. [1902?]
[291] Anglo-Boer War Album, Cape Town, D. Edwards & co. [1902?]
[292] Anglo Boer War - Images from 'On the Transvaal border' published by W H Gill, 1900 - Boer War Forum Credit Dr. David Biggins. PF: My guess is that that's a Basuto pony.
[293] Hillegas, Boers in War, 1900. Takhaar is a pejorative term for a boorish rustic.
[294] File:The Second Boer War, 1899-1902 Q72213.jpg - Wikimedia Commons Author: William Skeoch Cumming
[295] File:The Second Boer War, 1899-1902 Q72211.jpg - Wikimedia Commons
[296] File:The Second Boer War, 1899-1902 Q72217.jpg - Wikimedia Commons Author: William Skeoch Cumming
[297] Image: File:Mauser C96 M1916 Red 9 7.JPG - Wikimedia Commons
[298] File:The Second Boer War, 1899-1902 Q72246.jpg - Wikimedia Commons Author: William Skeoch Cumming
[299] File:The Second Boer War, 1899-1902 Q71974.jpg - Wikimedia Commons Author: William Skeoch Cumming
[300] Anglo-Boer War Album, Cape Town, D. Edwards & co. [1902?]
[301] This picture taken at Frere Camp. Black & White Weekly Vol 1-2 1899-1900 Photo by D. Barnett.
[302] Navy & Army Illustrated Vol 10
[303] Hillegas, Boers in War, 1900.
[304] File:The Second Boer War, 1899-1902 Q71956.jpg - Wikimedia Commons
[305] Creswicke, Louis South Africa and the Transvaal War, Vol. 2 1900 Photo by Cribb, Southsea.
[306] Repository: Library of Congress Prints and Photographs Division Washington, D.C. 20540 USA https://www.loc.gov/pictures/item/2006684574/resource/
[307] File:The Second Boer War, 1899-1902 Q72196.jpg - Wikimedia Commons Author: William Skeoch Cumming
[308] With Roberts to the Transvaal. C.N. Robinson [1900]
[309] Navy & Army Illustrated Vol 11
[310] Anglo Boer War - Small Arms and Ammunition - Page 15 - Boer War Forum_Credit: Neville Constantine
[311] Courtesy of Neville Constantine: Anglo Boer War - Artillery and Ammunition - Page 17 - Boer War Forum
[312] Courtesy of Neville Constantine: Anglo Boer War - Artillery and Ammunition - Page 17 - Boer War Forum
[313] File:The Second Boer War, 1899-1902 Q72153.jpg - Wikimedia Commons
[314] Courtesy of Neville Constantine: Anglo Boer War - Artillery and Ammunition - Page 22 - Boer War Forum
[315] Image Source: With the Flag to Pretoria, H. W. Wilson Vol 1
[316] Webster Davis, John Bull's crime; or, Assaults on Republics. 1901 New York The Abbey Press
[317] With the Inniskilling Dragoons: the record of a cavalry regiment during the Boer War, 1899-1902 by Lieut.-Colonel J. Watkins Yardley. Longmans, Green, and co., 1904. Harvard Copy: #11 - With the Inniskilling Dragoons : the record of a cavalry regiment ... - Full View | HathiTrust Digital Library
[318] File:The Second Boer War, 1899-1902 Q72198.jpg - Wikimedia Commons Author: William Skeoch Cumming
[319] Creswicke SA War Vol 5 Map page VIII
[320] Anglo-Boer War Album, Cape Town, D. Edwards & co. [1902?]
[321] Anglo-Boer War Album, Cape Town, D. Edwards & co. [1902?]
[322] File:The Second Boer War, 1899-1902 Q72296.jpg - Wikimedia Commons Author: William Skeoch Cumming
[323] File:Waterfall, Witwatersrand National Botanical Gardens.jpg - Wikimedia Commons Author: Chris Eason from London
[324] File:Klipriviersberg Nature Reserve.jpg - Wikimedia Commons Author: Pietsnor
[325] Vredefort crater cross section 2 - Vredefort crater - Wikipedia
[326] South Africa; its History, Heroes and Wars. Mackenzie [1899]
[327] Image source: britishbattles.com
[328] War with the Boers, H. Brown, London [1902?] Vol 2
[329] File:Viljoensdrif station.jpg - Wikipedia
[330] File:South Africa; English tourists at the battlefield of Doornko Wellcome V0037949.jpg - Wikimedia Commons File from: Wellcome Images.
[331] Ian Hamilton's March 1900 The Copp, Clark Co. Toronto. And Frontiers & Wars.
[332] File:Doornkop Battle.jpg - Wikimedia Commons Author: AxelRavera

[333] 155 mm Creusot Long Tom - Wikipedia Image Source: 155 mm Creusot Long Tom - Wikipedia File:Long tom on platform at Mafeking.jpg - Wikimedia Commons
[334] Anglo-Boer War Album, Cape Town, D. Edwards & co. [1902?]
[335] https://www.angloboerwar.com/unit-information/imperial-units/557-gordon-highlanders?start=2
[336] https://www.angloboerwar.com/unit-information/imperial-units/557-gordon-highlanders?start=2
[337] Cunliffe, History of the Boer War Vol 1 (NY Library Scan)
[338] File:Grass fire in Zakarpattia Oblast, 2022 (01).jpg - Wikimedia Commons Author: State Emergency Service of Ukraine. dsns.gov.ua
[339] File:Boer War; searching for the wounded from the battlefield at Wellcome V0015550.jpg - Wikimedia Commons Watercolour sketch by F. Craig, 1899.
[340] Image source: With the Flag to Pretoria, H. W. Wilson Vol 1 1900 and britishbattles.com (Original Caption: On Spion Kop the day after the battle.)
[341] Heroes of the Boer War. Review of Reviews, 1903.
[342] ABW Source
[343] War in SA, Harding 1899
[344] This is what Joburg looked like in 1900 | Southern africa, Johannesburg, Cape town holidays (pinterest.com) Credit Andre Hoffman & Friedel Hansen
[345] Anglo-Boer War Album, Cape Town, D. Edwards & co. [1902?]
[346] https://www.angloboerwar.com/unit-information/imperial-units/557-gordon-highlanders?start=2
[347] War with the Boers, H. Brown, London [1902?] Vol 1
[348] File:The Second Boer War, 1899-1902 Q68480.jpg - Wikimedia Commons And: Black & White Weekly Vol 1-2 1899-1900
[349] South Africa; its History, Heroes and Wars. Mackenzie [1899]
[350] picture of old johannesburg - Bing images
[351] After Pretoria: The Guerilla War Vol 3 March, 1901
[352] South Africa; its History, Heroes and Wars. Mackenzie [1899] The trees are mostly Blue Gums a type of Eucalyptus and brought from Australia."
[353] SA War Vol 8
[354] Boer PoWs | Saint Helena Island Info: All about St Helena, in the South Atlantic Ocean And: Second Boer War concentration camps - Wikipedia
[355] With the Inniskilling Dragoons: the record of a cavalry regiment during the Boer War, 1899-1902 by Lieut.-Colonel J. Watkins Yardley. Longmans, Green, and co., 1904. Harvard Copy: #11 - With the Inniskilling Dragoons : the record of a cavalry regiment ... - Full View | HathiTrust Digital Library
[356] Image Source: Boer PoWs | Saint Helena Island Info: All about St Helena, in the South Atlantic Ocean
[357] Image Source: Boer PoWs | Saint Helena Island Info: All about St Helena, in the South Atlantic Ocean
[358] Image Source: Boer PoWs | Saint Helena Island Info: All about St Helena, in the South Atlantic Ocean
[359] Black & White Weekly Vol 3 1900
[360] Image Source: Boer PoWs | Saint Helena Island Info: All about St Helena, in the South Atlantic Ocean
[361] Image Source: Boer PoWs | Saint Helena Island Info: All about St Helena, in the South Atlantic Ocean Credit: Domaines Francais de Sainte Helene boercampdeadwood2_ishe.jpg
[362] Image Source: Boer PoWs | Saint Helena Island Info: All about St Helena, in the South Atlantic Ocean
[363] Per Neville Constantine: Anglo Boer War - Artillery and Ammunition - Page 15 - Boer War Forum
[364] Black & White Weekly Vol 3, Sept, 1900. Photo by Biograph Co.
[365] Johannesburg, 24 April 1900. Explosion accident at the Begbies munition works. - Atom site for DRISA P1880.jpg (1400×1007) (drisa.co.za)
[366] After Pretoria: The Guerilla War Vol 3 Jan., 1901
[367] With the Flag to Pretoria, H. W. Wilson Vol 2 June, 1900.
[368] Illustrated Police News The Illustrated Police News - Wikipedia
[369] File:JamesStreet Secunderabad 1880.jpg - Wikimedia Commons Author: Creator: Deen Dayal
[370] War Impressions: Being a Record in Colour by Mortimer Menpes;[370] transcribed by Dorothy Menpes. Published by Adam & Charles Black, Soho Square, London 1901 Catalog Record 1: Catalog Record: War impressions : being a record in colour | HathiTrust Digital Library

[371] War Impressions: Being a Record in Colour by Mortimer Menpes;[371] transcribed by Dorothy Menpes. Published by Adam & Charles Black, Soho Square, London 1901 Catalog Record 1: Catalog Record: War impressions : being a record in colour | HathiTrust Digital Library
[372] File:Dorset Yeomanry Boer War.png - Wikimedia Commons Credit: Axel Ravera. Public Domain.
[373] Navy & Army Illustrated Vol 11 November 17th, 1900.
[374] File:Boer War Photograph Taken by B W Caney Q115366.jpg - Wikimedia Commons
[375] Black & White Weekly Vol 1-2 1899-1900
[376] War in SA, Harding 1899
[377] The well-known image of a Naval 12-pdr QF 12-cwt gun on "Scott" improvised carriage, taken by society photographer Horace Nicholls Anglo Boer War - Artillery and Ammunition - Page 31 - Boer War Forum Credit: Neville Constantine
[378] Cunliffe, History of the Boer War Vol 1 (NY Library Scan)
[379] File:Light horse walers.jpg - Wikimedia Commons
[380] This photo taken before the assault on Belmont 11/23/99 File:Australische soldaten te paard bij Belmont, 31 december Australian Mounted Rifles just before the taking of Belmont (Dec. 31st), South Africa (titel op object), RP-F-F08945, crop.jpg - Wikimedia Commons
[381] Haldane, How We Escaped from Pretoria.
[382] Richard Harding Davis With Both Armies in South Africa. Charles Scribner's Sons, New York 1900
[383] Creswicke, Louis South Africa and the Transvaal War, Vol. 1 Photo by Wilson, Aberdeen.
[384] Haldane, How We Escaped from Pretoria.
[385] Davitt, Boer Fight for Freedom, 3rd ed. 1902. (Duke Scan)
[386] Photo by Rosslyn and appearing in his work, Twice Captured.
[387] Richard Harding Davis With Both Armies in South Africa. Charles Scribner's Sons, New York 1900
[388] Earl of Rosslyn, James St Clair-Erskine, 5th Earl of Rosslyn Twice Captured: A Record of Adventure During the Boer War. William Blackwood and Sons, 1900. See: Twice Captured - Google Books and: James St Clair-Erskine, 5th Earl of Rosslyn - Wikipedia
[389] With the Flag to Pretoria, H. W. Wilson Vol 2 June, 1900.
[390] After Pretoria: The Guerilla War Vol 3 August, 1900
[391] Photo by Rosslyn and appearing in his work, Twice Captured.
[392] Rosslyn, James, 5th Earl of. Twice Captured: A Record Of Adventure During The Boer War William Black Wood & Sons, Edinburgh 1900
[393] Volume 14 at HathiTrust June 1901 to November 1901: Catalog Record: Catalog Record: The Windsor magazine : an illustrated monthly... | HathiTrust Digital Library Michigan Scan: #7 - The Windsor magazine : an illustrated monthly for ... v.14. - Full View | HathiTrust Digital Library
[394] File:The Second Boer War, 1899-1902 Q72136.jpg - Wikimedia Commons Author: William Skeoch Cumming
[395] File:The Second Boer War, 1899-1902 Q72151.jpg - Wikimedia Commons William Skeoch Cumming
[396] With the Flag to Pretoria, H. W. Wilson Vol 2 June, 1900.
[397] Richard Harding Davis With Both Armies in South Africa. Charles Scribner's Sons, New York 1900
[398] Library of Congress Prints and Photographs Division Washington, D.C.
https://www.loc.gov/pictures/item/00651051/resource/
[399] Anglo-Boer War Album, Cape Town, D. Edwards & co. [1902?]
[400] The Biograph in battle: its story in the South African War related with personal experiences. by William Kennedy Laurie Dickson, London: T. Fisher Unwin, 1901. Catalog Search:
https://archive.org/search?query=title%3A%28The+Biograph+in+Battle%29
[401] Richard Harding Davis With Both Armies in South Africa. Charles Scribner's Sons, New York 1900
[402] With the Flag to Pretoria, H. W. Wilson Vol 2 June, 1900.
[403] Davitt, Boer Fight for Freedom, 3rd ed. 1902. (Duke Scan)
[404] Hillegas, Boers in War, 1900.
[405] The Biograph in battle: its story in the South African War related with personal experiences. by William Kennedy Laurie Dickson, London: T. Fisher Unwin, 1901. Catalog Search:
https://archive.org/search?query=title%3A%28The+Biograph+in+Battle%29
[406] The Biograph in battle: its story in the South African War related with personal experiences. by William Kennedy Laurie Dickson, London: T. Fisher Unwin, 1901. Catalog Search:
https://archive.org/search?query=title%3A%28The+Biograph+in+Battle%29

[407] The Biograph in battle: its story in the South African War related with personal experiences. by William Kennedy Laurie Dickson, London: T. Fisher Unwin, 1901. Catalog Search: https://archive.org/search?query=title%3A%28The+Biograph+in+Battle%29
[408] The Biograph in battle: its story in the South African War related with personal experiences. by William Kennedy Laurie Dickson, London: T. Fisher Unwin, 1901. Catalog Search: https://archive.org/search?query=title%3A%28The+Biograph+in+Battle%29
[409] The Biograph in battle: its story in the South African War related with personal experiences. by William Kennedy Laurie Dickson, London: T. Fisher Unwin, 1901. Catalog Search: https://archive.org/search?query=title%3A%28The+Biograph+in+Battle%29
[410] Black & White Weekly Vol 3 1900
[411] After Pretoria: The Guerilla War Vol 3 April, 1901
[412] Ganna, circa 1902. Damage to rolling stock after a train was captured by Boer forces at Ganna siding near Nelspoort. - Atom site for DRISA
[413] Derailment on the Johannesburg - Volksrust line. - Atom site for DRISA N68524.jpg (1400×1029) (drisa.co.za)
[414] Hillegas, With the Boer forces, 1901 (Photograph by Leo Weinthal, Pretoria.)
[415] Anglo-Boer War Album, Cape Town, D. Edwards & co. [1902?]
[416] File:The Second Boer War, 1899-1902 Q72199.jpg - Wikimedia Commons Author: William Skeoch Cumming
[417] Anglo-Boer War Album, Cape Town, D. Edwards & co. [1902?]
[418] Anglo-Boer War Album, Cape Town, D. Edwards & co. [1902?]
[419] File:The Second Boer War, 1899-1902 Q72171.jpg - Wikimedia Commons Author: William Skeoch Cumming
[420] File:The Second Boer War, 1899-1902 Q72071.jpg - Wikimedia Commons Author: William Skeoch Cumming
[421] Times History War in S.A Vol 4 (1906) Photo by Duffus Bros., Cape Town See Also: Louis Botha - Wikipedia
[422] Times History War in S.A Vol 5 (1907) Photo by Duffus Bros., Cape Town See Also: Koos de la Rey - Wikipedia
[423] A Handbook of the Boer War - Gale and Polden Limited, London 1910 Source: Anonymous ISBN 978-1-37-497455-5. https://babel.hathitrust.org/cgi/pt?id=uc1.l0061277489&seq=7
[424] War with the Boers, H. Brown, London [1902?] Vol 4
[425] Image Source: With the Flag to Pretoria, H. W. Wilson Vol 1
[426] Frederick Woods, Young Winston's Wars. Viking Press. 1972.
[427] Ian Hamilton's March 1900 The Copp, Clark Co. Toronto. And Frontiers & Wars
[428] Anglo Boer War - Images from 'On the Transvaal border' published by W H Gill, 1900 - Boer War Forum Credit Dr. David Biggins.
[429] Images - Page 2 (angloboerwar.com)
[430] This sketch is of the Lancer Charge at Elandslaagte. Image by R. Caton Woodville File:Elandslaagte 1.jpg - Wikimedia Commons
[431] File:The Earl of Airlie in 1883.jpg - Wikimedia Commons
[432] Lieutenant colonel David Stanley William Ogilvy, 11th and 6th Earl of Airlie (20 January 1856 – 11 June 1900) was a Scottish peer. David Ogilvy, 11th Earl of Airlie - Wikipedia
[433] File:The Second Boer War, 1899-1902 Q71968.jpg - Wikimedia Commons Author: William Skeoch Cumming
[434] File:The Second Boer War, 1899-1902 Q71977.jpg - Wikimedia Commons Author: William Skeoch Cumming
[435] Frederick Woods, Young Winston's Wars. Viking Press. 1972.
[436] Ian Hamilton's March 1900 The Copp, Clark Co. Toronto. And Frontiers & Wars.
[437] File:The Second Boer War, 1899-1902 Q72027.jpg - Wikimedia Commons Author: William Skeoch Cumming
[438] Navy & Army Illustrated Vol 10
[439] File:The Second Boer War, 1899-1902 Q72183.jpg - Wikimedia Commons Author: William Skeoch Cumming
[440] Times History Vol 4. P. 293
[441] File:The Second Boer War, 1899-1902 Q71967.jpg - Wikimedia Commons Author: William Skeoch Cumming
[442] Times History Vol 4. P. 293
[443] File:The Second Boer War, 1899-1902 Q72300.jpg - Wikimedia Commons Author: William Skeoch Cumming
[444] This one "captured by Boers and retaken by British troops at Bethlehem" Courtesy of Neville Constantine: Anglo Boer War - Artillery and Ammunition - Page 4 - Boer War Forum
[445] With the Inniskilling Dragoons: the record of a cavalry regiment during the Boer War, 1899-1902 by Lieut.-Colonel J. Watkins Yardley. Longmans, Green, and co., 1904. Catalog Record: With the Inniskilling Dragoons : the record... |

HathiTrust Digital Library Harvard Copy: #11 - With the Inniskilling Dragoons : the record of a cavalry regiment ... - Full View | HathiTrust Digital Library

[446] With the Inniskilling Dragoons: the record of a cavalry regiment during the Boer War, 1899-1902 by Lieut.-Colonel J. Watkins Yardley. Longmans, Green, and co., 1904. Harvard Copy: #11 - With the Inniskilling Dragoons : the record of a cavalry regiment ... - Full View | HathiTrust Digital Library

[447] Digital ID: (b&w film copy neg.) cph 3c02309 http://hdl.loc.gov/loc.pnp/cph.3c02309
Reproduction Number: LC-USZ62-102309 (b&w film copy neg. Lord Roberts (to left) and Lord Kitchener (white horse) reviewing the troops entering Kroonstadt [sic], S. Africa (loc.gov)

[448] Listening for the Drums by General Sir Ian Hamilton Faber And Faber Ltd 24 Russell Square London 1944

[449] Listening for the Drums by General Sir Ian Hamilton Faber And Faber Ltd 24 Russell Square London 1944

[450] Morning Post, Protecting the Railway 29 June
https://www.britishnewspaperarchive.co.uk/viewer/bl/0000174/19000629/022/0005

[451] The Biograph in battle: its story in the South African War related with personal experiences. by William Kennedy Laurie Dickson, London: T. Fisher Unwin, 1901. Catalog Search:
https://archive.org/search?query=title%3A%28The+Biograph+in+Battle%29

[452] South Africa; its History, Heroes and Wars. Mackenzie [1899]

[453] Bethulie. Bridge over Orange River damaged during Anglo-Boer War Bethulie. Bridge over Orange River damaged during Anglo-Boer War. - Atom site for DRISA P2444_02.jpg (1400×1030) (drisa.co.za)

[454] Atom site for DRISA Page 29 (top). Temporary bridge over Sundays River. - Atom site for DRISA

[455] Armoured train on temporary trestle bridge during Boer War. - Atom site for DRISA P1812.jpg (1400×927) (drisa.co.za)

[456] The Biograph in battle: its story in the South African War related with personal experiences. by William Kennedy Laurie Dickson,, London: T. Fisher Unwin, 1901. Catalog Search:
https://archive.org/search?query=title%3A%28The+Biograph+in+Battle%29

[457] Cunliffe, History of the Boer War Vol 1 (NY Library Scan)

[458] SA National Archives: http://www.nationalarchives.gov.za/node/1236

[459] War with the Boers, H. Brown, London [1902?] Vol 1

[460] Heroes of the Boer War. Review of Reviews, 1903.

[461] Heroes of the Boer War. Review of Reviews, 1903.

[462] Langsnek Tunnel, July 1900. Damaged northern portal after being blown up. - Atom site for DRISA
P1628_01.jpg (1400×814) (drisa.co.za)

[463] South Africa; its History, Heroes and Wars. [1899]

[464] Creswicke, Louis South Africa and the Transvaal War, Vol. 2 1900 Photo by Knight, Aldershot.

[465] The Biograph in battle: its story in the South African War related with personal experiences. by William Kennedy Laurie Dickson, London: T. Fisher Unwin, 1901. Catalog Search:
https://archive.org/search?query=title%3A%28The+Biograph+in+Battle%29

[466] The Biograph in battle: its story in the South African War related with personal experiences. by William Kennedy Laurie Dickson, London: T. Fisher Unwin, 1901. Catalog Search:
https://archive.org/search?query=title%3A%28The+Biograph+in+Battle%29

[467] The Biograph in battle: its story in the South African War related with personal experiences. by William Kennedy Laurie Dickson, London: T. Fisher Unwin, 1901. Catalog Search:
https://archive.org/search?query=title%3A%28The+Biograph+in+Battle%29

[468] Proclamations in SA, https://babel.hathitrust.org/cgi/pt?id=hvd.32044089026009&seq=301&view=1up

[469] Proclamations in SA, https://babel.hathitrust.org/cgi/pt?id=hvd.32044089026009&seq=301&view=1up

[470] Proclamations in SA, https://babel.hathitrust.org/cgi/pt?id=hvd.32044089026009&seq=302&view=1up

[471] From: Cecil Rhodes: A Study of a Career", by Howard Hensman, William Blackwood and Sons, Edinburgh and London (1911) File:Col. f.w.rhodes (cropped).jpg - Wikimedia Commons

[472] War Impressions: Being a Record in Colour by Mortimer Menpes.by Adam & Charles Black, Soho Square, London 1901

[473] War Impressions: Being a Record in Colour by Mortimer Menpes.by Adam & Charles Black, Soho Square, London 1901

[474] South Africa; its History, Heroes and Wars. Mackenzie [1899]

[475] South Africa; its History, Heroes and Wars. Mackenzie [1899]

[476] The Biograph in battle: its story in the South African War related with personal experiences. by William Kennedy Laurie Dickson, London: T. Fisher Unwin, 1901. Catalog Search: https://archive.org/search?query=title%3A%28The+Biograph+in+Battle%29
[477] Rosslyn, Twice Captured.
[478] Library of Congress Prints and Photographs Division. https://www.loc.gov/pictures/resource/stereo.1s37241/
[479] South Africa; its History, Heroes and Wars. Mackenzie [1899]
[480] South Africa; its History, Heroes and Wars. Mackenzie [1899]
[481] After Pretoria: The Guerilla War Vol 3 Nov., 1900 Photo by J. Bowers, Pretoria.
[482] Burnham, Frederick Russell (1926). Scouting on Two Continents. Garden City, New York: Doubleday, Page & Company. ISBN 978-0-86920-126-8. OCLC 407686. Pages #439-441 of compilation. Pages 349-351 of bound book #11 - Scouting on two continents, by Major Frederick Russell Burnham ... - Full View | HathiTrust Digital Library. Second Source: RMS Dunottar Castle - Wikipedia File:Burnham churchill jul1900.jpg - Wikipedia Published by Ogden's as a cigarette card in 1900. Frederick Russell Burnham II (Major Burnham's grandson) and C. Atkinson, Burnham family members.
[483] Image File: Rms dunottar castle - RMS Dunottar Castle - Wikipedia
[484] The Sphere - Saturday 04 August 1900 p. 20 https://www.britishnewspaperarchive.co.uk/viewer/bl/0001861/19000804/019/0020
[485] File:Burnham 1902 loc collection.jpg - Wikimedia Commons
[486] Times History War in S.A Vol 3 Photo by Bassano.
[487] Lady Randolph Churchill's Engagement." [Philadelphia]Public Ledger, July 21, 1900. Newspaper clipping. Moreton Frewen Papers, Manuscript Division, Library of Congress (39) Bookmark this item: //www.loc.gov/exhibits/churchill/wc-youth.html#39
[488] The Sphere - Saturday 21 July 1900. P. 4: https://www.britishnewspaperarchive.co.uk/viewer/bl/0001861/19000721/005/0004
[489] https://commons.wikimedia.org/wiki/File:Churchills-in-Canada-768x564.jpg
[490] https://commons.wikimedia.org/wiki/File:Jack_Churchill_in_Calgary,_1929_(cropped).jpg
[491] File:Winston Churchill (1901).jpg - Wikimedia Commons
[492]Source: C. E. Bechhofer Roberts, Winston Churchill; being an account of the life of the Right Hon. Winston Leonard Spencer Churchill, P.C., C.H., T.D., M.P. by Ephesian [pseud.] With 11 illustrations Published 1927 London, Mills & Boon, limited [1927] Author Roberts, C. E. Bechhofer
[493] File:Winston Churchill 1899.jpg - Wikimedia Commons Also: With the Flag to Pretoria, H. W. Wilson Vol 1 Photo by Elliot & Fry.
[494] Listening for the Drums by General Sir Ian Hamilton Faber And Faber Ltd 24 Russell Square London 1944
[495] Listening for the Drums by General Sir Ian Hamilton Faber And Faber Ltd 24 Russell Square London 1944
[496] File:Positions round Ladysmith - November 1899 - Project Gutenberg eText 16466.jpg - Wikimedia Commons
[497] Creswicke SA War Vol 2
[498] War with the Boers, H. Brown, London [1902?] Vol 2
[499] Creswicke SA War Vol 2
[500] Times History Vol 3
[501] The Sphere - Saturday 24 November 1900 p. 13 https://www.britishnewspaperarchive.co.uk/viewer/bl/0001861/19001124/016/0013
[502] The Sphere - Saturday 24 November 1900 p. 13 https://www.britishnewspaperarchive.co.uk/viewer/bl/0001861/19001124/016/0013
[503] The Sphere - Saturday 24 November 1900 p. 13 https://www.britishnewspaperarchive.co.uk/viewer/bl/0001861/19001124/016/0013
[504] After Pretoria: The Guerilla War Vol 3 p. 84 1900 Photo by Bennett
[505] Cunliffe, History of the Boer War Vol 1 (NY Library Scan)
[506] File:Philip Rudolph Botha (1851-1901), Boer War general.jpg - Wikimedia Commons
[507] The Biograph in battle: its story in the South African War related with personal experiences. by William Kennedy Laurie Dickson, London: T. Fisher Unwin, 1901. Catalog Search: https://archive.org/search?query=title%3A%28The+Biograph+in+Battle%29
[508] Originally Included in Winston Spencer Churchill's book; Ian Hamilton's March, 1900 The Copp, Clark Company, Ltd. Toronto.

[509] Image credit: Francis L. H. de Souza (Mr. de Sozua would sneak in a bottle of whiskey at the bottom of a basket of fruit he would bring in for the prisoners.)
[510] Image source Dr. Jonathan de Souza
[511] Hofmeyr The Story of My Captivity.
[512] File:TradeCardMasonAndHamlinOrganCoCirca1880s.jpg - Wikimedia Commons
[513] Cunliffe, History of the Boer War Vol 1 (NY Library Scan)
[514] Haldane, How We Escaped from Pretoria. By: Robert Carnegie
[515] Images Credit: Rob Davidson
[516] Anglo-Boer War Album, Cape Town, D. Edwards & co. [1902?]
[517] Haldane, How We Escaped from Pretoria.
[518] Creswicke, Louis South Africa and the Transvaal War, Vol. 1 Photo by Wilson, Aberdeen.
[519] Rosslyn -Twice Captured.
[520] Richard Harding Davis With Both Armies in South Africa. Charles Scribner's Sons, New York 1900
[521] With the Flag to Pretoria, H. W. Wilson Vol 2 June, 1900.
[522] With the Flag to Pretoria, H. W. Wilson Vol 2 June, 1900.
[523] Richard Harding Davis With Both Armies in South Africa. Charles Scribner's Sons, New York 1900
[524] With the Inniskilling Dragoons: the record of a cavalry regiment during the Boer War, 1899-1902 by Lieut.-Colonel J. Watkins Yardley. Longmans, Green, and co., 1904.
[525] Image Credit QSAMike Anglo Boer War - Armored Train Ambush 11-15-99 Chuchill & Haldane - Page 17 - Boer War Forum
[526] Image per Neville Chamberlain. Anglo Boer War - Armored Train Ambush 11-15-99 Chuchill & Haldane - Page 17 - Boer War Forum
[527] Ian Hamilton's March 1900 The Copp, Clark Co. Toronto. And Frontiers & Wars.
[528] Hillegas, Boers in War, 1900.
[529] With the Flag to Pretoria, H. W. Wilson Vol 2 June, 1900. Photo by J. F. Archibald
[530] Hillegas, Boers in War, 1900.
[531] Richard Harding Davis With Both Armies in South Africa. Charles Scribner's Sons, New York 1900
[532] Hillegas, Boers in War, 1900.
[533] Black & White Weekly Vol 4 1901
[534] With the Flag to Pretoria, H. W. Wilson Vol 2 June, 1900.
[535] Anglo-Boer War Album, Cape Town, D. Edwards & co. [1902?]
[536] After Pretoria: The Guerilla War Vol 3 August, 1900
[537] The Biograph in battle: its story in the South African War related with personal experiences. by William Kennedy Laurie Dickson, London: T. Fisher Unwin, 1901. Catalog Search: https://archive.org/search?query=title%3A%28The+Biograph+in+Battle%29
[538] The Biograph in battle: its story in the South African War related with personal experiences. by William Kennedy Laurie Dickson, London: T. Fisher Unwin, 1901. Catalog Search: https://archive.org/search?query=title%3A%28The+Biograph+in+Battle%29
[539] Image Credit: QSAMIKE: Anglo Boer War - Armored Train Ambush 11-15-99 Chuchill & Haldane - Page 18 - Boer War Forum
[540] File:Sarah Wilson War Correspondent.JPG - Wikimedia Commons
[541] Unknown Source (Roberts)
[542] Unknown Source (Roberts)
[543] "The South African War: The lifting of the siege of Mafeking". South African History Online. Retrieved 15 June 2022. https://en.wikipedia.org/wiki/Siege_of_Mafeking#
[544] With the Flag to Pretoria, H. W. Wilson Vol 2 May 1900.
[545] https://en.wiktionary.org/wiki/maffick
[546] Navy & Army Illustrated Vol 10 1900 Photo by Bradley
[547] Navy & Army Illustrated Vol 10 1900 Photo by Bradley
[548] Navy & Army Illustrated Vol 10 1900 Photo by Stremack
[549] Navy & Army Illustrated Vol 10 – June 3th, 1900
[550] Anglo-Boer War Album, Cape Town, D. Edwards & co. [1902?]
[551] Anglo-Boer War Album, Cape Town, D. Edwards & co. [1902?]

[552] With the Flag to Pretoria, H. W. Wilson Vol 2 May 1900.
[553] With the Flag to Pretoria, H. W. Wilson Vol 2 Dec., 1899.
[554] With the Flag to Pretoria, H. W. Wilson Vol 2 May 1900.
[555] With the Flag to Pretoria, H. W. Wilson Vol 2 Feb. 1900.
[556] Image Source: With the Flag to Pretoria, H. W. Wilson Vol 2
[557] https://commons.wikimedia.org/wiki/File:Sol_Plaatje_002.jpg
[558] Anglo-Boer War Album, Cape Town, D. Edwards & co. [1902?]
[559] Anglo-Boer War Album, Cape Town, D. Edwards & co. [1902?]
[560] Webster Davis, John Bull's crime; or, Assaults on Republics. 1901 New York The Abbey Press.
[561] Richard Harding Davis With Both Armies in South Africa. Charles Scribner's Sons, New York 1900
[562] With the Flag to Pretoria, H. W. Wilson Vol 2 May 12 1900 p. 615.
[563] With the Flag to Pretoria, H. W. Wilson Vol 2 May 12 1900 p. 623.
[564] With the Flag to Pretoria, H. W. Wilson Vol 2 May 12 1900 p. 624.
[565] https://commons.wikimedia.org/wiki/File:Blowing_Mutinous_Sepoys_From_the_Guns,_September_8,_1857_-_steel_engraving.jpg And: https://en.wikipedia.org/wiki/Blowing_from_a_gun
[566] Rare British South African war song. Citation to the music and image is a screenshot of one used in the video. And: https://www.youtube.com/watch?app=desktop&v=geKCr-SYNM8.
[567] File:British Army, Boer War, Q 71952.jpg - Wikimedia Commons
[568] Image Credit: Rob Davidson. Anglo Boer War - Small Arms and Ammunition - Page 6 - Boer War Forum
[569] File:80-G-38559 (25601980716).jpg - Wikimedia Commons U.S. Navy photograph. National Archives. (2016/03/08).
[570] The Biograph in battle: its story in the South African War related with personal experiences. by William Kennedy Laurie Dickson, London: T. Fisher Unwin, 1901. Catalog Search: https://archive.org/search?query=title%3A%28The+Biograph+in+Battle%29
[571] The Biograph in battle: its story in the South African War related with personal experiences. by William Kennedy Laurie Dickson, London: T. Fisher Unwin, 1901. Catalog Search: https://archive.org/search?query=title%3A%28The+Biograph+in+Battle%29
[572] Anglo-Boer War Album, Cape Town, D. Edwards & co. [1902?]
[573] Davitt, Boer Fight for Freedom, 3rd ed. 1902. (Duke Scan)
[574] Hillegas, Boers in War, 1900.
[575] Dutch protected cruiser Gelderland. https://commons.wikimedia.org/wiki/File:HNLMS._Gelderland_(1897).jpg
[576] Vereeniging, 1904. President Kruger's funeral train on the bridge over the Vaal River. - Atom site for DRISA P2775_02.jpg (1400×1279) (drisa.co.za)

Made in United States
Orlando, FL
05 November 2026

90230222R00199